To Professor Peter Lax
this book
is
Affectionately dedicated

Mathematical Physics Studies

The series publishes original research monographs on contemporary mathematical physics. The focus is on important recent developments at the interface of Mathematics, and Mathematical and Theoretical Physics. These will include, but are not restricted to: application of algebraic geometry, D-modules and symplectic geometry, category theory, number theory, low-dimensional topology, mirror symmetry, string theory, quantum field theory, noncommutative geometry, operator algebras, functional analysis, spectral theory, and probability theory.

More information about this series at http://www.springer.com/series/6316

Hassan Emamirad

Scattering Theory for Transport Phenomena

Springer

Hassan Emamirad
Laboratoire de Mathématiques
Université de Poitiers
Chassneuil du Poitou, Cedex, France

ISSN 0921-3767 ISSN 2352-3905 (electronic)
Mathematical Physics Studies
ISBN 978-981-16-2372-1 ISBN 978-981-16-2373-8 (eBook)
https://doi.org/10.1007/978-981-16-2373-8

This Springer imprint is published by the registered company Springer Nature Singapore Pte Ltd.
The registered company address is: 152 Beach Road, #21-01/04 Gateway East, Singapore 189721, Singapore

Foreword by Peter Lax

Scattering theory is both a point of view and a powerful technical tool in mathematical physics, as well as in the more theoretical branches of mathematics.There are two formalisms; the classical one, initiated and developed by Wheeler, Heisenberg, Moller, Friedrichs, Kato, and others, has its roots in the relation between a self-adjoint operator and perturbations of it. The scattering operator defined in this way is extremely useful in analyzing the reflection of acoustic and electromagnetic waves from obstacles, as well as in studying the effect of potentials in quantum mechanics. The second formalism, introduced by Ralph Phillips and the undersigned, is based on the notion of translation representation. It has shown itself to be useful in scattering by obstacles, in Faddeev and Pavlov's theory of automorphic functions, and transport theory.

Transport theory falls within the province of statistical physics; this calls for some changes in the scattering theory for transport processes. It is the L^1 rather than the L^2 function class over phase space that is relevant here. The positivity preserving property of transport dynamics makes it possible to exploit the theory of positive operators, and the Banach lattice character of L^1.

There are many sources and inspirations for this book: neutron transport in nuclear reactors, both fission and fusion, computerized tomography, and semiconductor devices; it is to these applications that the book is devoted, after laying the foundations of the abstract theory of semigroups and its scattering theory.

A discussion of neutron transport with transparent and semitransparent boundary conditions is followed by a treatment of streams of charged particles in a plasma. In a noteworthy section, the author is able to establish a relation between the concept of albedo and scattering operators, and uses this connection to justify Hejtmanek's interpretation of computerized tomography.

There is an interesting discussion of scattering theory for the quantum Liouville equation, and its connection to the Schrödinger equation via the Wigner distribution. This is used to analyze semiconductor devices.

This book will be very useful to those who wish to learn the theory of scattering, and how it is used to study nonclassical, modern applications. The prerequisites for this book are the by now standard topics in functional analysis and operator theory.

New York, NY, US
October 2002

Peter Lax

Preface

The scattering theory for transport phenomena was the title of a paper by P. Lax and R. Phillips, which was published in 1967 in the proceedings of a conference. Since then, great progress has been made in this area. In this book, we try to develop a part of this progress which has been worked on for more than half a century.

This book is divided into seven chapters. The Chap. 1 treats some preliminaries on the theory of semigroups and C*-algebra, different types of semigroups, Schatten-von Neumann classes of operators, and some facts about ultraweak operator topology.

In Chap. 2, we go into the abstract scattering theory in a general Banach space X but we set the stage for use of $X = L^1(\mathbb{R}^n \times V)$, for example, the contribution of positivity. We define the wave and scattering operators and their basic properties. Some abstract methods such as smooth perturbations and the limiting absorption principle are presented.

Chap. 3 is devoted to the transport or linearized Boltzmann equation, which is the advection equation perturbed by the sum of absorption and production operators. We define the production cross-section $\sigma_p(x, v) = \int_V k(x, v, v')dv'$ and the finite Radon transform $F(\sigma)$, and we show that if $F(\sigma_p) < 1$ and $F(\sigma_a) < \infty$, then the scattering operator $S = W_+ W_-$ exists.

In Chap. 4, we introduce the Lax and Phillips formalism in scattering theory for the transport equation. In their seminal book, Lax and Phillips introduce the incoming and outgoing subspaces which verify their representation theorem for a dissipative hyperbolic system initially and match also for the transport problem. With means of these subspaces, we define the Lax and Phillips semigroup and we prove that this semigroup is eventually compact, hence hyperbolic. If we replace the L^1 by an admissible weighted space L^1_ω, this semigroup is also hypercyclic. The transport problem occurs also in plasma physics in a collision between an ionized atom and a neutral one. If we take $f(x, v, t)$ as the distribution of neutral particles and $g(x, v, t)$ as the distribution of ions into account balance equations give rise to two transport equations; by taking $h = f + g$, h satisfies an advection equation and f a transport equation in which h acts as a coefficient. But this equation will be nonautonomous, and for generating, we have to use the Howland semigroup and Howland's formalism in Chap. 5.

Chapter 6 is the highlight of the book in which we explain how the scattering operator for the transport problem can lead us to formulate the computerized tomography in medical science. Here, we use the albedo operator, which is a notion borrowed from astrophysics that relies on the association of incoming flux with outgoing flux in the transport problem. This notion occurs in the transport problems with transparent boundary conditions and also for semitransparent boundary conditions.

In the last chapter, we introduce the Wigner function and see how this function connects the Schrödinger equation to statistical physics and the Husimi distribution function. We use the ultraweak operator topology to analyze the semiclassical limit of the Wigner and Husimi distributions. We see the relationship between the Wigner function and Quantum Dynamical Semigroup (QDS), and we end by establishing the existence of the scattering operator for the QDS.

It gives me pleasure to thank Profs. J. Goldstein and F. Golse for their constructive comments, and especially Prof. C. Bardos for reading the manuscript carefully and making numerous suggestions and corrections. I cannot end this preface without thanking Prof. P. Lax, who kindly accepted in 2002 to write the Foreword.

Chassneuil du Poitou, Cedex, France Hassan Emamirad

Introduction

When we investigate a specific domain of physics, we are generally faced with different evolutionary systems such that, in a certain sense, each of them is a perturbation of the others. A common feature of an nonstationary physical problem can be formulated as a dynamical system. A rough simplification of such a system leads to an initial value problem with well-defined mathematical properties. Such a system is called an *unperturbed system*, whereas the original problem is called a *perturbed system*.

Scattering theory is concerned with the asymptotic behavior in the future and in the past of two different dynamics. To be more precise, let us consider two initial value problems on a Banach space X:

$$(P_0) \qquad \begin{cases} \dfrac{\partial u_0}{\partial t} = T_0 u_0 \\ u_0(0) = f_0 \in X, \end{cases}$$

and

$$(P) \qquad \begin{cases} \dfrac{\partial u}{\partial t} = T u \\ u(0) = f \in X. \end{cases}$$

We suppose that T_0 and T generate two strongly continuous one-parameter groups ((C_0) groups) on X, $\{e^{tT_0}\}_{t\in\mathbb{R}}$ and $\{e^{tT}\}_{t\in\mathbb{R}}$ respectively. Thus the mild solutions of (P_0) and (P) are given by $u_0(t) = e^{tT_0} f_0$ and $u(t) = e^{tT} f$.

By the asymptotic behavior in the future and in the past we mean that, given an initial datum f_0^+ $[f_0^-]$, does there exist an initial datum $f \in X$ such that

$$\lim_{t\to+\infty} \left\| u(t) - u_0^+(t) \right\| = 0 \ \ [\lim_{t\to-\infty} \left\| u(t) - u_0^-(t) \right\| = 0]\,? \tag{1}$$

It is clear that if e^{tT} is uniformly bounded on $\mathbb{R}^{\pm}$ (i.e. $\sup_{t\in\mathbb{R}^{\pm}} \|e^{tT}\| \leq M$), then this consideration reduces to the existence of the so-called *wave operators*. In fact, if the following strong limits

$$W_{\pm}(T, T_0) = s - \lim_{t\to\pm\infty} e^{-tT} e^{tT_0} \tag{2}$$

exist, then for $f = W_{\pm}(T, T_0) f_0^{\pm}$ we have

$$\left\|e^{tT} f - e^{tT_0} f_0^{\pm}\right\| \leq \left\|e^{tT}\right\| \left\|f - e^{-tT} e^{tT_0} f_0^{\pm}\right\| \to 0, \text{ as } t \to \pm\infty.$$

Conversely, given the initial datum $f \in X$, does there exist $f_0^{\pm} \in X$ such that (1) holds? By a similar argument, the answer to this question can be guaranteed by the existence of two more wave operators, namely

$$W_{\pm}(T_0, T) = s - \lim_{t\to\pm\infty} e^{-tT_0} e^{tT}, \tag{3}$$

provided e^{-tT_0} is equibounded on $\mathbb{R}^{\pm}$.

We remark that (2) provides us two independent problems, depending on whether $t \to +\infty$ or $t \to -\infty$. These problems can be connected to each other via, so-called *scattering operator*,

$$S : f_0^- \longrightarrow f_0^+ .$$

Note that we can decompose the above picture in the following way:

$$f_0^- \overset{W_-(T,T_0)}{\longrightarrow} f \overset{W_+(T_0,T)}{\longrightarrow} f_0^+ .$$

So, we have

$$S = W_+(T_0, T) W_-(T, T_0). \tag{4}$$

In the following chapters, we confine our attention to the existence of the wave operators and scattering operator for linear transport equations.

Linear transport theory deals with equations that describe linear transport phenomena in matter. These phenomena may relate to neutron transport in a nuclear reactor, radiative transfer in a stellar or planetary atmosphere, penetration of x-rays or γ-rays through scattering media, charge-exchange dynamics in ionized particles of plasma and similar processes. The domain of application of transport phenomena covers a broad area, ranging from intergalactic matter to semiconductor devices and quantum transport theory.

The object of investigation in transport theory is the distribution of particles in phase space. *Phase space* is the direct product of configuration space Ω and velocity space V; generally, each is a three-dimensional space and in many cases of our study $\Omega = \mathbb{R}^3$ and the velocity space is a spherical shell $V = \{v \in \mathbb{R}^3 \big| 0 \le a \le |v| \le b \le \infty\}$. The phase space distribution function, which we denote by u, is a function of variables x, v and t. It is defined on $\Omega \times V \times \mathbb{R}$ and satisfies an equation of the form

$$\frac{\partial u}{\partial t} = -v.\nabla_x u + Au, \tag{5}$$

where the perturbation term Au has different expressions according to the physical context in which the transport problem appears. In the neutron transport problem Au is the rate of change of u due to collisions. It is expressed as

$$Au := -\sigma_a(x, v)u + \int_V k(x, v', v)u(x, v', t)dv',$$

which is the difference between loss and gain terms. Here σ_a is the absorption rate and k is the production rate after collisions. In the computerized tomography

$$Au := \sigma(x)u,$$

where $\sigma(x)$ is the human body attenuation coefficient. In the transport theory for the charged particles $Au := A(t)u$ is an evolutive operator defined by

$$A(t)u := -\left(\int_V w(x - tv', v')dv'\right)u + w(x - tv, v)\int_V u(x, v')dv',$$

where $w = u+v$, u being the distribution of neutral particles and v the distribution of ions. Finally, in quantum transport theory A is a pseudo-differential operator defined by the Weyl's transformation of the potential.

The common feature of all these theories is the unperturbed operator $-v \cdot \nabla_x := -\sum_{i=1}^{n} v_i \frac{\partial}{\partial x_i}$ which is called the *advection operator*. This operator generates a unitary group of operators $\{U(t)\}_{t\in\mathbb{R}}$, defined by

$$U(t)f(x, v) := f(x - tv, v) \tag{6}$$

in any $L^p(\mathbb{R}^n \times V)$, $(1 \le p \le \infty)$, which is strongly continuous for any $1 \le p < \infty$.

Motivated by the above results, the Chap. 1 is devoted to the theory of the one-parameter semigroup of linear operators. In this chapter, we are not going to give a complete synopsis of this subject, but rather give a set of results which will be of use in the subsequent chapters. Although the basic concepts of the theory of semigroups of operators date back to the first half of the last century, there are some update notions which will be the subject of our discussion. Seminal notions of hypercyclic or chaotic semigroups or the widely disseminated concept of *Howland semigroups* were instrumental in creating a flourishing new area for the operator-theoretic contributors.

Almost all the results of the Chap. 1 are standard and we are not going to prove them. However, at the end of each chapter, a section is devoted to notes and comments. In this section, we give the precise references of the unproved results. However, if in this section some of Propositions have a proof, this is perhaps a new or an abridged version of the standard proof.

In Introduction to his monograph [104], G. F. Roach says that "Scattering theory means different things to different people". In Chap. 2, we indicate what we mean by scattering theory. Our approach is essentially based on the existence of wave operators, which, in turn, implies the similarity between perturbed and free dynamics. In this chapter, we place ourselves in a Banach space and we formulate different methods to prove the existence of wave operators in an abstract formalism. Most of these methods are based on Cook's Lemma, which roughly says that the asymptotic boundedness of a differentiable function follows from integrability of its derivative. We can say that this lemma finds itself in the heart of the chapter and methods like *smooth perturbation, limiting absorption principle* are the corollaries. One of the main features of transport problems or more generally in statistical physics, the unknown solution is the probability density of the particles and this function lies in the positive cone $L^1(\mathbb{R}^n \times V)_+$ of Banach lattice $L^1(\mathbb{R}^n \times V)$. This property prompts us to study the contribution of positivity and see how the theory of Banach lattices can alleviate the assumptions in Cook's Lemma. Finally, in this chapter, we indicate some simple results concerning the contribution of the existence of the wave operators on the spectrum of perturbed operators.

In Chap. 3, we apply the materials of the preceding chapter to the linear Boltzmann equation. In classical transport systems, the evolution of a classical probability density imposes as a natural setting the Banach lattice L^1_+. Within this framework, the scattering theory for the linear Boltzmann equation was initiated by Hejtmanek [60] and Simon [108].

The physical situation is the following: a finite convex scattering body, D is surrounded by a vacuum in which, if there is no collision, the classical particles move freely according to their initial velocities. This generates the free evolution operator

$$[U_0(t)f](x, v) \equiv f(x - tv, v)$$

as the dynamic of the following advection system:

When the particles encounter the body, the effect of a collision is an instantaneous change of velocity and therefore trajectory. Depending on the physical properties of the medium, pure scattering may be accompanied by net absorption or creation. The effect of all these changes is usually described by the linear Boltzmann equation.[1]

---

[1]The terminology linear Boltzmann or linearized Boltzmann equation is given by the pioneers J. Hejtmanek and B. Simon, but it is somehow misleading. Since the Boltzmann equation is the

$$\begin{aligned}\frac{\partial u}{\partial t} &= -v.\nabla_x u + \chi_D(x)\Big[-\sigma_a(x, v)u + \int_V k\big(x, v', v\big)u\big(x, v', t\big)d\mu\big(v'\big)\Big] \\ &= (T_0 + A_a + A_p)u(x, v, t) \\ &= Tu(x, v, t),\end{aligned} \tag{7}$$

where χ_D is the characteristic function of the domain D. Here the collision term admits a natural decomposition into an absorptive part described by the absorption cross-section $\sigma_a(x, v)$ and a re-emitting part, described by the differential cross-section $k(x, v\prime, v)$.

In the scattering situation, which will be analyzed in Chap. 3, the surface of the body ∂D is perfectly transparent for both incident and emergent distributions. This physical particularity yields the time reversibility of the system, which, in turn, implies the questions of the existence of the wave operators $W_\pm(T_0, T)$ and $W_\pm(T, T_0)$. The chapter is devoted to the answer to these questions. The positivity of the dynamics e^{tT_0} for $t \in \mathbb{R}$ and e^{tT} for $t \geq 0$, the isometry of e^{tT_0} in $L^1(\mathbb{R}^n \times V)$ and nonproliferation of e^{tT}, i.e., $\sup_{t\geq 0}\|e^{tT}\| < \infty$ play a crucial role in establishing some optimal conditions for the existence of the wave operators. We study different methods such as smooth perturbation or local the decaying property of scattering theory in transport problems and we give the example of the finite collisions system.

The Lax–Phillips scattering theory is especially well-suited to scattering problems of classical physics that obey Huygens' principle rather than to quantum physics, i.e., to scattering problems involving the wave equation, Maxwell's equations, or other symmetric hyperbolic systems. The basic idea which was isolated and developed in the Lax–Phillips theory is that of incoming and outgoing subspaces and their representation theorem. Let $U(t)$ be a unitary (C_0) group on the Hilbert space $\mathcal{H}$. The subspace $\mathcal{D}_+$ [resp. $\mathcal{D}_-$] is called an *outgoing subspace* [resp. *incoming subspace*] for $U(t)$ if it satisfies the following *representation theorem*:

(i) $U(t)\mathcal{D}_\pm \subset \mathcal{D}_\pm$for $\pm t \in \mathbb{R}_+$;
(ii) $\bigcap_{t\in\mathbb{R}} U(t)\mathcal{D}_\pm = \{0\}$;
(iii) $\overline{\bigcup_{t\in\mathbb{R}} U(t)\mathcal{D}_\pm} = \mathcal{H}$.

In [76], Lax and Phillips took $U(t)$ to be the (C_0) group which governs the following symmetric first order system[2]:

---

paradigm of nonlinear kinetic theory and if we replace the nonlinear perturbation by a linear one the physical meaning of the problem changes. So perhaps the name neutron transport equation would be more appropriate. That is what M. Mokhtar-Kharroubi used in [92]. But despite this inconsistency, we continue to use the standard terminology of J. Hejtmanek and B. Simon.

[2]In general this system contains the wave, Maxwell or Dirac equations; and in general leads to the ill-posed problem in $L^p(\mathbb{R}^n), p \neq 2$. In 1966, Ph. Brenner showed in [21] that the well-posedness of this system in $L^p(\mathbb{R}^n), p \neq 2$ is equivalent to an algebraic relation $[A_j, A_k] = 0, \forall j, k$, where $[.,.]$ is the Lie bracket. In my opinion this paper is one of the most beautiful papers in the field which gives an equivalence between the well-posedness and a pure algebraic property of the system. Then W. Littman, C. McCarthy and N. Rivière, proved in [82] the non-existence of L^p estimates for certain translation-invariant operators. This was extended in [112] by R. S. Strichartz, leading to the eponym Strichartz estimates. Perhaps this motivates the notion of the distribution semigroup

$$\frac{\partial u}{\partial t} = \sum_{j=1}^{n} A_j(x) \frac{\partial u}{\partial x_j} + B(x)u,$$

where A_j is a real symmetric $m \times m$ matrix-valued function on $\mathbb{R}^n$ such that for $|x| \geq \rho$, $A_j(x) = A_j^0$ is a constant matrix and $B(x) = 0$. The matrix $B(x)$ is chosen such that $U(t)$ is unitary. The corresponding "free" unitary (C_0) group $U_0(t)$ governs

$$\frac{\partial u}{\partial t} = \sum_{j=1}^{n} A_j^0 \frac{\partial u}{\partial x_j}.$$

Let $c > 0$ be the smallest propagation speed of $U_0(t)$ and let

$$\mathcal{D}_{\pm} := \{f \in \mathcal{H} | U(t)f(x) = 0 \text{ for } |x| \leq \rho \pm ct, \ \pm t \in \mathbb{R}_{-}\}.$$

They proved that $\mathcal{D}_{+}$ [resp. $\mathcal{D}_{-}$] are outgoing [resp. incoming] for $U(t)$. In an earlier publication [77], they constructed the outgoing and incoming subspaces for the streaming free group of the transport problem in $L^2(\mathbb{R}^3 \times S^2)$. To our knowledge, this was the first paper on the *scattering theory for transport phenomena*, which was the title of their paper. Here, we have chosen the same title for the present book to demonstrate the development of this topic for more than the last five decades.

Another important aspect of this theory is the introduction of the Lax–Phillips semigroup. In the case of a hyperbolic system, one can define by means of the orthogonal projections $P_{\pm}$ onto $\mathcal{D}_{\pm}^{\perp}$ the semigroups $Z(t) := P_{+}U(t)P_{-}$ on $\mathcal{K} := \mathcal{H} \ominus (\mathcal{D}_{+} \oplus \mathcal{D}_{-})$ which, roughly speaking, contain the behavior of $U(t)$ near the perturbed region. In contrast to $U(t)$, $Z(t)$ is not a group and is defined only for $t \geq 0$, and is eventually compact, hence its generator has only pure point spectra which are the poles of meromorphic extension of the scattering operator.

In Chap. 4, we will see that all these notions have their own meanings in the transport theory. We prove the representation theorem for full transport group e^{tT}. The terminology of outgoing and incoming subsets arose naturally in applications and correspond to the framework of statistical physics. The set of all outgoing [resp. incoming] points from interaction region is called the *outgoing points set* [resp. *incoming points set*] and the set of L^1 functions which have their supports on this set is called the *outgoing* [resp.*incoming*] *subset*. By multiplication of the characteristic function of these sets, one can define the projections $P_{\pm}$ and in the same manner the Lax–Phillips semigroups $\mathcal{Z}(t)$ can be defined. The eventual compactness of $\mathcal{Z}(t)$ leads to the generalized eigenfunction expansion of $U(t)$.

---

which was introduced by J. L. Lions in [80] and in [9, 10] used to prove the well-posedness of the Schrödinger equation in $L^p(\mathbb{R}^n)$.

Hypercyclicity is the study of linear operators that possess a dense orbit. More precisely if T is a bounded linear operator on a Banach space X, an element $x \in X$ is called a *hypercyclic element* for T, when its orbit $\{T^n x | n \in \mathbb{N}\}$ is dense in X. In this case the operator T is called *hypercyclic*. Similarly, the strongly continuous semigroup of operators $\{T(t)\}_{t \geq 0}$ is *hypercyclic* if there exists $x \in X$ such that the orbit $\{T(t)x | t \geq 0\}$ is dense in X, and is *chaotic* if, in addition, there is a dense set of *periodic points*: x such that $T(t)x = x$ for some $t > 0$. A prototype example of a (C_0) semigroup is the translation semigroup $T(t)f(s) := f(s+t)$ on a weighted space $L^p([0, \infty), \rho)$, where the weight function $\rho : [0, \infty) \mapsto [0, \infty)$ is a measurable and admissible, i.e. $\frac{\rho(s)}{\rho(s+t)} \leq Me^{\omega t}$, for some $M \geq 1, \omega > 0$ for all $t \geq 0$, and all $s \geq 0$. Since the streaming free semigroup in transport theory is a kind of translation, this inspired us to look at the material of hypercyclic and chaotic operator theory in the framework of Lax–Phillips representation theorem. In fact, it turn out that the density property (iii) of Lax–Phillips representation theorem can be generated by a single hypercyclic element if the ambient space is weighted with an adequate weight function. In Chap. 4, a section is devoted to this point of view.

The scattering theory occurs also in the context of the charged particle transport problem. Among the large number of different physical processes, in the dynamics of an ionized gas (plasma) in a tokamak, the contact with a material wall is an extremely difficult problem (see [25]). In particular it has been realized that charge exchange, near the wall, is extremely important. We recall that charge exchange is the collisional process by which two atoms exchange an electron, so that, in the case of a neutral atom and a singly-ionized ion, the only effect of the collision is that the neutral particle is transformed into an ion, and the ion into a neutral particle. Let $f := f(x, v, t)$ be the density distribution of neutral particles and $g := g(x, v, t)$ of those of ions. We will see that in the one-dimensional case the system of equations describing our problem is

$$\frac{\partial h}{\partial t} + v \cdot \nabla_x h = 0, \tag{8}$$

$$\frac{\partial f}{\partial t} + v \cdot \nabla_x f = -\sigma f \int_V h dv' + \sigma h \int_V f dv', \tag{9}$$

where $h = f + g$ is the distribution function of all the heavy particles (ions and neutrals), and $\sigma := \sigma(x, t)$ and σ the collision rate for charge exchange, which we assume to equal 1. Even for a constant σ the equation (9) is a nonautonomous equation which can be put in the following abstract form:

$$\begin{cases} \dfrac{df}{dt} = [T_0 + A(t) + K(t)]f(t) \\ f(s) = \varphi \in X, \quad s \in \mathbb{R}, \end{cases}$$

for which the propagator $U(t, s)$ is defined such that the solution of the above problem is given by $f(t) = U(t, s)f(s)$. The theory of scattering for the nonautonomous equations was initiated by J. S. Howland in [66]. In order to model that process in our case, we needed to introduce the *evolution* or Howland semigroup, which is defined in the Chap. 1 (Section 13).

One the most prominent applications of the scattering theory for the transport equation is to describe the functionality of computerized tomography. In computerized tomography, if there is no absorbing obstacle between the X-ray source and the detection cell, the system obeys an advection dynamic, which generates the free system

$$\frac{\partial u}{\partial t} + \omega.\nabla_x u = 0, \tag{10}$$

$$u(0) = f,$$

but if the particles meet an absorbing obstacle (human body), with an absorption rate $\sigma(x)$ at the point x, then the term $-\sigma(x)u$ must be added to the right-hand side of (10), so we retrieve the linear Boltzmann equation (7) in $\mathbb{R}^2$, with D a circular disk, $k = 0$ and $\sigma_a(x, v) = \sigma(x)$. In computerized tomography we emit from a point x_0 of the boundary of D an X-ray into D with density $f(x_0, \omega)$ in direction $x_0 + t\omega$ and we detect it on the other side of the circle with intensity g and the formula

$$[R\sigma](x, w) = \text{In}(f(x_0, x)/g(x, w)) \tag{11}$$

gives the *Radon transformation* of σ that is

$$[R\sigma](x, \omega) = \int_{-\infty}^{+\infty} \sigma(x + \omega t)dt.$$

In the Chap. 6, we calculate explicitly the scattering operator S for this problem and we show why we can replace the outgoing flux g in the formula (11) by S. In order to achieve this result, we will use the notion of an *albedo operator*, which is a notion that we have imbued from astrophysics. This operator, denoted $\mathcal{A}$, acts on the boundary of our region ∂D, which associates the incoming flux with outgoing flux,

$$\mathcal{A} : f \mapsto g.$$

With the help of this operator, we show why in formula (11), we can replace g by the scattering operator S defined in (4).

In this chapter, we discuss also the relationship between the albedo and scattering operators for the semitransparent boundary condition. This is a new aspect of dynamics.

Chapter 7 is devoted to analysis of the quantum transport equation. That begins with the prominent paper by E. Wigner [124] in 1932, in which he constructed a bridge between quantum and statistical physics. This was done by introducing the so-called Wigner function. Nowadays, it lies at the core of the phase space formulation of quantum mechanics and has a variety of applications in statistical mechanics, quantum optics, and signal analysis, to name a few. This is a trendy subject for the prominent mathematicians who have made very high level investigations into it, see for example [85, 86, 81, 7, 78, 12, 13] and the references therein.

Equip the classical phase space with coordinates $(x, \xi) \in \mathbb{R}^{2n}$. The Wigner is quasiprobability distribution on $\mathbb{R}^{2n}$, associated with a wave function $\psi_\hbar \in L^2(R^n)$ and its complex conjugate $\psi^*_\hbar$, is defined by

$$W_\hbar(x,\xi) = (2\pi)^{-n/2} \int_{\mathbb{R}^n} e^{-i\xi y} \psi_\hbar(x + \frac{\hbar y}{2}) \psi^*_\hbar(x - \frac{\hbar y}{2}) dy. \tag{12}$$

One can prove that if the function $\psi_\hbar(t)$ is the solution of Schrödinger's equation for an adequate potential V, then the correspondent Wigner function $W_\hbar(t)$ satisfies an equation called the quantum Liouvillle equation which is a perturbation of the advection equation whose solution is the free transport. The function $W_\hbar$ possesses many of the properties of a phase space probability distribution; in particular, it is real. However, $W_\hbar$ is not a genuine probability distribution as it can assume negative values for certain values of (x, ξ) in phase space. To address this inconvenience K. Husimi has proposed in [67] a modification of this distribution. By taking the convolution product of $W_\hbar$ with the Gaussian he presents a function which carries out all the properties of a density function. This function is called the Husimi function. In this chapter, we are interested in studying the semiclassical limit of all these functions when the Planck constant $\hbar \to 0$. In [85], Markowich and Ringhofer have shown that if $\hbar \to 0$ the weak solution of the quantum Liouville equation converges weakly to the weak solution of the Liouville–Vlasov equation

$$\frac{\partial v}{\partial t} + \xi \cdot \nabla_x v = \nabla_x V \cdot \nabla_\xi v. \tag{13}$$

In this chapter, we generalize this result in the following manner: We use the notion of ultraweak topology which is introduced in the Chap. 1 and we prove that this convergence is not only in the weak sense, but also in the sense of ultraweak topology which is stronger than the weak topology. This notion also helps us to show that the Husimi function converges in the space $L^1(\mathbb{R}^n_x \times \mathbb{R}^n_\xi)$ to a v solution of the Liouville–Vlasov equation.

The Wigner function is intimately related with the quantum dynamical semigroup which acts on the Banach space of trace class operators which are tackled in the Chap. 1. This semigroup is denoted by $e^{t\mathbb{L}}\rho := e^{-\mathrm{i}tH}\rho e^{\mathrm{i}tH}$, where ρ is a trace class operator. If H is a self-adjoint operator in a Hilbert space $\mathcal{H}$, then the mapping $\rho \mapsto e^{t\mathbb{L}}\rho$ is not only a semigroup, but also a (C_0) group in $\mathscr{I}_1(\mathcal{H})$, the space of a trace class operator with ultraweak topology, and is generated by

$$\frac{d}{dt}\rho = -\mathrm{i}[H, \rho], \tag{14}$$

which is called the *Heisenberg–von Neumann* or *quantum transport* equation. So, in the last section, we study the scattering theory for this equation.

Contents

Chapter 1
Semigroups of Linear Operators

1.1 Closed Operators

Let X and Y be two real or complex Banach spaces. Let $\|.\|$ (or $\|.\|_X$ when we need to specify) denote the norm in X. The space of all bounded linear operators from X to Y will be denoted by $\mathcal{L}(X, Y)$. It is a unit Banach algebra with the norm

$$\|A\| := \sup_{f \neq 0} \frac{\|Af\|_Y}{\|f\|_X},$$

which is equivalent to

$$\|A\| := \sup_{\|f\|_X \leq 1} \|Af\|_Y. \tag{1.1}$$

We denote $\mathcal{L}(X, X)$ by $\mathcal{L}(X)$.

Let the ambient space be a complex separable Hilbert space $\mathcal{H}$, endowed with the inner product $(\cdot, \cdot)$ and the norm $\|\cdot\| = (\cdot, \cdot)^{1/2}$. The set of all bounded operators acting on $\mathcal{H}$ is denote by $\mathcal{L}(\mathcal{H})$; it is an algebra with respect to the composition of operators and has an *involution mapping* $* : \ A \mapsto A^*$ satisfying

(i) $(A^*\phi, \psi) = (\phi, A\psi)$;
(ii) $(AB)^* = B^*A^*$ and $(A^*)^* = A$;
(iii) $(\alpha A + \beta B)^* = \overline{\alpha} A^* + \overline{\beta} B^*$;
(iv) $\|A^*A\| = \|A\|^2 = \|A^*\|^2$.

An operator $A \in \mathcal{L}(\mathcal{H})$ is *self-adjoint* if $A = A^*$ and *positive* if $(A\phi, \phi) \geq 0$; in this case we write $A \geq 0$.

When a linear operator A is not defined in the whole of space X, we need to specify its domain denoted by $D(A) \subseteq X$. In this case the operator A is usually referred to as an *unbounded operator*.

H. Emamirad, *Scattering Theory for Transport Phenomena*, Mathematical Physics Studies, https://doi.org/10.1007/978-981-16-2373-8_1

Definition 1.1.1 An operator A on X to Y is *closed* if its graph $\mathcal{G}(A) := \{(f, Af) \mid f \in D(A)\}$ is a closed subspace of $X \times Y$, or equivalently, if $f_n \in D(A)$ converges in X to f and $Af_n \to g$ then $f \in D(A)$ and $Af = g$.

The following lemma gives the basic facts about closed operators.

Lemma 1.1.2 *Let A on X to Y with domain $D(A)$ be a closed operator. Then:*

(i) *$A + B$ with $D(A + B) := D(A)$ is closed for each $B \in \mathcal{L}(X, Y)$.*
(ii) *If A is one-to-one, then A^{-1} is closed.*
(iii) (**Closed graph theorem**) *If $D(A)$ is closed in X then A is bounded;*
(iv) *a bounded operator A is closed iff $D(A)$ is closed.*

Definition 1.1.3 We say that $\mathcal{D} \subset D(A)$ is a *core* for an arbitrary closed operator A if $\mathcal{D}$ is dense in $D(A)$ for the norm

$$\|f\|_{D(A)} = \|Af\| + \|f\|, \tag{1.2}$$

which is called a *graph norm*.

Definition 1.1.4 A linear operator A on X to Y is called *closable* if the closure of its graph $\overline{\mathcal{G}(A)}$ is a graph (i.e. $(0, g) \in \overline{\mathcal{G}(A)}$ implies $g = 0$) or equivalently, if $D(A) \ni f_n \to 0$ and $Af_n \to g$, then $g = 0$.

Definition 1.1.5 Let A and B be operators on X to Y. If $\mathcal{G}(A) \subset \mathcal{G}(B)$, then B is said to be an *extension* of A and we write $A \subset B$. Equivalently, $A \subset B$ if and only if $D(A) \subset D(B)$ and $Bf = Af$ for all $f \in D(A)$.

If A is a closable operator defined on X, then we may define an extension of A, denoted by $\overline{A}$ and called the closure of A, in the following manner:

$$D(\overline{A}) := \{f \in X \mid \exists \{f_n\} \subset D(A) \text{ such that } f_n \to f \text{ and } \{Af_n\} \text{ is a Cauchy sequence}\}.$$

For $f \in D(\overline{A})$, $\overline{A}f$ is defined by $\overline{A}f := \lim_{n\to\infty} Af_n$, which exists and is unique, since Y is complete.

Definition 1.1.6 Let X be a Banach space and X^* its dual ($X^* := \mathcal{L}(X, \mathbb{C})$). The elements of X^* are the bounded linear functionals in X. Application of a functional $f^* \in X^*$ to an element f of X will be denoted by $\langle f^*, f\rangle$. Let A be a densely defined operator in X. Let $D(A^*)$ be the set of $f^* \in X^*$ for which there is an element $g^* \in X^*$ such that

$$\langle f^*, Af\rangle = \langle g^*, f\rangle \quad \text{for all } f \in D(A). \tag{1.3}$$

For each such $f^* \in D(A^*)$, we define $A^* f^* = g^*$. A^* is called *adjoint* of A.

Since $D(A)$ is dense in X, g^* as in (1.3) is unique and thus A^* is well-defined. However, it may be the case that $D(A^*)$ reduces to $\{0\}$. The following lemma is a standard result for unbounded operators (see e.g. [69]).

Lemma 1.1.7 *Let A be a densely defined operator in a Banach space X, then A^* is closed in X^*. Furthermore if X is reflexive, then:*

(i) *if A is closable, its closure $\overline{A} = A^{**}$;*
(ii) *if A is closed, then $D(A^*)$ is dense in X^*.*

One of the important consequences of the closed operator is the following lemma:

Lemma 1.1.8 *Let B be a closed operator on X, f a X-valued continuous function on $[a, b]$ such that $f(t) \in D(B)$ for all $t \in [a, b]$ and $Bf \in C([a, b], X)$. Then the Riemann integral $\int_a^b f(t)dt \in D(B)$ and*

$$B \int_a^b f(t)dt = \int_a^b Bf(t)\, dt. \tag{1.4}$$

Proof If $\sum_{i=1}^n f(\tau_i)(t_i - t_{i-1})$ is the X-valued Riemann sum of $\int_a^b f(t)dt$ and $\sum_{i=1}^n Bf(\tau_i)(t_i - t_{i-1})$ is the X-valued Riemann sum of $\int_a^b Bf(t)dt$, then

$$B \sum_{i=1}^n f(\tau_i)(t_i - t_{i-1}) = \sum_{i=1}^n Bf(\tau_i)(t_i - t_{i-1})$$

and (1.4) follows from closedness of B. □

Here we introduce some spectral properties of a linear operator $A : X \to X$. With A we associate the operator $(\lambda I - A) : \ D(A) \to X$, where $\lambda \in \mathbb{C}$ and I is the identity operator.

If $(\lambda I - A)$ is invertible then we denote by $(\lambda - A)^{-1}$ its inverse and we refer to it as the *resolvent* of A.

Definition 1.1.9 Let $A : X \to X$, $D(A) \subseteq X$ be a linear operator. The set of all complex numbers such that $(\lambda I - A) : \ D(A) \to X$ is bijective and $(\lambda - A)^{-1} \in \mathcal{L}(X)$, denoted $\rho(A)$, is called the *resolvent set* of A. The *spectrum* of A, denoted $\sigma(A)$, is the complement in the complex plane of $\rho(A)$, that is $\sigma(A) = \mathbb{C} \setminus \rho(A)$.

If there exists $\lambda_0 \in \rho(A)$, then $(\lambda_0 I - A)$ is bijective, hence we have:

Lemma 1.1.10 *If $\rho(A) \neq \emptyset$, then the operator A is closed.*

1.2 Strongly Continuous Semigroups

Definition 1.2.1 A *strongly continuous semigroup* of bounded linear operators defined on a Banach space X (called a *(C_0) semigroup*), is a family of operators $\{T(t)\}_{t\geq 0}$, $T(t) \in \mathcal{L}(X)$, such that:

(i) $T(0) = \text{Id}$;
(ii) for any $f \in X$ and any $t, s \geq 0$, $T(t)T(s)f = T(t+s)f$;
(iii) for any $f \in X$, $\lim_{t\to 0_+} T(t)f = f$.

Definition 1.2.2 The *infinitesimal generator* (or simply the *generator*) of a semigroup is its *derivative at zero*, that is, the operator A defined by

$$Af = \lim_{t\to 0_+} \frac{T(t)f - f}{t} \qquad \text{(in } X\text{)} \tag{1.5}$$

$D(A)$, which is called the *maximal domain of* A, is the set of all $f \in X$ such that the limit (1.5) exists.

In order to highlight the generator, in this book we denote by $\{e^{tA}\}_{t\geq 0}$ the (C_0) semigroup generated by A.

Remark 1.2.3 For any (C_0) semigroup $\{e^{tA}\}_{t\geq 0}$, there exist two constants $M \geq 1$ and $\omega \in \mathbb{R}$ such that $\|e^{tA}\| \leq Me^{\omega t}$ for all $t \geq 0$. This exponential estimate can be used to define an important characteristic of the semigroup.

Definition 1.2.4 By the *growth bound* or *type* of the semigroup $\{e^{tA}\}_{t\geq 0}$ we understand the number

$$\begin{aligned}
\omega(A) &:= \inf\{\omega \in \mathbb{R} \mid \text{there exists } M \geq 1 \text{ such that } \|e^{tA}\| \leq Me^{\omega t}\}. \\
&= \lim_{t\to\infty} \frac{1}{t} \ln \|e^{tA}\| \\
&= \inf_{t>0} \frac{1}{t} \ln \|e^{tA}\|.
\end{aligned}$$

In [69], T. Kato used the following notation: $A \in \mathcal{G}(M, \omega)$ if A generates a (C_0) semigroup e^{tA} such that

$$\|e^{tA}\| \leq Me^{\omega t}.$$

Particularly if $\omega \leq 0$, then we have

$$\sup_{t\geq 0} \|e^{tA}\| \leq M. \tag{1.6}$$

Such a semigroup is called a *bounded semigroup* and in the case $M = 1$ is called a *contraction semigroup*.

Remark 1.2.5 Note that the type of a (C_0) semigroup can be equal to zero, without e^{tA} satisfying (1.6), for example if there exist constants $C > 0$ and $N \in \mathbb{N}$ such that

$$\|e^{tA}\| \leq C(1+t)^N \qquad \text{for all } t \geq 0. \tag{1.7}$$

In this case $\omega(A) = 0$ and such a semigroup is called a *temperate semigroup*. In Chap. 3 we will give an example for (1.7).

A (C_0) semigroup $\{e^{tA}\}_{t\geq 0}$ can be extended to a (C_0) group $\{e^{tA}\}_{t\in\mathbb{R}}$, iff $-A$ generates also a (C_0) semigroup $\{e^{-tA}\}_{t\geq 0}$. In this case $\{e^{tA}\}_{t\in\mathbb{R}}$ is obtained as

$$e^{tA} := \begin{cases} e^{tA} & \text{for } \ t \geq 0; \\ e^{(-t)A} & \text{for } \ t \leq 0. \end{cases}$$

In [49, Proposition 1.7] one can find the following result.

Lemma 1.2.6 *If $\mathcal{D} \subseteq X$ is dense in X and invariant under the action of (C_0) semigroup e^{tA}, then $\mathcal{D}$ is a core for A.*

Remark 1.2.7 It is not hard to verify that if $\mathcal{D}$ is a core of the generator of a (C_0) semigroup, then $\overline{A\mid_{\mathcal{D}}} = A$ and $D(\overline{A\mid_{\mathcal{D}}})$ is equal to the maximal domain of A.

Remark 1.2.8 In Definition 1.2.1, the condition (iii) can be replaced by a weaker form:

(iii)' $\qquad \langle \phi, e^{tA} f \rangle \in C(\mathbb{R}_+, \mathbb{C}), \quad$ for each $f \in X$ and each $\phi \in X^*$,

where X^* is the dual space of X and $\langle \cdot, \cdot \rangle$ denotes the pairing between X and X^*. In other words, weak continuity implies strong continuity for semigroups. But strong continuity does not imply uniform continuity. In fact, if a (C_0) semigroup e^{tA} is uniformly continuous, then its generator would be a bounded operator.

1.3 Dissipative Operators

Definition 1.3.1 Let X be a Banach space and X^* its dual. A multivalued function $\mathcal{J}$ defined by

$$\mathcal{J}(f) := \{\phi \in X^* \ : \ \|\phi\|^2 = \|f\|^2 = \langle \phi, f \rangle\}$$

is called a *duality map* of X. A function $J : X \mapsto X^*$ such that for each $f \in X$, $J(f) \in \mathcal{J}(f)$ will be called a *duality section*. Now, an operator A on X is called *dissipative* if there exists a duality section J such that for each $f \in D(A)$, $\text{Re}\langle J(f), Af \rangle \leq 0$. A dissipative operator A is called *m-dissipative* provided $\rho(A) \cap]0, \infty[\neq \emptyset$.

An equivalent definition for a dissipative operator can be deduced from the following lemma.

Lemma 1.3.2 *A linear operator* A *on* X *is dissipative iff, for any* $\lambda > 0$ *and any* $f \in D(A)$,

$$\|(\lambda I - A)f\| \geq \lambda \|f\|.$$

An important consequence of this lemma is the celebrated Lumer–Phillips theorem:

Theorem 1.3.3 (Lumer–Phillips Theorem) *If* A *is an m-dissipative densely defined linear operator on* X*, then it generates a* (C_0) *semigroup of contractions. Conversely, if* A *generates a* (C_0) *semigroup of contractions, then:*

(i) $\overline{D(A)} = X$;
(ii) A *is dissipative for some* $J \in \mathcal{J}$;
(iii) $]0, \infty[\subset \rho(A)$.

1.4 Abstract Cauchy Problem

Let A be a closed operator on a Banach space X and consider the abstract Cauchy problem

$$\textbf{(ACP)} \qquad \begin{cases} \dfrac{\partial u}{\partial t}(t) = Au(t) \quad (t \geq 0), \\ u(0) = \phi \in X. \end{cases}$$

Definition 1.4.1 The problem **(ACP)** is called *well-posed* if $\rho(A) \neq \emptyset$ and for each $\phi \in D(A)$ there is a unique solution $u : \mathbb{R}_+ \mapsto D(A)$ of **(ACP)** in $C^1(\mathbb{R}_+, X)$. Such a solution is called a *classical solution*.

The following theorem gives the relationship between the well-posedness and the generation of a (C_0) semigroup.

Theorem 1.4.2 *The problem* **(ACP)** *is well-posed iff* A *generates a* (C_0) *semigroup* $\{e^{tA}\}_{t \geq 0}$.

This theorem implies that the existence and the uniqueness alone are not sufficient to guarantee the well-posedness. For the well-posedness we need also to have the continuous dependence of the initial data. In this case the classical solution of **(ACP)** is given by $u(t) = e^{tA}\phi$, for any $\phi \in D(A)$. Now this expression still makes sense if ϕ belongs to X. But in this case $u(t)$ is not differentiable (unless e^{tA} has a regularization effect, e.g. if e^{tA} is an *analytic semigroup*). Such a solution is called a *mild solution*.

The notion of a mild solution is more appropriate when we are dealing with an inhomogeneous initial value problem of the form

$$\textbf{(IP)} \qquad \begin{cases} \dfrac{\partial u}{\partial t}(t) = Au(t) + f(t) \quad (t \geq 0) \\ u(0) = \phi \in X. \end{cases}$$

It can be proved that (see e.g. [55]), if A generates a (C_0) semigroup e^{tA}, $\phi \in D(A)$ and if the known function f has a certain amount of regularity (e.g.; if $f \in C^1(\mathbb{R}_+, X)$ or $f \in C(\mathbb{R}_+, X)$ takes values in $D(A)$ and $Af \in C(\mathbb{R}_+, X)$), then the problem **(IP)** has a unique solution $u \in C^1(\mathbb{R}_+, X)$ with values in $D(A)$. This solution is a classical solution for **(IP)** and one can even write down the solution by means of the *variation of constants formula*:

$$u(t) = e^{tA}\phi + \int_0^t e^{(t-s)A} f(s)\, ds. \tag{1.8}$$

This formula still makes sense if for any $t \geq 0$, $f(t) \in X$. In this case, $u(t)$ given by (1.8) is referred to as a mild solution of **(IP)**.

1.5 Hille–Yosida Theorem

Thanks to Theorem 1.4.2, in order to prove the well-posedness of an abstract Cauchy problem we are led to verify if the operator A generates a C_0–semigroup. On the one hand, there is a two–way relationship between the resolvent of A and e^{tA}. In fact, $(\lambda - A)^{-1}$ can be regarded as the Laplace transform of e^{tA}

$$(\lambda - A)^{-1} f = \int_0^\infty e^{-\lambda t} e^{tA} f\, dt \qquad \text{for all } f \in X. \tag{1.9}$$

More precisely:

Lemma 1.5.1 *Let e^{tA} be a (C_0) semigroup of type ω. Then for any $\lambda > \omega$, $\lambda \in \rho(A)$ and for any $B \in \mathcal{L}(X)$*

$$B(\lambda - A)^{-1} f = \int_0^\infty e^{-\lambda t} B e^{tA} f\, dt \qquad \textit{for all } f \in X. \tag{1.10}$$

For $B = I$, we obtain (1.9), which is called the *resolvent formula*. Conversely, if Γ is a piecewise smooth curve surrounding the spectrum of A and $(\lambda - A)^{-1}$ satisfies certain conditions which ensure the convergence of the *Dunford–Taylor integral*

$$T(t) = \frac{1}{2\pi i} \int_\Gamma e^{\lambda t} (\lambda - A)^{-1}\, d\lambda \qquad \text{for all } t \geq 0, \tag{1.11}$$

then $T(t) = e^{tA}$. On the other hand, it is much easier to obtain information about the resolvent than to verify the well-posedness of a Cauchy problem. Therefore, since the Hille–Yosida theorem provides us the generation conditions by means of the estimates on the powers of the resolvent, this theorem is central in the theory of one-parameter semigroups.

First we apply this theorem to a (C_0) semigroup of contractions which is an obvious consequence of Lemma 1.3.2 and the Lumer–Phillips Theorem. This result is referred to in some treatises as the Hille–Yosida theorem.

Theorem 1.5.2 *A linear operator A on X generates a (C_0) semigroup of contractions iff:*

(i) *$D(A)$ is dense in X;*
(ii) *A is closed;*
(iii) *$]0, \infty[\subset \rho(A)$; and*
(iv)

$$\|(\lambda I - A)^{-1}\| \leq \frac{1}{\lambda} \quad \textit{for all } \lambda > 0.$$

But we refer to the generalization of this theorem for $A \in \mathcal{G}(M, \omega)$ as the Hille–Yosida theorem, which is stated as follows.

Theorem 1.5.3 (Hille–Yosida Theorem) *A linear operator A on X generates a (C_0) semigroup, iff:*

(i) *$D(A)$ is dense in X;*
(ii) *A is closed; and*
(iii) *there exist two constants $M \geq 1$ and $\omega \in \mathbb{R}$ such that $\lambda \in \rho(A)$ for each $\lambda > \omega$ and*

$$\|(\lambda I - A)^{-n}\| \leq \frac{M}{(\lambda - \omega)^n} \quad \textit{for all } \lambda > \omega \textit{ and all } n \in \mathbb{N}.$$

In this case $\|e^{tA}\| \leq Me^{\omega t}$, for all $t \geq 0$.

1.6 The Spectral Properties of the Generator of a (C_0) Semigroup

Definition 1.6.1 If $\lambda \in \sigma(A)$ then three situations could occur. Either $(\lambda I - A)$ is not injective, in which case we say that λ is a *point spectrum* of A and we write $\lambda \in \sigma_p(A)$; or $(\lambda I - A)$ is injective but $R(\lambda I - A)$, where the range of $(\lambda I - A)$ is not dense in X, then we say that λ is in *residual spectrum* of A and we write $\lambda \in \sigma_r(A)$; or finally, $(\lambda I - A)$ is injective and has a dense range but it is not surjective. In this case we say that λ belongs to the *continuous spectrum* of A and we write $\lambda \in \sigma_c(A)$. In this way $\sigma(A)$ can be partitioned into three disjoint sets,

$$\sigma(A) = \sigma_p(A) \cup \sigma_r(A) \cup \sigma_c(A). \tag{1.12}$$

Lemma 1.6.2 *Let A be a densely defined operator on X. Then:*

(i) *if λ is in $\sigma_r(A)$, then $\overline{\lambda}$ is in $\sigma_p(A^*)$;*
(ii) *if $\overline{\lambda}$ is in $\sigma_p(A^*)$, then λ is in either $\sigma_p(A)$ or in $\sigma_r(A)$.*

The proof is given in [117, Theorem 2.3]. If A is a bounded linear operator and $e^{tA} = \sum_{n=0}^{\infty} \frac{t^n A^n}{n!}$, then

$$\sigma(e^{tA}) = e^{t\sigma(A)} := \{e^{\lambda t} \mid \lambda \in \sigma(A)\}. \tag{1.13}$$

Example 1.6.3 Take a (C_0) semigroup $(T(t))_{t\geq 0}$ that cannot be extended to a group (e.g., $T(t)f(x) = f(x-t)$) on $C_0(\mathbb{R}_+)$. Then $0 \in \sigma(T(t))$ for all $t > 0$, although evidently 0 is never contained in $exp(t\sigma(A))$.

Hence the relation (1.13) becomes

(SMT) $$\sigma(e^{tA}) \setminus \{0\} = e^{t\sigma(A)}, \text{ for all } t \geq 0.$$

This relation is referred to as the *spectral mapping theorem* (SMT for short). In general, such a straightforward relationship between the spectrum of a (C_0) semigroup and the spectrum of its generator does not exist. In fact, one can find in Davies [34, Theorem 2.17] a counterexample in which for the generator of a (C_0) group e^{tA}, one has $\sigma(A) \subseteq i\mathbb{R}$ and $e^{|t|} \in \sigma(e^{tA})$ for all $t \in \mathbb{R}$. However, we have some relationships between the different parts of the spectrum of a (C_0) semigroup and the correspondent parts of its generator. For example, (**SMT**) holds for some special classes of (C_0) semigroups as indicated in the following lemma.

Lemma 1.6.4 *The identity* (**SMT**) *holds for the following classes of (C_0) semigroups:*

(a) eventually compact semigroups;
(b) eventually norm continuous semigroups;
(c) eventually differentiable semigroups;
(d) analytic semigroups.

Some of these classes will be discussed in the following sections. In each of these cases we have the spectral bound equal growth bound condition

(SBeGB) $$s(A) = \omega(A)$$

(see [49, Corollary 2.10] in which paper the abbreviation (SBeGB) is used).

Remark 1.6.5 It is worthwhile to observe that the growth bound is always greater than the spectral bound $s(A) \leq \omega(A)$ and in [49] one can find many examples in which $s(A) < \omega(A)$. This observation leads us to return to the identity (1.12) and remark that (**SMT**) holds for σ_p, hence if $\sigma_a \neq \emptyset$, the presence of a continuous spectrum causes the gap between $s(A)$ and $\omega(A)$.

Lemma 1.6.6 (Spectral mapping theorem) *Let A be the generator of a (C_0) semigroup $\{e^{tA}\}_{t\geq 0}$. For any $t \geq 0$,*

(a) $e^{t\sigma(A)} \subseteq \sigma(e^{tA})$;
(b) $\lambda \in \sigma_p(A)$ *iff* $e^{t\lambda} \in \sigma_p(e^{tA})$;
(c) $e^{t(\sigma_p(A)\cup\sigma_r(A))} = (\sigma_p(e^{tA}) \cup \sigma_r(e^{tA})) \setminus \{0\}$.

1.7 (C_0) Semigroups in a Hilbert Space

Let $\mathcal{H}$ be a Hilbert space endowed with the inner product $(\cdot,\cdot)$ and the norm $\|\cdot\| = \sqrt{(\cdot,\cdot)}$.

According to Definition 1.1.5, $A \subset B$ implies that $B^* \subset A^*$.

Definition 1.7.1 An operator A on $\mathcal{H}$ is *symmetric* or *Hermitian* if $A \subset A^*$, that is, A is densely defined and $(Af, g) = (f, Ag)$ for all $f, g \in D(A)$. Equivalently, A is symmetric iff

$$(Af, g) = (f, Ag) \qquad \text{for all} \quad f, g \in D(A).$$

We say that A is *skew-symmetric* if $A \subset -A^*$. A is *self-adjoint* iff $A = A^*$, that is, iff A is symmetric and $D(A) = D(A^*)$. Similarly, A is *skew-adjoint* if $A = -A^*$. An invertible bounded operator $U \in \mathcal{L}(\mathcal{H})$ is called *unitary* if $U^* = U^{-1}$ and a (C_0) *unitary group* is a (C_0) group of unitary operators.

Remark 1.7.2 (a) When $\mathcal{H}$ is a complex Hilbert space, A is skew-adjoint iff iA is self-adjoint.
(b) A (C_0) semigroup e^{tA} is self-adjoint for each $t \in \mathbb{R}_+$ iff its generator A is self-adjoint.
(c) An operator U is unitary iff U is isometric and surjective.

Lemma 1.7.3 *For a self-adjoint or skew-adjoint operator A in a Hilbert space, $\sigma_r(A) = \emptyset$.*

For a self-adjoint operator, this lemma is proved in [104, Theorem 4.23]. For a skew-adjoint operator A, we remark that iA is self-adjoint and $\sigma_p(iA) = i\sigma(A)$.

Theorem 1.7.4 (Stone's Theorem) *Let $\mathcal{H}$ be a Hilbert space. An operator A is the generator of a (C_0) unitary group, if and only if A is skew-adjoint.*

1.8 Perturbation Theory

As we have mentioned in the introduction, the central aim of the scattering theory is to compare the solution of an abstract Cauchy problem **(ACP)** which was designated as an unperturbed problem with a perturbed Cauchy problem

(PCP)
$$\begin{cases} \dfrac{\partial u}{\partial t}(t) = Cu(t) := Au + Bu \quad (t \geq 0), \\ u(0) = \phi \in X. \end{cases}$$

Here the operator C is an additive perturbation of A by an operator B. The main preoccupation of the perturbation theory is providing the sufficient conditions in order that C generates a (C_0) semigroup. Once these conditions are fulfilled the C_O-semigroup e^{tC} can be represented by *Duhamel formulae*. More precisely:

Theorem 1.8.1 (i) *Let A generate a (C_0) semigroup e^{tA} and B a linear operator with $D(B) \supset D(A)$ such that $C = A + B$ generates a (C_0) semigroup e^{tC}. Then there are two ways to represent e^{tC}:*

$$e^{tC} = e^{tA} + \int_0^t e^{(t-s)A} B e^{sC} \, ds, \quad t \in \mathbb{R}_+, \tag{1.14}$$

which is the first Duhamel formula, or

$$e^{tC} = e^{tA} + \int_0^t e^{(t-s)C} B e^{sA} \, ds, \quad t \in \mathbb{R}_+, \tag{1.15}$$

which is the second Duhamel formula

(ii) *By taking $r = t - s$ we can write also*

$$e^{tC} = e^{tA} + \int_0^t e^{rA} B e^{(t-r)C} \, dr, \quad t \in \mathbb{R}_+ \tag{1.16}$$

and

$$e^{tC} = e^{tA} + \int_0^t e^{rC} B e^{(t-r)A} \, dr, \quad t \in \mathbb{R}_+. \tag{1.17}$$

Corollary 1.8.2 *By iterating the first Duhame formula, one obtains the so-called Dyson–Phillips expansion of the (C_0) semigroup e^{tC}, given by*

$$e^{tC} = \sum_{n=0}^{\infty} S_n(t), \tag{1.18}$$

where $S_0(t) = e^{tA}$, and

$$S_n(t) = \int_0^t e^{(t-s)A} B S_{n-1}(s) \, ds \qquad (n \geq 1). \tag{1.19}$$

The series (1.18) converges in $\mathcal{L}(X)$ uniformly in any bounded interval $[t_1, t_2]$, $0 \leq t_1 < t_2 < \infty$.

To obtain the Duhamel formula and Dyson–Phillips expansion we do not need to assume that $B \in \mathcal{L}(X)$. In the framework of dissipative operators we have the following result:

Theorem 1.8.3 *Let $\{e^{tA}\}_{t\geq 0}$ be a (C_0) semigroup of contractions on a Banach space X. Let B be a dissipative operator with $D(B) \supset D(A)$. Assume there are constants $0 \leq a < 1$ and $b \geq 0$ such that*

$$\|Bf\| \leq a\|Af\| + b\|f\| \quad \text{for all } f \in D(A). \tag{1.20}$$

Then $C := A + B$ generates also a (C_0) contraction semigroup.

If A and B satisfy the domination condition (1.20), we say that *B is A-bounded with constants a and b*. In [90] Miyadera gave some more general conditions. He shows how A-boundedness of the operator B suffices to get the generation of an extension of $A + B$. Here we cite a performed version of this theorem.

Theorem 1.8.4 (Miyadera's Perturbation Theorem) *Let $\{e^{tA}\}_{t\geq 0}$ be a (C_0) semigroup on a Banach space X. Let B be a linear operator in X such that the following conditions are satisfied:*

(i) *B is A-bounded.*
(ii) *There exist constants $0 < a \leq +\infty$, $0 \leq \alpha < 1$, such that for all $f \in \mathcal{D} := D(A)$*

$$\int_0^a \|Be^{tA}f\|dt \leq \alpha\|f\|. \tag{1.21}$$

Then, there exists a unique (C_0) semigroup $\{e^{tC}\}_{t\geq 0}$ on X satisfying the following equivalent conditions:

(a) *C is an extension of $(A + B)\,|_{\mathcal{D}}$.*
(b) *For all $f \in \mathcal{D}$ and all $t \geq 0$ we have*

$$e^{tC}f = e^{tA}f + \int_0^t e^{(t-s)A}Be^{sC}f\,ds, \quad \text{for all } f \in \mathcal{D} \text{ and all } t \in \mathbb{R}_+. \tag{1.22}$$

(c) *There exists $a > 0$ such that (1.22) holds for all $f \in \mathcal{D}$ and all $t \in [0, a)$.*

Definition 1.8.5 We say that B is a *Miyadera perturbation* of A, if the assumptions of the above theorem are satisfied.

Corollary 1.8.6 *Under the assumptions of the above theorem, if $\|e^{tA}\| \leq M$ for all $t \geq 0$, then the Dyson–Phillips expansion formula (1.18), converges uniformly on a bounded interval $[0, \tau]$ in $\mathcal{L}(X)$ and we have the estimate*

$$\|e^{tC}\| \leq \frac{M}{1-\alpha}\exp\left(\frac{1}{\tau}\ln\frac{M}{1-\alpha}\right)t \quad \text{for all} \quad t \geq 0. \tag{1.23}$$

Proof Fix $\tau > 0$ and define e^{tC} for $0 \le t < \tau$ by (1.18). By induction one can show that for any $n \in \mathbb{N}$, S_n which is defined in Corollary 1.8.2 satisfies $\|S_n(t)\| \le M\alpha^n$ and consequently $\|e^{tC}\| \le \sum_{n=0}^{\infty} \|S_n(t)\| \le \frac{M}{1-\alpha}$. Let $t \ge 0$ and $n \in \mathbb{N}$ be such that $\frac{t}{n} < \tau \le \frac{t}{n-1}$. This implies that $n \le \frac{t}{\tau} + 1$ and

$$\|e^{tC}\| = \| \left[e^{\frac{t}{n}C} \right]^n \| \le \left(\frac{M}{1-\alpha} \right)^n \le \frac{M}{1-\alpha} \left(\frac{M}{1-\alpha} \right)^{t/\tau}.$$

□

Remark 1.8.7 If $B \in \mathcal{L}(X)$ and Be^{tA} is strongly continuous at 0_+, then B is A-bounded and for $a > 0$ small enough, one can find $\alpha < 1$ such that the assumptions (i) and (ii) of Theorem 1.8.4 hold. Hence we have the following corollary:

Corollary 1.8.8 (a) *Suppose A is the generator of a (C_0) semigroup on a Banach space X and $B \in \mathcal{L}(X)$, then the closed operator $C = A + B$ with domain $D(C) = D(A)$ also generates a (C_0) semigroup and the (C_0) semigroup e^{tC} can be expressed by the Duhamel formulae.*

(b) *In the case where A is the generator of a (C_0) group such that*

$$\|e^{tA}\| \le Me^{\omega|t|} \quad \text{for all} \quad t \in \mathbb{R},$$

$C = A + B$ also generates a (C_0) group and we have

$$\|e^{tC}\| \le Me^{(\omega + M\|B\|)|t|} \quad \text{for all} \quad t \in \mathbb{R}. \tag{1.24}$$

Theorem 1.8.9 (Chernoff's Theorem) *Let $\{F(t)\}_{t\ge 0}$ be a family of contractions in $\mathcal{L}(X)$, such that $F(0) = I$ and let $\mathcal{D}$ be a core of the generator A of a (C_0) semigroup e^{tA} of contractions on X. If*

$$\lim_{t\to 0_+} \frac{F(t)f - f}{t} = Af \quad \text{for all } f \in \mathcal{D},$$

then

$$\lim_{n\to\infty} \{F(\frac{t}{n})\}^n f = e^{tA} f \quad \text{for all } f \in X,$$

uniformly for t in bounded intervals.

Corollary 1.8.10 (Trotter product formula) *Suppose A, B and $C := \overline{A + B}$ generate (C_0) semigroups e^{tA}, e^{tB} and e^{tC} on X. Then*

$$\lim_{n\to\infty} \{e^{\frac{t}{n}A} e^{\frac{t}{n}B}\}^n f = e^{tC} f \quad \text{for all } f \in X,$$

uniformly for t in compact intervals.

Proof This can be obtained by taking $F(t) := e^{tA}e^{tB}$ and $\mathcal{D} := D(A + B) = D(A) \cap D(B)$. □

Corollary 1.8.11 *Suppose $A \in \mathcal{G}(1, \omega)$ generates a (C_0) semigroup e^{tA} and $B \in \mathcal{L}(X)$. Then, if $C = A + B$,*

$$\lim_{n\to\infty} \{e^{\frac{t}{n}A}e^{\frac{t}{n}B}\}^n f = e^{tC} f \quad \textit{for all } f \in X, \tag{1.25}$$

uniformly for t in compact intervals.

Proof If $A \in \mathcal{G}(1, \omega)$, then it suffices to renorm the space X with an equivalent norm

$$|||f||| := \sup_{t\geq 0} \|e^{-t\omega}e^{tA} f\|,$$

which makes A the generator of a (C_0) contraction semigroup. □

1.9 Positive Semigroups

When we talk about the positivity in the theory of operators, the first concept that comes to mind is the Krein–Rutman theorem, which is a generalization of the Perron–Frobenius theorem to infinite-dimensional Banach spaces, which was proved by Krein and Rutman in 1948.

In fact in linear algebra, the Perron–Frobenius theorem [75, Chap. 16], asserts that a real square matrix with positive entries has a unique largest real eigenvalue and that the corresponding eigenvector can be chosen to have strictly positive components, and also asserts a similar statement for certain classes of nonnegative matrices. And the Krein–Rutman theorem asserts that if X is a Banach lattice, which will be defined in Definition 1.9.4, and we let $K \subset X$ be a positive convex cone such that $K - K$ is dense in X and let $T \in \mathcal{L}(X)$ be a non-zero compact operator which is positive, that is, $T(K) \subset K$ and assume that its spectral radius $r(T)$ is strictly positive, then $r(T)$ is an eigenvalue of T with a positive eigenvector.

In this section we briefly outline the basic concepts necessary for understanding the essentials of theory of positive operators. Those desiring more detailed treatments should refer to [106] or [89].

We start with the definition of an ordered vector space and a vector lattice.

Definition 1.9.1 An *ordered vector space* is a real vector space X endowed with an ordering, usually written $\leq$, such that $(X, \leq)$ is compatible with the linear structure of X, i.e.,

$$f \leq g \quad \text{implies} \quad f + h \leq g + h \quad \text{for all } f, g, h \in X \tag{1.26}$$

and

$$f \geq 0 \quad \text{and} \quad \lambda \in \mathbb{R}^+ \quad \text{implies} \quad \lambda f \geq 0. \tag{1.27}$$

A *vector lattice* is an ordered vector space X, such that $f \vee g \equiv \sup\{f, g\}$ and $f \wedge g \equiv \inf\{f, g\}$ exist for all $f, g \in X$. Vector lattices are also called *Riesz spaces* or *linear lattices*.

Letting $(\Omega, \mathscr{B}, \mu)$ be a probability space, we say that a function $f : \Omega \mapsto \mathbb{R}$ is positive iff $f(x) \geq 0$ for almost every $x \in \Omega$ and we write $f \geq 0$. Now, let X be a vector space of all measurable functions endowed with an ordering $f \leq g$ iff $g - f \geq 0$. We denote by $X_+ \equiv \{f \in X; f \geq 0\}$ the *positive part* of X.

From the properties (1.26) and (1.27) it is not hard to see that, if X is an ordered vector space, X_+ is a convex set and forms a cone with vertex 0, which is usually called the *positive cone* of X. Conversely, given a proper cone C with vertex 0 in a vector space X, the ordering defined by

$$f \leq g \quad \text{iff} \quad g - f \in C$$

builds an ordered vector space on X.

Let $\{f_\alpha\}_{\alpha \in A}$ be a family in a vector lattice X. It is readily seen from the definition that whenever $\sup_\alpha f_\alpha$ exists then $\inf_\alpha(-f_\alpha)$ exists and

$$\sup_\alpha f_\alpha = -\inf_\alpha(-f_\alpha). \tag{1.28}$$

Moreover, for each $f \in X$ we have

$$f + \sup_\alpha f_\alpha = \sup_\alpha(f + f_\alpha), \tag{1.29}$$

$$f + \inf_\alpha f_\alpha = \inf_\alpha(f + f_\alpha).$$

Definition 1.9.2 Let X be a vector lattice. For any $f \in X$, $f^+ \equiv f \vee 0$, $f^- \equiv (-f) \vee 0$ and $|f| \equiv f \vee (-f)$ are called respectively the *positive part*, the *negative part*, and the *modulus* of f. We say also that a subset C of a vector lattice is *generating* when

$$X = C - C.$$

Lemma 1.9.3 *The positive cone X_+ of a vector lattice X is always generating.*

Proof For any $f, g, u, v \in X$,

$$\begin{aligned}(u + v) - (f \wedge g) &= (u + v) + (-f) \vee (-g) && \text{(by 1.28)}\\ &= (u + v - f) \vee (u + v - g) && \text{(by 1.29)}\end{aligned}$$

and for $u = f$ and $v = g$, we get

$$f + g = f \vee g + f \wedge g,$$

which yields, for $g = 0$,

$$f = f^+ - f^-$$

or,

$$X = X_+ - X_+.$$

□

Definition 1.9.4 A real Banach space X is called a *Banach lattice* if X is a vector lattice for some ordering $\leq$ and furthermore

$$0 \leq f \leq g \quad \text{implies} \quad \|f\| \leq \|g\|. \tag{1.30}$$

In this case we say that the norm $\|\cdot\|$ is *absolutely monotone.*

For a measure space (Ω, Σ, μ) all the spaces $L^p(\Omega, \mu)$ $(1 \leq p < \infty)$ of all (real) measurable functions on (Ω, μ) equipped with

$$\|f\|_p := \left(\int_\Omega |f|^p \, d\mu \right)^{1/p}$$

and the space $L^\infty(\Omega, \mu)$ of all μ–essentially bounded (real) functions equipped with the sup-norm

$$\|f\|_\infty := \text{ess-sup}\{|f(x)| \; : \; x \in \Omega\}$$

are Banach lattices for usual ordering "$\leq$" ($f \geq 0$ iff $f(x) \geq 0$ for almost all $x \in \Omega$).

If $X_\mathbb{R}$ is a Banach lattice over $\mathbb{R}$, one can define its complexification $X_\mathbb{C} := X_\mathbb{R} + iX_\mathbb{R}$ equipped with the norm

$$\|u\|_\mathbb{C} := \| \, |u| \, \|_\mathbb{R}, \tag{1.31}$$

where the absolute value of $u = f + ig \in X_\mathbb{C}$ $(f, g \in X_\mathbb{R})$ is defined by $|u| = \sup\{(\cos\theta) f + (\sin\theta) g \; : \; 0 \leq \theta \leq 2\pi\}$. The space $(X_\mathbb{C}, \|\cdot\|_\mathbb{C})$ is called a *complex Banach lattice*. Every real linear operator $A_\mathbb{R}$ on $X_\mathbb{R}$ may by uniquely extended to a complex linear operator $A_\mathbb{C}$ on $X_\mathbb{C}$ by

$$A_\mathbb{C} u := A_\mathbb{R} f + i A_\mathbb{R} g$$

whenever $u = f + ig \in X_\mathbb{C}$ with $f, g \in X_\mathbb{R}$. If $X_\mathbb{R}$ is the real Banach lattice $L^p(\Omega, \mu)$ $(1 \leq p < \infty)$, then $X_\mathbb{C}$ can be endowed with

$$\|u\| = \frac{1}{4}\int_0^{2\pi} \|(\cos\theta)f + (\sin\theta)g\|\,d\theta, \tag{1.32}$$

which is equivalent to the norm defined in (1.31) and we have:

Lemma 1.9.5

$$\|A_{\mathbb{R}}\| = \|A_{\mathbb{C}}\|. \tag{1.33}$$

Proof Let $p = 1$. The proof follows from the identity

$$\sqrt{a^2+b^2}\int_0^{2\pi} |\cos\theta|\,d\theta = \int_0^{2\pi} |a\cos\theta + b\sin\theta|\,d\theta.$$

In fact, for any $u = f + ig \in X_{\mathbb{C}}$

$$\begin{aligned}
\|A_{\mathbb{C}}u\|_1 &= \frac{1}{4}\int_0^{2\pi} \|(\cos\theta)A_{\mathbb{R}}f + (\sin\theta)A_{\mathbb{R}}g\|_1\,d\theta \\
&= \frac{1}{4}\int_\Omega\int_0^{2\pi} |(\cos\theta)A_{\mathbb{R}}f(x) + (\sin\theta)A_{\mathbb{R}}g(x)|\,d\theta\,d\mu(x) \\
&= \frac{1}{4}\int_\Omega \sqrt{|A_{\mathbb{R}}f(x)|^2 + |A_{\mathbb{R}}g(x)|^2}\int_0^{2\pi} |\cos\theta|\,d\theta\,d\mu(x) \\
&= \|\,|A_{\mathbb{R}}f + iA_{\mathbb{R}}g|\,\|_1,
\end{aligned}$$

which implies (1.33) in $L^1(\Omega,\mu)$. The proof for $1 < p < \infty$ is similar (see for example [34, Lemma 7.5]). □

Unless explicitly stipulated otherwise, we consider always that a Banach lattice X is complex.

Definition 1.9.6 A bounded linear operator A on a Banach lattice X is called *positive* or *positivity preserving*, and we write $A \geq 0$, iff $AX_+ \subseteq X_+$. We write $A \geq B$ iff $A - B \geq 0$.

The following lemma follows readily from the above Definition.

Lemma 1.9.7 *Let A, B and C be bounded linear operators on a Banach lattice.*

(a) *If $A \geq B$ and $C \geq 0$, then $AC \geq BC$ and $CA \geq CB$.*
(b) *If $A \geq B \geq 0$, then $A^n \geq B^n \geq 0$ for any integer $n \geq 1$.*

Remark 1.9.8 If X is a Banach lattice then the operator norm on $\mathcal{L}(X)$ is also absolutely monotone; that is,

$$0 \leq A \leq B \quad \text{implies that} \quad \|A\| \leq \|B\|.$$

In [14, Theorem 1.3.2 and Corollary 1.7.5] it is shown that this can always be arranged by choosing an equivalent norm.

Definition 1.9.9 Let X be a Banach lattice and $\{e^{tA}\}_{t\geq 0}$ a (C_0) semigroup in X. The semigroup $\{e^{tA}\}_{t\geq 0}$ is called *positive* if $e^{tA}X_+ \subseteq X_+$ for all $t \geq 0$.

Definition 1.9.10 Let X be a Banach lattice. An operator A on X is called *resolvent positive* if there exists $w \in \mathbb{R}$ such that $(w, \infty) \subseteq \rho(A)$ and $(\lambda - A)^{-1} \geq 0$ for all $\lambda > w$. For a resolvent positive operator A the *spectral bound* $s(A)$ is defined by

$$s(A) := \inf\{w \in \mathbb{R} \mid (w, \infty) \subseteq \rho(A) \text{ and } (\lambda - A)^{-1} \geq 0 \text{ for all } \lambda > w\}.$$

In the following theorem we characterize the resolvent positive operators which are automatically generators of a positive (C_0) semigroup.

Theorem 1.9.11 *Let A be a densely defined resolvent positive operator. If there exist $\lambda_0 > s(A)$ and $c > 0$ such that*

$$\|(\lambda_0 - A)^{-1} f\| \geq c\|f\| \quad \text{for all } f \in X_+,$$

then A generates a positive (C_0) semigroup and the spectral bound of A is equal to its growth bound ($s(A) = \omega(A)$).

In the forthcoming chapters we will dealing essentially with $L^1(\Omega, \mu)$ spaces. In spite of the fact that the Banach lattice $L^1(\Omega, \mu)$ is not reflexive, this space is an AL-space which is extremely relevant for dealing with positive operators.

Definition 1.9.12 If the norm on a Banach lattice X satisfies

(L) $\qquad \|f + g\| = \|f\| + \|g\| \quad$ for all $\quad f, g \in X_+$

then X is called an *abstract L-space* or an *AL-space*.

The property **(L)** can be generalized to an X_+-valued Riemann integrable function $f : [a, b] \mapsto X_+$ and we have

$$\left\| \int_a^b f(t)dt \right\| = \int_a^b \|f(t)\|dt. \tag{1.34}$$

The dual of an AL-space is an *AM-space*. This is a Banach lattice satisfying

(M) $\qquad \|f \vee g\| = \|f\| \vee \|g\| \quad$ for all $\quad f, g \geq 0$

and infinite-dimensional AL- or AM-spaces are never reflexive.

The space $L^1(\Omega, \mu)$ possesses another important property, the so-called Dunford–Pettis property.

Definition 1.9.13 A Banach space X is said to have the *Dunford–Pettis property* if for every Banach space Y, any weakly compact operator $K : X \mapsto Y$ maps a weakly compact subset of X onto a compact subset of Y.

Remark 1.9.14 An important consequence of the Dunford–Pettis property is the fact that if K is a weakly compact operator in $\mathcal{L}(X)$, then K^2 is compact in $\mathcal{L}(X)$. Every AL-space and every AM-space has the Dunford–Pettis property (see [106, Theorem II.9.9]) and consequently the spaces $L^1(\Omega, \mu)$, $L^\infty(\Omega, \mu)$ and $C(\Omega)$ have the Dunford–Pettis property. As AL- or AM-spaces, a reflexive space with the Dunford–Pettis property is finite dimensional.

We can gather all the important consequences of the Dunford–Pettis property in the following lemma.

Lemma 1.9.15 *Let X be a Banach lattice having the Dunford–Pettis property. Let S, T be two bounded linear operators on X.*

(i) *The set of all weakly compact operators is norm-closed subset of $\mathcal{L}(X)$.*
(ii) *If T is weakly compact and $0 \leq S \leq T$, then S is weakly compact.*
(iii) *If S and T are weakly compact, then ST is compact.*

In the following theorem we will see that if the Banach lattice X is an AL-space the Miyadera Perturbation Theorem can be considerably enlarged.

Theorem 1.9.16 *Let X be an AL-space, and let A be the generator of a positive C_0-semigroup on X. Let $B : D(A) \mapsto X$ be a positive operator, and assume that $A + B$ is resolvent positive. Then $A + B$ is the generator of a positive C_0-semigroup.*

Lemma 1.9.17 *Let $\{e^{tA}\}_{t\geq 0}$ be a bounded positive (C_0) semigroup on a AL–space X and let $B : D(A) \mapsto X$ be a A–bounded positive operator on X. Then the following assertions are equivalent:*

(i) $\lim_{t\to\infty} B \int_0^t e^{sA} f\, ds$ *exist for all* $f \in X$,
(ii) $\lim_{\lambda\to 0_+} B(\lambda - A)^{-1} f$ *exist for all* $f \in X$,

and any of the above assertions implies

$$B \int_0^\infty e^{sA} f\, ds = B(0_+ - A)^{-1} f \quad \textit{for all } f \in X. \tag{1.35}$$

Proof Assume (i) and write

$$\int_0^\infty Be^{sA} ds := \mathrm{s} - \lim_{t\to\infty} \int_0^t Be^{sA} ds,$$

then for $f \in X_+$ and $\lambda > 0$

$$\int_0^\infty Be^{sA} f ds = B \int_0^\infty e^{sA} f ds \geq B \int_0^\infty e^{-\lambda s} e^{sA} f ds = B(\lambda - A)^{-1} f.$$

The first equality is a consequence of Lemma 1.1.8, since B is A–bounded and the last equality shows that $\lambda \longrightarrow B(\lambda - A)^{-1} f$ is decreasing on $\mathbb{R}_+$. Hence by the monotone convergence theorem it follows that $B(0_+ - A)^{-1}f := \lim_{\lambda\to 0_+} B(\lambda - A)^{-1}f$ exists and

$$B(0_+ - A)^{-1}f \le B\int_0^\infty e^{sA}f\,ds.$$

Conversely, assume (ii), then for all $f \in X_+$, $\lambda > 0$ and $t > 0$

$$B\int_0^\infty e^{-\lambda s}e^{sA}f\,ds \le B(\lambda - A)^{-1}f \le B(0_+ - A)^{-1}f.$$

As $\lambda \to 0_+$, we get

$$B\int_0^\infty e^{sA}f\,ds \le B(0_+ - A)^{-1}f.$$

Once more by applying the monotone convergence theorem we obtain the reverse inequality:

$$B(0_+ - A)^{-1}f \ge B\int_0^\infty e^{-sA}f\,ds = \int_0^\infty Be^{-sA}f\,ds.$$

Since X_+ is generating, the above inequalities hold for all $x \in X$. □

Remark 1.9.18 If $-A$ generates a bounded positive (C_0) semigroup the formula (1.35) can be replaced by

$$\int_{-\infty}^0 Be^{sA}f\,ds = \int_0^\infty Be^{-sA}f\,ds = B(0_+ + A)^{-1}f = -B(0_- - A)^{-1}f, \qquad f \in X.$$

Another consequence of the positive semigroups is the following:

Lemma 1.9.19 *Let A generate a positive (C_0) semigroup e^{tA} and B be a negative bounded operator such that $B + \|B\|Id \ge 0$. Then, if $C = A + B$, we have $e^{tC} \le e^{tA}$ for all $t \ge 0$.*

Considerably stronger results are available if one considers that a positive semigroup is irreducible. For example, if $\{e^{tA}\}_{t\ge 0}$ is irreducible then the spectrum $\sigma(A)$ is always non-empty, positive eigenfunctions are strictly positive and if the spectral bound $s(A)$ is a pole of A, it is algebraically and geometrically simple. In the next section we will give these properties in the context of eventually compact semigroup. First we give the definition of irreducibility.

Definition 1.9.20 A positive (C_0) semigroup $\{e^{tA}\}_{t\ge 0}$ on a Banach lattice X is called *irreducible* if, given $0 < f \in X, 0 < f^* \in X^*$, there exists $t_0 > 0$ such that $\langle e^{t_0 A}f, f^*\rangle > 0$.

Lemma 1.9.21 *A (C_0) semigroup $\{e^{tA}\}_{t\ge 0}$ on a Banach lattice X is irreducible iff for some, hence for every, $\lambda > s(A)$ we have, $(\lambda - A)^{-1}f > 0$, whenever $f > 0$.*

1.10 Compact Semigroups

Definition 1.10.1 Let X, Y be two Banach spaces. An operator $A \in \mathscr{L}(X, Y)$ is called *compact* or *completely continuous* if it maps bounded sets of X into compact sets of Y. Note that A is compact if and only if it maps bounded sets of X onto relatively compact sets of Y.

The following lemma enumerates deferent properties of compact operators. The proofs can be found in any standard functional analysis manuscript.

Lemma 1.10.2 *Let A be a compact operator in $\mathscr{L}(X, Y)$,*

(i) *A maps weakly convergent sequences of X into convergent sequences of Y.*
(ii) *If $\{B_n\}$ is a sequence of compact operators which converges uniformly to $B \in \mathscr{L}(X, Y)$, then B is also compact.*
(iii) *If $A \in \mathscr{L}(X)$, $B \in \mathscr{L}(X)$ and A is compact then AB and BA are also compact.*
(iv) *$\sigma(A)$ is a discrete set with no other accumulation point than zero.*
(v) *Any $\lambda \in \sigma(A)$ is an eigenvalue of A with finite multiplicity, that is the corresponding eigenspace is finite dimensional.*

Definition 1.10.3 A (C_0) semigroup $\{e^{tA}\}_{t\geq 0}$ on a Banach space X is called *compact* if for every $t > 0$, e^{tA} is compact. It is called *eventually compact* if there exists $t_0 > 0$ such that $e^{t_0 A}$ is compact.

If $e^{t_0 A}$ is compact, then for $t \geq t_0$, $e^{tA} = e^{t_0 A} e^{(t-t_0)A}$, thus e^{tA} is the product of a bounded operator and a compact operator and therefore e^{tA} is compact.

Lemma 1.10.4 *Let $\{e^{tA}\}_{t\geq 0}$ be a (C_0) semigroup on a Banach space X. e^{tA} is compact for $t \geq t_0$, iff e^{tA} is uniformly continuous for $t \geq t_0$ and $(\lambda - A)^{-1} e^{t_0 A}$ is compact for some $\lambda \in \rho(A)$.*

Definition 1.10.5 A (C_0) semigroup $\{e^{tA}\}_{t\geq 0}$ on a Banach space X is called *eventually norm continuous* if there exists $t_0 > 0$ such that the function $t \mapsto e^{tA}$ is norm continuous from (t_0, ∞) into $\mathcal{L}(X)$.

Lemma 1.10.6 *Any eventually compact (C_0) semigroup on a Banach space X is eventually normcontinuous.*

The eventual compactness property allows us to describe the long-term behavior of the semigroup by using the results on the spectrum of its generator.

Lemma 1.10.7 (a) *Let $\{e^{tA}\}_{t\geq 0}$ be a (C_0) semigroup on a Banach space X which is eventually compact. Then, the generator A of $\{e^{tA}\}_{t\geq 0}$ has only the point spectrum, which is a countable set (possibly finite or empty) and contains only poles of finite multiplicity. Furthermore, the set $\{\mu \in \sigma(A) \ : \ \operatorname{Re} \mu \geq r\}$ is finite for every $r \in \mathbb{R}$ and $\sigma(A) = \{\lambda_0, \lambda_1, \cdots\}$ with $\operatorname{Re} \lambda_{n+1} \leq \operatorname{Re} \lambda_n$, for all $n \in \mathbb{N}$ and $\lim_{n\to\infty} \operatorname{Re} \lambda_n = -\infty$ if $\sigma(A)$ is infinite. Denoting the pole order at λ_n by*

N_n and the corresponding residue by P_n (i.e., $P_n := \frac{1}{2\pi i}\int_{|\lambda-\lambda_n|=\varepsilon}(\lambda - A)^{-1}d\lambda$), for any $m \in \mathbb{N}$ we have

$$\|e^{tA} - \sum_{n=0}^{m} e^{\lambda_n t}\sum_{k=0}^{N_n-1}\frac{t^k}{k!}(A-\lambda_n)^k P_n\| \le C_{\epsilon,m}e^{(\epsilon+\mathrm{Re}\,\lambda_m)t}, \quad \textit{for } t \ge 0 \textit{ and } \epsilon > 0.$$

(b) *If in addition, X is a Banach lattice and $\{e^{tA}\}_{t\ge 0}$ is irreducible, then the spectral bound of A is a first-order pole of $(\lambda - A)^{-1}$ with geometric multiplicity one. In other words, the corresponding spectral projection P_0 is of rank one, given by $P_0 f = \langle f, f_0^*\rangle f_0$, for some $0 < f_0 \in X$ and strictly positive functional $f_0^* \in X^*$ satisfying $Af_0 = \lambda_0 f_0$ and $\langle f_0, f_0^*\rangle = 1$.*

1.11 Hyperbolic and Chaotic Semigroups

Definition 1.11.1 A strongly continuous semigroup (or (C_0) semigroup) $\{e^{tA}\}_{t\ge 0}$ of linear operators on a Banach space X is called *hyperbolic* if X can be written as a direct sum $X = X_s \oplus X_u$ of e^{tA}–invariant closed stable subspace X_s and unstable subspace X_u such that the restrictions $T_s(t) := e^{tA}\big|_{X_s}$ and $T_u(t) := e^{tA}\big|_{X_u}$ satisfy the following properties:

(i) $\{T_s(t)\}_{t\ge 0}$ is uniformly exponentially stable; that is, there exists $M \ge 1$ and $\varepsilon > 0$ such that $\|T_s(t)\| \le Me^{-\varepsilon t}$ for all $t \ge 0$;
(ii) $\{T_u(t)\}_{t\ge 0}$ extends to a (C_0) group such that $\|T_u^{-1}(t)\| \le Me^{-\varepsilon t}$ for all $t \ge 0$.

Lemma 1.11.2 *For a (C_0) semigroup $\{T_s(t)\}_{t\ge 0}$ on a Banach space X, the following assertions are equivalent:*

(i) $\{e^{tA}\}_{t\ge 0}$ is hyperbolic;
(ii) $\sigma(\{e^{tA}\}_{t\ge 0}) \cap \Gamma = \emptyset$ for one (hence for all) $t > 0$, where Γ is the unit circle.

According to the spectral mapping theorem (SMT) one can express the following lemma in terms of the spectrum of $\{e^{tA}\}_{t\ge 0}$.

Lemma 1.11.3 *For a (C_0) semigroup $\{T_s(t)\}_{t\ge 0}$ on a Banach space X, consider that the following assertions are equivalent:*

(i) $\{e^{tA}\}_{t\ge 0}$ is hyperbolic;
(ii) $\sigma(e^{tA}) \cap \Gamma = \emptyset$ for one, hence for all $t > 0$, where Γ is the unit circle in $\mathbb{C}$.

This lemma is proved in [49, Proposition 3.12].

From Lemma 1.6.4 we have:

Corollary 1.11.4 *If the (C_0) semigroup $\{e^{tA}\}_{t\ge 0}$ is one of the following types:*

(a) eventually compact semigroup;
(b) eventually norm continuous semigroup;

(c) eventually differentiable semigroup;
(d) analytic semigroup,

then it is hyperbolic iff

$$\sigma(A) \cap i\mathbb{R} = \emptyset.$$

Definition 1.11.5 Let X be a separable infinite-dimensional Banach space over the complex field. A strongly continuous semigroup (or (C_0) semigroup) $\{e^{tA}\}_{t\geq 0}$ of linear operators on X is called *hypercyclic* if there exists a vector $x \in X$ such that its orbit $\{e^{tA}x \mid t \geq 0\}$ is dense in X, and it is called *chaotic* if in addition the set of periodic points of e^{tA}, $\{x \in X \mid \text{there exists } t_0 > 0 \text{ such that } e^{t_0 A}x = x\}$, is dense in X.

Here we give a *Hypercyclicity Criterion.*

Theorem 1.11.6 *Let $\{e^{tA}\}_{t\geq 0}$ be a (C_0) semigroup on a Banach separable space X. Let $X_0 = \{x \in X \mid \lim_{t\to\infty} e^{tA}x = 0\}$ and $X_\infty = \{x \in X \mid \forall\epsilon > 0, \exists\, u \in X \text{ and } t > 0 \text{ such that } \|u\| < \epsilon \text{ and } \|e^{tA}u - x\| < \epsilon\}$. If both X_0 and X_∞ are dense subsets of X, then $\{e^{tA}\}_{t\geq 0}$ is hypercyclic.*

1.12 Family of Strongly Continuous Propagators

We consider a nonautonomous Cauchy problem

$$\textbf{(NCP)} \qquad \begin{cases} \dfrac{du}{dt} = A(t)u(t) \\ u(s) = f \in X, \quad t, s \in \mathbb{R},\ t \geq s, \end{cases}$$

on a Banach space X with possibly unbounded operator $A(t)$, $t \in \mathbb{R}$, on X. The solutions of **(NCP)** can be expressed under appropriate conditions, by a family $\{U(t,s)\}_{(t,s)\in\mathbb{R}^2, t\geq s}$ in the space $\mathcal{L}(X)$ of bounded linear operators on X satisfying the following properties:

(P1) $(t,s) \mapsto U(t,s)$ from $\{(t,s) \in \mathbb{R}^2 \;:\; t \geq s\}$ into $\mathcal{L}(X)$ is strongly continuous;
(P2) $U(t,t) = I, \quad \forall t \in \mathbb{R}$;
(P3) the Chapman–Kolmogorov equation holds, i.e.

$$U(t,s)U(s,r) = U(t,r), \qquad \forall (t,s,r) \in \mathbb{R}^3, t \geq s \geq r;$$

(P4) there is a constant $M \geq 1$ and $\omega \in \mathbb{R}$ such that $\|U(t,s)\| \leq Me^{\omega(t-s)}$ for all $(t,s) \in \mathbb{R}^2, t \geq s$.

We call such a family of operators a *family of exponentially bounded strongly continuous propagators* on X. If $\omega = 0$ (resp. $M = 1$) we say that the family of propagators are *uniformly bounded* (resp.*of contractions*). In what follows, by a solution we mean a differentiable function $u(t)$ with respect to t such that it satisfies **(NCP)**. In the following theorem we give a sufficient condition for the existence of a family of strongly continuous propagators of contractions.

Theorem 1.12.1 *Suppose that for each* $t \in \mathbb{R}$, $A(t)$ *generates a* (C_0) *contraction semigroup on a Banach space* X *such that* $\{e^{sA(t)}\}$ *and* $\{e^{\sigma A(\tau)}\}$ *commute for all* $s, t, \sigma, \tau \in \mathbb{R}$. *Let* $\mathcal{D} := \bigcap_{t\in\mathbb{R}} D(A(t))$ *be a dense subset subspace of* X *such that* $A(t)f \in C(\mathbb{R}, X)$ *for each* $f \in \mathcal{D}$. *Then there exists a family of strongly continuous propagators of contractions* $\{(t, s), t \geq s\}$ *such that for all* $f \in \mathcal{D}, t \geq s$,

$$\partial U(t, s)f/\partial t = A(t)U(t, s)f = U(t, s)A(t)f,$$
$$\partial U(t, s)f/\partial s = -A(s)U(t, s)f = -U(t, s)A(s)f,$$

and $u(\cdot) = U(\cdot, s)f$ *is the unique solution of* **(NCP)**.

By renorming the space we can replace in the above theorem, the (C_0) semigroup by a uniformly bounded semigroup; then the family $\{U(t, s)\}$ becomes a family of uniformly bounded strongly continuous propagators.

At this stage we are going back over the perturbation theory and we consider a nonautonomous perturbation of **(NCP)** as follows:

$$\textbf{(PNCP)} \qquad \begin{cases} \dfrac{du}{dt} = [A(t) + B(t)]u(t) \\ u(s) = f \in X, \quad s \in \mathbb{R}, t \geq s. \end{cases}$$

The perturbation theory for evolution families is a touchy matter. Since in contrast to the autonomous case, even if $A(t)$ generates a family of strongly continuous propagators and if $B(t)$ were a bounded operator for any $t \in \mathbb{R}$, $A(t) + B(t)$ cannot generate a family of strongly continuous propagators. However, we have the following result.

Theorem 1.12.2 *Let* $\{U(t, s)\}_{(t,s)\in\mathbb{R}^2, t\geq s}$ *be a family of exponentially bounded strongly continuous propagators and let* $\mathcal{L}_s(X)$ *be the space of all bounded linear operators on* X *endowed with the strong operator topology. If* $B(\cdot)$ *is a bounded continuous* $\mathcal{L}_s(X)$*-valued mapping (we write* $B(\cdot) \in C_b(\mathbb{R}, \mathcal{L}_s(X))$*), then there exists a unique family of exponentially bounded strongly continuous propagators* $\{V(t, s)\}_{(t,s)\in\mathbb{R}^2, t\geq s}$ *on* X *such that*

$$V(t, s) = U(t, s) + \int_s^t V(t, \tau)B(\tau)U(\tau, s)d\tau \tag{1.36}$$

for all $t \geq s$.

If A generates a (C_0) semigroup e^{tA} the corresponding exponentially bounded strongly continuous propagator is $U(t, s) = e^{(t-s)A}$. In a forthcoming chapter we will use a straightforward consequence of the above Theorem in the following form.

Corollary 1.12.3 *Suppose A generates a (C_0) semigroups e^{tA} and $B(\cdot) \in C_b(\mathbb{R}, \mathcal{L}_s(X))$. Then there exists a unique family of exponentially bounded strongly continuous propagators $\{V(t,s)\}_{(t,s)\in\mathbb{R}^2, t\geq s}$ on X given implicitly by*

$$V(t,s) = e^{(t-s)A} + \int_s^t V(t,\tau)B(\tau)e^{(\tau-s)A}d\tau \tag{1.37}$$

such that $u(\cdot) = U(\cdot, s)f$ is the unique solution of the nonautonomous Cauchy problem

$$\begin{cases} \dfrac{du}{dt} = [A + B(t)]u(t) \\ u(s) = f \in X, \quad t, s \in \mathbb{R}, \ t \geq s. \end{cases}$$

1.13 The Howland Semigroup

Let $E := C_0(\mathbb{R}, X)$ the space of continuous X-valued functions vanishing at $\pm\infty$, with the supremum norm. On E we define a bounded multiplication operator as follows: For each $\phi \in C_0(\mathbb{R})$, the operator $\mathcal{M}(\phi)$ defined by

$$[\mathcal{M}(\phi)\psi](t) = \phi(t)\psi(t), \qquad \text{for all} \qquad \psi \in E$$

is called *bounded scalar multiplication operator* on E. Let $\{A(t)\}_{t\in\mathbb{R}}$ be a family of strongly continuous bounded operators on X such that $\|A(.)\|_\infty := \sup_{t\in\mathbb{R}} \|A(t)\| < \infty$. With such a family we can associate an element $\mathcal{A}$ of $\mathcal{L}(E)$ by

$$[\mathcal{A}\psi](t) = A(t)\psi(t) \qquad \text{for all} \qquad \psi \in E.$$

Such an operator commutes with $\mathcal{M}(\phi)$, for all $\phi \in C_0(\mathbb{R})$, and is called a *bounded multiplication operator* on E. We have $\|\mathcal{A}\| = \|A(.)\|_\infty$. It is not hard to verify that if the operators $A(t)$ are invertible in $\mathcal{L}(X)$, then $\mathcal{A}$ is also invertible and $\mathcal{A}^{-1}$ is given by $[\mathcal{A}^{-1}\psi](t) = A(t)^{-1}\psi(t)$, for any $\psi \in E$. Given an exponentially bounded strongly continuous propagator $U(.,.)$, the formula

$$e^{t\mathcal{G}}\psi(.) := U(., . - t)\psi(. - t), \qquad \text{for} \qquad \psi \in E, \ t \geq 0 \tag{1.38}$$

defines a strongly continuous semigroup on E and is called the *Howland* or *evolution semigroup* associated with the propagator $U(.,.)$.

It is clear that at the point $t = 0$, $e^{t\mathcal{G}}$ is the identity of $\mathcal{L}(E)$. The group property $e^{(t+s)\mathcal{G}} = e^{t\mathcal{G}}e^{s\mathcal{G}}$ is equivalent to the Chapman–Kolmogorov equation (P3). Denote by $C_c(\mathbb{R}, X)$ the space of continuous functions $\psi : \mathbb{R} \mapsto X$ with compact support and norm $\|\psi\| = \sup_{s\in\mathbb{R}} \|\psi(s)\|_X$. For $\psi \in C_c(\mathbb{R}, X)$ we obtain $e^{t\mathcal{G}}\psi \in C_c(\mathbb{R}, X)$, and since

$$\|e^{t\mathcal{G}}\psi - \psi\| = \sup_{s\in\mathbb{R}} \|U(s, s-t)\psi(s-t) - \psi(s)\|_X$$

the properties (P1) and (P4) of $U(t, s)$ imply that $\lim_{t\to 0} \|e^{t\mathcal{G}}\psi - \psi\| = 0$ and the strong continuity follows from the density of $C_c(\mathbb{R}, X)$ in E.

The operator $\mathcal{G}$ is the generator of $e^{t\mathcal{G}}$; it is an unbounded operator with domain

$$D(\mathcal{G}) := \{\psi \in E \quad | \quad \lim_{t\to 0} \frac{1}{t}[U(\cdot, \cdot - t) - I]\psi(\cdot)\},$$

which is defined by

$$[\mathcal{G}\psi](\cdot) := \lim_{t\to 0} \frac{1}{t}[U(\cdot, \cdot - t) - I]\psi(\cdot).$$

The right translation $\{\mathbf{T}_0(t)\}_{t\in\mathbb{R}}$, acting on E, defined by

$$\mathbf{T}_0(t)\psi(s) := \psi(s-t), \qquad \text{for} \qquad s, t \in \mathbb{R}, \tag{1.39}$$

is a (C_0) group on E. It is clear that this is a Howland group corresponding to the family $U(t, s) := I$ and its generator is $\mathbf{d}f = -f'$ with $D(\mathbf{d}) = \{f \in C^1(\mathbb{R}, X) : f, f' \in E\}$.

These considerations allow us to give a new characterization of the Howland group $e^{t\mathcal{G}}$. Let $U(t, s)$ be a strongly continuous propagator on $\mathbb{R}^2$. Define an invertible bounded multiplication operator $\mathcal{U}$ on E which acts in the following way. For any $t \in \mathbb{R}$ and any $\psi \in E$,

$$[\mathcal{U}\psi](t) \equiv U(t, 0)\psi(t),$$

and write

$$e^{t\mathcal{G}} = \mathcal{U}\mathbf{T}_0(t)\mathcal{U}^{-1}. \tag{1.40}$$

This formula shows that the Howland group is always similar to the right translation group. In fact

$$\begin{aligned} e^{t\mathcal{G}}\psi(.) &= [\mathcal{U}\mathbf{T}_0(t)]U(0, \cdot)\psi(\cdot) \\ &= [\mathcal{U}]U(0, \cdot - t)\psi(\cdot - t) \\ &= U(\cdot, 0)U(0, \cdot - t)\psi(\cdot - t) = U(\cdot, \cdot - t)\psi(\cdot - t), \end{aligned}$$

and once more (1.40) shows that $\mathbf{T}_0(t)$ is a Howland group corresponding to the family $U(t, s) = I$.

Let $\{A(t)\}_{t\in\mathbb{R}}$ be a family of unbounded linear operators with domain $D(A(t))$. We define a multiplication operator $\mathcal{A}$ by

$$[\mathcal{A}\psi](t) := A(t)\psi(t), \qquad \text{for all } \psi \in D(\mathcal{A}),$$

where

$$D(\mathcal{A}) := \{\psi \in E \quad | \quad \psi(t) \in D(A(t)),\ A(t)\psi(t) \in E \quad \text{for a. a. } t \in \mathbb{R}\}$$

is the maximal domain of $\mathcal{A}$.

Lemma 1.13.1 *Suppose $\{U(t,s)\}_{(t,s)\in\mathbb{R}^2, t\geq s}$ be a family of exponentially bounded strongly continuous propagators and $e^{t\mathcal{G}}$ its corresponding Howland semigroup. This family solves the Cauchy problem* **(NCP)** *if and only if there is a core $\mathcal{D} \subseteq C_0^1(\mathbb{R}, X) \cap D(\mathcal{A})$ of $\mathcal{G}$ such that $\mathcal{G}f = \mathbf{d}f + \mathcal{A}f$ for all $f \in \mathcal{D}$.*

By means of this result one can constitute a one-to-one correspondence between the nonautonomous Cauchy problem **(NCP)** and the autonomous Cauchy problem

$$\begin{cases} \dfrac{df}{dt} = [\mathbf{d} + \mathcal{A}]f(t) \\ f(0) = \psi \in E. \end{cases}$$

1.14 Schatten–von Neumann Classes

In the theory of von Neumann algebra, $\mathcal{L}(\mathcal{H})$ denotes the algebra of linear bounded operators on the separable Hilbert space $\mathcal{H}$ equipped with the uniform norm $\|A\| := \sup_{\|x\|\leq 1} \|Ax\|$. Let us denote by $\mathscr{I}_\infty(\mathcal{H})$ the subset of all compact operators in $\mathcal{L}(\mathcal{H})$. From Lemma 1.10.2[(iii)], and the fact that A is compact if and only if its adjoint A^* is compact, it follows that $\mathscr{I}_\infty(\mathcal{H})$ is a two-sided $^*-$ideal of $\mathcal{L}(\mathcal{H})$. On the other hand, if $\mathcal{H}$ is separable and if $\mathscr{I}$ is a two-sided ideal in $\mathcal{L}(\mathcal{H})$ containing an operator which is not compact, then $\mathscr{I} = \mathcal{L}(\mathcal{H})$ (see [109, Theorem 2.1]).

Definition 1.14.1 Let A be a compact operator in $\mathcal{L}(\mathcal{H})$. Then there exist orthonormal sets (not necessarily complete) $\{\varphi_n\}$ and $\{\psi_n\}$ and positive real numbers $\{s_n\}$ called *singular values of* A such that

$$Au = \sum_{n=1}^{N} s_n(u, \psi_n)\varphi_n. \tag{1.41}$$

The integer N in (1.41) may be finite or infinite. For $N = \infty$ the series (1.41) converges in a norm operator. The expression (1.41) is called the *canonical representation of* A.

The Hilbert–Schmidt Theorem (see [102, Theorem VI16]) shows that if the operator A is self-adjoint and compact, then there is a complete orthonormal basis $\{\varphi_n\}$ such that $A\varphi_n = \lambda_n\varphi_n$ and $\lambda_n \to 0$ as $n \to \infty$. This is shows that for A^*A which is compact, self-adjoint and positive there exist a complete orthonormal basis $\{\psi_n\}$ and a sequence of positive numbers μ_n such that $A^*A\psi_n = \mu_n\psi_n$. By taking $\varphi_n = A\psi_n/s_n$, where $s_n = \sqrt{\mu_n}$, for any $u = \sum_{n\geq 1}(u, \psi_n)\psi_n \in \mathcal{H}$, we get

$$Au = \sum_{n\geq 1}(u, \psi_n)A\psi_n = \sum_{n\geq 1} s_n(u, \psi_n)\varphi_n,$$

which gives (1.41). This shows that the singular values of A are precisely the eigenvalues of $|A| := \sqrt{A^*A}$.

Definition 1.14.2 We call a bounded operator A in $\mathcal{L}(\mathcal{H})$ of *finite rank* if the dimension of $RanA$ called the *rank* of A is finite (dim(Ran A) $< \infty$).

In a Hilbert space the following proposition can be used as the definition of a compact operator.

Proposition 1.14.3 *A bounded operator A in $\mathcal{L}(\mathcal{H})$ is compact if and only if it is a norm limit of finite rank operators.*

If we denote by $\mathcal{F}(\mathcal{H})$ the set of all finite rank operators of $\mathcal{L}(\mathcal{H})$, by taking $A \in \mathcal{F}(\mathcal{H})$ and the sequence of finite rank operators $\{A, A, A, \cdots\}$, Proposition 1.14.3 shows that $A \in \mathscr{I}_\infty(\mathcal{H})$. Hence the canonical representation of A would be

$$Au = \sum_{n=1}^{r} s_n(u, \psi_n)\varphi_n,$$

where r is the rank of A. Knowing that $\{\varphi_n\}$ and $\{\psi_n\}$ are two orthonormal bases we obtain easily the following estimation:

$$\|Au\| = \sum_{n=1}^{r} s_n|(u, \psi_n)|\|\varphi_n\| \leq \|u\|\sum_{n=1}^{r} s_n,$$

Consequently $\|A\|_\infty \leq \sum_{n=1}^{r} s_n$, so we can endow $\mathcal{F}(\mathcal{H})$ not only with a $\|\cdot\|_\infty$ -norm but also with any

$$\|A\|_p = \left(\sum_{n=1}^{r} s_n^p\right)^{1/p}, \qquad 1 \leq p < \infty, \tag{1.42}$$

which are all equivalent on $\mathcal{F}(\mathcal{H})$.

This formalism is generalized in infinite-dimensional spaces and gives the notion of *Schatten–von Neumann classes* $\mathscr{I}_p(\mathcal{H})$. An operator A belongs to an $\mathscr{I}_p(\mathcal{H})$ class if it satisfies (1.42), with $r = \infty$. $\mathscr{I}_p(\mathcal{H})$ classes form an increasing sequence of subsets; that means for $1 < p < q < \infty$ we have

$$\mathscr{I}_1(\mathcal{H}) \subset \mathscr{I}_p(\mathcal{H}) \subset \mathscr{I}_q(\mathcal{H}) \subset \mathscr{I}_\infty(\mathcal{H}).$$

Among these classes there are three which are popular.

First, $\mathscr{I}_1(\mathcal{H})$ is compact, it follows that $*-$ideal of $\mathcal{L}(\mathcal{H})$ which is called a *trace class* ideal with the norm

$$\|A\|_1 := \sum_{i=1}^{\infty} s_i, \tag{1.43}$$

where s_i are the singular values of A, or eigenvalues of $|A| = \sqrt{AA^*}$. It is clear that if $A \in \mathscr{I}_1(\mathcal{H})$ then the trace of A

$$\mathrm{Tr}(A) := \sum_{i=1}^{\infty} \lambda_i < \infty,$$

where λ_i are the eigenvalues of A. Note that $\mathrm{Tr}A$ can be any value of $\overline{\mathbb{C}}$ whereas $\|A\|_1$ is always positive. In the following lemma we gather together all the properties of the trace and the norm $\|\cdot\|_1$.

Lemma 1.14.4 *For any* $A, B \in \mathscr{I}_1(\mathcal{H})$

(i) $\mathrm{Tr}(A+B) = \mathrm{Tr}A + \mathrm{Tr}B$ *and* $\|A+B\|_1 \le \|A\|_1 + \|B\|_1$;
(ii) $\mathrm{Tr}(\lambda A) = \lambda Tr A$ *and* $\|\lambda A\|_1 = |\lambda| \|A\|_1$ *for any* $\lambda \in \mathbb{C}$;
(iii) $\mathrm{Tr}(U^{-1}AU) = \mathrm{Tr}A$, *for any unitary operator* U;
(iv) *If* $0 \le A \le B$, *then* $\mathrm{Tr}A \le \mathrm{Tr}B$.

Lemma 1.14.5 ($\mathscr{I}_p(\mathcal{H})$-**Hölder inequality**). *Let* $A \in \mathscr{I}_p(\mathcal{H})$ *and* $B \in \mathscr{I}_q(\mathcal{H})$, *with* $p > 1$ *and* $\frac{1}{p} + \frac{1}{q} = 1$, *then* $AB \in \mathscr{I}_1(\mathcal{H})$ *and* $\|AB\|_1 \le \|A\|_p \|B\|_q$.

One can also characterize $\mathscr{I}_p(\mathcal{H})^*$, the dual of $\mathscr{I}_p(\mathcal{H})$. For any $A \in \mathcal{L}(\mathcal{H})$, the map $F_A : B \in \mathscr{I}_p(\mathcal{H}) \mapsto \mathrm{Tr}(AB)$ belongs to $\mathscr{I}_p(\mathcal{H})^*$. So, there are also two more important properties of trace class operators.

Lemma 1.14.6 (i) $\mathscr{I}_1(\mathcal{H}) = \mathscr{I}_\infty(\mathcal{H})^*$, *that is, the map* F_A *is an isometric isomorphism of* $\mathscr{I}_1(\mathcal{H})$ *onto* $\mathscr{I}_\infty(\mathcal{H})^*$.
(ii) $\mathscr{I}_1(\mathcal{H})^* = \mathcal{L}(\mathcal{H})$, *that is, the map* F_B *is an isometric isomorphism of* $\mathcal{L}(\mathcal{H})$ *onto* $\mathscr{I}_1(\mathcal{H})^*$.

For the proof we refer to [107, Theorem IV, 1, 3] or [102, Theorem VI.26].

The second popular class between Schatten–von Neumann classes is the Hilbert–Schmidt two-sided $*-$ideal of $\mathcal{L}(\mathcal{H})$. This class occurs for $p = 2$ and if $A \in \mathscr{I}_2(\mathcal{H})$, so is $A^* \in \mathscr{I}_2(\mathcal{H})$, hence according to the above lemma $B^*A \in \mathscr{I}_1(\mathcal{H})$. It is not hard to see that $\mathrm{Tr}(B^*A)$ defines an inner product in $\mathscr{I}_2(\mathcal{H})$ which is denoted by

$$\langle A, B\rangle_2 := \mathrm{Tr}(B^*A).$$

There are two more arguments which make this ideal more discernible among the Schatten–von Neumann classes. First $\mathcal{I}_2(\mathcal{H})$ endowed with the inner product is a Hilbert space; the second is that any $A \in \mathcal{I}_2(\mathcal{H})$ is an integral operator, as is shown in the following proposition (see [102, Theorem VI.23].

From now on we denote an integral operator with the kernel $\rho(x, y)$ by the bold character $\boldsymbol{\rho}$, such that $\boldsymbol{\rho} f(x) = \int \rho(x, y) f(y) dy$. and the following Proposition shows that all Hilbert–Schmidt operators have this property.

Proposition 1.14.7 *Let (Ω, μ) be a measure space and $\mathcal{H} = L^2(\Omega, d\mu)$. Then a bounded operator $\boldsymbol{\rho}$ in $\mathcal{L}(\mathcal{H})$ is Hilbert–Schmidt if and only if there is a function $\rho \in L^2(\Omega \times \Omega, d\mu \otimes d\mu)$ such that $(\boldsymbol{\rho} f)(x) = \int_\Omega \rho(x, y) f(y) d\mu(y)$. In this case the $\mathcal{I}_2(\mathcal{H})$-norm of $\boldsymbol{\rho}$ is given by*

$$\|\boldsymbol{\rho}\|_2 = \left(\iint |\rho(x, y)|^2 d\mu(x) d\mu(y) \right)^{1/2}.$$

The third popular Schatten–von Neumann class is $\mathcal{I}_\infty(\mathcal{H})$, which has been widely discussed previously.

1.15 Tensor Product Semigroup

Let X be a separable Banach space. It is formally clear that if the (C_0) semigroups $G_1(t)$ and $G_2(t)$ are generated by the operators A_1 and A_2, then the application $\mathcal{G}(t)$ defined by

$$\mathcal{L}(X) \ni \boldsymbol{\rho} \mapsto G_2(t) \boldsymbol{\rho} G_1(t) \in \mathcal{L}(X)$$

forms a family of semigroups on $\mathcal{L}(X)$ in the following sense:

(i) $\mathcal{G}(0)\boldsymbol{\rho} = \boldsymbol{\rho}$;

(ii)

$$\begin{aligned} \mathcal{G}(t)\mathcal{G}(s)\boldsymbol{\rho} &= \mathcal{G}(t)[G_2(s)\boldsymbol{\rho} G_1(s)] \\ &= G_2(t)G_2(s)\boldsymbol{\rho} G_1(s)G_1(t) \\ &= G_2(t+s)\boldsymbol{\rho} G_1(t+s) \\ &= \mathcal{G}(t+s)\boldsymbol{\rho}, \quad t, s \geq 0. \end{aligned}$$

(iii) $\mathcal{G}(t)\boldsymbol{\rho}$ converges strongly to $\boldsymbol{\rho}$, as $t \to 0_+$, for any $\boldsymbol{\rho} \in \mathcal{L}(X)$.

In general the above convergence is not in the sense of uniform topology of $\mathcal{L}(X)$, but if X is a finite- dimensional space or if $\mathcal{G}(t)$ acts on an ideal of $\mathcal{L}(X)$, then the convergence can be considered to be uniform in $\mathcal{L}(X)$.

The semigroup $\mathcal{G}(t)$ has Δ_{A_1,A_2} as its infinitesimal generator, where

$$\Delta_{A_1,A_2} = A_2\rho + \rho A_1. \tag{1.44}$$

This is strictly true when the generators A_1 and A_2 are bounded. This semigroup occurs in the study of the Lyapunov operator equation

$$A_2\rho + \rho A_1 = -I,$$

which itself occurs in the study of the stability of the semigroups.

By applying the tensor product semigroup $\mathcal{G}(t) = G_2 \otimes G_1(t)^*$ to an operator of rank one $u \otimes v^*(.) = \langle ., v^* \rangle u$ $(u \in X, v^* \in X^*)$, it turns out that

$$\begin{aligned}\mathcal{G}(t)(u \otimes v^*) &= G_2(t)u \otimes G_1^*(t)v^* \\ &= \langle \cdot, G_1^*(t)v^* \rangle G_2(t)u \\ &= G_2(t)(u \otimes v^*)G_1(t).\end{aligned}$$

The same manipulation holds for finite rank operators. Hence $\mathcal{G}(t)$ is an extension to $\mathcal{L}(X)$ of the tensor product $G_2 \otimes G_1(t)^*$ acting on the space of finite rank, $X \otimes X^*$. Therefore this $\mathcal{G}(t)$ is referred to as a *tensor product semigroup*. In Chap. 7, we will use a particular form of this semigroup which is called *quantum dynamic semigroup*.

Another application of the tensor product semigroup is the relationship of its infinitesimal generator (1.44) with the notion of hyperbolic semigroups (see [29, Theorem 4.40]).

1.16 Some Facts on Ultraweak Operator Topology

Here we begin to recall some useful facts on ultraweak operator topology. For more complete information we refer to Dixmier [40].

Definition and Theorem 1.16.1 *We say that the sequence of the bounded operators* $\{A_\alpha\}$ *converges weakly to* A *and we write* $A_\alpha \stackrel{w}{\rightarrow} A$, *if and only if*

$$\lim_\alpha ((A_\alpha - A)x, y) = 0, \qquad \textit{for any } \ x, y \in \mathcal{H}. \tag{1.45}$$

We say that the sequence of the bounded operators $\{A_\alpha\}$ *converges ultraweakly to* A *and we write* $A_\alpha \stackrel{uw}{\rightarrow} A$, *if and only if*

$$\lim_\alpha \sum_{i=1}^{\infty} ((A_\alpha - A)x_i, y_i) = 0, \tag{1.46}$$

for any pair of sequences (x_i), (y_i) *in* $\mathcal{H}$ *satisfying* $\sum_{i=1}^{\infty} \|x_i\|^2 + \|y_i\|^2 < \infty$, *which is equivalent to saying that*

$$\operatorname{Tr}(A_\alpha \rho) \to \operatorname{Tr}(A\rho), \tag{1.47}$$

for any $\rho \in \mathscr{I}_1(\mathcal{H})$.

Proof Assume (1.46) holds. Any ρ can be represented in its orthonormal eigenfunctions basis (ϕ_i) as $\rho\phi_i = \lambda_i\phi_i$ and $\operatorname{Tr}(A_\alpha\rho) = \sum_{i=1}^{\infty}(A_\alpha\lambda_i\phi_i, \phi_i)$. If $\lambda_i = |\lambda_i|e^{i\theta_i}$, by taking $x_i = \sqrt{|\lambda_i|}e^{i\theta_i/2}\phi_i$ and $y_i = \sqrt{|\lambda_i|}e^{-i\theta_i/2}\phi_i$, we get

$$\sum_{i=1}^{\infty} \|x_i\|^2 + \|y_i\|^2 \le 2\|\rho\|_1.$$

Since $\operatorname{Tr}(A_\alpha\rho) = \sum_{i=1}^{\infty}(A_\alpha\lambda_i\phi_i, \phi_i)$ and $\operatorname{Tr}(A\rho) = \sum_{i=1}^{\infty}(A\lambda_i\phi_i, \phi_i)$, we have (1.47).

Conversely, if we suppose that (1.47) is true, then given (x_i) and (y_i) satisfying $\sum_{i=1}^{\infty} \|x_i\|^2 + \|y_i\|^2 < \infty$, for ρ defined by $\rho x = \sum_{i=1}^{\infty}(x, y_i)x_i$, we have

$$\begin{aligned}
\|\rho x\| &\le \sum_{i=1}^{\infty} |(x, y_i)|\|x_i\| \\
&\le \sum_{i=1}^{\infty} \|x\|\|y_i\|\|x_i\| \le C\|x\|,
\end{aligned}$$

where $C^2 = \left(\sum_{i=1}^{\infty} \|x_i\|^2\right)\left(\sum_{i=1}^{\infty} \|y_i\|^2\right)$, so we see that ρ is bounded. To show that ρ is a trace class, let (ψ_j) be any orthonormal set in $\mathcal{H}$; by using the Bessel's inequality we have

$$\begin{aligned}
\sum_{j=1}^{n} |(\rho\psi_j, \psi_j)| &= \sum_{j=1}^{n} \left| \left(\sum_{i=1}^{\infty} (\psi_j, y_i)x_i, \psi_j \right) \right| \\
&\le \sum_{i=1}^{\infty}\sum_{j=1}^{n} |(\psi_j, y_i)||(x_i, \psi_j)| \\
&\le \sum_{i=1}^{\infty}\sum_{j=1}^{n} \frac{1}{2}\left(|(\psi_j, y_i)|^2 + |(x_i, \psi_j)|^2\right) \\
&\le \sum_{i=1}^{\infty} \frac{1}{2}\left(\|x_i\|^2 + \|y_i\|^2\right),
\end{aligned}$$

which is finite and independent of n, which implies that ρ is a trace class operator on $\mathcal{H}$.

Now if we replace in (1.46) $\rho\psi_j$ by $\sum_{i=1}^{\infty}(\psi_j, y_i)x_i$, we get

$$\lim_{\alpha} \sum_{j=1}^{\infty} \left((A_{\alpha} - A) \sum_{i=1}^{\infty} (\psi_j, y_i) x_i, \psi_j \right) = 0,$$

which implies that

$$\begin{aligned}
\lim_{\alpha} \sum_{i=1}^{\infty} \sum_{j=1}^{\infty} (\psi_j, y_i)((A_{\alpha} - A)x_i, \psi_j) &= \lim_{\alpha} \sum_{i=1}^{\infty} \left((A_{\alpha} - A)x_i, \sum_{j=1}^{\infty} (y_i, \psi_j)\psi_j \right) \\
&= \lim_{\alpha} \sum_{i=1}^{\infty} (A_{\alpha} - A)x_i, y_i) = 0.
\end{aligned}$$

□

Remark 1.16.2 In [34], (1.47) is taken as the definition of ultraweak convergence.

Lemma 1.16.3 *If the family $\{A_{\alpha}\}$ is uniformly bounded, then the ultraweak topology is equivalent to weak topology.*

Proof It is clear that if we choose the sequences $\{x_i\}$ and $\{y_i\}$ such that $x_1 = x$, $y_1 = y$, where x and y are arbitrary and $x_i = y_i = 0$ for any $i \geq 2$, then (1.46) can be written as $\lim_{\alpha}((A_{\alpha} - A)x, y) = 0$ which implies the weak convergence of $\{A_{\alpha}\}$. Conversely, if $L := \sup_{\alpha} \|A_{\alpha}\|$, for any $\varepsilon > 0$ we can find an integer $N \gg 1$ such that

$$\sum_{i \geq N} \|x_i\|^2 + \|y_i\|^2 < \frac{\varepsilon}{L + \|A\|}$$

and also saying $\lim_{\alpha}((A_{\alpha} - A)x, y) = 0$ is equivalent to saying that for any finite set Λ, if $\alpha \notin \Lambda$, then $((A_{\alpha} - A)x_i, y_i) < \frac{\varepsilon}{2(N-1)}$ for all $i = 1, \cdots, N-1$. So we can write for any $\alpha \notin \Lambda$,

$$\begin{aligned}
\sum_{i=1}^{\infty} ((A_{\alpha} - A)x_i, y_i) &= \sum_{i \leq N-1} ((A_{\alpha} - A)x_i, y_i) + \sum_{i \geq N} ((A_{\alpha} - A)x_i, y_i) \\
&\leq \frac{\varepsilon}{2} + (L + \|A\|) \sum_{i \geq N} \|x_i\| \|y_i\| \\
&\leq \frac{\varepsilon}{2} + \frac{L + \|A\|}{2} \sum_{i \geq N} \|x_i\|^2 + \|y_i\|^2 < \varepsilon.
\end{aligned}$$

□

Remark 1.16.4 We have seen that the ultraweak convergence implies the weak convergence, but the converse is not true. Taking an example of a sequence of linear bounded operators $\{A_{\ell}\}_{\ell \geq 0} \subset \mathcal{L}(L^2(\mathbb{R}))$ such that $A_{\ell} \xrightarrow{w} A$, but the convergence is not ultraweak, we use some facts of *wavelet theory*. Let $\{\psi_{j,k}\}_{j,k \in \mathbb{Z}}$ be *Shannon's* *orthonormal basis*, $\{\psi_{j,k}\}_{j,k\in\mathbb{Z}}$, where $\psi_{j,k}(x) = 2^{\frac{j}{2}}\psi(2^j x - k)$ and the function ψ is such that the Fourier transform of ψ denoted by $\widehat{\psi}(\xi) = \frac{1}{\sqrt{2\pi}}\int_{\mathbb{R}} e^{-ix\xi}\psi(x)dx$ satisfies $\widehat{\psi}(\xi) = \chi_I(\xi)$, where $\chi_I(\xi)$ is the characteristic function of the interval $I = [-1, -\frac{1}{2}] \cup [\frac{1}{2}, 1]$.

We define the operator A_ℓ by $\widehat{A_\ell f}(\xi) := \chi_\ell(\xi)\widehat{f}(\xi)$, where

$$\chi_\ell(\xi) = \begin{cases} 2^\ell & \text{if, } \ 2^\ell \le \xi \le 2^\ell + 2^{-2\ell} \\ 0 & \text{elsewhere.} \end{cases}$$

Note that the family $\{A_\ell\}_{\ell\ge 0}$ converges weakly to zero as $\ell \to \infty$. In fact for any $f, g \in L^2(\mathbb{R}) \cap L^1(\mathbb{R})$, by the Plancherel identity and Riemann–Lebesgue lemma $\widehat{f}, \widehat{g} \in C_0(\mathbb{R}^n)$

$$(A_\ell f, g) = (\widehat{A_\ell f}, \widehat{g}) = 2^\ell \int_{2^\ell}^{2^\ell + 2^{-2\ell}} \widehat{f}(\xi)\widehat{g}(\xi)d\xi \le f_M g_M 2^{-\ell}, \tag{1.48}$$

where $f_M = \sup_{x\in\mathbb{R}} |\widehat{f}(\xi)|$ and $g_M = \sup_{x\in\mathbb{R}} |\widehat{g}(\xi)|$. Since $f_M g_M 2^{-\ell} \to 0$, as $\ell \to \infty$. Consequently, the family $\{A_\ell\}_{\ell\ge 0}$ converges weakly to zero. But this family is not ultraweakly convergent. To see this we replace in the formula (1.46) the families (x_i) and (y_i) in $\mathcal{H} = L^2(\mathbb{R})$ by $\{\psi_{j,k}\}_{j,k\in\mathbb{Z}}$. Since $\widehat{\psi_{j,k}}(\xi) = 2^{-j/2}e^{-i2^{-j}k\xi}\widehat{\psi}(2^{-j}\xi)$ and $|\widehat{\psi_{j,k}}(\xi)|^2 = 2^{-j}|\widehat{\psi}(2^{-j}\xi)|^2$. In

$$\sum_{j,k\in\mathbb{Z}} (A_\ell \psi_{j,k}, \psi_{j,k}) = \sum_{j,k\in\mathbb{Z}} 2^{\ell-j} \int_{2^\ell}^{2^\ell+2^{-2\ell}} |\widehat{\psi}(2^{-j}\xi)|^2 d\xi,$$

we see that the second term is independent of k in which only one term is non-zero, that is, when $j = \ell$, so this series is not convergent.

Remark 1.16.5 Let A_n and B_n be two sequences such that $A_n \xrightarrow{w} A$ and $B_n \xrightarrow{w} B$, then it is not true that $A_n B_n \xrightarrow{w} AB$ (see [102, Problem VI 6]). But by decomposing

$$A_n B_n - AB = (A_n - A)B_n + A(B_n - B)$$

it is easy to show that if $A_n \xrightarrow{uw} A$ and $B_n \xrightarrow{uw} B$, then $A_n B_n \xrightarrow{uw} AB$.

1.17 Quantum Dynamical Semigroup

The main motivation for this section is to provide an easy access to the notion of quantum dynamical semigroups and studying their properties. For this aim I prefer to provide this section with Dirac's notations. In the early days of quantum theory, P. A. M. Dirac created a powerful and concise formalism which is now referred to as Dirac notation (see [39]). In quantum mechanics, a quantum state is typically represented as an element of a complex Hilbert space $\mathcal{H}$. In Dirac notation this element is denoted by $|\varphi\rangle \in \mathcal{H}$ a *ket*. With this ingenious notation one feels, at first glance, that something is missing and the ket $|\cdot\rangle$ needs to be complemented. This complement is an element of its dual and is called a *bra* denoted by $\langle\cdot|$. If the ket is a complex number $z = \alpha + \mathrm{i}\beta$ the bra is its conjugate $z^* = \alpha - \mathrm{i}\beta$; if the ket is a column vector

$$|\hat{v}\rangle = \begin{pmatrix} z_1 \\ z_2 \\ \vdots \\ z_n \end{pmatrix} \in \mathbb{C}^n, \quad \text{then the bra would be a row vector } \langle\hat{v}^*| = (\overline{z}_1, \overline{z}_2, \cdots, \overline{z}_n) \in \mathbb{C}^n.$$

If $|\varphi\rangle \in L^2$ is a state vector (quadratically integrable function), then the bra $\langle\varphi|$ is the complex conjugate of φ. Completing a ket by its bra or applying the bra on the ket gives $\langle\cdot|\cdot\rangle$ called a *braket* which acts as the inner product $(\cdot,\cdot)$ of the Hilbert space analysis. Whereas by commuting these two identities one gets *ket-bra* which can be interpreted as an operator. This operator can be obtained as the outer product of two vectors. For example $|\varphi\rangle\langle\psi|$ is an operator of rank 1, which acting on $|v\rangle$ from the right gives a multiple of the ket $|\varphi\rangle$; $|\varphi\rangle\langle\psi|v\rangle = \langle v|\psi\rangle|\varphi\rangle$. $|\varphi\rangle\langle\psi|$ can also act from left; $\langle v|\varphi\rangle\langle\psi|$ is a multiple of the bra $\langle\psi|$ that is $\langle v|\varphi\rangle\langle\psi| = (\langle v|\varphi\rangle)\langle\psi|$.

If $\{\varphi_i\}$ is an orthonormal basis in $\mathcal{H}$, then for any finite number n and any vector $(\lambda_1\lambda_2, \cdots, \lambda_n) \in \mathbb{C}^n$, $A = \sum_{i=1}^n \lambda_i|\varphi_i\rangle\langle\varphi_i|$ is an operator of rank n, thus any compact operator is a uniform limit of finite rank operators. A linear operator A transforms a ket $|\varphi\rangle$ to another ket $A|\varphi\rangle = |A\varphi\rangle$, whereas if we denote by $A^\dagger$ the adjoint of A, $A^\dagger$ transforms a bra (from right to left) to a bra $\langle\psi|A^\dagger = \langle\psi^*|$; the bra $\langle\psi^*|$ is the image of $\langle\psi|$ by $A^\dagger$. So, we can write

$$\langle\psi|A^\dagger = \langle A^\dagger\psi|. \tag{1.49}$$

Let $\mathscr{I}_1(\mathcal{H})$ be the Banach space of trace class operator and V the subspace of $\mathscr{I}_1(\mathcal{H})$ of all self-adjoint operators called the *state space* of $\mathcal{H}$ and V_+ the positive cone of state space.

Definition 1.17.1 A *dynamical semigroup* is a one-parameter family of linear operators $\{\Phi(t)\}_{t\geq 0}$ on V satisfying

(i) $\Phi(t)(V_+) \subset V_+$, for all $t \geq 0$;
(ii) $\mathrm{Tr}(\Phi(t)\rho)$=$\mathrm{Tr}(\rho)$, for all $\rho \in V$ and all $t \geq 0$;
(iii) $\Phi(t)\Phi(s)\rho = \Phi(t+s)\rho$, for all $\rho \in V$ and all $t, s \geq 0$;
(iv) $\Phi(t) \to I$ ultraweakly as $t \to 0$.

If

$$\Phi(t)\rho = e^{-\mathrm{i}tH}\rho e^{\mathrm{i}tH} \tag{1.50}$$

and $e^{-\mathrm{i}tH}$ is the Schrödinger group, $\Phi(t)$ satisfies all the assumptions of the above definition and will be call a *quantum dynamical semigroup*, (**QDS**, for short). Concerning (**QDS**), we have to state that, in any $\mathscr{I}_p(\mathcal{H})$, $p \in [1, \infty]$ Schatten operators class

$$K(0) \mapsto K(t) = e^{-\mathrm{i}tH} K(0) e^{\mathrm{i}tH} \tag{1.51}$$

is a unitary group in $\mathscr{I}_p(\mathcal{H})$, which is the solution of *Heisenberg's motion equation*

$$\textbf{(HME)} \qquad \frac{\partial}{\partial t} K(t) = \mathrm{i}(HK(t) - K(t)H),$$

which in $\mathscr{I}_1(\mathcal{H})$ is nothing but (**QDS**).

If the Hamiltonian H in (1.50) is a bounded operator, then the infinitesimal generator of $\Phi(t)$ will be $-\mathrm{i}(ad_H)$, where

$$ad_H \rho = [H, \rho] = H\rho - \rho H \tag{1.52}$$

is the Lie bracket of H and ρ. In [33, Lemma 5.5.1], E. B. Davies gives a nice representation of the domain of ad_H when H is an unbounded operator and we take it as the definition of $D(ad_H)$.

Definition 1.17.2 Let $D(H)$ be the domain of H in $\mathcal{H}$. The domain of ad_H denoted by $D(ad_H)$ is an ultraweakly dense subset of V such that if $\rho \in D(ad_H)$, $\rho(D(H)) \subset D(H)$ and $H\rho - \rho H$ is norm bounded on $D(H)$.

Example 1.17.3 Let $H := H_0 = -\frac{\hbar}{2}\Delta^0$ be a free potential Schrödinger operator with domain $D(H) = \{\varphi \in \mathcal{H} : |\xi|^2 \widehat{\varphi}(\xi) \in L^2(\mathbb{R}^n_\xi)\}$, where $\widehat{\varphi}(\xi) = 1/(2\pi)^{n/2} \int \varphi(x) e^{-ix.\xi} dx$ is the Fourier transform of φ. Now take φ as a unitary element of $L^2(\mathbb{R}^n)$ and $\rho(x, y) = \varphi(x)\varphi^*(y)$. It is clear that $\rho \in \mathscr{H} := L^2(\mathbb{R}^n_x \times \mathbb{R}^n_y)$, thus,

$$[\rho f](x) = \int_{\mathbb{R}^n_y} \varphi(x)\varphi^*(y) f(y) dy. \tag{1.53}$$

This operator is of *Hilbert–Schmidt class*, that is, $\rho \in \mathscr{I}_2(\mathcal{H})$ and since $\rho^2 = \rho$ or equivalently

$$\begin{aligned}
[\rho^2 f](x) &= \int_{\mathbb{R}^n_y} \rho^2(x, y) f(y) dy \\
&= \int_{\mathbb{R}^n_y} \rho(x, y) \left[\int_{\mathbb{R}^n_z} \rho(y, z) f(z) dz \right] dy \\
&= \int_{\mathbb{R}^n_y} \varphi(x)\varphi^*(y) \left[\int_{\mathbb{R}^n_z} \varphi(y)\varphi^*(z) f(z) dz \right] dy \\
&= \int_{\mathbb{R}^n_z} \varphi(x)\varphi^*(z) f(z) dz = [\rho f](x),
\end{aligned}$$

it belongs also to $\mathcal{I}_1(\mathcal{H})$, the space of trace class operators, since an operator $A \in \mathcal{I}_1(\mathcal{H})$, if and only if $A = B_1 B_2$ with $B_i \in \mathcal{I}_2(\mathcal{H})$ for $i = 1, 2$ (see [102, Theorem VI. 22]).

Another important property of ρ is that it defines a positive self-adjoint operator. In fact,

$$\begin{aligned}\langle \rho f | f \rangle &= \int_{\mathbb{R}^n_x} \left[\int_{\mathbb{R}^n_y} \varphi(x)\varphi^*(y) f(y) dy \right]^* f(x) dx \\ &= \left(\int_{\mathbb{R}^n_x} \varphi^*(x) f(x) dx \right) \left(\int_{\mathbb{R}^n_y} \varphi(y) f^*(y) dy \right) = |\langle \varphi | f \rangle|^2,\end{aligned}$$

and since $\langle \rho f | f \rangle$ is positive, ρ is automatically self-adjoint.

If we take the quantum dynamical semigroup defined as (1.50) in which the operator ρ has the form (1.53), then

$$\Phi(t)\rho = e^{-\mathrm{i}tH} \int_{\mathbb{R}^n_y} \psi(x)\psi^*(y) e^{\mathrm{i}} t H dy = \psi_\hbar(x, t) \int_{\mathbb{R}^n_y} \psi^*(y) e^{\mathrm{i}} t H dy,$$

where $\psi_\hbar(x, t) = e^{-\mathrm{i}tH}\psi(x)$. According to (1.49) in Dirac notation we can write

$$\Phi_\hbar(t)\rho = |\psi_\hbar(x, t)\rangle\langle\psi^*_\hbar(y, t)|, \tag{1.54}$$

where $\psi^*_\hbar(y, t) = e^{\mathrm{i}tH}\psi^*(y)$.

Now, let ϕ be a unitary element of $\mathcal{S}(\mathbb{R}^n)$. Let us show that the operator ρ_ϕ defined as in (1.53) in which we replace φ by ϕ belongs to $D(ad_{H_0})$. First, we remark that ρ_ϕ is a positive self-adjoint operator of trace class, hence $\rho_\phi \in V_+$. To prove that $\rho_\phi(D(H_0)) \subseteq D(H_0)$, we take $u \in D(H_0)$, then $\rho_\phi u = \phi(x) \int_{\mathbb{R}^n_y} \phi^*(y) u(y) dy \in \mathcal{S}(\mathbb{R}^n) \subset D(H_0)$ independent of the choice of $u \in D(H_0)$. Furthermore for any $u \in D(H_0)$, let us write $\mu = \int_{\mathbb{R}^n_y} \phi^*(y) u(y) dy$, then $\|H_0 \rho_\phi u\|_\infty = \frac{\mu\hbar}{2} \|\Delta^0 \phi\|_\infty < \infty$ and for $\nu = \int_{\mathbb{R}^n_y} \phi^*(y) \Delta^0 u(y) dy$ we have $\|\rho_\phi H_0 u\|_\infty = \frac{\nu\hbar}{2} \|\phi\|_\infty < \infty$, which implies that $H_0\rho_\phi - \rho_\phi H_0$, is a norm bounded on $D(H_0)$.

Finally, when $\rho_\phi \in D(ad_{H_0})$ let us calculate the derivative of the quantum dynamical semigroup $\Phi(t)\rho_\phi$ with respect to t:

$$\begin{aligned}\frac{d}{dt}\Phi(t)\rho_\phi &= -\mathrm{i}(ad_{H_0})\Phi(t)\rho_\phi = \frac{\mathrm{i}\hbar}{2} \left(\Delta^0 \Phi(t)\rho_\phi - \Phi(t)\rho_\phi \Delta^0 \right) \\ &= e^{\frac{\mathrm{i}\hbar t}{2}\Delta^0} \left[\frac{\mathrm{i}\hbar}{2} (\Delta^0 \rho_\phi - \rho_\phi \Delta^0) \right] e^{-\frac{\mathrm{i}\hbar t}{2}\Delta^0}.\end{aligned}$$

So for any state $\rho \in D(ad_H)$

$$\frac{d}{dt}\Phi(t)\rho = e^{-\mathrm{i}tH} \left[-\mathrm{i}(H\rho - \rho H) \right] e^{\mathrm{i}tH}. \tag{1.55}$$

1.18 Notes and Comments

The theory of strongly continuous semigroups of bounded linear operators generalizes to Banach spaces the properties of the exponential of matrices in $\mathbb{C}^n$. There is a wealth of literature on semigroup theory, starting from the monumental treatises by Hille and Philips [64] and Dunford and Schwartz [42], many research articles and many expository books. The main objective of this theory is dealing with the abstract Cauchy problem. Seemingly almost all evolutionary linear partial differential equations can be formulated as an abstract Cauchy problem and for its well-posedness one uses systematically the theory of semigroups. This theory is also generalized in various directions: The theories of distribution semigroups of Lions [80], C-regularized semigroups of Da Prato [32], integrated semigroups of Arendt [4, 5] and the existence families of deLaubenfels [35] are introduced for studying the different types of ill-posed Cauchy problems. In the following chapters we will have the opportunity to discuss some of them.

In the present chapter, we restrict ourselves to those notions and results directly connected with our needs. In fact, in the abstract scattering theory we are faced with the study of perturbation theory, and this is a hard subject in the theory of semigroups. Many researchers in this area are preoccupied with this topic and create an evolving character for it.

In the first section we deal with the closed operators. The closedness of an operator is the main property of the generator of a strongly continuous semigroup. All the results of this section are proved in the classical monographs by Lax [74], Goldstein [55] and Pazy [96], and the proof of Lemma 1.2.6 can be found in the book by Davies [34, Theorem 1.9].

How can we recognize if an unbounded operator is the infinitesimal generator of a particular semigroup? Any result for answering this question is called a *generation* result. The Lumer–Phillips Theorem 1.3.3 enables us to recognize if an operator is the generator of a contraction semigroup in a Banach space, while Hille-Yosida Theorem generalizes this theorem for an arbitrary strongly continuous semigroup. Lumer–Phillips Theorem, as we have cited in 1.3.3 is proved in [55, p. 26] and our version of Hille–Yosida Theorem is proved in [96, Theorem 1.5.3]. This result has been extended to equi-continuous semigroups in locally convex linear spaces. An extensive bibliography in this area was compiled by J. Goldstein in his book [55]. Theorem 1.4.2 is a fundamental result which links up the theory of semigroups with abstract Cauchy problems. This theorem is due to Hille [63], see also Pazy [96, Theorem 4.1.3]. Concerning the spectral properties of an unbounded operator, the proof of Lemma 1.6.2 is similar to those given by Reed and Simon [102, Proposition of p. 196]. The spectral mapping Lemma 1.6.6 is proved in Clément et al. [30, Proposition 8.3]. Stone's Theorem is due to Stone [111], and one can find a nice proof by Goldstein [55] which is modelled on Lax and Phillips's treatise [77]. In Sect. 1.8 we mention some old and new results on perturbation theory which will be used in the construction of wave operators. It was Phillips [97] who proved that if A is the generator of a (C_0) semigroup on a Banach space X and $B \in \mathcal{L}(X)$, then the closed operator $C = A + B$ with domain $D(C) = D(A)$ generates also a (C_0) semigroup (see Corollary 1.8.8).

In Chap. II of K. J. Engel and R. Nagel's book [49] there is a complete and updated discussion including the proof of Theorem 1.8.1 and Corollary 1.8.2. Theorem 1.8.4 is a generalization of *Miyadera's Perturbation Theorem* which is performed by Voigt [122].

In 1914, S. Lie proved the following result. If A and B are two n-dimensional matrices, then

$$\lim_{n\to\infty} \left[e^{A/n}e^{B/n}\right]^n = e^{A+B}$$

(see [49, Theorem VIII 29] for the proof).

This result was extended by Trotter [116] to the generator of the contraction semigroups, when $\overline{A + B}$ generates also a contraction semigroup. In [28], P. Chernoff gave a remarkable result which may include the Trotter Theorem.

In Chap. 2, we dedicate a section to describing the contribution of positivity in the abstract scattering theory. That is why here we have to introduce the preliminaries for this topic. Almost all the results of Sect. 1.9 can be found in [94].

In 1892, A. M. Liapunov proved in his thesis "Stability of Motion" the following stability theorem. The semigroup $\{e^{tA}\}_{t\geq 0}$ generated by a matrix $A \in \mathbf{M}_n(\mathbb{C})$ is stable (i.e. $\lim_{t\to\infty} \|e^{tA}\| = 0$) iff all eigenvalues of A have negative real parts. From this result one can infer that the hyperbolicity of $\{e^{tA}\}_{t\geq 0}$ in the sense of Definition 1.11.1 is equivalent to the fact that the matrix A has no eigenvalue of modulus 1 and by the spectral mapping theorem is equivalent to the fact that the matrix A has no purely imaginary eigenvalue. This result is generalized in the case of (C_0) semigroup in terms of Lemma 1.11.2. The proofs can be found in the book by Engel and Nagel [49]. Concerning the hyperbolic semigroup one can find many examples in the book by Chicone and Latushkin [29].

Hypercyclicity is the study of linear operators that possess a dense orbit. Although the first example of hypercyclic operators dates back to the first half of the last century with widely disseminated papers of Birkhoff [19] and MacLane [84], a systematic study of this concept has only been undertaken since the mid–eighties. In his survey [59], K. Grosse-Erdmann provided a complete synopsis of hypercyclicity and related areas such as universality and chaos. There are several concepts of chaos for not necessarily linear mappings in the literature. Since many of them require, among others, the existence of a dense orbit, it seems natural to introduce a notion of chaos into the theory of hypercyclicity. The first to consider this problem were Godefroy and Shapiro [52] who took up a definition of chaos due to Devaney [38] and showed that the right notion of a linear chaotic operator on a topological vector space is an operator with a dense orbit and dense set of periodic points- let us recall that a point x in a topological vector space X is called *periodic* for a mapping $T : X \mapsto X$ if $T^N x = x$ for some $N \in \mathbb{N}$. The notion of chaotic (C_0) semigroups (see Definition 1.11.5) was introduced independently by MacCluer [83] and Protopopescu and Azmy [100]; the first systematic study of this concept is due to Desch, Schappacher and

Webb [37] in which one can find a Hypercyclicity Criterion (Theorem 1.11.6). In [36], this notion is generalized to the families of unbounded operators.

The area of nonautonomous infinite-dimensional dynamical systems is vast. For further reading and more detailed investigation we refer to monographs like [51, 72, 73, 96, 113, 114]. When the family $\{A(t)\}_{t\in\mathbb{R}}$ commutes, the corresponding propagator $U(t,s)$ can be formulated by

$$U(t,s) = \exp\left[\int_s^t A(\tau)d\tau\right].$$

In the non-commutative case one can replace the integral $\int_s^t A(\tau)d\tau$ by its Riemann sum and take

$$U(t,s) = \lim_{\mu\to 0}\prod_{j=1}^{n}\left[\exp(\tau_j-\tau_{j-1})A(\xi_j)\right],$$

where the mesh μ is defined by $s=\tau_0<\tau_1<\cdots<\tau_n=t$, $\xi_j\in[\tau_{j-1},\tau_j]$, $j=1,\cdots,n$ and $\mu:=\max_j(\tau_j-\tau_{j-1})$. In [54], Goldstein used the above approximation in order to prove Theorem 1.12.1. The above formula is called *product integration* and there is a substantial literature by Dollard and Friedmann on this topic, which is gathered in their book [41]. Theorem 1.12.2 is taken from Engel and Nagel's book [49, Corollary VI, 9.20].

In Sect. 1.13 we study the evolution semigroups. Originally this formalism was used by Howland [66] in the Hilbert space context and by Evans [50] in the Banach space context to indicate how the stationary theory of scattering can be applied to problems in the theory of transitions of a quantum mechanical system.

Recently, motivated by the Lyapunov Stability Theorem for differential equations on Banach spaces and Mather's Hyperbolicity Theorem for hyperbolic diffeomorphisms on smooth manifolds, a substantial literature drew attention to the properties of this semigroup. In [29], Chicone and Latushkin describe in their monograph a complete synthesis on the evolution semigroups.

The results of this section are given in $E=C_0(\mathbb{R},X)$. But by a slight modification all these results can be carried over to $E:=L^p(\mathbb{R},X)$ (see [50]).

As we have said, the concept of the tensor product semigroup comes through the notion of quantum dynamic semigroups in Sect. 1.15. In this section we define the notion of of the tensor product semigroup $\mathcal{G}(t)\rho$ and its relationship with $G_2\otimes G_1(t)^*$. This semigroup is already used in remediability theory in [47] and in [17] in which we have investigated its chaotic character of this semigroup.

In Sect. 1.16 we recall some useful facts on ultraweak operator topology. For more complete information we refer to Dixmier [40]. With this notion we generalize the result by Markowich and Ringhofer [85] in Chap. 7 and we also use this notion to show that the solution of the Husimi equation converges in the space $L^1(\mathbb{R}^{2n})$ to the solution of Liouville–Vlasov equation as $\hbar\to 0$.

Finally, Sect. 1.17 is a preparation for introducing the quantum dynamic semigroup (QDS). First let us define for a self adjoint operator H the family of operators $K(t) = e^{itH} K(0) e^{-itH}$. The mapping $K(0) \mapsto K(t)$ is a unitary group in any Schatten–von Neumann class $\mathscr{I}_p(\mathcal{H})$ for any $p \in [1, \infty]$, but among all $\mathscr{I}_p(\mathcal{H})$, the trace class ideal $\mathscr{I}_1(\mathcal{H})$ is physically the most meaningful, since the state space V is a subset of $\mathscr{I}_1(\mathcal{H})$ and there are many more properties in quantum physics which deal with $\mathscr{I}_1(\mathcal{H})$. That is why E. B. Davies dedicates the whole of his book [33] to this space and studies the quantum dynamic semigroup in $\mathscr{I}_1(\mathcal{H})$. We choose the definition of QDS given by Lindbladt [79] in 1.17.1. This definition brings the notion of ultraweak operator topology into the hart of the concept.

Trace class operators are in particular a Hilbert–Schmidt class of operators which can be constructed by the kernel functions. For this reason, in the process of investigating this subject we confuse the concepts of the trace class operator with its kernel function. In order to perform this analysis properly we use the Dirac notation, which was introduced by Dirac in [39] and became a standard language in quantum mechanics. The general theory of QDS is followed in [33] with some classical examples, but Arnold and Sparber in [8] have expanded it to broader examples.

Chapter 2
Wave and Scattering Operators

2.1 Basic Properties of the Wave and Scattering Operators

In the Introduction of this book we gave an intuitive point of view of the abstract scattering theory in a Banach space X. With the same notations we suppose that T_0 and T generate two (C_0) groups in X.

Definition 2.1.1 We call the *wave operator* the strong limit

$$W_{\pm}(T, T_0) = s - \lim_{t \to \pm\infty} e^{-tT} e^{tT_0} \tag{2.1}$$

and

$$W_{\pm}(T_0, T) = s - \lim_{t \to \pm\infty} e^{-tT_0} e^{tT}, \tag{2.2}$$

whenever each exists. The existence of the wave operators $W_{+}(T_0, T)$ and $W_{-}(T, T_0)$ imply the existence of a new operator

$$S = W_{+}(T_0, T) W_{-}(T, T_0). \tag{2.3}$$

which is called the *scattering operator.*

Remark 2.1.2 In the definition of the scattering operator S, we use only the wave operators $W_{+}(T_0, T)$ and $W_{-}(T, T_0)$. Since

$$W_{-}(T, T_0) = s - \lim_{t \to +\infty} e^{tT} e^{-tT_0}$$

these operators invoke only the positive values of t in e^{tT}. Thus, for the existence of S it is enough that e^{tT} be only a semigroup of linear operators (instead of being a group, which is a strong assumption).

In the subsequent chapter we will use this aspect of the theory.

H. Emamirad, *Scattering Theory for Transport Phenomena*, Mathematical Physics Studies, https://doi.org/10.1007/978-981-16-2373-8_2

Lemma 2.1.3 *The existence of each wave operator implies the correspondent identity in the following list:*

(a)
$$e^{tT}W_\pm(T,T_0)f = W_\pm(T,T_0)e^{tT_0}f; \qquad f \in X.$$

(b)
$$TW_\pm(T,T_0)f = W_\pm(T,T_0)T_0 f; \qquad f \in D(T_0).$$

(c)
$$e^{tT_0}W_\pm(T_0,T)f = W_\pm(T_0,T)e^{tT}f; \qquad f \in X.$$

(d)
$$T_0 W_\pm(T_0,T)f = W_\pm(T_0,T)Tf; \qquad f \in D(T).$$

(e)
$$W_\pm(T,T_0)W_\pm(T_0,T)f = W_\pm(T_0,T)W_\pm(T,T_0)f = f; \qquad f \in X.$$

Each one of the above identities is called an *intertwining identity* of the wave operators.

Proof (a) Fix t in $\mathbb{R}$ and let s tend to $\pm\infty$ in

$$e^{tT}e^{-sT}e^{sT_0}e^{-tT_0}g = e^{-(s-t)T}e^{(s-t)T_0}g; \qquad g \in X,$$

then one gets

$$e^{tT}W_\pm(T,T_0)e^{-tT_0}g = W_\pm(T,T_0)g.$$

The assertion follows by taking $f = e^{-tT_0}g$.

(b) Take the Laplace transform of (a) at the point $\lambda \in \mathbb{R}_+$, for λ sufficiently large,

$$\int_0^\infty e^{-\lambda t}e^{tT}W_\pm(T,T_0)f\,dt = \int_0^\infty W_\pm(T,T_0)e^{-\lambda t}e^{tT_0}f\,dt.$$

Since $W_\pm(T,T_0) \in \mathcal{L}(X)$, by Lemma 1.5.1 it follows that

$$(\lambda - T)^{-1}W_\pm(T,T_0)f = W_\pm(T,T_0)(\lambda - T_0)^{-1}f,$$

thus

$$W_\pm(T,T_0)D(T_0) \subseteq D(T). \tag{2.4}$$

This allows us to take the derivative of (a) with respect to t, for $f \in D(T_0)$ and get

$$Te^{tT}W_\pm(T,T_0)f = W_\pm(T,T_0)T_0e^{tT_0}f.$$

At the point $t = 0$, this is (b).

For (c) and (d) use the same arguments.

For (e), we have even more: Let $e^{tT_1}, e^{tT_2}, e^{tT_3}$ be three (C_0) group on X such that $W_{\pm}(T_1, T_2) = s - \lim_{t\to\pm\infty} e^{-tT_1}e^{tT_2}$ and $W_{\pm}(T_2, T_3) = s - \lim_{t\to\pm\infty} e^{-tT_2}e^{tT_3}$ exist. Then, it is clear that

$$W_{\pm}(T_1, T_2)W_{\pm}(T_2, T_3) = W_{\pm}(T_1, T_3). \tag{2.5}$$

The above identity is called the *chain rule for wave operators* and since $W_{\pm}(T, T) = I$ for any generator of a (C_0) group, the identities (e) follow. □

Remark 2.1.4 By taking $g \in D(T_0)$ and $f = W_{-}(T, T_0)g$, the inclusion (2.4) and (d) of Lemma 2.1.3 imply that $T_0W_{+}(T_0, T)W_{-}(T, T_0)g = W_{+}(T_0, T)TW_{-}(T, T_0)g$. This, together with (b), leads to the commutative relation of S with T_0:

$$T_0S = ST_0 \quad \text{on} \quad D(T_0).$$

Remark 2.1.5 The relation (e) of Lemma 2.1.3 implies that

$$W_{\pm}(T, T_0)^{-1} = W_{\pm}(T_0, T) \qquad \text{and} \qquad W_{\pm}(T_0, T)^{-1} = W_{\pm}(T, T_0).$$

Hence the existence of the pair $(W_{-}(T, T_0), W_{-}(T_0, T))$ or $(W_{+}(T, T_0), W_{+}(T_0, T))$ conveys the similarity of two operators T and T_0 and the correspondent (C_0) groups in the following sense:

$$W_{\pm}(T_0, T)TW_{\pm}(T, T_0)f = T_0f; \quad f \in D(T_0),$$

$$W_{\pm}(T, T_0)T_0W_{\pm}(T_0, T)f = Tf; \quad f \in D(T).$$

$$W_{\pm}(T_0, T)e^{tT}W_{\pm}(T, T_0)f = e^{tT_0}f; \quad f \in X,$$

$$W_{\pm}(T, T_0)e^{tT_0}W_{\pm}(T_0, T)f = e^{tT}f; \quad f \in X.$$

2.2 Theorems of Existence for the Wave Operators

Unless explicitly stipulated otherwise, we suppose in the sequel that T is a bounded perturbation of T_0 (i.e. there is bounded operator $A \in \mathcal{L}(X)$ such that $T = T_0 + A$), then $D(T) = D(T_0)$ and if T_0 generates a (C_0) group e^{tT_0}, then by Theorem 1.8.1, e^{tT} also generates a (C_0) group.

Concerning the conditions which lead to the existence of the wave operators, the most common result which fits perfectly to an arbitrary Banach space is *Cook's*

Lemma. This lemma is based essentially on the following observation: Let g be a X-valued function which is, for some $a > 0$, strongly differentiable on $(a, +\infty)$ with $g' \in L^1((a, \infty); X)$, then $\lim_{t\to\infty} g(t)$ exists, for X is a complete space and

$$\|g(t) - g(s)\| \le \| \int_s^t g'(r)\,dr\| \le \int_s^t \|g'(r)\|dr \longrightarrow 0,$$

as s and t tend to $+\infty$.

Lemma 2.2.1 (Cook's Lemma) *Let T and T_0 be the infinitesimal generators of the (C_0) groups $\{e^{tT}\}_{t\in\mathbb{R}}$ and $\{e^{tT_0}\}_{t\in\mathbb{R}}$. Suppose $D = D(T)\cap D(T_0)$ is dense in X and there exist three constants $a \ge 0$, M_0 and M such that:*

(a) $\sup_{t\le -a} \|e^{tT}\| \le M$ *(resp.* $\sup_{t\ge a}\|e^{tT}\| \le M$*),*
(b) $\sup_{t\ge a} \|e^{tT_0}\| \le M_0$ *(resp.* $\sup_{t\le -a}\|e^{tT_0}\| \le M_0$*) and*
(c) $\int_a^\infty \|(T - T_0)e^{tT_0}u\|dt < \infty$ *(resp.* $\int_{-\infty}^{-a} \|(T - T_0)e^{tT_0}u\|dt < \infty$*), for all* $u \in D$.

Then $W_+(T, T_0)$ (resp. $W_-(T, T_0)$) exists.

Proof Let $u \in D$, and $g(t) = e^{-tT}e^{tT_0}u$, for $t > a$. Then g is strongly differentiable on (a, ∞) and $g'(t) = -e^{-tT}(T - T_0)e^{tT_0}u$. Hence

$$\begin{aligned}\|g(t) - g(s)\| &\le \int_s^t \|g'(r)\|\,dr\\ &= \int_s^t \|e^{-rT}(T - T_0)e^{rT_0}u\|dr\\ &\le M\int_s^t \|(T - T_0)e^{rT_0}u\|dr.\end{aligned}$$

By the assumption (c), the limit for $t \to \infty$ of $e^{-tT}e^{tT_0}u$ exist for all $u \in D$ which is dense in X. Furthermore, by (a) and (b) $\sup_{t\in[a,\infty)}\|e^{-tT}e^{tT_0}\| \le M\,M_0$, and the uniform boundedness principle implies the existence of $W_+(T, T_0)$ (similarly the existence of $W_-(T, T_0)$). □

Theorem 2.2.2 *Let e^{tT_0} be a uniformly bounded (C_0) group on $\mathbb{R}$, and $A \in \mathcal{L}(X)$. Suppose*

$$\|e^{tT_0}\| \le 1 \quad \textit{for} \quad t \ge 0, \tag{2.6}$$

and there exists $\alpha < 1$ such that

$$\int_{-\infty}^0 \|Ae^{tT_0}f\|dt \le \alpha\|f\| \tag{2.7}$$

for all f in a dense subset of X. Then the wave operators $W_\pm(T, T_0)$ and $W_\pm(T_0, T)$ exist.

To prove the above theorem we need the following lemmas.

Lemma 2.2.3 *Let e^{tT_0} be a bounded (C_0) group on $\mathbb{R}$, and $A \in \mathcal{L}(X)$. Then, if for some $\alpha < 1$ one has*

$$\int_0^\infty \|Ae^{tT_0} f\|dt \le \alpha\|f\|, \quad (resp. \int_{-\infty}^0 \|Ae^{tT_0} f\|dt \le \alpha\|f\|), \tag{2.8}$$

for all $f \in X$, then the (C_0) group e^{tT} is uniformly bounded on $\mathbb{R}_+$ (resp. $\mathbb{R}_-$).

Proof Let us show that the (C_0) group e^{tT} generated by $T = T_0 + A$ is uniformly bounded on $\mathbb{R}_+$. For this we use the second Duhamel formula (1.15), which asserts that

$$e^{tT} = e^{tT_0} + \int_0^t e^{(t-s)T} Ae^{sT_0}\, ds, \quad t \in \mathbb{R}.$$

Let $\tau > 0$. For $t \in [0, \tau]$ and $f \in X$, from (2.8) we have

$$\begin{aligned} \|e^{tT} f\| &\le \|e^{tT_0} f\| + \sup_{s\in[0,\tau]} \|e^{sT}\| \int_0^\tau \|Ae^{sT_0} f\|\, ds \\ &\le M_0\|f\| + \alpha \sup_{s\in[0,\tau]} \|e^{sT}\|\, \|f\|, \end{aligned}$$

and therefore,

$$\sup_{t\in[0,\tau]} \|e^{tT}\| \le \frac{M_0}{1-\alpha}.$$

Since τ is chosen arbitrarily, we conclude that e^{tT} is uniformly bounded on $\mathbb{R}_+$.

For $\mathbb{R}_-$ the proof is similar. □

Lemma 2.2.4 *Let e^{tT_0} be a uniformly bounded (C_0) group on $\mathbb{R}$, and $A \in \mathcal{L}(X)$. Then* (2.6) *and*

$$\int_{-\infty}^0 \|Ae^{tT_0} f\|dt \le \alpha\|f\|, \quad f \in X, \tag{2.9}$$

imply

$$\int_0^\infty \|Ae^{tT_0} f\|dt \le \alpha\|f\|, \quad f \in X. \tag{2.10}$$

Proof Take $r > 0$ and $s = t - r$, then

$$\begin{aligned} \int_0^r \|Ae^{tT_0} f\|dt &= \int_{-r}^0 \|Ae^{(s+r)T_0} f\|ds \\ &\le \alpha\|e^{rT_0} f\| \qquad \text{(by (2.9))} \end{aligned}$$

$$\leq \alpha \|f\|. \qquad \text{(by (2.6))}$$

Since r is an arbitrary number, we thus get (2.10). □

Lemma 2.2.5 *Under the assumptions of Theorem 2.2.2, e^{tT} is uniformly bounded on $\mathbb{R}$ and we have one of the following equivalent assertions:*

(a)

$$\int_0^{\infty} \|Ae^{tT} f\| dt < \infty; \qquad f \in X. \tag{2.11}$$

(b)

$$\int_{-\infty}^{0} \|Ae^{tT} f\| dt < \infty; \qquad f \in X. \tag{2.12}$$

Proof To prove that under the assumptions of Theorem 2.2.2, e^{tT} is uniformly bounded on $\mathbb{R}$, first we apply Lemma 2.2.4 to get (2.10), then we use Lemma 2.2.3.

The uniform boundedness of e^{tT} on $\mathbb{R}$ implies the equivalence between (a) and (b). In fact, assume (a) and apply the principle of uniform boundedness, then there exists a constant $C \geq 0$, such that

$$\int_0^{\infty} \|Ae^{tT} f\| dt \leq C\|f\| \quad f \in X.$$

Let $r > 0$ and $s = t + r$, then

$$\begin{aligned}\int_{-r}^{0} \|Ae^{tT} f\| dt &= \int_0^{r} \|Ae^{(s-r)T} f\| ds \\ &\leq C\|e^{-rT} f\| \\ &\leq CM\|f\|,\end{aligned}$$

r being arbitrary, one gets (b). For the converse we may argue in a similar way.

Thus, for the proof of this lemma we have only to show one of the statements of this lemma. Let us prove (2.12). For this, we present the (C_0) group e^{tT} by its Dyson–Phillips expansion (see (1.18)) given by

$$e^{tT} = \sum_{n=0}^{\infty} S_n(t),$$

where $S_0(t) = e^{tT_0}$, and

$$S_n(t) = \int_0^{t} e^{(t-s)T_0} A S_{n-1}(s)\, ds \qquad (n \geq 1).$$

For any $f \in X$,

$$\int_{-\infty}^{0} \|Ae^{tT} f\| dt \le \int_{-\infty}^{0} \|A \sum_{n=0}^{\infty} S_n(t) f\| \, dt,$$

hence it is enough to show that

$$\sum_{n=0}^{\infty} \int_{-\infty}^{0} \|AS_n(t) f\| \, dt \le C\|f\|, \tag{2.13}$$

where, $C = \alpha(1-\alpha)^{-1}$. We shall prove by induction that

$$\int_{-\infty}^{0} \|AS_n(t) f\| \, dt \le \alpha^{n+1} \|f\|, \tag{2.14}$$

which is given in the assumptions for $n = 0$. Suppose

$$\int_{-\infty}^{0} \|AS_{n-1}(t) f\| \, dt \le \alpha^{n} \|f\|, \tag{2.15}$$

then,

$$\begin{aligned}
\int_{-\infty}^{0} \|AS_n(t) f\| \, dt &= \int_{-\infty}^{0} \|A\{-\int_{t}^{0} e^{(t-s)T_0} AS_{n-1}(s) f \, ds\}\| \, dt \\
&\le \int_{-\infty}^{0} \int_{-\infty}^{0} \|Ae^{tT_0} \left[e^{-sT_0} AS_{n-1}(s) f\right] \| \, dt \, ds \\
&\le \alpha \int_{-\infty}^{0} \|e^{-sT_0} AS_{n-1}(s) f\| \, ds && \text{(by (2.7))} \\
&\le \alpha \int_{-\infty}^{0} \|AS_{n-1}(s) f\| \, ds && \text{(by (2.6))} \\
&\le \alpha^{n+1} \|f\|. && \text{(by (2.15))}
\end{aligned}$$

This implies (2.14) and consequently (2.13). □

Proof of the Theorem 2.2.2 For the existence of $W_+(T, T_0)$, according to the Lemma 2.2.1, we need to show that for some $a > 0$:

(i) $\sup_{t \le -a} \|e^{tT}\| \le M$,
(ii) $\sup_{t \ge a} \|e^{tT_0}\| \le M_0$, and
(iii) $\int_a^{\infty} \|Ae^{tT_0} u\| dt < \infty$, for all $u \in D(T)$.

Statement (ii) is already in the assumptions and the point (iii) follows from the uniform boundedness of e^{tT} on $\mathbb{R}_+$ which is shown in Lemma 2.2.5.

Cook's Lemma ensures also the existence of $W_-(T, T_0)$. In fact we need to show that for some $a > 0$:

(i) $\sup_{t\geq a} \|e^{tT}\| \leq M$,
(ii) $\sup_{t\leq -a} \|e^{tT_0}\| \leq M_0$, and
(iii) $\int_{-\infty}^{-a} \|Ae^{tT_0}u\|dt < \infty$, for all $u \in D(T)$.

The statements (ii) and (iii) are already in the assumptions and the point (i) comes directly from Lemma 2.2.5. The existence of $W_{\pm}(T_0, T)$ can be proved in a similar way. □

The existence of the wave operators can also imply the boundedness of the perturbed semigroup.

Lemma 2.2.6 *Let e^{tT_0} be a (C_0) group such that $\sup_{t\leqslant 0} \|e^{tT_0}\| \leqslant C$. If one of the wave operators $W_-(T, T_0)$, $W_+(T_0, T)$ exists then the (C_0) semigroup e^{tT} is uniformly bounded for $t \geqslant 0$.*

Proof If $W_-(T, T_0)$ exists, then $\{e^{-tT}e^{tT_0} \mid t \leqslant 0\}$ is bounded. It follows from the principle of uniform boundedness that $\sup_{t\leqslant 0} \|e^{-tT}e^{tT_0}\| \leqslant M$, hence $\sup_{t\leqslant 0} \|e^{-tT}\| = \sup_{t\leqslant 0} \|e^{-tT}e^{tT_0}e^{-tT_0}\| \leqslant MC$. The statement from the existence of $W_+(T_0, T)$ is similar. □

2.3 Smooth Perturbations

In the previous section we assumed that the perturbation operator A is bounded. Even if the operator A is not bounded, there is the possibility to obtain the existence of the wave operators. For this we have to define the notion of T-smoothness.

Definition 2.3.1 Let T be the generator of a (C_0) group e^{tT} on a Banach space X and $\mathcal{D}$ a dense subset of X. A linear operator A is called T*-smooth* with constant $\alpha > 0$, if

$$\int_{-\infty}^{\infty} \|Ae^{tT}f\|dt \leq \alpha\|f\| \qquad \text{for all } f \in \mathcal{D} \tag{2.16}$$

(and hence for all f in X). Furthermore, if $\{e^{tT}\}_{t\geq 0}$ is only a (C_0) semigroup, we say that A is *positively* T*-smooth* when

$$\int_{0}^{\infty} \|Ae^{tT}f\|dt \leq \alpha\|f\|, \qquad f \in \mathcal{D}, \tag{2.17}$$

(and hence for all f in X). When

$$\int_{-\infty}^{0} \|Ae^{tT}f\|dt \leq \alpha\|f\|, \qquad f \in \mathcal{D}, \tag{2.18}$$

instead of saying that A is negatively T-smooth, we say that A is positively $(-T)$-smooth, when $\{e^{-tT}\}_{t\geqslant 0}$ is a (C_0) semigroup.

Notice that all the results of the previous section are valid if we replace $A \in \mathcal{L}(X)$ by A is T_0-smooth; For example if $A : D(T) \mapsto X$, then (2.16) can be satisfied for all $f \in D(T)$. When $\{e^{tT}\}_{t\in\mathbb{R}}$ is a uniformly bounded (C_0) group by a similar argument of the proof of Lemma 2.2.5, one can show that (2.17) and (2.18) are equivalent, thus in this case saying that A is positively T-smooth is equivalent to saying that A is T-smooth.

In the applications, generally the (C_0) group e^{tT_0} is isometric (i.e. $\|e^{tT_0} f\| = \|f\|$ for all $f \in X$). In this case we can weaken the hypotheses of Theorem 2.2.2 in the following way.

Theorem 2.3.2 *Let e^{tT_0} be an isometric (C_0) group on $\mathbb{R}$. If A is T_0-smooth with constant $\alpha < 2$, then the wave operators $W_\pm(T, T_0)$ and $W_\pm(T_0, T)$ exist.*

Proof Let us verify that (2.16) with $\alpha < 2$ implies simultaneously (2.17) and (2.18) with $\alpha < 1$. In fact, one of those is necessarily satisfied. The contrary leads to contradicting (2.16) with $\alpha < 2$. Now, if (2.18) is satisfied, since $\|e^{tT_0}\| = 1$ for $t \geq 0$, by Lemma 2.2.4 one gets (2.17) and if (2.17) is satisfied, since $\|e^{tT_0}\| = 1$ also for $t \leq 0$, by repeating the proof of Lemma 2.2.4 one gets (2.18). Hence we are in the framework of Theorem 2.2.2. □

The notion of T-smoothness is strong enough to involve not only the existence theorem for the wave operators, but also a generation result for the perturbed operator. In fact in Theorem 1.8.4 we have mentioned that (1.21), which is weaker condition than (2.17) can imply a generation result for $T_0 + A$.

2.4 Contribution of Positivity

In this section we assume that X is a Banach lattice, X_+ is its generating positive cone and A is a linear positive operator on X. We suppose

$$Af \geqslant 0, \qquad \text{for all } f \in X_+. \tag{2.19}$$

We will see how the positivity of the operators e^{tT_0} and A contribute to prove the existence of the wave operators $W_+(T_0, T)$ and $W_+(T, T_0)$ without using Cook's Lemma directly.

Theorem 2.4.1 *Suppose T_0 generates a (C_0) group $\{e^{tT_0}\}_{t\in\mathbb{R}}$ on X, $A \in \mathcal{L}(X)$ satisfying* (2.19) *and $T = T_0 + A$ are such that:*

(i) $e^{tT_0} \geqslant 0$ *for all $t \in \mathbb{R}_+$;*
(ii) $\|e^{tT_0}\| \leqslant M_0$ *for all $t \in \mathbb{R}$;*
(iii) $\|e^{tT}\| \leqslant M$ *for all $t \in \mathbb{R}_-$;*

(iv) *A is T_0-smooth on the positive cone X_+; that is*

$$\int_{-\infty}^{+\infty} \|Ae^{tT_0} f\| dt < +\infty \quad \textit{for all} \quad f \in X_+. \tag{2.20}$$

Then the wave operators $W_+(T_0, T)$ and $W_+(T, T_0)$ exist.

For the proof of this theorem we need the following lemmas.

Lemma 2.4.2 *Suppose the assumptions* (i) *and* (ii) *of the above theorem hold. Then*

$$e^{tT} \geqslant e^{tT_0} \geqslant 0 \quad \textit{for all} \quad t \geqslant 0. \tag{2.21}$$

Proof By Corollary 1.8.6 and Remark 1.8.7 the Dyson–Phillips expansion of e^{tT} (see (1.18)) converges in $\mathcal{L}(X)$. The operators A and $S_0(t) = e^{tT_0}$ being positive for all $t \geqslant 0$, Lemma 1.9.7 implies that $S_1(t) = \int_0^t [\exp(t-s)T_0] A S_0(s) ds$ is also positive. By induction one can prove that for any $n \geqslant 1$ and for any $t \geqslant 0$, $S_n(t)$ given by (1.19) is a positive operator and the lemma follows from the closedness of the positive cone X_+ in X. □

Lemma 2.4.3 *Suppose the assumptions* (i), (ii) *and* (iii) *of the above theorem hold. Then*

$$\int_{-\infty}^{0} \|Ae^{tT_0} f\| dt < \infty, \qquad \textit{for all} \quad f \in X_+, \tag{2.22}$$

implies

$$\int_{0}^{\infty} \|Ae^{tT} f\| dt < \infty, \qquad \textit{for all} \quad f \in X_+. \tag{2.23}$$

Proof The fact that X_+ is generating in X implies that every $f \in X$ can be written in the form $f = g_1 - g_2 + i(h_1 - h_2)$ with $g_j, h_j \in X_+$, where $j = 1, 2$. Thus, (2.22) holds for all $f \in X$. Then, applying the principle of uniform boundedness, we find a constant $C > 0$ such that

$$\int_{-\infty}^{0} \|Ae^{tT_0} f\| dt \leqslant C\|f\| \qquad \text{for all} \quad f \in X. \tag{2.24}$$

Now, let $r > 0$ and $0 \leqslant s \leqslant r$. By taking $t = r - s$ in (2.21) and applying Lemma 1.9.7 (i), we get

$$e^{rT} \geqslant e^{(r-s)T_0} e^{sT} \geqslant 0.$$

By using (i), the positivity of A implies that

$$Ae^{(s-r)T_0} e^{rT} f \geqslant Ae^{sT} f \geqslant 0, \qquad \text{for all} \quad f \in X_+.$$

Hence we have by Remark 1.9.8,

$$\|Ae^{sT}f\| \leqslant \|Ae^{(s-r)T_0}e^{rT}f\| \qquad \text{for all } f \in X_+.$$

Integrating with respect to s, a change of variable gives

$$\int_0^r \|Ae^{tT}f\|dt \leqslant \int_{-r}^0 \|Ae^{tT_0}e^{rT}f\|dt.$$

Due to (2.24) and the assertion (iii), the right-hand side of this inequality is bounded by $CM\|f\|$. Since r was arbitrary, the conclusion follows. □

Proof of Theorem 2.4.1 Replace in the first Duhamel formula (1.14) A, B and C by T_0, A and T. The action of e^{-tT_0} on this formula yields

$$e^{-tT_0}e^{tT}f = f + \int_0^t e^{-sT_0}Ae^{sT}f\,ds. \tag{2.25}$$

The formula (2.25) together with Lemma 2.4.3 implies the existence of the wave operator $W_+(T_0, T)$.

For the existence of $W_+(T, T_0)$ the second Duhamel formula (1.15) yields that

$$e^{-tT}e^{tT_0}f = f - \int_0^t e^{-sT}Ae^{sT_0}f\,ds, \tag{2.26}$$

and the convergence of the integral $\int_0^\infty e^{-sT}Ae^{sT_0}f\,ds$ follows directly from the assumptions. □

Remark 2.4.4 In Theorem 2.4.1, we can replace the condition (2.20) by the condition (2.22), since by a similar proof of Lemma 2.2.5, one can show that these two conditions are equivalent.

Corollary 2.4.5 *Suppose T_0 generates a bounded positive (for $t \geqslant 0$) (C_0) group $\{e^{tT_0}\}_{t\in\mathbb{R}}$ on X such that for some $\alpha < 1$,*

$$\int_{-\infty}^0 \|Ae^{tT_0}f\|dt < \alpha\|f\| \quad \text{for all } f \in X. \tag{2.27}$$

Then the wave operator $W_+(T_0, T)$ exists.

Proof It suffices to remark that by virtue of Lemma 2.2.3 (ii), the assumption (2.27) implies (iv) of Theorem 2.4.1. □

2.5 Limiting Absorption Principle

The notion of the limiting absorption principle (l.a.p. for short) concerns the trace of a resolvent operator on its spectrum. Roughly speaking, if Γ is a line in the complex plane and T_0 is an operator such that $\sigma(T_0)$, the spectrum of T, lies in Γ, we say that T_0 satisfies the l.a.p. if $(\lambda - T_0)^{-1}$ admits continuous boundary values as $\lambda \to \lambda_0 \in \sigma(T_0)$ on both side of the line Γ. Such a limit does not exist in a general setting. For example if T_0 generates a (C_0) group $\{e^{tT_0}\}_{t\in\mathbb{R}}$ in $X = L^1(\Omega, \mu)$, such that $\|e^{tT_0}\| = 1$, for all $t \in \mathbb{R}$, then $\sigma(T_0) \subset i\mathbb{R}$, and as we have seen in Lemma 1.9.17 (by taking $B := I$), the existence of $\lim_{\lambda\to 0_+}(\lambda - T_0)^{-1}$ is equivalent to the existence of $\int_0^\infty e^{tT_0}dt$. From the property (1.34) of $L^1(\Omega, \mu)$ space, for any $f \in L^1(\Omega, \mu)_+$, $\|\int_0^\infty e^{tT_0} f dt\| = \int_0^\infty \|e^{tT_0} f\| dt$, which diverges due to the isometric character of e^{tT_0}. So to remedy this problem we have either to change the topology of the underlying space X, or to introduce an operator A on X, which plays the role of a mollifier and require that the strong limit of $A(\lambda - T_0)^{-1}$ as $\lambda \to \lambda_0^{\pm}$ exists. The latter consideration brings the following definition.

Definition 2.5.1 Let T_0 (resp.$-T_0$) be the generator of a bounded (C_0) semigroup on a Banach lattice X and $A \in \mathcal{L}(D(T_0), X)$ be positive. We say that the pair (A, T_0) satisfies the *limiting absorption principle as* $\lambda \to 0_+$ (resp. $\lambda \to 0_-$) on X, if the limit operator

$$A(0_+ - T_0)^{-1} f := \lim_{\lambda\to 0_+} A(\lambda - T_0)^{-1} f \quad (\text{resp. } A(0_- - T_0)^{-1} f := \lim_{\lambda\to 0_-} A(\lambda - T_0)^{-1} f) \tag{2.28}$$

exist for all $f \in X$. Or equivalently,

$$A(0_+ - T_0)^{-1} f = \int_0^\infty A e^{sT_0} f ds \quad (\text{resp.} \; - A(0_- - T_0)^{-1} f = \int_{-\infty}^0 A e^{sT_0} f ds) \tag{2.29}$$

exists for all $f \in X$.

The equivalence between (2.28) and (2.29) follows from the following lemma

Lemma 2.5.2 *Assume that the pair* (A, T_0) *satisfies the limiting absorption principle. Then the assertions* (2.28) *and* (2.29) *are equivalent.*

Proof Since the positive cone X_+ is generating in the Banach lattice X, we may restrict ourselves to $f \in X_+$. Assume that (2.28) holds, that is, the strong limit $s - \lim_{\lambda\to 0_+} A(\lambda - T_0)^{-1} f$ exists. Since $(\lambda - T_0)^{-1}$ is nonincreasing for $\lambda > 0$, it follows that

$$A \int_0^\infty e^{-\lambda s} e^{sT_0} f ds = A(\lambda - T_0)^{-1} f \le A(0_+ - T_0)^{-1} f.$$

Since the operator A is closed,

$$A\int_0^\infty e^{-\lambda s}e^{sT_0}f\,ds = s-\lim_{t\to\infty} A\int_0^t e^{-\lambda s}e^{sT_0}f\,ds$$
$$= s-\lim_{t\to\infty}\int_0^t e^{-\lambda s}Ae^{sT_0}f\,ds = \int_0^\infty e^{-\lambda s}Ae^{sT_0}f\,ds.$$

Thus,

$$\int_0^\infty e^{-\lambda s}Ae^{sT_0}f\,ds = A\int_0^\infty e^{-\lambda s}e^{sT_0}f\,ds \le A(0_+ - T_0)^{-1}f.$$

Letting $\lambda \to 0_+$, we get

$$\int_0^\infty Ae^{sT_0}ds \le A(0_+ - T_0)^{-1}.$$

Conversely, suppose the positive part of (2.29) holds, then

$$A(\lambda - T_0)^{-1}f = \int_0^\infty e^{-\lambda s}Ae^{sT_0}f\,ds \le \int_0^\infty Ae^{sT_0}f\,ds < \infty$$

Letting $\lambda \to 0_+$, we get the reverse inequality, which proves the equivalence of the positive part of the lemma. For the negative part of (2.28) and (2.29), let $\widetilde{T}_0 = -T_0$, e^{tT_0} being a (C_0) group, so $\{e^{t\widetilde{T}_0}\}_{t\ge 0}$ would be a (C_0) semigroup. According to previous analysis, $\int_0^\infty Ae^{t\widetilde{T}_0}dt$ exists if and only if $A(0_+ - \widetilde{T}_0)^{-1}$ exists and these strong limits are equal. Hence by a change of variable t to $-s$ we get

$$\int_{-\infty}^0 Ae^{sT_0}ds = -A(0_- - T_0)^{-1}.$$

□

Theorem 2.5.3 *Let X be an AL-space, and let T_0 be the generator of a bounded positive (C_0) group on X. Let $A : D(T_0) \mapsto X$ be a positive operator.*

(a) *If the pair (A, T_0) satisfies the limiting absorption principle as $\lambda \to 0_+$ on X and*

$$\|A(0_- - T_0)^{-1}\| < 1, \tag{2.30}$$

then the wave operator $W_+(T, T_0)$ exists.

(b) *If the pair (A, T_0) satisfies the limiting absorption principle as $\lambda \to 0_-$ on X and*

$$\|A(0_+ - T_0)^{-1}\| < 1. \tag{2.31}$$

Then the wave operator $W_-(T, T_0)$ exists.

(c) *If $\|e^{tT_0}\| \leqslant 1$ for all $t \geqslant 0$ and the estimation* (2.30) *holds, then $W_+(T_0, T)$ exists.*

Proof (a) According to formula (2.26), the existence of $W_+(T, T_0)$ is conditioned by the existence of

$$\int_0^\infty e^{-sT} A e^{sT_0} f \, ds \qquad \text{for all} \quad f \in X \tag{2.32}$$

By using Remark 1.9.18 and the property (1.34), for any $f \in X_+$ we have

$$\begin{aligned} \int_{-\infty}^0 \|Ae^{tT_0} f\| dt &= \left\| \int_{-\infty}^0 Ae^{tT_0} f dt \right\| \\ &= \|A(0_- - T_0)^{-1} f\| \\ &\leq \alpha \|f\|, \end{aligned}$$

where $\alpha = \|A(0_- - T_0)^{-1}\| < 1$. By Lemma 2.2.3 (ii), this implies that $\sup_{t\in\mathbb{R}_-} \|e^{tT}\| \leqslant M_-$ and

$$\begin{aligned} \int_0^\infty \|e^{-sT} Ae^{sT_0} f\| ds &\leqslant M_- \int_0^\infty \|Ae^{sT_0} f\| ds \\ &= M_- \left\| \int_0^\infty Ae^{sT_0} f \, ds \right\| \\ &= M_- \|A(0_+ - T_0)^{-1} f\| \end{aligned}$$

and we obtain (2.32), since $X = X_+ - X_+$.

(b) For the existence of $W_-(T, T_0)$ the procedure is the same. We use the formula (1.15) but for $t < 0$. For the existence of $\int_{-\infty}^0 e^{-sT} Ae^{sT_0} f \, ds$, we apply Lemma 2.2.3 (i) to get the boundedness of $\{e^{tT}\}_{t\geqslant 0}$ and the existence of $A(0_- - T_0)^{-1}$ is equivalent to the existence of $\int_{-\infty}^0 Ae^{sT_0} f \, ds$.

(c) According to (2.25), for the existence of $W_+(T_0, T)$, it suffices that the integral $\int_0^\infty e^{-sT_0} Ae^{sT} ds$ converges. From $\alpha = \|A(0_- - T_0)^{-1}\| < 1$ we get $\int_{-\infty}^0 \|Ae^{sT_0} f\| ds \leqslant \alpha \|f\|$, for all $f \in X$. Hence all the assumptions of Lemma 2.4.3 are satisfied and by (2.23) we have

$$\int_0^\infty \|e^{-sT_0} Ae^{sT} f\| \, ds \leq \int_0^\infty \|Ae^{sT}\| ds < \infty \qquad \text{for all } f \in X_+$$

thus for all $f \in X$. □

2.6 What Can We Obtain from the Existence of the Wave Operators?

First, there exists some relationships between different wave operators. In the following lemma we will see that, under some assumptions, the existence of some wave operators implies the existence of others.

Lemma 2.6.1 *Let e^{tT_0} be a positive (for all $t \in \mathbb{R}$) and uniformly bounded for $t \leqslant 0$, (C_0) group. Let A and A' be two bounded positive operators in a Banach lattice X such that $A' - A \geqslant 0$. If e^{tT} and $e^{tT'}$ are two positive (C_0) groups generated by $T = T_0 + A$ and $T' = T_0 + A'$, then the existence of $W_-(T', T_0)$ implies the existence of $W_-(T, T_0)$.*

Proof By Cook's Lemma for the existence of $W_-(T, T_0)$ it suffices to show that e^{tT} is uniformly bounded on $\mathbb{R}_+$ and

$$\int_{-\infty}^{0} \|Ae^{sT_0} f\| \, ds < \infty. \tag{2.33}$$

By virtue of Lemma 2.2.6 the existence of $W_-(T', T_0)$ implies that $e^{tT'}$ is uniformly bounded on $\mathbb{R}_+$ and consequently the existence of the integral $\int_{-\infty}^{0} \|e^{-sT'} A' e^{sT_0} f\| \, ds$ for all $f \in X$. But for $f \in X_+$ we have the domination relation $Ae^{tT_0} f \leq A' e^{tT_0} f$ which implies that

$$\int_{-\infty}^{0} \|Ae^{sT_0} f\| \, ds \leqslant \int_{-\infty}^{0} \|A' e^{sT_0} f\| \, ds.$$

We conclude the lemma by remarking that $X = X_+ - X_-$. □

The wave operators have also the transitive properties in the following sense.

Lemma 2.6.2 *Let T_0, T' and T be the generators of the (C_0) groups e^{tT_0}, $e^{tT'}$ and e^{tT}. If the wave operators $W_\pm(T_0, T')$ and $W_\pm(T', T)$ exist, then wave operator $W_\pm(T_0, T)$ exists also.*

Proof The proof is rather trivial. Since $e^{-tT_0} e^{tT'} e^{-tT'} e^{tT} = e^{-tT_0} e^{tT}$, the existence of $\lim_{t \to \pm\infty} e^{-tT_0} e^{tT'} f$ and $\lim_{t \to \pm\infty} e^{-tT'} e^{tT} f$ imply the existence of $\lim_{t \to \pm\infty} e^{-tT_0} e^{tT} f$, for all $f \in X$. □

Another interesting consequence of the existence of the wave operators is the geometrical configuration of the spectrum $\sigma(T)$.

Lemma 2.6.3 *Suppose that $\{e^{tT_0}\}_{t \in \mathbb{R}}$ is a bounded $C_0)$ group (that is $\|e^{tT_0}\| \leq M$ for all $t \in \mathbb{R}$) and $\{e^{tT}\}_{t \in \mathbb{R}}$ is a (C_0) group satisfying $\|e^{tT}\| \leq M e^{\omega |t|}$ for all $t \in \mathbb{R}$.*

(a) *If $W_+(T_0, T)$ or $W_-(T, T_0)$ exists, then e^{tT} is uniformly bounded for $t \geq 0$ and $\sigma(T) = \{\lambda \in \mathbb{C} \, : \, -\omega \leq \operatorname{Re} \lambda \leq 0\}$.*
(b) *If $W_-(T_0, T)$ or $W_+(T, T_0)$ exists, then e^{tT} is uniformly bounded for $t \leq 0$ and $\sigma(T) = \{\lambda \in \mathbb{C} \, : \, 0 \leq \operatorname{Re} \lambda \leq \omega\}$.*

Proof The existence of $W_+(T_0, T)$ (resp. $W_-(T, T_0)$) implies upon application of the principle of uniform boundedness that $\|e^{-tT_0} e^{tT}\| \leq C$ (resp. $\|e^{-tT} e^{tT_0}\| \leq C$) for all $t \geq 0$ (resp. for all $t \leq 0$), therefore we get $\|e^{tT}\| \leq \|e^{tT_0}\| \|e^{-tT_0} e^{tT}\| \leq MC$ for all $t \geq 0$ (resp. $\|e^{-tT}\| \leq \|e^{-tT} e^{tT_0}\| \|e^{-tT_0}\| \leq MC$ for all $t \leq 0$). Applying the

Hille–Yosida Theorem (see Theorem 1.5.3), we have $\{\lambda \in \mathbb{C} : \operatorname{Re}\lambda \geq 0\} \subset \rho(T)$. Similarly, $\|e^{tT}\| \leq Me^{-\omega t}$ for all $t \leq 0$ implies that $\{\lambda \in \mathbb{C} : \operatorname{Re}\lambda \leq -\omega\} \subset \rho(T)$ from which statement (a) follows. The proof of statement (b) is similar. □

2.7 Notes and Comments

One of the earliest accounts of *Abstract Scattering Theory* is to be found in Taylor's book [115]. A fine introduction to this theory is given by Read and Simon [103], Baumgärtel and Wollenberg [15] and Roach [104] and a more advanced text is Yafaev's monograph [126]. In all these texts the *intertwining identity* has a predominant role in the scattering theory. That is why we opened the first section with the proof of this identity. Surely, the application of this identity needs the proof of the existence of wave operators. As we have indicated in the Introduction, our methods to prove the existence of the wave operators are based essentially on Cook's Lemma. This lemma was proved by Cook [31], which is indicated in the second section. We also gave a scale of results which revolves around this lemma. In Sect. 2.3 we introduced the notion of the smooth perturbation in an abstract manner. This theory was developed by T. Kato in two remarkable papers [70, 71] (see also [103, Sect. XIII. 7].

The notion of the *Limiting Absorption Principle* is similar. Roughly speaking, it is similar to that in Chap. 1 (1.10), taking $\lambda \to 0$. This notion was introduced by Eidus [43]. In Sect. 2.5 we introduced this notion as given by Mokhtar-Kharroubi [91] (see also [92]).

Section 2.4 is the abstract part of [118]. In this paper T. Umeda shows how heavily positivity can contribute to the existence of wave operators if we work in a Banach lattice like L^1. The application of this contribution to the transport problem appears in the next chapter.

Chapter 3
Existence of the Wave Operators for the Transport Equation

3.1 Statement of the Time-Dependent Problem

In the transport phenomena, the dynamical object is a positive function $u(x, v)$ representing the probability density function of neutrons at a point (x, v) in the phase space $\mathbb{R}^n \times V$. Thus

$$\int_{\mathbb{R}^n \times V} u(x, v) dx dv < \infty \tag{3.1}$$

and represents the total number of neutrons in the entire space $\mathbb{R}^n$. Such a framework is physically justifiable, since we consider a bounded convex open subset Ω of $\mathbb{R}^n$, with C^1-boundary and we assume that the particles can enter or leave Ω freely. In this situation we say that the transport phenomena is taken with *transparent boundary condition* and this assumption renders the dynamics invertible. Note that at the end of the Chap. 6 we will consider also the *semi-transparent boundary conditions* which has also some new peculiarity. However, we will prove that for transparent boundary condition the transport operator generates a one-parameter (C_0) group of operators.

The finiteness of (3.1) stipulates that the natural space of the states is not a Hilbert space as is customary in scattering theory for the wave or Schrödinger equations but the positive cone $L^1(\mathbb{R}^n \times V)_+$ in the Banach lattice $L^1(\mathbb{R}^n \times V)$. These characteristics illustrate the natural use of the theory of the positive semigroups of operators in Banach lattices.

In the phase space $\mathbb{R}^n \times V$, if there is no collision, the classical particles move freely according to their initial velocities. This generates the free evolution operator

$$[U_0(t) f](x, v) := f(x - tv, v) \tag{3.2}$$

H. Emamirad, *Scattering Theory for Transport Phenomena*, Mathematical Physics Studies, https://doi.org/10.1007/978-981-16-2373-8_3

as the dynamic of the following system:

(P0)
$$\begin{cases} \dfrac{\partial u}{\partial t} = -v.\nabla_x u = T_0 u \\ u(x, v, 0) = f(x, v). \end{cases}$$

which is called an *advection system* or *streaming free problem*.

We suppose that Ω is a measurable subset of $\mathbb{R}^n$ such that its closure $\overline{\Omega}$ is bounded, taking a finite convex domain with Lebesgue measure dx and

$$V = \{v \in \mathbb{R}^3 \quad ; 0 \leq v_{min} \leq |v| \leq v_{max} \leq +\infty\}$$

to be the velocity space with Lebesgue measure dv. We denote by V^* when the velocity space is a shell in $\mathbb{R}^3$, that is when $v_{min} > 0$ and $v_{max} < \infty$. The particles encounter the body only in Ω and outside Ω there is no collision. The effect of a collision is an instantaneous change of velocity and therefore trajectory. Depending on the physical properties of the medium, pure scattering may be accompanied by net absorption or creation. The effect of all these changes is usually described by the linear Boltzmann equation:

$$\begin{aligned} \frac{\partial u}{\partial t} &= -v.\nabla_x u + \chi_\Omega(x)\left[-\sigma_a(x, v)u + \int_V k(x, v', v)u(x, v', t)dv'\right] \\ &= (T_0 + A_a + A_p)u(x, v, t) \\ &= Tu(x, v, t), \end{aligned} \tag{3.3}$$

where χ_Ω is the characteristic function of the domain Ω. Here the collision term admits a natural decomposition into an absorptive part described by the absorption cross-section $\sigma_a(x, v)$ and a re-emitting part, described by the differential cros-section $k(x, v', v)$.

We analyze the scattering situations where the surface of the body $\partial\Omega$, is perfectly transparent for both incident and emergent distributions.

In the sequel we shall need the following conditions on σ_a and k appearing in (3.3).

Definition 3.1.1 The pair (k, σ_a), is called *admissible* if:

(i) $k(x, v, v')$ is a nonnegative measurable function on $\mathbb{R}^n \times V \times V$ and $\sigma_a(x,v)$ is a nonnegative measurable function on $\mathbb{R}^n \times V$;
(ii) for each (x, v) , $k(x, v,.)$ is in $L^1(V)$;
(iii) $\sigma_a(x, v)$ and

$$\sigma_p(x, v) := \int_V k(x, v, v')dv' \tag{3.4}$$

are uniformly bounded on $\mathbb{R}^n \times V$.

An admissible pair (k, σ_a) is *regular* if furthermore:

(iv) $k(x, v, v')$ and $\sigma_a(x, v)$ vanish for $x \notin \Omega$.

It is not hard to see that if (k, σ_a) is an admissible pair,

$$A_p f(x, v) := \int_V k(x, v', v) f(x, v') dv' \quad \text{and} \quad A_a f(x, v) = -\sigma_a(x, v) f(x, v) \tag{3.5}$$

define bounded operators on $L^1(\mathbb{R}^n \times V)$ and we have

$$\|A_i\| = \|\sigma_i\|_\infty \qquad (i = a, p). \tag{3.6}$$

We will also write

$$A := A_a + A_p \qquad T_a := T_0 + A_a \text{ and } T := T_a + A_p. \tag{3.7}$$

Definition 3.1.2 If we denote by

(MFP) $$M(\sigma) := \operatorname*{ess\,sup}_{(x,v)\in\mathbb{R}^n\times V^*} |v|^{-1}\sigma(x, v) < \infty$$

the essential bound of the macroscopic cross-section $|v|^{-1}\sigma(x, v)$, $M(\sigma)$ is called the *mean free path* **(MFP)**. Here and in the sequel σ stands for σ_a or σ_p or $\sigma_a + \sigma_p$.

Physically, $M(\sigma)$ represents the mean distance between two consecutive collisions before a production and the condition **(MFP)** is appropriate for the solvability of the transport problem.

Instead of **(MFP)**, one can impose some weaker conditions. Let us define

$$F(\sigma) := \operatorname*{ess\,sup}_{(x,v)\in\mathbb{R}^n\times V} \int_{-\infty}^{\infty} \sigma(x + tv, v)\, dt,$$

$$F_+(\sigma) := \operatorname*{ess\,sup}_{(x,v)\in\mathbb{R}^n\times V} \int_{0}^{\infty} \sigma(x + tv, v)\, dt,$$

$$F_-(\sigma) := \operatorname*{ess\,sup}_{(x,v)\in\mathbb{R}^n\times V} \int_{-\infty}^{0} \sigma(x + tv, v)\, dt.$$

Since $\int_{-\infty}^{\infty} \sigma(x + tv, v)\, dt$ is the Radon transform of σ, the condition

(FRT) $$F(\sigma) = \operatorname*{ess\,sup}_{(x,v)\in\mathbb{R}^n\times V} \int_{-\infty}^{\infty} \sigma(x + tv, v)\, dt < \infty$$

stands for *finite Radon transform*.

Lemma 3.1.3 *Let (k, σ_a) be an admissible pair, then*

$$\int_{-\infty}^{\infty} \|A_i U_0(t) f\| dt \leq F(\sigma_i) \|f\|, \quad \text{for all } f \in L^1(\mathbb{R}^n \times V) \;\; i = a, p. \tag{3.8}$$

Proof

$$\begin{aligned}
\int_{-\infty}^{\infty} \|A_i U_0(t) f\| dt &= \int_{-\infty}^{\infty} \iint_{\mathbb{R}^n \times V} |\sigma_i(x, v) f(x - tv, v)| dx dv dt \\
&= \int_{-\infty}^{\infty} \iint_{\mathbb{R}^n \times V} |\sigma_i(y + tv, v) f(y, v)| dy dv dt \\
&\leq F(\sigma_i) \iint_{\mathbb{R}^n \times V} |f(y, v)| dy dv.
\end{aligned}$$

□

Lemma 3.1.4 *Let (k, σ_a) be an admissible regular pair,*

(a) If σ satisfies **(FRT)**, *then $F(\sigma) = F_+(\sigma) = F_-(\sigma)$.*
(b) If σ satisfies **(MFP)**, *then $F(\sigma) \leqslant (\text{diam}\, \Omega) M(\sigma)$.*

Proof (a) Since σ is a positive function, it is clear that $F(\sigma) \geq F_+(\sigma)$. By definition of $F(\sigma)$, for any $\varepsilon > 0$, there exists $(x_\varepsilon, v_\varepsilon)$ with $v_\varepsilon \neq 0$ such that $F(\sigma) - \varepsilon \leqslant \int_{-\infty}^{\infty} \sigma(x_\varepsilon + s v_\varepsilon, v_\varepsilon) ds$. The regularity of σ implies that

$$\int_{-\infty}^{\infty} \sigma(x_\varepsilon + s v_\varepsilon, v_\varepsilon) ds = \int_{t_-(x_\varepsilon, v_\varepsilon)}^{t_+(x_\varepsilon, v_\varepsilon)} \sigma(x_\varepsilon + s v_\varepsilon, v_\varepsilon) ds, \tag{3.9}$$

where

$$t_-(x, v) = \min\{t \in \mathbb{R} \mid x + tv \in \Omega\} \quad \text{and} \quad t_+(x, v) = \max\{t \in \mathbb{R} \mid x + tv \in \Omega\}.$$

Taking $t = s - t_-(x_\varepsilon, v_\varepsilon)$ in (3.9), one obtains

$$\begin{aligned}
\int_{-\infty}^{\infty} \sigma(x_\varepsilon + s v_\varepsilon, v_\varepsilon) ds &= \int_0^{t_+(x_\varepsilon, v_\varepsilon) - t_-(x_\varepsilon, v_\varepsilon)} \sigma(x_\varepsilon + t_-(x_\varepsilon, v_\varepsilon) v_\varepsilon + t v_\varepsilon, v_\varepsilon) dt \\
&\leqslant \int_0^{\infty} \sigma(x_\varepsilon + t_-(x_\varepsilon, v_\varepsilon) v_\varepsilon + t v_\varepsilon, v_\varepsilon) dt \\
&\leqslant F_+(\sigma).
\end{aligned}$$

ε being arbitrary, one concludes that $F(\sigma) \leq F_+(\sigma)$. Hence $F(\sigma) = F_+(\sigma)$. The same argument works for $F(\sigma) = F_-(\sigma)$.

(b) Taking $s = \tau|v|^{-1}$,

$$\begin{aligned} F_+(\sigma) &= \operatorname*{ess\,sup}_{(x,v)\in\mathbb{R}^n\times V} \int_0^{\infty} \sigma(x+sv, v)\,ds \\ &= \operatorname*{ess\,sup}_{(x,v)\in\mathbb{R}^n\times V} \int_0^{\mathrm{diam}\Omega} |v|^{-1}\sigma(x+\tau\omega, v)\,d\tau \\ &\leqslant M(\sigma)(\mathrm{diam}\Omega), \end{aligned}$$

where $\omega = \frac{v}{|v|}$. Hence from part (a) we obtain

$$F(\sigma) = F_+(\sigma) \leqslant M(\sigma)(\mathrm{diam}\ \Omega).$$

□

3.2 Streaming Free Group

Let $C_0^{\infty}(\mathbb{R}^n \times V)$ be the space of infinitely many times differentiable complex-valued functions with compact support in $\mathbb{R} \times V$. The evolution operator $U_0(t)$ can be defined on any $L^p(\mathbb{R}^n \times V)(1 \leqslant p < \infty)$, but $U_0(t)f$ is not differentiable with respect to t, unless $f \in C^1(\mathbb{R}^n \times V)$. In the following theorem we prove that the advection system **(P0)** generates a (C_0) group on any $L^p(\mathbb{R}^n \times V)(1 \leqslant p < \infty)$ defined by (3.2) is called a *streaming free group*.

Theorem 3.2.1 *The family $\{U_0(t)\}_{t\in\mathbb{R}}$ is an one-parameter positive (C_0) group of isometries on $L^p(\mathbb{R}^n \times V)$, i.e.*

$$\|U_0(t)f\|_p = \|f\|_p \qquad \text{for all } 1 \leqslant p < \infty, \tag{3.10}$$

of positive operators for all $t \in \mathbb{R}$. Moreover $\mathcal{S}(\mathbb{R}^n \times V)$, the Schwartz space is a core of its generator T_0 which satisfies

$$T_0 f = -v \cdot \nabla_x f \tag{3.11}$$

for all $f \in \mathcal{S}(\mathbb{R}^n \times V)$. Hence the generator of $U_0(t)$ is the closure of T_0. Henceforth in order to highlight the generator of $U_0(t)$ we denote it by e^{tT_0}.

Proof The fact that e^{tT_0} is an Abelian group of isometries on $L^p(\mathbb{R}^n \times V)$ and its positivity follows readily from (3.2). For the strong continuity of e^{tT_0}, since $\mathcal{S}(\mathbb{R}^n \times V)$ is dense in $L^p(\mathbb{R}^n \times V)$ one can first establish this property for the $\mathcal{S}$ functions. Let $\varphi \in \mathcal{S}(\mathbb{R}^n \times V)$, for any $\epsilon > 0$ we can take t so small that

$$\iint_{\mathbb{R}^n\times V} |\varphi(x-tv,v)-\varphi(x,v)|^p dxdv$$
$$= \iint_{\mathbb{R}^n\times V} \left|\frac{(1+|x|^2+|v|^2)(\varphi(x-tv,v)-\varphi(x,v))}{1+|x|^2+|v|^2}\right|^p dxdv$$
$$\le C \sup_{(x,v)\in\mathbb{R}^n\times V} |(1+|x|^2+|v|^2)(\varphi(x-tv,v)-\varphi(x,v))|^p \le \frac{\epsilon}{3},$$

where $C = \iint_{\mathbb{R}^n\times V}(1+|x|^2+|v|^2)^{-p}dxdv$. Since e^{tT_0} is uniformly bounded we can use an $\frac{\epsilon}{3}$ argument to show this for all $f \in L^p(\mathbb{R}^n \times V)$. Furthermore, if $\mathcal{S}(\mathbb{R}^n \times V)$ is invariant under e^{tT_0}, then according to Lemma 1.2.6, $\mathcal{S}(\mathbb{R}^n \times V)$ is a core for T_0 and for all $f \in C_0^\infty(\mathbb{R}^n \times V)$

$$\frac{d}{dt} f(x-tv,v) = -v\cdot\nabla_x f(x-tv,v),$$

which implies (3.11) for $t = 0$. □

From this theorem it follows that the generator of e^{tT_0} is $T_0 := \overline{T_0|_{\mathcal{S}(\mathbb{R}^n\times V)}}$ and its maximal domain $D(T_0) = \mathcal{W}^p(\mathbb{R}^n \times V)$, where

$$\mathcal{W}^p(\mathbb{R}^n\times V) := \{f \in L^p(\mathbb{R}^n\times V) \mid v\cdot\nabla_x f \in L^p(\mathbb{R}^n\times V)\}.$$

Thus, T_0 is a closed densely defined operator in $L^p(\mathbb{R}^n \times V)$, but since $\mathcal{S}(\mathbb{R}^n \times V)$ is not dense in $L^\infty(\mathbb{R}^n \times V)$, one has to avoid the case $p = \infty$.

Here we confine our attention to the spectral properties of the operator T_0. From Hille–Phillips Theorem it comes out that for any $\lambda \in \mathbb{C}\setminus i\mathbb{R}$ the operator $(\lambda I - T_0)$ is invertible, thus $\sigma(T_0) \subseteq i\mathbb{R}$. Here we will prove a surprising result that the spectrum of T_0 in $L^p(\mathbb{R}^n \times V)$ change its nature with respect to $p \in [1,\infty]$.

Theorem 3.2.2 *(a)* $\sigma(T_0) = \sigma_r(T_0) = i\mathbb{R}$ *in* $L^1(\mathbb{R}^n \times V)$*;*
(b) $\sigma(T_0) = \sigma_c(T_0) = i\mathbb{R}$ *in* $L^2(\mathbb{R}^n \times V)$*;*
(c) for any $g \in C^1(\mathbb{R}^n \times V)$ *and any* $\lambda \in i\mathbb{R}$, ϕ *defined by*

$$\phi(x,v) = g(x_\perp,v)e^{\lambda x\cdot v/|v|^2},$$

where $x_\perp$ *is the component of vector* x *orthogonal to the direction* v*, belongs to* $L^\infty(\mathbb{R}^n \times V)$ *and satisfies* $-v\cdot\nabla_x\phi = \lambda\phi$*. In other word, if* T_0^* *is the adjoint of* T_0*, acting in* $L^\infty(\mathbb{R}^n \times V)$*, then* $\sigma_p(T_0^*) = i\mathbb{R}$*.*

Proof First we prove that $\sigma_p(T_0) = \emptyset$ in $L^p(\mathbb{R}^n \times V)$ for any $1 \leqslant p < \infty$. Assume that there exists a real number β such that $i\beta \in \sigma_p(T_0)$, then there exists $0 \neq \phi \in D(T_0)$ such that $T_0\phi = i\beta\phi$. By the spectral mapping Theorem 1.6.6 (b), it is equivalent to say that for all $t \in \mathbb{R}$, $e^{i\beta t}f(x,v) = e^{tT_0}f(x,v) = f(x-tv,v)$ for some $f \in L^p(\mathbb{R}^n \times V)$. Thus,

$$|f(x-tv,v)|^p = |f(x,v)|^p \qquad \text{for all } t \geqslant 0. \tag{3.12}$$

For $f \neq 0$, there exist $\rho > 0$, $v_0 > 0$ and $V_0 > 0$ such that

$$\int_{|x|\leqslant \rho} \int_{v_0 \leqslant |v| \leqslant V_0} |f(x,v)|^p dx dv > 0.$$

Now, let us choose the sequence $0 < t_1 < t_2 < \cdots$ such that $\rho < -\rho + v_0 t_1 < \rho + V_0 t_1 < -\rho + v_0 t_2 < \rho + V_0 t_2 < \cdots$, then from (3.12)

$$\begin{aligned} \|f\|_p^p &\geqslant \sum_{i=1}^{\infty} \int_{-\rho+v_0 t_i \leqslant |x| \leqslant \rho + V_0 t_i} \int_{v_0 \leqslant |v| \leqslant V_0} |f(x,v)|^p \, dx dv \\ &= \sum_{i=1}^{\infty} \int_{|x|\leqslant \rho} \int_{v_0 \leqslant |v| \leqslant V_0} |f(x - t_i v, v)|^p \, dx dv \\ &\geqslant \sum_{i=1}^{\infty} \int_{|x|\leqslant \rho} \int_{v_0 \leqslant |v| \leqslant V_0} |f(x,v)|^p dx dv = \infty, \end{aligned}$$

which leads to a contradiction.

(a) Since $\|e^{-tT_0}\|_1 = 1$ for all $t \in \mathbb{R}$, it follows that $\sigma(T_0) \subset i\mathbb{R}$. For $\mu \in \mathbb{R}$ define $\varphi_\mu \in L^\infty(\mathbb{R}_x \times \mathbb{R}_v)$ by

$$\varphi_\mu(x,v) := \exp(i\mu x.v/|v|^2).$$

One can easily check that for T_0^* adjoint of T_0, $\varphi_\mu \in D(T_0^*)$ and $T_0^* \varphi_\mu = -i\mu\varphi_\mu$, which means that, the $-i\mu \in \sigma_p(T_0^*)$ and according to Lemma 1.6.2 (ii) $i\mu$ is either in $\sigma_p(T_0^*)$ or in $\sigma_r(T_0^*)$. Now by taking $p = 1$ in the above analysis we find that $\sigma_p(T_0^*) = \emptyset$ in $L^1(\mathbb{R}^n \times V)$ and consequently by summing up, we have shown

$$\sigma(T_0) \subset i\mathbb{R} \subset \sigma_r(T_0),$$

which with (1.12) proves (a).

(b) Theorem 3.2.1 implies that in the Hilbert space $L^2(\mathbb{R}^n \times V)$, the (C_0) group e^{tT_0} is unitary and according to Stone's Theorem 1.7.4, the operator T_0 is skew-adjoint. Thus, by Lemma 1.7.3, $\sigma_r(T_0) = \emptyset$. On the other hand, from $\sigma_p(T_0) = \emptyset$, the formula (1.12) implies that $\sigma_c(T_0) = \sigma(T_0) = i\mathbb{R}$.

(c) Let $x_\perp = x - \frac{x\cdot v}{|v|^2} v$. Hence for $\phi(x,v) = g(x - \frac{x\cdot v}{|v|^2} v, v) e^{\lambda x\cdot v/|v|^2}$,

$$v_i \frac{\partial \phi}{\partial x_i} = v_i \left[\sum_{k\neq i} \frac{\partial g}{\partial x_k} \left(-\frac{v_i v_k}{|v|^2} \right) + \frac{\partial g}{\partial x_i} \left(1 - \frac{v_i^2}{|v|^2} \right) + \lambda \frac{v_i}{|v|^2} g \right] \exp\left(\lambda \frac{x\cdot v}{|v|^2} \right)$$

and

$$\sum_{i=1}^{n} v_i \frac{\partial \phi}{\partial x_i} = \left\{ \sum_{i=1}^{n} \left[\left(- \sum_{k=1}^{n} \frac{\partial g}{\partial x_k} \frac{v_i^2 v_k}{|v|^2} \right) + v_i \frac{\partial g}{\partial x_i} \right] + \lambda \frac{v_i^2}{|v|^2} g \right\} \exp\left(\lambda \frac{x \cdot v}{|v|^2} \right)$$
$$= \lambda \phi. \tag{3.13}$$

Notice that $\overline{\sigma_r(T_0)} = \sigma_r(T_0) = i\mathbb{R}$. Hence by Lemma 1.6.2 we have

$$\sigma_r(T_0) \subseteq \sigma_p(T_0^*) \subseteq \sigma_r(T_0) \cup \sigma_p(T_0).$$

Since $\sigma_p(T_0) = \emptyset$, it follows that $\sigma_p(T_0^*) = i\mathbb{R}$. □

3.3 Absorption Group

Let $\sigma_a(x, v)$ be a bounded positive function on $\mathbb{R}^n \times V$. It not hard to see that the multiplication operator

$$A_a f(x, v) := -\sigma_a(x, v) f(x, v) \tag{3.14}$$

is bounded on $L^1(\mathbb{R}^n \times V)$ and we have $\|A_a\| = \|\sigma_a\|_\infty$.

Definition 3.3.1 The operator $T_a := T_0 + A_a$ defined on $D(T_a) := D(T_0)$ is called the *absorption operator* or *penetration operator* and in the case where σ_a depends only on x we call it the *tomography operator*, as we have seen in the Introduction. In the next theorem we will see that T_a generates a (C_0) group which is called the *absorption group* or *penetration group* in $L^1(\mathbb{R}^n \times V)$.

Theorem 3.3.2 *The absorption operator T_a generates a positive (C_0) group $\{e^{tT_a}\}_{t\in\mathbb{R}}$ in $L^1(\mathbb{R}^n \times V)$ satisfying*

(a)

$$e^{tT_a} \le e^{tT_0} \quad \textit{for all} \quad t \geqslant 0, \tag{3.15}$$

(b)

$$\|e^{tT_a}\| \leqslant e^{-\|\sigma_a\|_\infty t} \quad \textit{for all} \quad t \leqslant 0. \tag{3.16}$$

(c) If $\sigma_a(x, v)$ vanish for $x \notin \Omega$ then

$$\|e^{tT_a}\| \leqslant e^{F_+(\sigma_a)} \quad \textit{for all} \quad t \leqslant 0. \tag{3.17}$$

Proof For the proof of this theorem we will use the fact that the (C_0) group e^{tT_a} has an explicit form given by

$$[e^{tT_a} f](x, v) = E(x, v, t) f(x - tv, v) \quad \text{where} \quad E(x, v, t) := e^{-\int_0^t \sigma_a(x - sv, v) ds} \tag{3.18}$$

for all $t \in \mathbb{R}$. Using this expression of e^{tT_a}, one can see that $e^{tT_a} f \geqslant 0$ whenever $f \in L^1(\mathbb{R}^n \times V)_+$. Arguing as in Theorem 3.2.1 proves that e^{tT_a} is a strongly continuous group on $L^1(\mathbb{R}^n \times V)$. The positivity of σ_a implies (a) and the boundedness of σ_a implies (3.16) for $t \leqslant 0$. In this case, since σ_a vanishes out of $\Omega \times V$, $F_+(\sigma) < \infty$ and (3.17) holds. □

Remark 3.3.3 (a) Notice that if we write $M = \exp(F(\sigma_a))$, then

$$\begin{cases} \dfrac{1}{M} \le E(x, v, t) \le 1 & \text{for} \quad t \ge 0, \\ 1 \le E(x, v, t) \le M & \text{for} \quad t \le 0. \end{cases} \tag{3.19}$$

(b) From Lemma 3.1.4, (3.17) implies that

$$\|e^{tT_a}\| \leqslant e^{M(\sigma_a)\text{diam}\Omega} \qquad \text{for all} \quad t \in \mathbb{R}. \tag{3.20}$$

Similarly to Theorem 3.2.2 we have:

Theorem 3.3.4 *Suppose that $M(\sigma_a) < \infty$.*

(a) $\sigma(T_a) = \sigma_r(T_a) = i\mathbb{R}$ in $L^1(\mathbb{R}^n \times V)$;
(b) $\sigma(T_a) = \sigma_c(T_a) = i\mathbb{R}$ in $L^2(\mathbb{R}^n \times V)$;
(c) if T_a^* is the adjoint of T_a, acting in $L^\infty(\mathbb{R}^n \times V)$, then $\sigma_p(T_a^*) \supseteq i\mathbb{R}$.

Proof We shall use the relation

$$\sigma_p(T_a^*) \subseteq \sigma_r(T_a) \cup \sigma_p(T_a),$$

which follows from Lemma 1.6.2. From (3.20) it follows that T_a is the generator of a uniformly bounded group, thus we have $\sigma(T_a) \subseteq i\mathbb{R}$. To prove (a) and (c), we have only to check the properties

$$\sigma_p(T_a) = \emptyset \tag{3.21}$$

and

$$i\mathbb{R} = \sigma_p(T_a^*). \tag{3.22}$$

Let $T_a f = isf$, for some $s \in \mathbb{R}$. Then $e^{tT_a} f = e^{ist} f$, or $f(x - tv, v) = e^{ist} E^{-1}(x, v, t) f(x, v)$. Hence

$$|f(x - tv, v)| = |E^{-1}(x, v, t_0) f(x, v)| \quad \text{for all} \quad t \in \mathbb{R}.$$

Since $|E^{-1}(x, v, t_0)| > 1$ and

$$\int_{\mathbb{R}_x^n \times V} |f(x - tv, v)| dx dv = \|f\|_1.$$

This proves that f cannot be an L^1-function and yields the assertion (3.21). To prove (3.22), let ρ be a positive number such that $\Omega \subset \{x \in \mathbb{R}^n \mid |x| \leq \rho\}$ and define the function $\varphi(x, v) := g(x_\perp, v) \exp isx.v/|v|^2$, $s \in \mathbb{R}$, where $x_\perp = x - (x \cdot v)v/|v|^2$ and g is C_0^∞-function such that $g(x_\perp, v) = 0$ for $|x_\perp| \leq 2\rho$. Since $\varphi(x, v) = 0$ for $|x| \leq \rho$, we obtain as in (3.13), $T_a^*\varphi = is\varphi$, which proves (3.22). This proves the theorem, since

$$i\mathbb{R} = \sigma_p(T_a^*) \subseteq \sigma(T_a) \subseteq i\mathbb{R}.$$

□

A trivial case of the existence of the wave operators is the following.

Theorem 3.3.5 *If the absorption rate σ_a obeys all the conditions of Definition 3.1.1 and if $F(\sigma_a) < \infty$, then all the wave operators $W_\pm(T_a, T_0)$ and $W_\pm(T_0, T_a)$ exist in $L^1(\mathbb{R}^n \times V)$.*

Proof For $f \in L^1(\mathbb{R}^n \times V)$ and $t \in \mathbb{R}$, according to the formulae (3.2) and (3.18) we have

$$[e^{-tT_a} e^{tT_0} f](x, v) = \exp\left(-\int_0^{-t} \sigma_a(x - sv, v)ds\right) f(x, v)$$

$$[e^{-tT_0} e^{tT_a} f](x, v) = \exp\left(-\int_{-t}^{0} \sigma_a(x - sv, v)ds\right) f(x, v).$$

Then, the fact that $\exp\left(-\int_{-\infty}^{\infty} \sigma_a(x - sv, v)ds\right) \leqslant \exp(F(\sigma_a))$ together with the Lebesgue dominated convergence theorem shows the existence of the wave operators $W_\pm(T_a, T_0)$ and $W_\pm(T_0, T_a)$ in $L^1(\mathbb{R}^n \times V)$. □

3.4 Transport Group

Now we consider the Cauchy problem (3.3):

$$(\mathbf{P}) \qquad \begin{cases} \dfrac{\partial u}{\partial t}(x, v, t) = Tu(x, v, t), & (x, v, t) \in \mathbb{R}^n \times V \times \mathbb{R}, \\ u(x, v, 0) = f(x, v). \end{cases}$$

The next theorem proves the well-posedness of **(P)** in $L^1(\mathbb{R}^n \times V)$

Theorem 3.4.1 *If the pair (k, σ_a) is admissible, then the transport operator T defined in* (3.7) *generates a* (C_0) *group $\{e^{tT}\}_{t\in\mathbb{R}}$ in $L^1(\mathbb{R}^n \times V)$ and furthermore:*

(a) *For all* $t \in \mathbb{R}_+$, e^{tT} *is positive and satisfies*

$$\|e^{tT}\| \leq \begin{cases} e^{\|\sigma_p\|_\infty t} & t \geq 0 \\ e^{-(\|\sigma_a\|_\infty + \|\sigma_p\|_\infty)t} & t \leq 0. \end{cases} \tag{3.23}$$

(b) *The spectrum* $\sigma(T)$ *is included in the strip* $\Lambda := \{\lambda \in \mathbb{C} \mid -\lambda_1 \leq Re\,\lambda \leq \lambda_2\}$, *where* $\lambda_1 = \|\sigma_a\|_\infty + \|\sigma_p\|_\infty$ *and* $\lambda_2 = \|\sigma_p\|_\infty$.

(c) *If* σ_a *vanishes out of* $\Omega \times V$, *then*

$$\|e^{tT}\| \leq \begin{cases} e^{\|\sigma_p\|_\infty t} & t \geq 0 \\ \min\left\{Me^{-Mt\|\sigma_p\|_\infty}, e^{-(\|\sigma_a\|_\infty + \|\sigma_p\|_\infty)t}\right\} & t \leq 0, \end{cases} \tag{3.24}$$

where $M := e^{F_+(\sigma_a)}$. *In this case the spectrum* $\sigma(T)$ *is included in the strip* $\Lambda := \{\lambda \in \mathbb{C} \mid -\lambda_1 \leq Re\,\lambda \leq \lambda_2\}$, *where* $\lambda_1 = \min\left\{M\|\sigma_p\|_\infty, \|\sigma_a\|_\infty + \|\sigma_p\|_\infty\right\}$ *and* $\lambda_2 = \|\sigma_p\|_\infty$.

(d) *For* $t \geq 0$, $f \in L^1(\mathbb{R}^n \times V)_+$, *we have* $e^{tT} f \geq e^{tT_a} f$.

(e) *For* $t \leq 0$, $f \in L^1(\mathbb{R}^n \times V)_+$, *we have* $e^{tT} f \leq e^{tT_a} f$.

Proof (a) T is the perturbing operator of T_a by the bounded operator A_p; by virtue of Corollary 1.8.8 (b), T generates a (C_0) group e^{tT}. e^{tT_a} and e^{tA_p} being positive for $t \geqslant 0$, Trotter's Product Theorem

$$\text{s}-\lim_{n\to\infty} (e^{\frac{t}{n}T_a} e^{\frac{t}{n}A_p})^n = e^{tT}$$

implies the positivity of e^{tT} and also together with (3.15) implies $\|e^{tT}\| \leqslant e^{\|A_p\|t}$ for $t \geq 0$. For (3.23) it is enough to remark that from (3.5) and (3.4) we get

$$\|A_p f\| \leqslant \int_{\mathbb{R}\times V} |\sigma_p(x, v')f(x, v')|dxdv' \leqslant \|\sigma_p\|_\infty \|f\|.$$

For $t \leq 0$ the estimation follows from (3.16) and Corollary 1.8.8 (b).

(b) It suffices to apply the Hille–Yosida theorem to the semigroups $\{e^{tT}\}_{t\geq 0}$ and $\{e^{-tT}\}_{t\geq 0}$.

(c) When σ_a vanishes out of $\Omega \times V$, (3.24) follows from (3.17) and Corollary 1.8.8 (b).

(d) Since e^{tT} is a positive group for all $t \in \mathbb{R}$ and $A_p \geq 0$, for any a and b in $\mathbb{R}$, $a < b$ and $f \in L^1(\mathbb{R}^n \times V)_+$

$$\int_a^b e^{-sT_a} A_p e^{sT} f\, ds \geq 0. \tag{3.25}$$

Thus, from Duhamel's formula

$$e^{tT} f = e^{tT_a} f + \int_0^t e^{(t-s)T_a} A_p e^{sT} f \, ds \geq e^{tT_a} f,$$

whenever $t \geq 0$ and $f \in L^1(\mathbb{R}^n \times V)_+$.

(e) For $t \leq 0$, Duhamel's formula written as

$$e^{tT} f = e^{tT_a} f - \int_t^0 e^{(t-s)T_a} A_p e^{sT} f \, ds \geq e^{tT_a} f,$$

whenever $t \geq 0$ and $f \in L^1(\mathbb{R}^n \times V)_+$. □

In the sequel, we call $\{e^{tT}\}_{t \in \mathbb{R}}$ the *transport group* generated by T. In the sequel we will show that for $t \leqslant 0$, the semigroup e^{tT} is not necessarily positive.

Another interesting feature of e^{tT} is that on the functions with support in the interaction region $\Omega \times V$ of the phase space e^{tT} and e^{tT_0} act similarly for $t \geqslant 0$ $[t \leqslant 0]$. More precisely, if for any Borel set S we denote by χ_S the characteristic function of S, we have

$$e^{tT} \chi_{\Omega \times V} f = e^{tT_0} \chi_{\Omega \times V} f, \quad \text{for all } t \in \mathbb{R}. \tag{3.26}$$

In the forthcoming lemma we will prove a more general version of the above identity.

Definition 3.4.2 Let $r_+(x, v)$ and $r_-(x, v)$ denote the *forward* and *backward rays* coming from $(x, v) \in \mathbb{R}^n \times V$:

$$r_+(x, v) := \{x + tv \ : \ t > 0\} \qquad r_-(x, v) := \{x + tv \ : \ t < 0\}.$$

With these rays we can associate the four disjoint sets as follows:

$$\begin{aligned}
\Sigma_+ &:= \{(x, v) \in \mathbb{R}^n \times V \ : \ r_+(x, v) \cap \Omega = \emptyset, r_-(x, v) \cap \Omega \neq \emptyset\}, \\
\Sigma_- &:= \{(x, v) \in \mathbb{R}^n \times V \ : \ r_+(x, v) \cap \Omega \neq \emptyset, r_-(x, v) \cap \Omega = \emptyset\}, \\
\Sigma_0 &:= \{(x, v) \in \mathbb{R}^n \times V \ : \ r_+(x, v) \cap \Omega = \emptyset, r_-(x, v) \cap \Omega = \emptyset\}, \\
\Sigma &:= \mathbb{R}^n \times V \setminus (\Sigma_+ \cup \Sigma_- \cup \Sigma_0).
\end{aligned}$$

The set Σ_+ consists of all those points of phase space from which a rectilinear trajectory coming from Ω pass through and stay in Σ_+ at all later times. This set is called the *free forward and trapped backward point set*. Similarly the set Σ_- is called the *free backward and trapped forward point set* and the set Σ_0 is called the *streaming free point set*. We remark that a particle at the point $(x, v) \in \Omega \times V$ is in Σ, but the converse is not true, one can find a particle $(x, v) \in \Omega^c \times V$ such that $(x, v) \in \Sigma$ (for example if Ω has a concave region on its surface). Furthermore, a particle at the point $(x, v) \in \Sigma_0$ has always been and always remains in Σ_0.

Lemma 3.4.3 *Let (k, σ_a) be a regular pair and $f \in L^1(\mathbb{R}^n \times V)$, then:*

(a) $e^{tT}\chi_{\Sigma+}f = e^{tT_0}\chi_{\Sigma+}f$, *for all* $t \geqslant 0$;
(b) $e^{tT}\chi_{\Sigma-}f = e^{tT_0}\chi_{\Sigma-}f$, *for all* $t \leqslant 0$;
(c) $e^{tT}\chi_{\Sigma 0}f = e^{tT_0}\chi_{\Sigma 0}f$, *for all* $t \in \mathbb{R}$.

Proof Since the functions $\sigma_a(\cdot, v)$ and $k(\cdot, v, v')$ have their supports in Ω, if $x \notin \Omega$, then $\sigma_a(x, v)f(x, v) = \int_V k(x, v', v)f(x, v')dv' = 0$ for any $f \in L^1(\mathbb{R}^n \times V)$ and furthermore, if $x \in \Omega$ and supp $h \subset \Sigma_0$ we have once more $\sigma_a(x, v)h(x - sv, v) = \int_V k(x, v', v)h(x - sv, v')dv' = 0$ for all $s \in \mathbb{R}$. Thus by the second Duhamel formula (1.15),

$$e^{tT}h(x, v) = e^{tT_0}h(x, v) + \int_0^t e^{(t-s)T}Ah(x - sv, v)ds \tag{3.27}$$

and consequently we obtain (c). Now if $x \in \Omega$, then $r_+(x - sv, v) \cap \Omega \neq \emptyset$ for $s \geqslant 0$, hence either $(x - sv, v) \in \Sigma_-$ or $(x - sv, v) \in \Sigma$. In any case, if supp $h \subset \Sigma_+$, then $Ah(x - sv, v) = 0$ for all $s \in \mathbb{R}_+$ and once more (3.27) implies (a). Similarly for $t \leqslant 0$ one has (b). □

In a similar way we can define the outgoing and incoming point set:

Definition 3.4.4 Let ρ denote the radius of a ball B_ρ around the origin which contains all the points of Ω. The sets

$$E_\rho^+ := \{(x, v) \, : \, x \cdot v \geqslant \rho|v|\} \quad \text{and} \quad E_\rho^- := \{(x, v) \, : \, x \cdot v \leqslant -\rho|v|\}$$

are called respectively the *outgoing* and *incoming point set*. If we denote

$$E_\rho^0 := \{(x, v) \, : \, -\rho|v| \leqslant x \cdot v \leqslant \rho|v|\},$$

it is not hard to see that we can have the following partition of the phase space;

$$\mathbb{R}^n \times V = E_\rho^+ \cup E_\rho^- \cup E_\rho^0.$$

In the sequel we denote by χ_ρ^+, χ_ρ^- and χ_ρ^0 the characteristic functions of E_ρ^+, E_ρ^- and E_ρ^0.

The following corollary is a consequence of Lemma 3.4.3.

Corollary 3.4.5 *Let (k, σ_a) be a regular pair. For any $f \in L^1(\mathbb{R}^n \times V)$, we have:*

(i) $E_\rho^\pm \subset \Sigma_\pm \cup \Sigma_0$;
(ii) $e^{tT}\chi_\rho^\pm f = e^{tT_0}\chi_\rho^\pm f$, *for all* $\pm t \geqslant 0$.

Proof (i) This follows from

$$|x + tv| \geqslant (x + tv) \cdot \frac{v}{|v|} \geqslant \rho + t|v| \geqslant \rho \quad \text{for} \quad t \geqslant 0$$

and

$$|x + tv| \geqslant -(x + tv) \cdot \frac{v}{|v|} \geqslant \rho - t|v| \geqslant \rho \quad \text{for} \quad t \leqslant 0.$$

(ii) This assertion is inferred from (i) and Lemma 3.4.3. □

Remark that $B_\rho \times V \subset E^0_\rho$, hence the assertion (ii) of the above corollary implies (3.26) in which we replace $\Omega \times V \subset E^0_\rho$ by $B_\rho \times V \subset E^0_\rho$, which is more general.

3.5 Nonproliferating Systems

Definition 3.5.1 We say that a regular pair (k, σ_a) defines a *nonproliferating system*, if and only if the transport semigroup $\{e^{tT}\}_{t \geq 0}$, satisfies

$$\sup_{t \geqslant 0} \|e^{tT}\| < \infty.$$

In [108], Simon used the term *subcritical*. This terminology does not quite coincide with the corresponding notion in reactor theory, when a reactor problem has a strictly dominant eigenvalue which is negative (see [18]). For this reason in [68], the term nonproliferating system is used.

Several physical conditions can be found in the literature that ensure the nonproliferating of e^{tT}. Among them we cite:

Definition 3.5.2 We say that the pair (k, σ_a) defines an *absorptive* or *pure scatterer* system according that $\sigma_a > \sigma_p$ or $\sigma_a = \sigma_p$.

Theorem 3.5.3 *Let (k, σ_a) be an admissible pair.*

(a) If the system is absorptive then e^{tT} is a semigroup of contractions for all $t \geqslant 0$.
(b) If the system is pure scatterer then e^{tT} is isometric on $L^1_+(\mathbb{R}^n \times V)$ and $\|e^{tT}\| = 1$ for any $t \geqslant 0$.
(c) Let $(k, \sigma_a)$and (k', σ'_a) be two admissible pairs such that $k' \leqslant k$ and $\sigma_a \geqslant \sigma'_a$. Then if e^{tT} and $e^{tT'}$ are the correspondent transport groups and if e^{tT} is non-proliferating, then so is $e^{tT'}$. Furthermore for any $f \in L^1(\mathbb{R}^n \times V)_+$,

$$e^{tT} f(x, v) \geqslant e^{tT'} f(x, v), \qquad \textit{for any} \quad (x, v) \in \mathbb{R}^n \times V \quad \textit{and} \quad t \geqslant 0. \tag{3.28}$$

Proof By integrating Duhamel's formula

$$e^{tT}f(x,v)=e^{tT_0}f(x,v)+\int_0^t e^{(t-s)T_0}(A_a+A_p)e^{sT}f(x,v)ds,$$

for $f\in L^1_+(\mathbb{R}^n\times V)$, over $\mathbb{R}^n\times V$ we obtain

$$\int_{\mathbb{R}^n\times V}e^{tT}f(x,v)dxdv=\int_{\mathbb{R}^n\times V}f(x,v)dxdv$$
$$+\int_0^t\left[\int_{\mathbb{R}^n\times V}(-\sigma_a(x,v)+\sigma_p(x,v))e^{sT}f(x,v)dxdv\right]ds.$$

In the absorptive case the positivity of e^{tT} for $t\geqslant 0$ implies that e^{tT} is a semigroup of contractions on $L^1_+(\mathbb{R}^n\times V)$ and according to Lemma 1.9.3, this is enough to get (a) and (b) simultaneously. For the last part, we may write Duhamel's formula as

$$e^{tT}=e^{tT_p}+\int_0^t e^{(t-s)T_p}(A-A_p)e^{sT}ds,$$

where $A_p:=A'_a+A'_p$ and $T'=T-A_a-A_p+A'_a+A'_p$. From assumption it follows that $(A'-A)f\geq 0$ and consequently $\int_0^t e^{(t-s)T'}(A-A')e^{sT}fds\leqslant 0$ for any $f\in L^1(\mathbb{R}^n\times V)_+$. This implies (3.28) and from positivity of $e^{tT'}$, one gets $|e^{tT'}f|\leqslant e^{tT'}|f|\leqslant e^{tT}|f|$ and $\|e^{tT}f\|\leqslant\|e^{tT'}|f|\|$ for all $f\in L^1(\mathbb{R}^n\times V)$, which implies (c). □

Corollary 3.5.4 *Let (k,σ_a) be an admissible pair and e^{tT} its correspondent transport group and let e^{tT_a} be the absorption transport group correspondent to the pair $(0,\sigma_a)$, then for any $f\in L^1(\mathbb{R}^n\times V)_+$, and $t>0$,*

$$e^{tT}f(x,v)\geq e^{-\int_0^t\sigma_a(x-sv,v)ds}f(x-tv,v). \tag{3.29}$$

Proof This is an obvious consequence of Theorem 3.5.3 (c) and (3.18). □

Remark 3.5.5 In Lemmas 2.2.3 and 2.6.3 (a), we gave some abstract conditions which yield the nonproliferation of the dynamic e^{tT} for $t\geq 0$. In Corollary 3.6.5 we will give a concrete form of these conditions.

3.6 Existence of the Scattering Operator

For the existence of the scattering operator $S=W_+(T_0,T)W_-(T,T_0)$, one has to justify the existence of the wave operators $W_+(T_0,T)$ and $W_-(T,T_0)$. In the sequel we suppose that the pair (k,σ_a) is admissible and the dynamic e^{tT} is nonproliferating.

Theorem 3.6.1 *Let (k, σ_a) be a admissible pair. If e^{tT} is nonproliferating and one of the assumptions* **(MFP)** *or* **(FRT)** *of Definition 3.1.2 is satisfied, then $W_+(T_0, T)$ exists in $L^1(\mathbb{R}^n \times V)$.*

Proof Let $f \in L^1(\mathbb{R}^n \times V)_+$. In order to use Cook's Lemma, we have to change the assumptions of Lemma 2.2.1, which was given for the existence of $W_+(T, T_0)$, to:

(a) $\sup_{t\le 0} \|e^{tT_0}\| \le M_0$, which is true since $\|e^{tT_0}\| = 1$ for all $t \in \mathbb{R}$.
(b) $\sup_{t\ge 0} \|e^{tT}\| \le M$, which is true since e^{tT} is nonproliferating.
(c) $\int_0^\infty \|(T - T_0)e^{tT}u\|dt < \infty$ for all $u \in D(T)$. Since $T - T_0 = A_a + A_p$ it suffices to prove that,

$$\int_0^\infty \|A_i e^{sT} f\|ds < \infty \quad i = a, p. \tag{3.30}$$

Taking in (3.29) $y = x - tv$, we get

$$f(y, v) \leqslant \exp(F_+(\sigma_i))e^{tT} f(y + tv, v)$$

Thus replacing f by $e^{sT} f$ and t by $t - s$

$$e^{sT} f(y, v) \leqslant \exp(F_+(\sigma_i))e^{tT} f(y + (t - s)v, v).$$

Therefore, for f positive,

$$\int_0^t \|A_i e^{sT} f\|ds \leqslant \int_0^t \iint_{\mathbb{R}^n\times V} \sigma_i(y, v)e^{sT} f(y, v)dydvds$$

$$\leqslant \int_0^t \iint_{\mathbb{R}^n\times V} \sigma_i(y, v) \exp(F_+(\sigma_i))e^{tT} f(y + (t - s)v, v)dydvds.$$

Finally, we get

$$\int_0^t \|A_i e^{sT} f\|ds \leqslant \exp(F_+(\sigma_i))F_+(\sigma_i)\|e^{tT}\|,$$

which implies (3.30).

□

For the existence of the scattering operator S, we need also to prove the existence of the wave operator $W_-(T, T_0)$. To do this, first we prove the existence of the wave operator $W_-(T_p, T_0)$, where e^{tT_p} is the transport group corresponding to the pair $(k, 0)$.

Theorem 3.6.2 *Let* (k, σ_a) *be a admissible pair. If* e^{tT} *is nonproliferating and one of the assumptions* **(MFP)** *or* **(FRT)** *of Definition 3.1.2 is satisfied for* $\sigma_i, i = a, p$, *then* $W_-(T, T_0)$ *exists in* $L^1(\mathbb{R}^n \times V)$.

In order to prove this theorem, first we define $T_p = T_0 + A_p$, which corresponds to the pair $(k, 0)$ and we prove the existence of the wave operator $W_-(T_p, T_0)$.

Lemma 3.6.3 *If the function* k *obeys all the conditions of Definition 3.1.1 with* $F(\sigma_p) < \infty$ *and if the pair* $(k, 0)$ *is nonproliferating, then* $W_-(T_p, T_0)$ *exists in* $L^1(\mathbb{R}^n \times V)$.

Since e^{tT_p} is uniformly bounded on $\mathbb{R}_+$, to use the version of Cook's Lemma 2.2.1 for the existence of $W_-(T_p, T_0)$, it suffices to show that

$$\int_{-\infty}^{0} \|A_p e^{tT_0} f\| dt < \infty. \tag{3.31}$$

For $f \in L^1(\mathbb{R}^n \times V)_+$

$$\begin{aligned}
\int_{-\infty}^{0} \|A_p e^{tT_0} f\| dt &= \int_{-\infty}^{0} \iint_{\mathbb{R}^n \times V} \int_V k(x, v', v) f(x - tv', v') dv' dx dv dt \\
&= \int_{-\infty}^{0} \iint_{\mathbb{R}^n \times V} \sigma_p(x, v') f(x - tv', v') dx dv' dt \\
&\leqslant \left(\sup_{(y, v') \in \mathbb{R}^n \times V} \int_{-\infty}^{0} \sigma_p(y + tv', v') \right) \iint_{\mathbb{R}^n \times V} f(y, v') dy dv' \\
&= F_-(\sigma_p)\|f\| = F(\sigma_p)\|f\|, .
\end{aligned} \tag{3.32}$$

which implies (3.31).

In the above lemma we have assumed that the pair $(k, 0)$ is nonproliferating, but this property can be obtained as follows:

Corollary 3.6.4 *If the function* k *satisfies the condition* **(MFP)** *with* $F(\sigma_p) < 1$, *then* e^{tT_p} *is uniformly bounded on* $\mathbb{R}_+$.

Proof By Lemma 2.2.4 and a calculation similar to (3.32), it is implied also that

$$\int_0^{\infty} \|A_p e^{tT_0} f\| dt \leq F_+(\sigma_p)\|f\|, \quad f \in L^1(\mathbb{R}^n \times V)_+.$$

Since $F_-(\sigma_p) \leq F(\sigma_p) < 1$, Lemma 2.2.3 implies the uniform boundedness of e^{tT_p} on $\mathbb{R}_+$. □

Proof of Theorem 3.6.2. The operator T_p corresponds to the pair $(k, 0)$, thus according to Theorem 3.5.3 (c), the (C_0) group e^{tT} is nonproliferating and Lemma 2.6.1 implies that $W_-(T, T_0)$ exists in $L^1(\mathbb{R}^n \times V)$.

□

In spite of the fact that in Theorem 3.6.1 the condition $F(\sigma_p) < 1$ is not imposed, we need this condition in order to recover that the dynamic e^{tT} is nonproliferating. In fact the following corollary shows that Theorems 3.6.1 and 3.6.2 have the same feature.

Corollary 3.6.5 *Let (k, σ_a) be a admissible pair. If*

$$F(\sigma_p) < 1 \quad \text{and} \quad F(\sigma_a) < \infty, \tag{3.33}$$

then e^{tT} is nonproliferating and the wave operators $W_+(T_0, T)$ and $W_-(T, T_0)$ exist in $L^1(\mathbb{R}^n \times V)$. Thus we have the existence of $S = W_+(T_0, T)W_-(T, T_0)$.

Proof In Corollary 3.6.4 we have proved that this condition implies the uniform boundedness of e^{tT_p}. The fact that e^{tT} is also nonproliferating follows from Theorem 3.5.3 (c) and the fact that

$$e^{tT} f = e^{tT_p} f - \int_0^t e^{(t-s)T_p} \sigma_a e^{sT} f ds \leq e^{tT_p} f$$

for all $t \geq 0$ and $f \in L^1(\mathbb{R}^n \times V)_+$. □

Remark 3.6.6 In Theorem 3.6.2, if in addition the pair (k, σ) is regular, then according to Lemma 3.1.4, we can replace the condition $F(\sigma_p) < 1$ by $(\text{diam } \Omega)M(\sigma_p) < 1$, which is physically much more eloquent.

3.7 Existence of the Other Wave Operators

For the existence of $W_+(T_0, T)$ and $W_-(T, T_0)$ in the previous section we have used in many points the fact that the dynamic e^{tT} preserves the positive cone $L^1(\mathbb{R}^n \times V)_+$ for $t \geq 0$. For the lack of this property for $t \leq 0$ we are constrained to applying Theorem 2.3.2 of the previous chapter.

Theorem 3.7.1 *If*

$$F(\sigma_a) + F(\sigma_p) < 2. \tag{3.34}$$

Then the wave operators $W_\pm(T, T_0)$ and $W_\pm(T_0, T)$ exist.

Proof An estimate similar to (3.32) shows

$$\int_{-\infty}^{\infty} \|Ae^{tT_0} f\| dt \leqslant \int_{-\infty}^{\infty} \|A_a e^{tT_0} f\| dt + \int_{-\infty}^{\infty} \|A_p e^{tT_0} f\| dt \leqslant \left(F(\sigma_a) + F(\sigma_p)\right) \|f\|,$$

which means that the operator A is T_0-smooth with a constant < 2. Thus, Theorem 2.3.2 is applicable. □

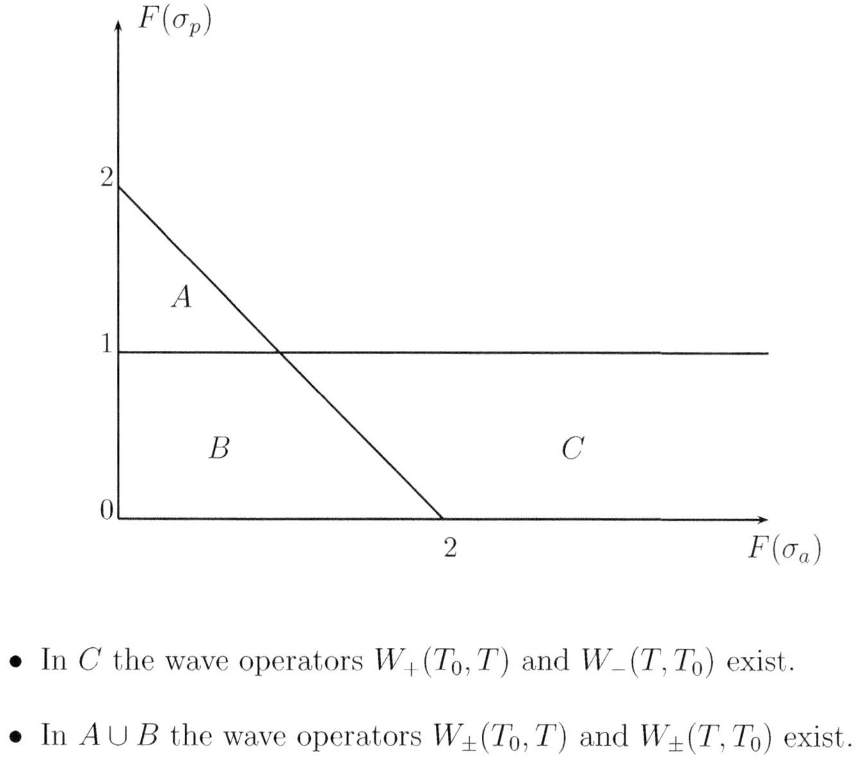

Fig. 3.1 .

In Fig. 3.1, we will use the conditions (3.33) and (3.34) for comparing the existence of the wave operators in different regions of the $(F(\sigma_a), F(\sigma_p))$-plane.

3.8 Locally Decaying Property

In [123], Voigt has introduced the notion of local decay for the existence of $W_+(T_0, T)$.

Definition 3.8.1 An admissible pair (k, σ_a) and its subsequent system are called *locally decaying* if for any $f \in L^1(\mathbb{R}^n \times V)$ and each compact set $K \subset \mathbb{R}^n$

$$\iint_{K\times V} |e^{tT} f(x, v)|dxdv \longrightarrow 0 \quad (\text{as } t \to \infty). \tag{3.35}$$

Remark 3.8.2 Let $V^* = V \setminus \{0\}$ and $C_c^\infty(\mathbb{R}^n \times V^*)$ be the space of all C^∞-functions on $\mathbb{R}^n \times V$ with compact support that vanish in a neighborhood of the manifold $\{(x, 0) \ : \ x \in \Omega\}$. $C_c^\infty(\mathbb{R}^n \times V^*)$ is dense in $L^1(\mathbb{R}^n \times V)$.

Remark 3.8.3 The streaming free system e^{tT_0} is locally decaying. In fact, let K be a compact subset of $\mathbb{R}^n$ and $\chi_{K\times V}$ the characteristic function of $K \times V$. If $\phi \in C_c^\infty(\mathbb{R}^n \times V^*)$, then $\chi_{K\times V} e^{tT_0}\phi = 0$ for t large enough (if the particles have no zero velocity, the streaming free group moves out all the particles in K after a finite time). Since $C_c^\infty(\mathbb{R}^n \times V^*)$ is dense in $L^1(\mathbb{R}^n \times V)$ and $\|e^{tT_0}\| \leqslant 1$, for any $\varepsilon > 0$ and any $f \in L^1(\mathbb{R}^n \times V)$ there exist $\phi \in C_c^\infty(\mathbb{R}^n \times V^*)$ and $t^* > 0$ such that

$$\iint_{K\times V} |e^{tT_0} f(x, v)| dx dv \leqslant \|f - \phi\| + \iint_{K\times V} |e^{tT_0}\phi(x, v)| dx dv < \varepsilon.$$

for $t \geqslant t^*$.

Remark 3.8.4 Let (k, σ_a) and (k', σ_a') be two admissible pairs such that $k' \leqslant k$ and $\sigma_a \leqslant \sigma_a'$. Then if (k', σ_a') is locally decaying, then so is (k, σ_a). In fact, according to Theorem 3.5.3 (c), $|e^{tT} f| \leqslant e^{tT}|f| \leq e^{tT'}|f|$, which implies the local decaying of e^{tT}.

In the following lemma we use the notations of Definition 3.4.4 and Remark 3.8.2.

Lemma 3.8.5 *If (k, σ_a) is regular, then:*

(a) For any $f \in L^1(\mathbb{R}^n \times V)$,

$$\chi_\rho^- e^{tT} \chi_\rho^0 f = 0 \ \ \textit{for all} \ \ t \geqslant 0.$$

(b) For any $f \in C_c^\infty(\mathbb{R}^n \times V^*)$ there exists a time $t^* \gg 1$ such that $\chi_\rho^- e^{t^*T} \chi_\rho^- f = 0$.

(c) If e^{tT} has the local decay property then for any $a > 0$

$$\lim_{t\to\infty} \int_{-a \leqslant x\cdot\frac{v}{|v|} \leqslant a} \int_V |e^{tT} f(x, v)| dx dv = 0.$$

Proof (a) Here we will use the first version of Duhamel's formula (1.14). For $(x, v) \in E_\rho^-$ we have $x \cdot v \leqslant -\rho|v|$ and

$$e^{tT}\chi_\rho^0 f(x, v) = \chi_\rho^0(x - tv, v) f(x - tv, v) + \int_0^t e^{(t-s)T_0} A e^{sT}\chi_\rho^0 f(x, v) ds. \tag{3.36}$$

Since $(x - tv).v \leqslant -\rho|v| - t|v|^2 \leqslant -\rho|v|$ for all $t \geqslant 0$. Thus, for this element $(x - sv, v)$ of the phase space $\chi_\rho^0(x - sv, v) = 0$ and consequently the first term of the right-hand side is identically zero. For the second term we take $g(x, v, t) = [e^{tT}\chi f](x, v)$, then

$$\int_0^t e^{(t-s)T_0} A e^{sT} \chi f(x,v) ds = -\int_0^t \sigma_a(x - tv + sv, v) g(x - tv + sv, v, s) ds$$
$$+ \int_0^t \int_V k(x - tv + sv, v', v) g(x - tv + sv, v', s) dv' ds.$$

Since $|x - tv + sv| \geqslant -(x - tv + sv) \cdot \frac{v}{|v|} \geqslant \rho + t - s \geqslant \rho$ for $t \geqslant s$, the lemma follows from the fact that σ_a and k have their supports in Ω with respect to x.

(b) f has compact support in configuration space and the minimal speed in the support of f is strictly positive $f(x - tv, v) = 0$ for some t^* large enough. Hence if we replace in the identity (3.36) $\chi_\rho^0 f$ by $\chi_\rho^- f$ and t by t^* first term of the right-hand side of this identity vanishes and the second term is identically zero just by mimicking the proof of part (a).

(c) For $x \in B_\rho = \{x \in \mathbb{R}^n \ : \ |x| \leq \rho\}$, $\pm x \cdot v \leqslant \rho|v|$, thus $B_\rho \subset E_\rho^0$ and

$$\iint_{E_\rho^0} |e^{tT} f(x,v)| dx dv = \iint_{B_\rho \times V} |e^{tT} f(x,v)| dx dv + \iint_{E_\rho^0 \setminus (B_\rho \times V)} |e^{tT} f(x,v)| dx dv.$$

By the local decay property of e^{tT}, the first integral of the right-hand side goes to zero as $t \to \infty$ and as previously using the first version of Duhamel's formula we can replace $\iint_{E_\rho^0 \setminus (B_\rho \times V)} |e^{tT} f(x,v)| dx dv$ by $\iint_{E_\rho^0 \setminus (B_\rho \times V)} |e^{tT_0} f(x,v)| dx dv$. This integral converges to zero as $t \to \infty$, when $f \in C_c^\infty(\mathbb{R}^n \times V^*)$ and the lemma follows from the density of $C_c^\infty(\mathbb{R}^n \times V^*)$ in $L^1(\mathbb{R}^n \times V)$. □

Theorem 3.8.6 *Let (k, σ_a) be an admissible and nonproliferating pair.*

(a) If $W_+(T_0, T)$ exists, then (k, σ_a) is locally decaying.
(b) If (k, σ_a) is regular and locally decaying, then $W_+(T_0, T)$ exists.

Proof (a) Letting $f \in L^1(\mathbb{R}^n \times V)$ and $g = W_+(T_0, T) f$,

$$\iint_{K \times V} |e^{tT} f(x,v)| dx dv \leqslant \|e^{tT} f - e^{tT_0} g\| + \iint_{K \times V} |e^{tT_0} g(x,v)| dx dv$$
$$= \|e^{-tT_0} e^{tT} f - g\| + \iint_{K \times V} |e^{tT_0} g(x,v)| dx dv$$

By definition of g and according to Remark 3.8.3 these terms tend to zero.

(b). We split a function $f \in L^1(\mathbb{R}^n \times V)$ into three parts; $f = \chi_\rho^+ f + \chi_\rho^0 f + \chi_\rho^- f$. According to Corollary 3.4.5 $e^{-tT_0} e^{tT}$, acts as the identity operator on $\chi_\rho^+ f$ for all $t \geqslant 0$, hence $W_+(T_0, T) \chi_\rho^+ f$ certainly exists. For the existence of $W_+(T_0, T) \chi_\rho^0 f$, we take t and t' as two positive numbers for which we have

$$\|e^{-tT_0}e^{tT}\chi_\rho^0 f - e^{-(t+t')T_0}e^{(t+t')T}\chi_\rho^0 f\| = \|e^{tT}\chi_\rho^0 f - e^{-t'T_0}e^{t'T}e^{tT}\chi_\rho^0 f\|$$
$$= \|(1 - e^{-t'T_0}e^{t'T})e^{tT}\chi_\rho^0 f\|.$$

By virtue of Lemma 3.8.5 (a) we can replace in the above identity $e^{tT}\chi_\rho^0 f$ by $\chi_\rho^0 e^{tT}\chi_\rho^0 f + \chi_\rho^+ e^{tT}\chi_\rho^0 f$ and since $e^{-t'T_0}e^{t'T}$ acts as the identity operator on $\chi_\rho^+ e^{tT}\chi_\rho^0 f$, we have

$$\|e^{-tT_0}e^{tT}\chi_\rho^0 f - e^{-(t+t')T_0}e^{(t+t')T}\chi_\rho^0 f\| = \|(1 - e^{-t'T_0}e^{t'T})\chi_\rho^0 e^{tT}\chi_\rho^0 f\|$$
$$\leq \left(1 + \sup_{t \geqslant 0}\|e^{t'T}\|\right)\iint_{E_\rho^0} |e^{tT}\chi_\rho^0 f(x, v)| dx dv. \quad (3.37)$$

By nonproliferation and local decay, Lemma 3.8.5 (c) implies that the upper bound in this estimate tends to zero as $t \to \infty$. Therefore, $e^{-tT_0}e^{tT}\chi_\rho^0 f$ forms a Cauchy net and $W_+(T_0, T)\chi_\rho^0 f$ exists.

Finally, for the existence of $W_+(T_0, T)\chi_\rho^- f$, we choose $t^* \gg 1$ as in Lemma 3.8.5 (b) and we remark that

$$e^{-tT_0}e^{tT}\chi_\rho^- f = e^{-tT_0}e^{(t-t^*)T}e^{t^*T}\chi_\rho^- f$$
$$= e^{-tT_0}e^{(t-t^*)T}\chi_\rho^0 e^{t^*T}\chi_\rho^- f + e^{-tT_0}e^{(t-t^*)T}\chi_\rho^+ e^{t^*T}\chi_\rho^- f$$
$$= e^{-t^*T_0}\left[e^{-(t-t^*)T_0}e^{(t-t^*)T}\chi_\rho^0 e^{t^*T}\chi_\rho^- f + e^{-(t-t^*)T_0}e^{(t-t^*)T}\chi_\rho^+ e^{t^*T}\chi_\rho^- f\right],$$

If we replace in (3.37) $\chi_\rho^0 f$ by $\chi_\rho^0 e^{t^*T}\chi_\rho^- f$ and t by $t - t^*$, this estimate shows that the first term inside the brackets forms a Cauchy net as $t \to \infty$. In the second term, the operator $e^{-(t-t^*)T_0}e^{(t-t^*)T}$ can be replaced by the identity operator. Thus the existence of $W_+(T_0, T)\chi_\rho^- f$ follows. This completes the proof of the theorem. □

The locally decaying property can also supply more information about the spectrum of operator T.

Theorem 3.8.7 *Let (k, σ_a) be an admissible regular and nonproliferating pair such that $M(\sigma_a) < \infty$. Then $\sigma_r(T) = i\mathbb{R}$ and $\sigma_p(T) \subset \{\lambda \in \mathbb{C} \mid -\lambda_1 \leq Re\,\lambda < 0\}$.*

Proof The nonproliferating property of e^{tT} implies that $\rho(T) \subset \{\lambda \in \mathbb{C} \mid \mathrm{Re}\,\lambda > 0\}$ and by Theorem 3.4.1 (b), $\sigma(T) \subset \{\lambda \in \mathbb{C} \mid -\lambda_1 \leq \mathrm{Re}\,\lambda \leq 0\}$.

Since (k, σ_a) is regular, $k(x, v, v') = 0$ whenever $|x| \geq \rho$ for some $\rho > 0$. Thus, the argument of (3.22), given for T_a works also for T and we have

$$i\mathbb{R} = \sigma_p(T') \subseteq \sigma_r(T) \cup \sigma_p(T) \subset \{\lambda \in \mathbb{C} \mid -\lambda_1 \leq \mathrm{Re}\,\lambda \leq 0\}.$$

To concludie the theorem it suffices to prove that $\sigma_p(T) \cap i\mathbb{R} = \emptyset$. By Lemma 3.1.4 (b), $F(\sigma_a) < \infty$. Hence Theorem 3.6.1 implies the existence of $W_+(T_0, T)$ and by Theorem 3.8.6 the local decaying of e^{tT}. Now, let $s \in \mathbb{R}$ and $\psi \in L^1(\mathbb{R}^n \times V)$ such that $T\psi(x, v) = is\psi(x, v)$. This implies that $|e^{tT}\psi(x, v)| = |\psi(x, v)|$ and for any compact set $K \subset \mathbb{R}^n$, $\iint_{K\times V} |e^{tT}\psi(x, v)|dxdv$ is constant. This contradicts the local decay property of e^{tT}. □

3.9 Finite Collision Systems

In this section we shall construct a transport system in which the Dyson–Phillips expansion of the system contains only a finite number of terms. We call such a system a *finite collision system*. Such construction is based on some physical assumptions on the collision kernel $k(x, v, v')$ and the velocity space V. Although these physical assumptions are far from reality, they help us to understand the mechanism of the collision processes and see how it can be plausible to obtain the wave operators for a transport system, without any further assumptions on the size of the target or on the height of the mean free path.

Let V the velocity space be a spherical shell in $\mathbb{R}^n$,

$$V := \{v \in \mathbb{R}^n \ : \ v_{min} \leqslant |v| \leqslant v_{max}\},$$

where $0 < v_{min} < v_{max} < \infty$. Let η_0 be a fixed positive number. Define the subsets

$$M_+ := \{(v, v') \in V \times V \mid |v'| \geqslant |v| + \eta_0\},$$
$$M_- := \{(v, v') \in V \times V \mid |v'| \geqslant |v| - \eta_0\}.$$

Definition 3.9.1 We say that an admissible system (k, σ_a) is *speeding-up* (resp. *slowing-down*) if the support of the function $k_x : (v, v') \longrightarrow k(x, v, v')$ is in M_+ (resp. in M_-) (Fig. 3.2).

In a speeding-up or in a slowing-down system, any particle can suffer only a finite number of collisions before becoming unable to get involved in any further collision process and leaking out through the boundary of the target Ω. In fact, in a speeding-up system, since $\operatorname{supp} k(x, \cdot, \cdot) \subset M_+$, in every collision process, a particle will gain at least the positive amount of velocity η_0 and for any $f \in L^1(\mathbb{R}^3 \times V)$, $\operatorname{supp}\,[A_p f](x, v) = \operatorname{supp} \int_V k(x, v', v) f(x, v')dv' \subset \mathbb{R}^n \times \{v \in V \ : \ v_{min} + \eta_0 \leqslant |v| \leqslant v_{max}\}$. After a finite number of collisions the velocity of any particle goes beyond the velocity v_{max}. For such a velocity the kernel k is inert to further collisions and the particle finds itself in a streaming free or penetration system. Conversely in a slowing-down model, a particle loses in every collision process an amount of velocity at least η_0 and particles with velocity $|v| < v_{min}$ cannot get involved in any further collision process, so every particle can suffer only a finite number of collisions. As a consequence we will have the following lemma.

Lemma 3.9.2 *(a) If the pair* $(k, 0)$ *defines a finite collision system, then* $\{e^{tT'}\}_{t\in\mathbb{R}}$ *is a temperate* (C_0) *group and*

$$\|e^{tT'}\| \leqslant \sum_{i=0}^{N} \frac{M^i |t|^i}{i!} \quad \textit{for all} \quad t \in \mathbb{R}. \tag{3.38}$$

(b) If the pair (k, p) defines a finite collision system and $F(\sigma_a) < \infty$, then $\{e^{tT}\}_{t\in\mathbb{R}}$ is a temperate (C_0) group in the sense of Remark 1.2.5.

Proof (a) The above statements imply that there exists an $N \in \mathbb{N}$ such that for all $t_1, t_2, \cdots, t_N$,

$$A_p e^{t_1 T_0} A_p e^{t_2 T_0} A_p \cdots A_p e^{t_N T_0} A_p = 0.$$

Consequently, the Dyson–Phillips expansion of $e^{tT'}$ contains only $N+1$ terms, that is $e^{tT'} = \sum_{i=0}^{N} S_i(t)$. Since $S_0(t) = e^{tT_0}$ is isometric on $L^1(\mathbb{R}^n \times V)$, an induction argument shows that if $\|S_i(t)\| \leqslant M^i |t|^i / i!$, then

$$\|S_{i+1}(t)\| \leqslant \int_0^{|t|} \|e^{(t-s)T_0} A_p S_i(s)\| \, ds \leqslant \int_0^{|t|} M \frac{M^i s^i}{i!} ds \leqslant \frac{M^{i+1}|t|^{i+1}}{(i+1)!},$$

where $M = \|A_p\| = \sup_{(x,v)\in\mathbb{R}^n\times V} \sigma_p(x, v)$. Hence we obtain (3.38).

(b) For the (C_0) group $\{e^{tT}\}_{t\in\mathbb{R}}$, we will use the Trotter product formula (1.25), by writing

$$e^{tT} f = \lim_{n\to\infty} \{e^{\frac{t}{n}A_a} e^{\frac{t}{n}T'}\}^n f \qquad \text{for all } f \in L^1(\mathbb{R}^n \times V). \tag{3.39}$$

Since $[e^{tA_a} f](x, v) = \exp\left(-\int_0^t \sigma_a(x - sv, v) ds\right) f(x, v)$, this (C_0) group is uniformly bounded by $e^{F(\sigma_a)}$, it follows from (3.39) that there exists a constant C such that $\|e^{tT}\| \leqslant C(1 + |t|)^N$. □

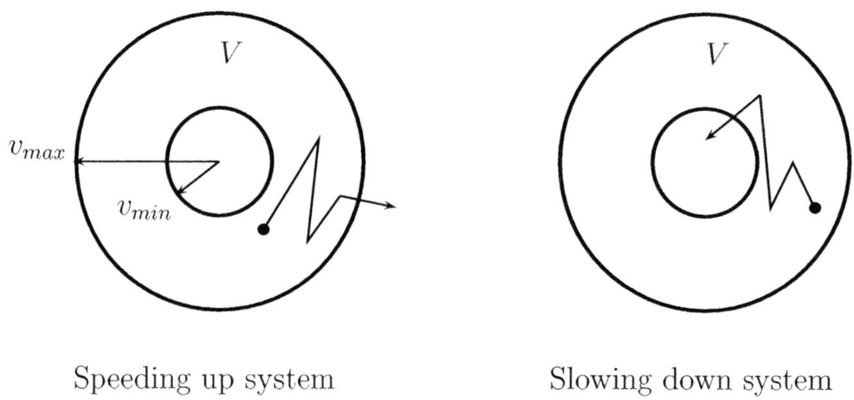

Fig. 3.2 .

In spite of the fact that such a system dies out after finite collisions (consequently, after finite time), we cannot conclude the existence of the wave operators in complete generality. But in some particular cases we shall prove in the following theorem that all the wave operators exist in $(F(\sigma_a), F(\sigma_p))$ plane defined in Fig. 3.1.

Theorem 3.9.3 *Suppose* $\eta_0 \geqslant (v_{max} - v_{min})/2$.

(i) *In this case* $N = 1$ *and*

$$e^{tT_p} = e^{tT_0} + \int_0^t e^{(t-s)T_0} A_p e^{sT_0}\, ds. \tag{3.40}$$

(ii) *If* $F(\sigma_p) < \infty$, *then* e^{tT_p} *is uniformly bounded on* $\mathbb{R}$.

(iii) *If* $F(\sigma_a) < \infty$ *and* $F(\sigma_p) < \infty$, *then* e^{tT} *is uniformly bounded on* $\mathbb{R}$ *and all wave operators* $W_\pm(T, T_0)$ *and* $W_\pm(T_0, T)$ *exist.*

Proof (i) Since $2\eta_0 \geqslant (v_{max} - v_{min})$, it follows that $A_p e^{tT_0} A_p = 0$ and $S_2(t) = 0$ for all $t \in \mathbb{R}$. This means that the Dyson–Phillips expansion of e^{tT_p} contains only two terms: $e^{tT'} = S_0(t) + S_1(t)$. This is exactly (3.40).

(ii) As (3.32), by (3.40) we have

$$\|e^{tT_p}\| = 1 + \int_0^t \|A_p e^{sT_0}\|\, ds \leqslant 1 + F(\sigma_p) \qquad \text{for all} \quad t \in \mathbb{R}.$$

(iii) For uniform boundedness of e^{tT} on $\mathbb{R}$, one can use once more the Trotter's product formula, as in Lemma 3.9.2 (b). For the existence of the wave operators $W_\pm(T, T_0)$, it suffices to use Cook's Lemma and the fact that

$$\int_{-\infty}^{\infty} \|Ae^{tT_0}\| dt \leqslant \big(F(\sigma_a) + F(\sigma_p)\big)\, \|f\|.$$

For the existence of the wave operators $W_\pm(T_0, T)$, one proves first the existence of $W_\pm(T_0, T_p)$, then uses the transitivity relation $W_\pm(T_0, T) = W_\pm(T_0, T_p) W_\pm(T_p, T)$. The existence of $W_\pm(T_0, T_p)$ comes from the fact that $e^{-tT_0} e^{tT} = I + \int_0^t e^{-sT} A_p e^{sT_0}\, ds$ and $\int_{-\infty}^{\infty} \|A_p e^{tT_0}\| dt \leqslant F(\sigma_p) \|f\|$. □

3.10 Notes and Comments

The story began in August 1974, in Oberwolfach (Germany), when J. Hejtmanek and B. Simon both attended a symposium on spectral and scattering theory. J. Hejtmanek presented his work [60] on scattering theory of the linear Boltzmann operator. Afterwards, B. Simon gave many suggestions, Lemma 1.9.5 is in particular due to him. He used the idea of [60] and wrote a paper [108], which was published before Hejtmanek's [60].

In the work by Hejtmanek [60] one can find two main theorems concerning the solubility of the transport problem for $t > 0$ and the existence of the wave operator $W_-(T, T_0)$ in the case of absorption or pure scattering. Remark 3.8.2 is also proved in this paper. In [108], B. Simon, showed the existence of the wave operator $W_+(T_0, T)$ under the assumption that the system is regular, admissible, nonproliferating (*subcritical* in term of [108]) and $\mathrm{diam}(\Omega)M(\sigma) < 1$ for $\sigma = \sigma_a + \sigma_p$ and he conjectured also Theorem 3.6.1 (see also [103]).

The materials and notions which we introduced in Sect. 3.1 are based on Reed and Simon's book [103]. Theorem 3.2.2 is partially proved independently in the papers by Hejtmanek [61], Umeda [117] and the author [44].

The notions of incoming and outgoing point sets of Definition 3.4.4 were introduced for the first time by Lax and Phillips [77]. Lemma 3.4.3 and Corollary 3.4.5 are given in [44].

As the preliminaries for Sect. 3.6 we introduced in Sect. 3.5 the sufficient conditions in order that the transport system was nonproliferating. In [118], T. Umeda proved that, if the pair (k, σ_a) is admissible and $F(\sigma_p)e^{F(\sigma_a)} < 1$, or (k, σ_a) is regular and $(\mathrm{diam}\ \Omega)M(\sigma_p)e^{(\mathrm{diam}\ \Omega)M(\sigma_a)} < 1$, then the wave operators $W_+(T_0, T)$ and $W_-(T, T_0)$ exist in $L^1(\mathbb{R}^n \times V)$. In Sects. 3.6 and 3.7 we follow his argument for the existence of $W_\pm(T_0, T)$ and $W_\pm(T, T_0)$. In his program Umeda used thoroughly the contribution of positivity in L^1 space. The existence of $W_+(T_0, T)$ under the locally decaying assumption (Theorem 3.8.6) was proved by Voigt [123]. This result has been commented on extensively in the books [68] and [103]. Finally, the notion of finite collision systems was introduced by Hejtmanek in [61] and the scattering formalism related to this notion by the author in [45].

Chapter 4
The Lax and Phillips Formalism for the Transport Problems

4.1 The Lax–Phillips Representation Theorem

Let $U(t)$ be a unitary group on a Hilbert space X and let $D_\rho^\pm \subset X$ be two distinguished subspaces designated as the *incoming* and *outgoing subspaces* which satisfy the following theorem.

Theorem 4.1.1 (Lax and Phillips representation theorem) *The subspaces D_ρ^+ and D_ρ^- satisfy the following properties:*

(i)$_+$ $\quad U(t)D_\rho^+ \subset D_\rho^+, \quad for\ t \geq 0;$

(i)$_-$ $\quad U(t)D_\rho^- \subset D_\rho^-, \quad for\ t \leq 0;$

(ii) $\quad \bigcap_{t\in\mathbb{R}_\pm} U(t)D_\rho^\pm = \{0\};$

(iii) $\quad \bigcup_{t\in\mathbb{R}} U(t)D_\rho^\pm$ *is dense in* X.

In a paper [77] earlier than their book, Lax and Phillips have applied this theorem to the transport phenomena. They took $U(t)$ as the streaming free group e^{tT_0}. To define the outgoing and incoming subspaces let us recall them as in Definition 3.4.4:

$$E_\rho^\pm = \{(x, v)\ :\ \pm x \cdot v \geqslant \rho|v|\}.$$

Corresponding to the subsets $E_\rho^\pm$ they define

$$D_\rho^\pm := \{f \in X\ :\ \mathrm{supp} f \subset E_\rho^\pm\}$$

and they show that Theorem 4.1.1 holds for $U(t) = e^{tT_0}$ in $X := L^2(\mathbb{R}^n \times S_{n-1})$. This theorem can be readily generalized for $X := L^1(\mathbb{R}^n \times V)$.

H. Emamirad, *Scattering Theory for Transport Phenomena*, Mathematical Physics Studies, https://doi.org/10.1007/978-981-16-2373-8_4

Throughout this chapter, the velocity space V is a spherical shell

$$V := \{v \in \mathbb{R}^n \ : \ v_{min} \leqslant |v| \leqslant v_{max}\},$$

with $0 < v_{min} \leq v_{max} < \infty$.

In this section we shall prove that the existence of the wave operators for the transport system implies the Lax and Phillips representation theorem for the transport group e^{tT}. To prove this result we need the following lemma. Let us define the projections $Q_\rho^+ f := \chi_\rho^+ f$, $Q_\rho^- f := \chi_\rho^- f$ and $Q_\rho^0 f := \chi_\rho^0 f$, where $\chi_\rho^\pm$ and χ_ρ^0 are defined in Definition 3.4.4.

Lemma 4.1.2 *Let (k, σ_a) be an admissible pair. If $W_\pm(T, T_0)$ exist, then*

$$\|e^{-tT} Q_\rho^0 e^{tT_0} f\| \to 0 \quad as \quad t \to \pm\infty \quad for\ all \quad f \in L^1(\mathbb{R}^n \times V). \tag{4.1}$$

Proof By the uniform boundedness theorem the existence of $W_+(T, T_0)$ implies that the upper bound M of $\|e^{-tT} e^{tT_0}\|$ is independent of t. Now by writing $\Pi_\rho^0 := e^{-tT_0} Q_\rho^0 e^{tT_0}$, since $(\mathrm{Im}\ \Pi_\rho^0) \subset X$ we have

$$\sup_{g \in \mathrm{Im}\ \Pi_\rho^0} \|e^{-tT} e^{tT_0} g\| \leq \sup_{f \in X} \|e^{-tT} e^{tT_0} f\| \leq M.$$

Since for any $f \in X$ and $g = e^{-tT_0} Q_\rho^0 e^{tT_0} f$,

$$e^{-tT} Q_\rho^0 e^{tT_0} f = e^{-tT} e^{tT_0} g,$$

the upper bound of $\|e^{-tT} Q_\rho^0 e^{tT_0}\|$ is also independent of t and consequently it suffices to prove (4.1) only for $f \in C_c^\infty(\mathbb{R}^n \times V)$ which is dense in $L^1(\mathbb{R}^n \times V)$. Let $f \in C_c^\infty(\mathbb{R}^n \times V)$ and R be large enough such that $\mathrm{supp}\ f \subset B_R \times V$, then for $|t| \geq (R + \rho)/v_{min}$ and for $(x, v) \in E_\rho^0$ we have $|x - tv| \geq (x - tv) \cdot (\frac{\mp v}{|v|}) \geq -\rho \pm t|v| \geq (-\rho \pm t)v_{min} \geq R$. Therefore for any $(x, v) \in \mathbb{R}^n \times V$, $\chi_\rho^0(x, v) f(x - tv, v) = 0$ and consequently $\|e^{-tT_0} Q_\rho^0 e^{tT_0} f\| = 0$. □

Theorem 4.1.3 *Let (k, σ_a) be a regular pair such that the wave operators $W_\pm(T, T_0)$ and $W_\pm(T_0, T)$ exist. Then Theorem 4.1.1 holds for $U(t) = e^{tT}$ and $X := L^1(\mathbb{R}^n \times V)$.*

Proof Statements (i) and (ii) of Theorem 4.1.1 are independent of the existence of the wave operators and can be shown by using Corollary 3.4.5. In fact, since for any $f \in D_\rho^\pm$, one has $\chi_\pm f = f$, Corollary 3.4.5 implies that

$$e^{tT} f = e^{tT_0} f \qquad \text{for all } f \in D_\rho^\pm \text{ and all } \pm t \geq 0. \tag{4.2}$$

Now, given the obvious fact that $e^{tT_0} D_\rho^+ \subset D_\rho^+$ for $t \geq 0$ and $e^{tT_0} D_\rho^- \subset D_\rho^-$ for $t \leq 0$, (4.2) implies (i)$_+$ and (i)$_-$.

In order to prove statement (ii) it suffices to make the following remark.

Remark 4.1.4 As we have said, the Lax–Phillips representation theorem is true for e^{tT_0} in $L^1(\mathbb{R}^n \times V)$ and we have

$$\bigcap_{t\in\mathbb{R}_\pm} e^{tT_0} D_\rho^\pm = \{0\} \tag{4.3}$$

and

$$\bigcup_{t\in\mathbb{R}_\mp} e^{tT_0} D_\rho^\pm \quad \text{is dense in } X. \tag{4.4}$$

The relations (ii) in Theorem 4.1.1 can be justified by replacement e^{tT_0} by e^{tT} in (4.3).

For (iii), first we prove that

$$\overline{\bigcup_{t\in\mathbb{R}_-} e^{tT} D_\rho^+} = X. \tag{4.5}$$

Take $f \in X$ and $g = W_+(T_0, T)f$. According to Lemma 2.1.3 (e) we have $W_+(T, T_0)g = f$, which means that for any $\epsilon > 0$ there exists $t_1 \geq 0$ such that for any $t \geq t_1$,

$$\|e^{-tT} e^{tT_0} g - f\| < \frac{\epsilon}{3}. \tag{4.6}$$

Since $g \in L^1(\mathbb{R}^n \times V)$, we can find r large enough (larger than ρ) such that

$$\int_{|x|\geq r} \int_V |g(x, v)| dx dv < \frac{\epsilon}{3}.$$

For such a constant r we consider χ_r^+, χ_r^- and χ_r^0, the characteristic functions of three disjoint sets E_r^+, E_r^- and E_r^0, and in (4.6) we replace $e^{tT_0}g$ by $(\chi_r^+ + \chi_r^- + \chi_r^0)e^{tT_0}g$. Then, for $h = \chi_r^+ e^{tT_0} g \in D_\rho^+$ we get

$$\|e^{-tT} h - f\| < \|e^{-tT} \chi_r^- e^{tT_0} g\| + \|e^{-tT} \chi_r^0 e^{tT_0} g\| + \frac{\epsilon}{3}. \tag{4.7}$$

Now, $\chi_r^- e^{tT_0} g \in D_r^- \subset D_\rho^-$ and $E_r^- \subset B_r^c$, the complement of the ball B_r in $\mathbb{R}^n$. For the first term of the right-hand side of (4.7), we have

$$\begin{aligned} \|e^{-tT} \chi_r^- e^{tT_0} g\| &= \|e^{-tT_0} \chi_r^- e^{tT_0} g\| \\ &= \iint \chi_r^-(x, v)|g(x - tv, v)| \, dx \, dv \\ &\leq \int_{|x|\geq r} \int_V |g(x, v)| \, dx \, dv < \frac{\epsilon}{3}. \end{aligned}$$

According to Lemma 4.1.2, the existence of $W_+(T, T_0)$ yields the existence of $t_2 > 0$ such that $\|e^{-tT}\chi_r^0 e^{tT_0} g\| < \frac{\epsilon}{3}$ for any $t \geq t_2$. Consequently, for any $\epsilon > 0$ we have found a function $h \in D_\rho^+$ and a time $t^* = \max\{t_1, t_2\}$ such that $\|e^{-tT}h - f\| < \epsilon$ for any $t \geq t^*$, which implies (4.5). Similarly, we can prove that the existence of $W_-(T_0, T)$ and $W_-(T, T_0)$ imply $\overline{\bigcup_{t\in\mathbb{R}_+} e^{tT} D_\rho^-} = X$. □

Corollary 4.1.5 *Let (k, σ_a) be a regular pair and assume that $W_+(T_0, T)$ exists. Then the range of $W_+(T_0, T)$ is dense in $L^1(\mathbb{R}^n \times V)$.*

Proof It suffices to prove that

$$\bigcup_{t\leq 0} e^{tT_0} D_\rho^+ \subset \operatorname{Ran}(W_+(T_0, T)). \tag{4.8}$$

Indeed, (4.4) concludes the lemma. To prove (4.8), take $f \in D_\rho^+$, it follows from (4.2) that $W_+(T_0, T)f = f$. If we fix $t \geq 0$ and $g = e^{tT_0} f$, since $e^{tT_0} D_\rho^+ \subset D_\rho^+$, then supp $g \subset E_\rho^+$, hence $e^{-tT_0} g = W_+(T_0, T)f$, which implies (4.8). □

4.2 The Lax–Phillips Semigroups

In the sequel we omit the index ρ and we replace $E_\rho^\pm$, E_ρ^0, $D_\rho^\pm$, D_ρ^0, $\chi_\rho^\pm$, χ_ρ^0, $Q_\rho^\pm$ and Q_ρ^0 by $E_\pm$, E_0, $D_\pm$, D_0, $\chi_\pm$, χ_0, $Q_\pm$ and Q_0. Let us define the projections $P_\pm$ by $P_\pm f := (I - Q_\pm)f = \chi'_\pm f$, where $\chi'_\pm = 1 - \chi_\pm$. Associated with the groups e^{tT_0}, e^{tT_1} and e^{tT}, we define the corresponding Lax–Phillips semigroups by

$$Z_i(t) := P_+ e^{tT_i} P_-,$$

where i can be replaced by 0, 1 or void.

Let us consider K a subspace of $L^1(\mathbb{R}^n \times V)$ consisting of functions f which are identically zero on $E_+ \cup E_-$. In the following theorem we will prove that $P_+ e^{tT_i} P_-$ defines a (C_0) semigroup on K. By respecting the Lax and Phillips notations, we denote by B_i the infinitesimal generator of this semigroup.

Theorem 4.2.1 *The operators $\{Z_i(t)\}_{t\geq 0}$ map K into itself and form the (C_0) semigroups on K.*

In order to prove this theorem we need the following lemma.

Lemma 4.2.2 *For any $f \in L^1(\mathbb{R}^n \times V)$:*

(i) supp $\{e^{tT_i} Q_+ f\} \cap$ supp $\{e^{tT_i} P_+ f\} = \emptyset$ for $t \geq 0$;
(ii) supp $\{e^{tT_i} Q_- f\} \cap$ supp $\{e^{tT_i} P_- f\} = \emptyset$ for $t \leq 0$.

Proof Since $Q_+ f \in D_+$, (4.2) implies that supp $\{e^{tT} Q_+ f\} \subset \{(x, v) \mid x \cdot v \geq (\rho + t|v|)|v|\}$. Let us show that supp $\{e^{tT} P_+ f\}$ cannot overlap with $\{(x, v) \mid x \cdot v \geq (\rho + t|v|)|v|\}$. Indeed, if $x \cdot v \geq (\rho + t|v|)|v|$, the estimation

$$|x + (t - s)v| \geqslant (x + (t - s)v) \cdot \frac{v}{|v|} \geqslant \rho + (t - s)|v| \geqslant \rho, \quad \text{for} \quad t \geqslant s,$$

implies that the last term of

$$e^{tT} P_+ f = e^{tT_0} P_+ f + \int_0^t e^{(t-s)T_0} A e^{sT} P_+ f \, ds$$

vanishes and since supp $\{e^{tT} Q_+ f\} \subset \{(x, v) \mid x \cdot v \leq (\rho + t|v|)|v|\}$, this proves the first part of this lemma for e^{tT}. We remark that e^{tT_0} and e^{tT_1} are the particular cases. The proof of (ii) is similar. □

Proof of Theorem 4.2.1. Let us define

$$D'_{\pm} := \{f \in L^1(\mathbb{R}^n \times V) \mid \text{supp } f \subset E^c_{\pm}\}.$$

The left factor in $Z_i(t)$ being P_+, it follows that the range of $Z_i(t)$ is in D'_+. Thus, in order to show that $Z_i(t)$ maps $K = D'_+ \cap D'_-$ into itself for $t \geq 0$, it suffices to show that $Z_i(t) f$ is in D'_- when f is in K. Writing $e^{tT_i} f = Q_- e^{tT_i} f + P_- e^{tT_i} f$, by using (4.2) for $t \leq 0$, since $Q_- e^{tT_i} f$ belongs to D_-, one gets

$$f = e^{-tT_0} Q_- e^{tT_i} f + e^{-tT_i} P_- e^{tT_i} f. \tag{4.9}$$

By virtue of Lemma 4.2.2 (ii), since $f(x, v) = 0$ when $x \cdot v \leq -\rho|v|$, (4.9) is not consistent unless $\chi_- e^{tT_i} f = 0$ for all $t \geq 0$. Consequently, $Z_i(t)$ maps K into itself.

We finish the proof of this theorem by remarking that $\{Z_i(t)\}_{t\geq 0}$ form the (C_0) semigroups on K, because $P_\pm$ acts as an identity on K and e^{-tT_i} are already the (C_0) semigroups on $L^1(\mathbb{R}^n \times V)$. □

Theorem 4.2.3 *For any* $t \geq 2\rho/v_{min}$, *one has* $Z_0(t) = Z_1(t) = 0$.

Proof From the definition of $Z_1(t)$ and (3.18) we conclude that

$$[Z_1(t) f](x, v) = \chi'_+(x, v) e^{-\int_0^t \sigma_a(x - sv, v) ds} \chi'_-(x - tv, v) f(x - tv, v).$$

Now, if $t \geq 2\rho/v_{min}$ we have either $x \cdot v \geq \rho|v|$ in which case $\chi'_+(x, v) = 0$ or $x \cdot v \leq \rho|v|$, which implies $x \cdot v - t|v|^2 \leq \rho|v| - t v_{min}|v| \leq -\rho|v|$ and consequently, $\chi'_-(x - tv, v) = 0$. Whatever the case we get $Z_1(t) f(x, v) = 0$. □

Since $\Omega \times V \subset E_0 = E^c_+ \cap E^c_-$, if we denote by $L^1(\Omega \times V)$ the subspace of $L^1(\mathbb{R}^n \times V)$ of the functions which are identically zero out of $\Omega \times V$ we get $L^1(\Omega \times V) \subset K$. In spite of this fact, $\{Z_i(t)\}_{t\geq 0}$ does not act as a semigroup on $L^1(\Omega \times V)$.

In [110], Stefanov has modified the projection operators $P_\pm$ in such a way that the resulting Lax–Phillips semigroup defines a (C_0) semigroup on $L^1(\Omega \times V)$. As we will see, the spectral properties of the generator of this type of semigroup are closely related to the spectral properties of T_i.

By recalling the sets $\Sigma_\pm$, Σ_0 and Σ defined in Definition 3.4.2, we write $\Sigma_\pm^0 := \Sigma_\pm \cup \Sigma_0$ and the corresponding projections $\Gamma_\pm f := \chi_{\Sigma_\pm^0} f$ and $\Pi_\pm f := (I - \Gamma_\pm) f$. We have shown in Corollary 3.4.5 that $E_\pm \subset \Sigma_\pm^0$, hence $\Pi_+ e^{tT_i} \Pi_-$ is a restriction of $Z_i(t)$ on $L^1(\Omega \times V)$ and if we replace the characteristic functions $\chi_\pm$ by $\chi_{\Sigma_\pm^0}$, all the preceding results concerning $Z_i(t)$ would also be valid for $\Pi_+ e^{tT_i} \Pi_-$. While

$$\text{supp } (\Gamma_\pm f) \cap \text{supp } (\Pi_\pm f) = \emptyset,$$

it becomes easy to show that:

Lemma 4.2.4 *For any $f \in L^1(\mathbb{R}^n \times V)$:*

(i) supp $\{e^{tT_i} \Gamma_+ f\} \cap$ supp $\{e^{tT_i} \Pi_+ f\} = \emptyset$ for $t \geq 0$;
(ii) supp $\{e^{tT_i} \Gamma_- f\} \cap$ supp $\{e^{tT_i} \Pi_- f\} = \emptyset$ for $t \leq 0$.

Now, since Σ_+, Σ_-, Σ_0 and Σ are four disjoint subsets of $\mathbb{R}^n \times V$ with $\Sigma = \Omega \times V$, we have:

(i) $(\Sigma_\pm^0)^c = \Sigma \cup \Sigma_\mp$;
(ii) $(\Sigma_-^0)^c \cap (\Sigma_+^0)^c = \Sigma$;
(iii) from Lemma 4.2.4, it follows that for any $f \in L^1(\mathbb{R}^n \times V)$,

$$\text{supp } e^{tT_i} \chi_\Sigma f \subset \Sigma \cup \Sigma_+.$$

Consequently,

$$\begin{aligned}\Pi_+ e^{tT_i} \Pi_- \chi_\Sigma f &= \chi_{\Sigma \cup \Sigma_-} e^{tT_i} \chi_{\Sigma \cup \Sigma_+} \chi_\Sigma f \\ &= \chi_{\Sigma \cup \Sigma_-} e^{tT_i} \chi_\Sigma f \\ &= \chi_{\Sigma \cup \Sigma_-} \chi_{\Sigma \cup \Sigma_+} e^{tT_i} \chi_\Sigma f \\ &= \chi_\Sigma e^{tT_i} \chi_\Sigma f.\end{aligned}$$

This proves that $\Pi_+ e^{tT_i} \Pi_-$ maps $L^1(\Omega \times V)$ into itself.

We denote by

$$\mathcal{Z}_i(t) := \Pi_+ e^{tT_i} \Pi_-, \quad i = 0, 1, \text{void}, \tag{4.10}$$

the corresponding Lax–Phillips semigroup.

In the sequel we suppose that the pair (k, σ_a) is regular and the kernel $k \in L^\infty(\Omega \times V \times V)$, then we have the following fundamental result.

Theorem 4.2.5 *$\mathcal{Z}(t)$ is an eventually compact semigroup in $L^1(\mathbb{R}^n \times V)$.*

For the proof of this theorem we need the following lemma. First we fix some notation. Let χ be the characteristic function of $\Omega \times V$. $\Omega \times V$ is bounded in $\mathbb{R}^{2n}$, thus χ is an element of $L^1 \cap L^\infty(\Omega \times V)$ and the operator $\chi \otimes \chi$ on $L^1(\Omega \times V)$ given by $[\chi \otimes \chi]f = \langle f, \chi \rangle \chi = (\int_{\Omega \times V} f dx dv)\chi$ is of rank 1.

Let us recall Corollary 1.8.2 and write the Dyson–Phillips expansion of e^{tT_1} by

$$e^{tT} = \sum_{j=0}^{\infty} S_j(t), \tag{4.11}$$

where $S_0(t) = e^{tT_1}$ and $S_j(t) = \int_0^t e^{(t-s)T_1} A_p S_{j-1}(s) ds$, for $j = 1, 2, \cdots$.

Lemma 4.2.6 *Suppose (k, σ_a) be a regular pair and $k \in L^\infty(\Omega \times V \times V)$, then for arbitrary $t > 0$, one has:*

(i) $e^{tT_1}(\chi \otimes \chi) \leq \chi \otimes e^{tT_0}\chi$.
(ii) $A_p e^{tT_1} A_p \leq (\|k\|_\infty^2 t^{-n})\chi \otimes \chi$, *which implies that the operator* $A_p e^{tT_1} A_p$ *is weakly compact.*
(iii) For $t > 0$ and $j \geq 2$ the operator $S_j(t)$ is weakly compact.
(iv) $[A_p(\lambda - T_1)^{-1}]^4$ *is a compact operator.*

Proof (i) Since $\sigma_a \geq 0$, it follows from (3.15) that $e^{tT_1}\chi \leq e^{tT_0}\chi$ and therefore $e^{tT_1}(\chi \otimes \chi) = \chi \otimes (e^{tT_1}\chi) \leq \chi \otimes e^{tT_0}\chi$.

(ii) For $0 \leq f \in L^1(\mathbb{R}^n \times V)$, we have

$$[A_p f](x, v) = \chi(x) \int_V k(x, v, v') f(x, v') dv' \leq \|k\|_\infty \chi(x) \int_V f(x, v') dv',$$

hence, from (3.15) and the substitution $x' = x - tv''$ we get

$$\begin{aligned}
[A_p e^{tT_1} A_p f](x, v) &\leq [A_p e^{tT_0} A_p f](x, v) \\
&= A_p e^{tT_0} \chi_\Omega(x) \int_V k(x, v, v') f(x, v') dv' \\
&= A_p \chi_\Omega(x - tv) \int_V k(x - tv, v, v') f(x, v') dv' \\
&= \chi_\Omega(x) \int_V k(x, v, v'') \chi_\Omega(x - tv'') \int_V k(x - tv'', v'', v') f(x, v') dv' \\
&\leq \|k\|_\infty^2 \int_V \int_V \chi_\Omega(x - tv") f(x - tv'', v') dv'' dv' \\
&\leq \|k\|_\infty^2 \int_\Omega \int_V \chi_\Omega(x') f(x', v') t^{-n} dx' dv' \\
&\leq \|k\|_\infty^2 t^{-n} \int_\Omega \int_V f(x', v') dx' dv' = (\|k\|_\infty^2 t^{-n}) \langle f, \chi \rangle.
\end{aligned}$$

The operator $A_p e^{tT_1} A_p$ is dominated by a rank 1 operator, hence by a compact operator and Lemma 1.9.15 (ii) implies that it is weakly compact.

(iii) Let us consider the case $j = 2$. For $t > 0$, the operator $S_2(t)$ is given by

$$S_2(t) = \int_0^t \int_0^s e^{(t-s)T_1} A_p e^{(s-r)T_1} A_p e^{rT_1} dr ds.$$

For $\epsilon > 0$ we define

$$S_2^\epsilon(t) = \int_\epsilon^t \int_0^{(s-\epsilon)} e^{(t-s)T_1} A_p e^{(s-r)T_1} A_p e^{rT_1} dr ds.$$

Using the fact that $\{e^{tT_1}\}_{t\geq 0}$ is a contraction semigroup and (3.6), a simple calculation shows that $\| S_2(t) - S_2^\epsilon(t)\| \leq \|\sigma_p\|_\infty^2 \epsilon t$. Thus by Lemma 1.9.15 (i) it suffices to show that $S_2^\epsilon(t)$ is weakly compact. Using (ii) we obtain

$$e^{(t-s)T_1} A_p e^{(s-r)T_1} A_p e^{rT_1} \leq (\|k\|_\infty^2 (s-r)^{-n}) \chi \otimes \chi,$$

which implies

$$S_2^\epsilon(t) \leq \left(\|k\|_\infty^2 \int_\epsilon^t \int_0^{(s-\epsilon)} (s-r)^{-n} dr ds \right) \chi \otimes \chi.$$

As in (ii), this estimate shows that each operator $S_2^\epsilon(t)$ is weakly compact.

For $j \geq 2$, in any $S_j(t)$ there appears a term $A_p e^{(s-r)T_1} A_p$ in its integral representation, therefore one can treat this case in a similar manner.

(iv) This assertion is proved directly by Stefanov [110], but here we give a short proof by using (ii). In fact, it follows from this assertion that $A_p e^{t_2 T_1} A_p e^{t_1 T_1}$ is weakly compact and the mapping $t_1 \mapsto A_p e^{t_4 T_1} A_p e^{t_3 T_1} A_p e^{t_2 T_1} A_p e^{t_1 T_1}$ maps $\mathbb{R}_+$ into the set $\mathscr{I}_\infty(X)$ of all compact operators on X, which is a norm closed ideal of $\mathcal{L}(X)$. This mapping is also strongly continuous with respect to t_1. Since it is an $\mathscr{I}_\infty(X)$-valued mapping it is also continuous in uniform norm topology (see [69, p.151]). Taking its Laplace transform we get

$$A_p e^{t_4 T_1} A_p e^{t_3 T_1} A_p e^{t_2 T_1} A_p (\lambda - T_1)^{-1} = \int_0^\infty e^{-\lambda t_1} A_p e^{t_4 T_1} A_p e^{t_3 T_1} A_p e^{t_2 T_1} A_p e^{t_1 T_1} dt_1.$$

Considering the integral as a limit of Riemann's sums we notice that $A_p e^{t_4 T_1} A_p e^{t_3 T_1} A_p e^{t_2 T_1} A_p (\lambda - T_1)^{-1}$ is the limit in norm of compact operators and therefore is compact. By repeating this argument successively for t_2, t_3 and t_4, one gets the compactness of $[A_p(\lambda - T_1)^{-1}]^4$. □

Proof of Theorem 4.2.5. By replacing (4.11) in (4.10) one gets

$$\mathcal{Z}(t) = \Pi_+ e^{tT_1} \Pi_- + \int_0^t \Pi_+ e^{(t-s)T_1} A_p e^{sT_1} \Pi_- ds + \sum_{n=2}^\infty \Pi_+ S_n(t) \Pi_-. \qquad (4.12)$$

Now let ρ_0 be the smallest real number such that the ball B_{ρ_0} contains Ω, then for $x \in \Omega$ we have $|x - tv| \geq tv_{min} - \rho_0 \geq \rho_0$, when $t \geq t_0 := 2\rho_0/v_{min}$. Hence for $t > t_0$ the first term of (4.12) is identically zero. The integral in (4.12) vanishes for $t > 2t_0$. Indeed, $A_p e^{sT_1}\Pi_-$ vanishes for $s > t_0$, while for $0 \leq s \leq t_0$ we have $t - s \geq t_0 \geq t_0$, hence $\Pi_+ e^{(t-s)T_1} A_p \equiv 0$. Thus, for $t > 2t_0$ we get $\mathcal{Z}(t) = \sum_{n=2}^{\infty} \Pi_+ S_n(t)\Pi_-$. Furthermore, this series converges in norm operator topology to $\Pi_+ V_2(t)\Pi_-$, where $V_2(t) := \int_0^t e^{(t-s)T_1} A_p e^{sT} ds$ (see Corollary 1.8.2). Each term of this series is weakly compact, hence by Lemma 1.9.15 (i) and (iii) $\mathcal{Z}(t)$, is weakly compact and its square $\mathcal{Z}(t) = \mathcal{Z}(t/2)^2$ is compact for $t > 4t_0$. □

4.3 Generalized Eigenfunction Expansion

In the preceding section we have proved that the semigroup $\mathcal{Z}(t)$ is eventually compact and thus, it has a series expansion given by Lemma 1.10.7 (a). This is called a *generalized eigenfunction expansion of* $\mathcal{Z}(t)$ with respect to the semigroup $e^{t\widetilde{T}} := e^{tT}\,|_{L^1(\Omega\times V)}$. Before giving the precise definition of these notations, let us give some details about $\sigma(T)$.

Theorem 4.3.1 *Let (k, σ_a) be a regular pair and assume that $W_+(T_0, T)$ exists. Then, if $\Lambda := \{\lambda \in \mathbb{C} \;:\; -(\|\sigma_a\|_\infty + \|\sigma_p\|_\infty) \leq Re\,\lambda < 0\}$:*

(i) $\sigma(T) = \sigma_r(T) \cup \sigma_p(T)$.
(ii) $\sigma_p(T) \subset \Lambda$.
(iii) The point spectrum $\sigma_p(T)$ of T contains a finite isolated points λ_k. Each λ_k is an eigenvalue of finite multiplicity and is a pole for the resolvent $(\lambda - T)^{-1}$.

Proof The assertions (i) and (ii) are already proved in Theorem 3.8.7. For (iii), let $\Phi(\lambda) := [A_p(\lambda - T_1)^{-1}]^2$. By Lemma 4.2.6 (iv), $\lambda \mapsto \Phi(\lambda)^2$ is an analytic compact operator-valued function in the half-planes $\mathrm{Re}\,\lambda > 0$ and $\mathrm{Re}\,\lambda < 0$. Moreover, $\|\Phi(\lambda)^2\| \leq M^4\|A_p\|^4/|\mathrm{Re}\,\lambda|^4 \to 0$, as $|\mathrm{Re}\,\lambda| \to \infty$. Hence, 1 is not an eigenvalue for all operators $\Phi(\lambda)^2$. By applying the analytic Fredholm theorem, we conclude that $(I - \Phi(\lambda)^2)^{-1}$ exists in the half-planes $\mathrm{Re}\,\lambda > 0$ and $\mathrm{Re}\,\lambda < 0$, except in S, a countable set of points λ_k, where the function $(I - \Phi(\lambda))^{-1}$ has a pole with finite multiplicity. The set S has no accumulation points in the open half-plane. From the equality

$$(\lambda - T)^{-1} = (\lambda - T_1)^{-1}(I + A_p(\lambda - T_1)^{-1})(I + \Phi(\lambda))(I - \Phi(\lambda)^2)^{-1}, \quad \lambda \notin S \text{ and } \mathrm{Re}\,\lambda \neq 0$$

we obtain that the resolvent $(\lambda - T)^{-1}$ has the same behavior as the function $(I - \Phi(\lambda)^2)^{-1}$. This proves that $S = \sigma_p(T)$ and by Theorem 3.8.7, $\sigma_p(T) \subset \Lambda$. The finiteness of $\sigma_p(T)$ follows from the following lemma. □

Lemma 4.3.2 *Let $\mathcal{B}$ be the generator of the semigroup $\mathcal{Z}(t)$. Under the assumptions of Theorem 4.3.1, $\sigma_p(T)$ is finite and we have*

$$\sigma_p(T) \subset \sigma(\mathcal{B}). \tag{4.13}$$

Proof According to Theorem 4.3.1, if $\lambda \in \sigma_p(T)$ and ϕ is the corresponding eigenfunction, then $\operatorname{Re}\lambda < 0$ and we have

$$e^{tT}[\chi_{\Sigma_+}\phi + \chi_{\Sigma_-}\phi + \chi_{\Sigma_0}\phi + \chi_{\Sigma}\phi] = e^{t\lambda}[\chi_{\Sigma_+}\phi + \chi_{\Sigma_-}\phi + \chi_{\Sigma_0}\phi + \chi_{\Sigma}\phi]. \quad (4.14)$$

For any point $(x, y) \in \Sigma_-^0$, it follows from Duhamel's formula that $|e^{\lambda t}\phi(x, v)| = |\phi(x - tv, v)|$. If $\phi(x, v) \neq 0$, taking $t \to -\infty$ in this equality, the left-hand side goes to $+\infty$, whereas the right-hand side goes to zero. This proves that $\chi_{\Sigma_-}\phi = \chi_{\Sigma_0} = 0$. By taking $t \to -\infty$ the same argument shows that $\chi_{\Sigma_+}\phi = 0$. Since Σ_+, Σ_-, Σ_0 and Σ are four disjoint subsets of $\mathbb{R}^n \times V$, (4.14) reduces to $e^{tT}[\chi_{\Sigma}\phi] = e^{t\lambda}[\chi_{\Sigma}\phi]$. This implies that the eigenfunctions of T belong to $L^1(\mathbb{R}^n \times V)$ and are the eigenfunctions of $\mathcal{B}$. Finally, the compactness of $\mathcal{Z}(t)$ together with Lemma 1.10.7 imply that there is a finite number of eigenvalues in the strip Λ and

$$\sigma(\mathcal{B}) \cap i\mathbb{R} = \emptyset. \quad (4.15)$$

□

Corollary 4.3.3 *The Lax and Phillips semigroup $\mathcal{Z}(t)$ is hyperbolic on $\mathcal{K}$.*

Proof Since $\mathcal{Z}(t)$ is an eventually compact semigroup. The proof follows from Corollary 1.11.4 and (4.15). □

According to Lax–Phillips terminology the eigenvalues of the generator $\mathcal{B}$ are called *resonances* of T. The above lemma asserts that any resonance of T in the strip Λ is an eigenvalue and vice versa. But, there can be extra resonances out of Λ (see Fig. 4.1). Throughout we denote by λ_i the eigenvalues of T and by μ_i its resonances.

As we have seen in Lemma 1.10.7, any eventually compact semigroup has an asymptotic eigenfunction expansion in term of its spectral elements. So the Lax–Phillips semigroup $\mathcal{Z}(t)$ has also such an expansion. The eigenfunction expansion for $\mathcal{Z}(t)$ is called a *generalized eigenfunction expansion* for $T(t)$. The following theorem can be considered as a corollary of Lemma 1.10.7.

Theorem 4.3.4 *Under the assumptions of Theorem ??, the operator T has only a countable set (possibly finite or empty) of resonances in the left half plane $\mathbb{C}_- := \{\mu \in \mathbb{C} \mid Re\,\mu > 0\}$, which are the poles of finite multiplicity of $(\mu - \mathcal{B})^{-1}$. Furthermore, for any $r > 0$ there is a finite number of μ in the strip $\{\mu \in \mathbb{C} \mid -r < Re\,\mu < 0\}$. So we can enumerate the resonances by $Re\,\mu_{n+1} \leq Re\,\mu_n$, for all $n \in \mathbb{N}$. Denoting the pole order at μ_n by N_n and the corresponding residue by P_n (i.e., $P_n := \frac{1}{2\pi i}\int_{|\mu-\mu_n|=\varepsilon}(\mu - \mathcal{B})^{-1}d\mu$), for any $m \in \mathbb{N}$ we have*

$$\Pi_+ e^{tT}\Pi_- = \sum_{n=0}^{m} e^{\mu_n t}\sum_{k=0}^{N_n-1}\frac{t^k}{k!}(\mathcal{B} - \mu_n)^k P_n + R_m(t), \quad (4.16)$$

where $\|e^{(-\epsilon+Re\,\mu_m)t}R_m(t)\| \to 0$ as $t \to \infty$.

Fig. 4.1 .

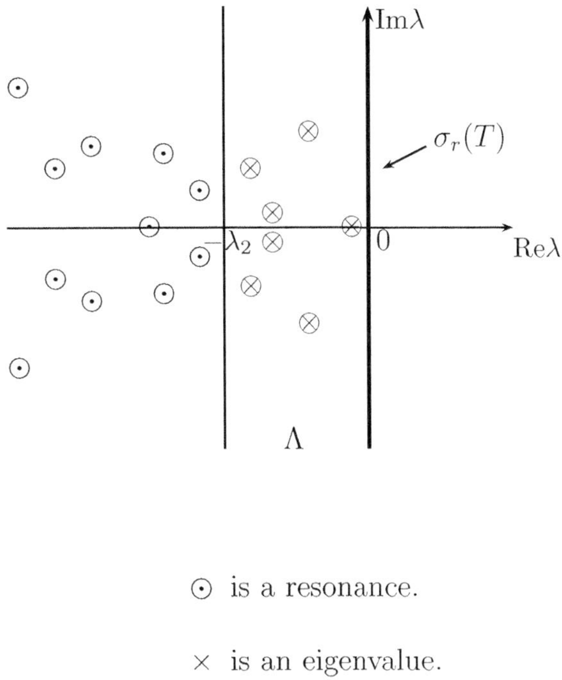

4.4 Hypercyclicity in the Lax–Phillips Formalism

In this section we will study the Lax and Phillips representation theorem in the weighted space $L^1_\omega(\mathbb{R}^n \times V)$. Let E be a measurable subset of $\mathbb{R}^n \times V$.

Definition 4.4.1 By an *admissible weight function* on E we mean a measurable positive function ω satisfying

$$\omega(x - tv, v) \leq r(t)\omega(x, v)$$

for all $(x, v) \in E$, where r is a positive function such that $\liminf_{t\to\infty} r(t) = 0$.

A prototype example of an admissible weight function is

$$\omega(x, v) = \exp(x \cdot v).$$

Let us denote by $L^1_\omega(E)$ the space of measurable functions $f : E \longmapsto \mathbb{C}$ such that

$$\|f\|_{E,\omega} = \int_E |f(x, v)|\omega(x, v)dxdv < \infty.$$

In this section we delete the subindex ρ and we denote by $E_\pm$ the outgoing and incoming point sets $E_\rho^\pm$ as in Definition 3.4.4. Let ω_+ and ω_- be two admissible weight functions defined respectively on E_+ and E_-. Let ω be a positive extension of $\omega_\pm$ on $\mathbb{R}^n \times V$. On $L^1_\omega(\mathbb{R}^n \times V)$ the streaming free group $\{e^{tT_0}\}_{t\in\mathbb{R}}$ defines the dynamics $V_0^\pm(t) = \chi_\pm e^{\mp tT_0}$ on $L^1_{\omega_\pm}(E_\pm)$, where $\chi_\pm$ are the characteristic functions of $E_\pm$.

Theorem 4.4.2 $\{V_0^\pm(t)\}_{t\geq 0}$ *are strongly continuous hypercyclic semigroups on* $L^1_{\omega_\pm}(E_\pm)$.

Proof The fact that $V_0^\pm(t)$ are two (C_0) semigroups can be proved by the standard argument of translation semigroup (see Theorem 3.2.1). Here we will prove the hypercyclicity of $V_0^-(t)$ by using the Hypercyclicity Criterion given in Theorem 1.11.6, the same argument will work for $V_0^+(t)$.

Let $X_0 = \left\{f \in L^1_{\omega_-}(E_-) \mid \lim_{t\to\infty} V_0^-(t)f = 0\right\}$ and $X_\infty = \left\{f \in L^1_{\omega_-}(E_-) \mid \forall\epsilon > 0, \exists\, u \in L^1_{\omega_-}(E_-) \text{ and } t>0 \text{ such that } \|u\|_{E_-,\omega_-} < \epsilon \text{ and } \|V_0^-(t)u - f\|_{E_-,\omega_-} < \epsilon\right\}$. Choose f an element of $L^1_{\omega_-}(E_-)$ with compact support in E_-. Thus, there is some $R > \rho$ such that for any $(x, v) \in \mathbb{R}^n \times V$ with $-R \geq x \cdot v$, $f(x, v) = 0$. Therefore for any $(x, v) \in E_-$ and $t \geq (R - \rho)/v^2_{min}$ one has $x \cdot v - t|v|^2 \leq -\rho - tv^2_{min} \leq -R$ and $f(x - tv, v) = 0$. This implies that X_0 contains all functions of compact support in $L^1_{\omega_-}(E_-)$, so that X_0 is dense.

For the density of X_∞, we choose an arbitrary function f of compact support in E_- and we define $u_t(x, v) = f(\tau_t(x, v))$, where $\tau_t(x, v) = (x + tv, v)$. For any $t \geq 0$ and $u_t \in L^1_{\omega_-}(E_-)$, we have

$$\begin{aligned}\|u_t\|_{E_-,\omega_-} &= \int_{E_-} |f(x + tv, v)|\omega_-(x, v)dxdv \\ &= \int_{E_-} |f(x, v)|\omega_-(x - tv, v)dxdv \\ &\leq r_-(t)\|f\|_{E_-,\omega_-}.\end{aligned}$$

Hence, from the vanishing property of $r_-(t)$, for any $\epsilon > 0$, we have $\|u_t\|_{E_-,\omega_-} < \epsilon$ for some t large enough, while $V_0^-(t)u_t = f$. Consequently, $f \in X_\infty$. Since f has been picked from a dense subset, X_∞ is dense in $L^1_{\omega_-}(E_-)$ and Theorem 1.11.6 concludes. □

Corollary 4.4.3 *The Lax and Phillips Representation Theorem holds for streaming free group* e^{tT_0} *in* $X := L^1_\omega(\mathbb{R}^n \times V)$.

Proof The properties (i) and (ii) of Theorem 4.1.1 can be obtained readily as in Remark 4.1.4. But for (iii) we have even more; in fact, we will prove that there exists $\varphi_+ \in L^1_{\omega_+}(E_+)$ such that instead of (4.4) we have

$$\overline{\bigcup_{t\le 0} e^{tT_0} D_+} = \overline{\bigcup_{t\le 0, s\ge 0} e^{tT_0} V_0^+(s)\{\varphi_+\}} = X. \tag{4.17}$$

In fact for any $\epsilon > 0$ and any $f \in L^1_\omega(\mathbb{R}^n \times V)$ there exists $g \in C_0^\infty(\mathbb{R}^n \times V)$ such that $\|f - g\|_{\mathbb{R}^n\times V,\omega} < \frac{\epsilon}{2}$. Let t be large enough such that $h = e^{tT_0} g \in L^1_{\omega_+}(E_+) = D_+$. From the hypercyclicity of $V_0^+(s)$ it follows that there exists $s \ge 0$ such that $\|V_0^+(s)\varphi_+ - h\|_{E_+,\omega_+} < \epsilon(r_+(t) + r_-(t))^{-1}/2$. Hence,

$$\|f - U_0(-t)V_0^+(s)\varphi_+\|_{\mathbb{R}^n\times V,\omega} \le \frac{\epsilon}{2} + \|g - U_0(-t)V_0^+(s)\varphi_+\|_{\mathbb{R}^n\times V,\omega}$$

$$\le \frac{\epsilon}{2} + (r_+(t) + r_-(t))\|e^{tT_0} g - V_0^+(s)\varphi_+\|_{\mathbb{R}^n\times V,\omega} \le \epsilon.$$

The same argument holds for E_- and we can write

$$\overline{\bigcup_{t\ge 0} e^{tT_0} D_-} = \overline{\bigcup_{t\ge 0, s\ge 0} e^{tT_0} V_0^-(s)\{\varphi_-\}} = X. \tag{4.18}$$

□

4.5 Notes and Comments

According to our notations, E_ρ^+ and E_ρ^- are the phase space subspaces which are free forward and free backward set points that P. Lax and R. Phillips used in their seminal treatise [76]. In Chap. 7, we will define the Koopman formalism for studying the scattering in quantum transport theory. In [76, Theorem I.6.2] the measure preserving of this formalism is used to show that the set of points trapped in one direction but not in the other has measure zero and by defining $D_\rho^\pm = \{f \in X \ : \ \text{supp}\ f \subset E_\rho^\pm\}$ they show their representation Theorem (4.1.1) for the dissipative hyperbolic systems.

In the introduction of their book (see also [77]), as an example, they have taken $U(t)$ as the free streaming group for the transport equation in $L^2(\mathbb{R}^3 \times S^2)$ and they have proved their representation theorem for transport phenomena. In [44] I have proved this representation theorem for full transport dynamics and for these dynamics I have constructed $Z(t)$, the Lax and Phillips semigroups as in [76], but unfortunately my arguments were not complete in some points. For example, to prove the finiteness of the point spectrum of B, the generator of $Z(t)$, I have claimed that $Z(t)$ is a differentiable semigroup as it holds for symmetric hyperbolic systems (see [76]). In [110] Stefanov remarked that this is not true and he proved the finiteness of $\sigma_p(B)$ by using the compactness of $\mathcal{Z}(t)$ for some large t, where $\mathcal{Z}(t)$ is a slight modification of $Z(t)$ as we have defined in Sect. 4.2. In this section we gave the proof of eventual compactness of $\mathcal{Z}(t)$ by using the simple and natural argument of Greiner [58] which works in $L^1(\mathbb{R}^n \times V)$. Section 4.3 is a rectification of the author's paper [44] which is augmented with some new results on the hyperbolicity of the Lax and Phillips semigroup. The results of Sect. 4.4 follow closely Emamirad [46].

Chapter 5
Scattering Theory for a Charged Particle Transport Problem

5.1 Derivation of the Charged Particle Transport Equation

The physical problem concerns the time evolution of the distributions of neutral and ionized atoms of equal atomic weights, taking into account only interactions giving rise to a charge exchange; that is, in a collision between an ionized atom and a neutral one, the first becomes neutral and the latter becomes ionized. This kind of phenomenon is important in mixtures, or plasmas, with low densities, so it is allowable to neglect the ion–ion and neutral–neutral interactions. This phenomenon is explained by the charge exchange effect; A_1 and A_2 being atoms of the same mass, a particle of type A_1 which is ionized and a particle of type A_2 which is neutral react producing a neutral particle of type A_1 and a charged particle of type A_2. This may be represented by the formula

$$A_1^+ + A_2 \to A_1 + A_2^+.$$

In [53], it is explained why the charge exchange without change in the sum of the internal energies of the two colliding particles influences the dynamics of the plasma and of its single components. Moreover, the sum of the kinetic energies of the two particles remains unchanged, so we can consider these collisions to be elastic. We assume also that the charge density vanishes everywhere, but we disregard the electron distribution. We denote by $f = f(x, v, t)$ the distribution of neutral particles and by $g = g(x, v, t)$ the distribution of ions; $x \in \mathbb{R}^3$ denotes the position of the particle in the configuration space, $v \in V \subset \mathbb{R}^3$ denotes the velocity of the particle, $t \in \mathbb{R}$ denotes the time. The balance equations are

$$\frac{\partial f}{\partial t} + v \cdot \nabla_x f = P_f - R_f, \tag{5.1}$$

$$\frac{\partial g}{\partial t} + v \cdot \nabla_x g = P_g - R_g, \tag{5.2}$$

H. Emamirad, *Scattering Theory for Transport Phenomena*, Mathematical Physics Studies, https://doi.org/10.1007/978-981-16-2373-8_5

where P_f, P_g, R_f, R_g denote the production and removal rates for neutral and ionized particles. For sake of simplicity, but allowing for preservations of mass, momentum, and energy during the single collision, under the assumption of equal mass for ions and neutral atoms, we put

$$P_f(x, v, t) = \sigma g(x, v, t) \int_V f(x, v', t) dv' = R_g(x, v, t), \tag{5.3}$$

$$P_g(x, v, t) = \sigma f(x, v, t) \int_V g(x, v', t) dv' = R_f(x, v, t). \tag{5.4}$$

The removal rate R_f of neutral particle (f) is modeled with a total cross-section $\sigma(x, t) = \sigma$ times the total number ionized particles (g) present in x at time t, and it is proportional to the ratio of the removal rate and the density of (f) particles. Clearly we have to introduce the dimensional factor σ that in the sequel will be put equal to one. During the collision between two particles of different types, the charged one yields its charge to the other (the neutral one) remaining left neutral. The neutral one obviously becomes charged. The velocities of the particles are assumed not to change. Therefore $P_f = R_g$ and $P_g = R_f$.

Here V is a measurable not necessarily bounded set of $\mathbb{R}^3$ whose measure $m(V)$ is finite: $m(V) \leq \infty$. The integro-differential equations are supplied with initial data for the two distributions. We can consider the corresponding abstract Cauchy problem in the space $L^1(\mathbb{R}^3 \times V)$. By putting $h = f + g$ we obtain the system

$$\frac{\partial h}{\partial t} + v \cdot \nabla_x h = 0, \tag{5.5}$$

$$\frac{\partial f}{\partial t} + v \cdot \nabla_x f = -\sigma f \int_V h dv' + \sigma h \int_V f dv', \tag{5.6}$$

showing that the total density h evolves as a pure streaming for $t \in \mathbb{R}$. Its evolution depends on the initial distributions of ionized and neutral atoms. Equation (5.5) is the advection equation and its corresponding Cauchy problem will be denoted by

$$\textbf{(AE)} \qquad \begin{cases} \dfrac{dh}{dt} = T_0 h := -v \cdot \nabla_x h, \\ h(., 0) = h_0 \in X. \end{cases}$$

Equation (5.6) for the neutral (or ion) density looks like a linear transport equation with time dependent absorption and production terms, which really are determined by the initial distributions.

Taking $\sigma = 1$, the Eq. (5.6) can be considered as a nonautonomous transport problem in the following abstract form:

$$\textbf{(NTP)} \qquad \begin{cases} \dfrac{df}{dt} = [T_0 + A(t) + K(t)] f(t) \\ f(s) = \varphi \in X, \quad s \in \mathbb{R}, \end{cases}$$

where

$$[A(t)f](x, v) := -\Big(\int_V h_0(x - tv', v')dv'\Big)f(x, v) \qquad \text{for all } (x, v) \in \mathbb{R}^3 \times V \tag{5.7}$$

and

$$[K(t)f](x, v) = h_0(x - tv, v)\int_V f(x, v')dv' \qquad \text{for all } (x, v) \in \mathbb{R}^3 \times V. \tag{5.8}$$

By writing $T(t) := T_0 + A(t) + K(t)$, the problem (**NTP**) can be reformulated as a nonautonomous Cauchy problem

$$\textbf{(NCP)} \qquad \begin{cases} \dfrac{df}{dt} = T(t)f(t) \\ f(s) = \varphi \in X, \quad s \in \mathbb{R}, \end{cases}$$

on a Banach space X.

5.2 Strongly Continuous Propagator for Charged Particle Transport Equation

Put $X := L^1(\mathbb{R}^3 \times V)$, $Y := UCB(\mathbb{R}^3 \times V)$; they are the complex Banach spaces of (classes of) Lebesgue summable functions and of uniformly continuous bounded functions respectively, with the usual supremum norm. They have the structure of Banach lattices with positive cones consisting of the nonnegative functions. Let $C_0^1(\mathbb{R}^3 \times V)$ be the space of continuously differentiable functions with compact support in $\mathbb{R}^3 \times V$, where $V \subset \mathbb{R}^3$ is the velocity space. In Sect. 3.2, we have seen that the streaming operator $T_0 h := -v \cdot \nabla_x h$, defined in $C_0^1(\mathbb{R}^3 \times V)$ has a closure in X denoted again by T_0, whose domain is $D(T_0) \subset C_0^1(\mathbb{R}^3 \times V)$. This closed operator generates a one-parameter strongly continuous group $\{e^{tT_0}\}_{t\in\mathbb{R}}$ of bounded operators in X acting as

$$(e^{tT_0}f)(x, v) = f(x - vt, v), \tag{5.9}$$

having the properties, for any $t \in \mathbb{R}$

$$\|e^{tT_0}f\|_X = \|f\|_X, \tag{5.10}$$

and it preserves the cone of positive functions. In the space Y we define the streaming operator $\widetilde{T_0}$ on $D(\widetilde{T_0})$

$$D(\widetilde{T_0}) = \{f \in Y : v \cdot \nabla_x f \in Y\},$$
$$(\widetilde{T_0}f)(x, v) = -v \cdot \nabla_x f(x, v).$$

It generates an isometric positive one parameter strongly continuous group of linear operators in Y acting as (5.9).

In either space X and Y the abstract Cauchy problem

$$\begin{cases} \dfrac{dh}{dt} = T_0 h, \\ h(s) = h_0 \in D(T_0), \end{cases}$$

where T_0 stands for T_0 itself or for $\widetilde{T}_0$, according to the space, has one and only one solution for any $t, s \in \mathbb{R}$, given by

$$h(t) = U_0(t,s)h(s) = e^{(t-s)T_0}h(s) = h_0(x - (t-s)v, v).$$

We recall that by a solution in X, or in Y, we mean a function with values in X, or in Y such that $h(t) \in D(T_0)$, h is strongly continuously differentiable for $t \in \mathbb{R}$, the differential equation and the initial conditions at the starting time s are satisfied. Some definitions are in order to present the analysis of evolution systems. Let $h_0 \in D(T_0) \cap Y^+$ (Y^+ being the positive cone in Y). For $t \in \mathbb{R}$, we define

$$H(x,t) := \int_V h_0(x - wt, w)dw, \tag{5.11}$$

then for the operators $A(t)$ and $K(t)$ defined by (5.7) and (5.8) we have

$$(A(t)f)(x,v) := -H(x,t)f(x,v)$$

and

$$(K(t)f)(x,v) := h_0(x - vt, v)\int_V f(x,w)dw.$$

The following lemma is necessary for the well-posedness of the problem (**NCP**).

Lemma 5.2.1 *Under the preceding assumptions and definitions, it follows that for any $f \in X$:*

(i) $H(x,t) = \int_V e^{t\widetilde{T}_0}h_0(x,w)dw$;

(ii) $\|A(t)f\|_X \leq m(V)\|h_0\|_Y\|f\|_X$ *and* $\|A(t)\| = \sup_x \int_V h_0(x - wt, w)dw$, *where $m(V)$ is the measure of V;*

(iii) $A(t)f$ *is continuously differentiable with respect to $t \in \mathbb{R}$,*

$$\frac{d}{dt}A(t)f(x,v) = -f(x,v)\int_V (e^{t\widetilde{T}_0}\widetilde{T}_0 h_0)(x,w)dw; \tag{5.12}$$

and

$$\left\|\frac{d}{dt}A(t)f\right\|_X \leq m(V)\|T_0 h_0\|\|f\|_X; \tag{5.13}$$

(iv) $[K(t)f](x, v) = [e^{t\widetilde{T}_0} h_0(x, v)] \int_V f(x, w)dw$;
(v) $\|K(t)f\|_X \le m(V)\|h_0\|_Y \|f\|_X$ *and* $\|K(t)\| = m(V) \sup_{(x,v)} h_0(x, v)$;
(vi) $K(t)f$ *is continuously differentiable with respect to* $t \in \mathbb{R}$

$$\frac{d}{dt} K(t)f(x, v) = \int_V f(x, w)dw(e^{t\widetilde{T}_0}\widetilde{T}_0 h_0)(x, v) \tag{5.14}$$

and

$$\left\| \frac{d}{dt} K(t)f \right\|_X \le m(V)\|T_0 h_0\| \|f\|_X. \tag{5.15}$$

Proof Since $h_0 \in Y$, parts (i) and (iv) are obvious consequences of the definitions. Parts (ii) and (v) are simple estimations.

For (iii), let $\tau \in \mathbb{R}$, by (i), we have

$$\begin{aligned} \tau^{-1}[H(x, t+\tau)f(x, v) - H(x, t)f(x, v)] - f(x, v) \int_V \left[e^{t\widetilde{T}_0}\widetilde{T}_0 h_0\right](x, w)dw \\ = f(x, v) \int_V e^{t\widetilde{T}_0} \left[\tau^{-1}\left(e^{\tau\widetilde{T}_0} - I\right) h_0 - \widetilde{T}_0 h_0\right](x, w)dw. \end{aligned}$$

Since $h_0 \in D(\widetilde{T}_0)$, for any $\varepsilon > 0$,

$$\left| e^{t\widetilde{T}_0} \left[\tau^{-1}\left(e^{\tau\widetilde{T}_0} - I\right) h_0 - \widetilde{T}_0 h_0\right](x, w) \right| < \varepsilon$$

if $0 < |\tau| < \delta(\varepsilon)$, uniformly with respect to (x, w). Thus we have

$$\begin{aligned} &\|\tau^{-1}[A(x, t+\tau)f - A(x, t)f] + \int_V \left[e^{t\widetilde{T}_0}\widetilde{T}_0 h_0\right](x, w)dwf\|_X \\ &\le \int_V \int_{\mathbb{R}^3 \times V} \left| e^{t\widetilde{T}_0} \left[\tau^{-1}\left(e^{\tau\widetilde{T}_0} - I\right) h_0 - \widetilde{T}_0 h_0\right](x, w) \right| \\ &\quad \times |f(x, w)|dxdwdv \\ &\le \varepsilon m(V)\|f\|_X, \end{aligned}$$

and consequently we get (5.12). The estimation (5.13) is a simple consequence of isometry of $e^{t\widetilde{T}_0}$ in Y. Lastly, by using the continuity of $e^{t\widetilde{T}_0}\widetilde{T}_0 h_0$ at $t = 0$ and the isometry of $e^{t\widetilde{T}_0}$ in Y, we see that for any $\epsilon > 0$, the X-norm of

$$\frac{d}{dt} A(t+\tau)f(x, v) - \frac{d}{dt} A(t)f(x, v) = -f(x, v) \int_V \left[e^{t\widetilde{T}_0}\left(e^{\tau\widetilde{T}_0} - I\right) \widetilde{T}_0 h_0\right](x, w)dw$$

is bounded by $\epsilon m(V)\|f\|_X$, for $|\tau|$ sufficiently small. This implies the continuity of $\frac{d}{dt}A(t)$.

For the proof of (vi), let $\delta \in \mathbb{R}$; by (iv), we have

$$\delta^{-1}[K(t+\delta)f(x,v) - K(t)f(x,v)] - \left[e^{t\widetilde{T}_0}\widetilde{T}_0 h_0\right](x,v)\int_V f(x,w)dw$$
$$= \left[\delta^{-1}\left(e^{(t+\delta)\widetilde{T}_0} - e^{t\widetilde{T}_0} - \delta e^{t\widetilde{T}_0}\widetilde{T}_0\right)h_0\right](x,v)\int_V f(x,w)dw.$$

Thus, for any $\varepsilon > 0$

$$\|\delta^{-1}[K(t+\delta)f - K(t)f] - \int_V f(\cdot,w)dw\left[e^{t\widetilde{T}_0}\widetilde{T}_0 h_0\right]\|_X$$
$$\leq \int_{\mathbb{R}^n\times V}\int_V |f(x,w)|dw\left|\delta^{-1}\left(e^{(t+\delta)\widetilde{T}_0} - e^{t\widetilde{T}_0} - \delta e^{t\widetilde{T}_0}\widetilde{T}_0\right)h_0(x,v)\right|dxdv$$
$$\leq \varepsilon m(V)\|f\|_X,$$

and as before we get (5.14) and (5.15).

Finally, for continuity of $\frac{d}{dt}K(t)$ on $\mathbb{R}$, we have

$$\frac{d}{dt}K(t+\delta)f(x,v) - \frac{d}{dt}K(t)f(x,v) = \left[\left(e^{(t+\delta)\widetilde{T}_0} - e^{t\widetilde{T}_0}\right)\widetilde{T}_0 h_0\right]\int_V f(x,w)dw,$$

whose X-norm is bounded by $\varepsilon m(V)\|f\|_X$ for any $\varepsilon > 0$, provided that δ is small enough. □

Theorem 5.2.2 *Under the preceding assumptions and definitions, it follows that:*

(i) $T_0 + A(t)$ *generates an exponentially bounded strongly continuous propagators* $U_a(t,s)$ *given by*

$$[U_a(t,s)f](x,v) = \exp\left\{-\int_s^t H(x-(t-r)v,r)dr\right\}f(x-(t-s)v,v), \tag{5.16}$$

preserves the order and satisfies

$$\|U_a(t,s)f\|_X \leq \begin{cases}\|f\|_X & \text{for } t \geq s\\ \|f\|_X\exp(\|h_0\|_Y m(V)|t-s|) & \text{for } t \leq s,\end{cases} \tag{5.17}$$

(ii) $T_0 + A(t) + K(t)$ *generates a unitary strongly continuous propagator* $U(t,s)$ *given by*

$$U(t,s) = U_a(t,s) + \int_s^t U_a(t,r)K(r)U(r,s)dr. \tag{5.18}$$

Proof Since $h \in Y$, (iii) of Lemma 5.2.1 implies that $A(\cdot) \in \mathcal{C}_b(\mathbb{R}, \mathcal{L}(X))$ and by virtue of Corollary 1.12.3 such a result is enough to prove that $T_0 + A(t)$ generates an exponentially bounded strongly continuous propagator $U_a(t,s)$ on X. To see that this propagator is defined by (5.16), we use the uniqueness part of Theorem 1.12.1 by verifying

$$\partial U_a(t,s)f/\partial t = [-v\cdot\nabla_x - H(x,t)]U_a(t,s)f = [T_0 + A(t)]U_a(t,s)f,$$
$$\partial U_a(t,s)f/\partial s = [v\cdot\nabla_x + H(x,t)]U_a(t,s)f = -[T_0 + A(s)]U_a(t,s)f.$$

Now we note that (5.16) implies the order preserving of $U_a(t,s)$ and (5.17) is a simple estimation of (5.16).

To verify that (ii) holds, we use once more the uniqueness part of Theorem 1.12.2. In fact, according to (vi) of the previous lemma and Theorem 1.12.2, $T(t) := T_0 + A(t) + K(t)$ generates another evolution system on X, say $U(t,s)$. Finally, Eq. (5.18) is the version (1.36) of Duhamel's formula. □

Theorem 5.2.3 *For any $t \geq s$ one has:*

(a) $U(t,s)$ preserves the order (i.e. $U(t,s)\phi \geq 0$ when $\phi \geq 0$);
(b)

$$\|U(t,s)\| = 1. \tag{5.19}$$

Proof (a) First we note that if we explicitly show the t-dependence of f, (5.18) implies

$$\begin{aligned} f(x,v,t) &= U(t,s)\phi(x,v) \\ &= U_a(t,s)\phi(x,v) + \left(\int_s^t U_a(t,r)K(r)U(r,s)\phi dr\right)(x,v) \\ &= \phi(x - v(t-s), v)\exp\left(-\int_s^t H(x - v(t-r), r)dr\right) \\ &+ h_0(x - vt, v)\int_s^t \exp\left(-\int_r^t H(x - v(t-\tau), \tau)d\tau\right)\int_V f(x \\ &- v(t-r), w, r)dwdr, \end{aligned}$$

whose solution is given by the series

$$f(t) = \sum_{n=0}^{\infty} U_n(t,s)\phi$$

with

$$U_0(t,s) = U_a(t,s) \quad \text{and} \quad U_{n+1}(t,s)\phi = \int_s^t U_n(t,s)\phi.$$

These formulae show that $\phi \geq 0$, $h_0 \geq 0$ imply $U_n(t,s)\phi \geq 0$ for $t \geq s$ for any $n \in \mathbb{N}$. Hence $U(t,s)$ preserves the order for $t \geq s$.

(b) We remark that for any $f \in D(T_0) \subset X$,

$$\int_{\mathbb{R}^n \times V} T_0 f dx dv = 0, \tag{5.20}$$

since this is true for $f \in C_0^1(\mathbb{R}^n \times V)$ (indeed $\int_{\mathbb{R}^n} v.\nabla_x f dx = 0, \forall v \in V$) and also for any $f \in D(T_0)$ because of the density of $C_0^1(\mathbb{R}^n \times V)$ in $D(T_0)$. Now, if $f \geq 0, f \in D(T_0 + A(t) + K(t)) = D(T_0)$, we have

$$\begin{aligned}\frac{d}{dt}\|U(t,s)f\|_X &= \frac{d}{dt}\int_{\mathbb{R}^n \times V} U(t,s)f(x,v)dxdv \\ &= \int_{\mathbb{R}^n \times V} \frac{\partial}{\partial t} U(t,s)f(x,v)dxdv \\ &= \int_{\mathbb{R}^n \times V} [T_0 + A(t) + K(t)]U(t,s)f(x,v)dxdv \\ &= \int_{\mathbb{R}^n \times V} T_0 U(t,s)f(x,v)dxdv \\ &+ \int_{\mathbb{R}^n \times V} [A(t) + K(t)]U(t,s)f(x,v)dxdv.\end{aligned}$$

According to (5.20) the first integral is null, since for $f \in D(T_0)$, $U(,s)f \in D(T_0)$. For the second integral

$$\begin{aligned}&\int_{\mathbb{R}^n \times V} [A(t) + K(t)]U(t,s)f(x,v)dxdv \\ &\quad = -\int_{\mathbb{R}^n \times V} \left(\int_V h_0(x - wt, w)dw\right) U(t,s)f(x,v)dxdv \\ &\quad + \int_{\mathbb{R}^n \times V} h_0(x - vt, v) \int_V U(t,s)f(x,w)dwdxdv \\ &\quad = \int_{\mathbb{R}^n} [-A(x)B(x) + A(x)B(x)]dx = 0,\end{aligned}$$

where

$$A(x) = \int_V h_0(x - vt, v)dv, \qquad B(x) = \int_V U(t,s)f(x,v)dv.$$

Thus $t \geq s$, $f \geq 0$, $f \in D(T_0)$ imply

$$\frac{d}{dt}\|U(t,s)f\|_X = 0$$

and

$$\|U(t,s)f\|_X = \|U(s,s)f\|_X = \|f\|_X.$$

From the density of the subset $D(T_0) \cap X_+$ in X_+, we deduce $\|U(t,s)f\|_X = \|f\|_X$ for any $f \in X_+$. Finally, since any $f \in X$ can be split as $f = f_1 - f_2 + if_3 - if_4$ with $f_j \geq 0$ for $j = 1, ..., 4$ and since the operator $U(t,s)$ is order preserving for $t \geq s$, we conclude the proof of (5.19). □

5.3 Wave Operators for Charged Particle Transport Equation

With these preliminaries it is possible to consider in the space $\mathcal{X} = X^2$ the Cauchy problem for the densities of neutral and ionized particles

$$\textbf{(PTP)} \qquad \begin{cases} \dfrac{df}{dt} = T_0 f + P_f - R_f, & f(0) = f_0, \\ \dfrac{dg}{dt} = T_0 g + P_g - R_g, & g(0) = g_0, \end{cases}$$

where P_f, P_g, R_f, R_g are the bilinear operators defined previously in (5.3) and (5.4), with domains $D(T_0)^2 \subset \mathcal{X}$ where $f_0, g_0 \in D(T_0) \cap X^+ \cap D(\widetilde{T}_0)$. If such a system admits a solution, that is a pair of functions f and g with values in X which are strongly continuously differentiable with respect to $t \in \mathbb{R}$, belong to $D(T_0)$, and satisfy the differential equations and initial conditions, then the pair f and $h := f + g$ satisfy

$$\begin{cases} \dfrac{\partial f}{\partial t} = T_0 f + h \int_V f dw - f \int_V h dw, & f(0) = f_0, \\ \dfrac{\partial h}{\partial t} = \widetilde{T}_0 h, \quad h(0) = h_0 = f_0 + g_0 \end{cases}$$

not only in $\mathcal{X} = X^2$, but in $X \times Y$. This consideration enables us to use Lemma 5.2.1 for $h_0 \in Y$. The results quoted above assure that $h(x, v, t) = h_0(x - vt, v)$ and consequently that the Cauchy problem (**PTP**) has a unique strongly continuously differentiable solution defined for any $t \in \mathbb{R}$, $f_0, g_0 \in D(T_0) \cap D(\widetilde{T}_0) \cap X^+$ given by

$$f(t) = U(t, 0) f_0,$$
$$g(t) = U(t, 0) g_0.$$

Since $\|f(t)\|_X = \|f_0\|_X$ and from (5.10) $\|h(t)\|_X = \|h_0\|_X$ and the norm of L^1 is positively additive, it follows that $\|g(t)\|_X = \|g_0\|_X$. Note that $U(t, 0)$ depends on $h_0 = f_0 + g_0$.

In order to define the time dependent wave operators we consider as the unperturbed dynamics those driven by the family $U_0(t, s) := e^{(t-s)T_0}$, $t, s \in \mathbb{R}$, and as the perturbed dynamics those driven by the family $U(t, s)$. The time dependent wave operators are defined for $s \in \mathbb{R}$, $h_0 \in D(T_0) \cap D(\widetilde{T}_0) \cap X^+$, $f \in X$ by

$$W_+(s) f := \lim_{t \to \infty} U_0(s, t) U(t, s) f,$$
$$W_-(s) f := \lim_{t \to -\infty} U(s, t) U_0(t, s) f,$$

provided the limits exist with respect to the norm of the space X. We note that the function $t \mapsto U(t, s) f$ behaves like the function $t \mapsto U_0(t, s) W_+(s) f$ for $t \to +\infty$, and the function $t \mapsto U_0(t,s)f$ behaves like the function $t \mapsto U(t,s)W_-(s)f$ for $t \to -\infty$. We are able to determine sufficient conditions on h_0 in order that the wave operators exist. For this we have to announce two lemmas.

Lemma 5.3.1 *When $f \geq 0$ and $t \geq s$,*

$$\left\| \int_s^t U_0(s,r)[-A(r)]U(r,s)f dr \right\|_X = \left\| \int_s^t U_0(s,r)K(r)U(r,s)f dr \right\|_X .$$

Proof It follows from positivity of the operators $U_0(s,r)$, $-A(r)$, and $K(r)$ for $s, r \in \mathbb{R}$, and $U(r,s)$ for $r \geq s$ that for $f \geq 0$ and $t \geq s$

$$\begin{aligned}
&\| \int_s^t U_0(s,r)[-A(r)]U(r,s)f dr \|_X \\
&\quad = \int_s^t \|U_0(s,r)[-A(r)]U(r,s)f\|_X \, dr \\
&\quad = \int_s^t \|[-A(r)]U(r,s)f\|_X \, dr \\
&\quad = \int_s^t dr \int_{\mathbb{R}^n \times V} H(x,r)[U(r,s)f](x,v) dx dv \\
&\quad = \int_s^t dr \int_{\mathbb{R}^n} \left\{ \int_V h_0(x - wr, w) dw \int_v [U(r,s)f](x,v) dx dv \right\} \\
&\quad = \int_s^t dr \int_{\mathbb{R}^n \times V} h_0(x - wr, w) \int_V [U(r,s)f](x,v) dv dx dw
\end{aligned}$$

and

$$\begin{aligned}
&\| \int_s^t U_0(s,r)K(r)U(r,s)f dr \|_X \\
&\quad = \int_s^t \|U_0(s,r)K(r)U(r,s)f\|_X dr \\
&\quad = \int_s^t \|K(r)U(r,s)f\|_X dr \\
&\quad = \int_s^t dr \int_{\mathbb{R}^n \times V} h_0(x - vr, v) \int_V [U(r,s)f](x,w) dw dx dv.
\end{aligned}$$

This shows that the two norms are equal. □

Lemma 5.3.2 *For any $f \in X$ and $t \geq s$,*

$$\int_s^t \|K(r)U(r,s)f\|_X dr \leq 4(e^{\alpha_+}-1)\|f\|_X.$$

Proof For $f \geq 0$ and $s \leq r \leq t$,

$$U(t,s)f = U(t,r)U(r,s)f \geq U_a(t,r)U(r,s)f.$$

We recall that $U_a(r,t) \geq 0$ even if $r \leq t$. So we obtain

$$\begin{aligned} U(r,s)f &\leq U(r,t)U(t,s)f \\ &= \exp\left\{-\int_t^r H(x-w(r-r'),r')dr')dr'\right\} U_0(r,t)U(t,s)f(x,w). \end{aligned}$$

By putting $x - w(r-t) = y$, $w = w'$, we obtain

$$\begin{aligned} &\int_s^t \|K(r)U(r,s)f\|_X dr \\ &\quad = \int_s^t \|A(r)U(r,s)f\|_X dr \\ &\quad \leq \int_s^t \int_{\mathbb{R}^n \times V} H(x,r) \exp\left\{-\int_t^r H(x-w(r-r'),r')dr'\right\} \\ &\qquad \times [U_0(r,t)U(t,s)f](x,w)dxdwdr \\ &\quad = \int_{\mathbb{R}^n \times V} dxdw \int_s^t H(x,r) \exp\left\{-\int_t^r H(x-w(r-r'),r')dr'\right\} \\ &\qquad \times [U_0(r,t)U(t,s)f](x,w)dr \\ &\quad = \int_{\mathbb{R}^n \times V} dydw' \int_s^t H(y+w'(r-t),r) \\ &\qquad \times \exp\left\{-\int_t^r H(y+w'(r-t)-w'(r-r'),r')dr'\right\} [U(t,s)f](y,w')dr \\ &\quad = \int_{\mathbb{R}^n \times V} \left[\exp\left\{\int_s^t H(y+(r'-t)w',r')dr'\right\} - 1\right] [U(t,s)f](y,w')dydw' \\ &\quad \leq (e^{\alpha_+}-1)\|U(t,s)f\|_X \leq (e^{\alpha_+}-1)\|f\|_X, \end{aligned}$$

where $\alpha_+ = \int_s^\infty \sup_{x\in\mathbb{R}^3} H(x,r)dr$. □

Theorem 5.3.3 *Let $h_0 \in D(T_0) \cap D(\widetilde{T}_0) \cap X_+$ and $H(x,t)$ be defined as in (5.11). Then for all $f \in X$, and all $s \geq 0$, the wave operator $W_+(s)f$ exists if*

$$\alpha_+ := \int_s^{+\infty} \sup_{x\in\mathbb{R}^3} H(x,r)dr < \infty, \tag{5.21}$$

and $W_-(s)f$ exists if

$$\alpha_- := \int_{-\infty}^{s} \sup_{x\in\mathbb{R}^3} H(x,r)dr < \infty. \tag{5.22}$$

Proof By using the Duhamel formula

$$U(t,s) = U_0(t,s) + \int_s^t U_0(t,r)[A(r)+K(r)]U(r,s)dr$$

for any $f \in X$ and any $t \geq s$, we can write

$$\begin{aligned} U_0(s,t)U(t,s)f &= U_0(s,t)U_0(t,s) \\ &\quad + U_0(s,t)\int_s^t U_0(t,r)[A(r)+K(r)]U(r,s)f dr \\ &= f + \int_s^t U_0(s,r)[A(r)+K(r)]U(r,s)f dr. \end{aligned}$$

From isometry of the propagators $U_0(t,s)$ for all $t, s \in \mathbb{R}$ and $U(t,s)$ for $t \geq s$ in X (see Theorems 3.2.1 and 5.2.3) it follows that for any $f \geq 0$

$$\|U_0(s,t)U(t,s)f\|_X = \|f\|_X.$$

This identity can be extended to the whole of X and gives

$$\|U_0(s,t)U(t,s)f\|_X \leq 4\|f\|_X, \tag{5.23}$$

which implies the uniform boundedness of $U_0(s,t)U(t,s)$ with respect to t. For the existence of $W_+(s)$ we use the time dependent version of Cook's Lemma and for the existence of $W_-(s)$ instead of (5.23) we can write

$$\|U(s,t)U_0(t,s)f\|_X \leq 4\|f\|_X. \tag{5.24}$$

for any $f \in X$ and any $t \leq s$ and Duhamel's formula gives

$$U(s,t)U_0(t,s)f = f + \int_t^s U_0(s,r)[A(r)+K(r)]U(r,t)U_0(t,s)f dr$$

and by the same strategy of the first part we obtain for any $f \geq 0$

$$
\begin{aligned}
&\| \int_t^s U_0(s,r)[-A(r)]U(r,t)U_0(t,s)f dr \|_X \\
&\quad = \| \int_t^s U_0(s,r)K(r)U(r,t)U_0(t,s)f dr \|_X \\
&\quad = \int_t^s \|U_0(s,r)[-A(r)]U(r,t)U_0(t,s)f\|_X dr \\
&\quad = \int_t^s \|U_0(s,r)K(r)U(r,t)U_0(t,s)f\|_X dr \\
&\quad \le \int_{\mathbb{R}^n \times V} \left[\exp\left\{ \int_t^s H(y+(r'-s)w', r')dr' \right\} - 1 \right] \\
&\qquad \times [U(s,t)U_0(t,s)f](y,w')dy dw'.
\end{aligned}
$$

This implies that

$$
\begin{aligned}
\int_{-\infty}^s \|U_0(s,r)K(r)U(r,t)U_0(t,s)f\|_X dr &\le (e^{\alpha_-}-1)\|U(s,t)U_0(t,s)f\|_X \\
&\le (e^{\alpha_-}-1)\|f\|_X.
\end{aligned}
$$

As in the preceding part, an extension to the whole of X gives

$$
\int_{-\infty}^s \|U_0(s,r)K(r)U(r,t)U_0(t,s)f\|_X dr \le 4(e^{\alpha_-}-1)\|f\|_X.
$$

which enables us to use the time dependent version of Cook's Lemma and achieve the proof. □

Remark 5.3.4 By the same argument we can show that

$$
\int_s^t \|K(r)U(r,s)f\|_X dr \le 4\beta\|f\|_X, \qquad \forall f \in X, \quad \forall t \ge s,
$$

where $\beta = e^{\alpha_+} - 1$.

5.4 Characterization of the Evolution Group

Now we are in a position to specify the particular properties of the evolution group $e^{t\mathcal{G}}$ when the correspondent $U(t,s)$ is the propagator for the charged particle transport problem (**NTP**) defined in the previous sections.

Let $E := C_0(\mathbb{R}, X)$ be the space of continuous X-valued functions vanishing at $\pm\infty$, with the supremum norm. On E we define the Howland semigroup $e^{t\mathcal{G}}$ as in Sect. 1.13.

Theorem 5.4.1 *Under the assumptions of Sect. 5.2, $e^{t\mathcal{G}}$ is a strongly continuous group of isometries in E.*

Proof By virtue of (5.19) and the fact that the right translation (1.39) is isometric, it follows that $e^{t\mathcal{G}}$ is also isometric in E. □

The operators $A(t)$, $K(t)$ and consequently $B(t) = A(t) + K(t)$ are uniformly bounded in X. For $T(t) = T_0 + A(t) + K(t)$ with $D(T(t)) = D(T_0)$ and for all $f \in D(T(t))$ we have

$$\frac{\partial}{\partial t}U(t,s)f = T(t)U(t,s)f \qquad \text{and} \qquad \frac{\partial}{\partial s}U(t,s)f = -U(t,s)T(s)f.$$

Let us denote by $\mathcal{B}$ the bounded multiplication operator on E corresponding to the family $\{B(t)\}_{t\in\mathbb{R}}$ on X and by $\mathcal{T}$ the unbounded multiplication operator corresponding to the family $\{(T(t), D(T_0))\}_{t\in\mathbb{R}}$ of unbounded operators on X. The operator $\mathcal{T}$ with domain

$$D(\mathcal{T}) := \{\psi \in E \mid \lim_{t\to 0}\frac{1}{t}[U(.,.-t) - I]\psi(.) \in E\} \tag{5.25}$$

is defined by

$$[\mathcal{T}\psi](.) := \lim_{t\to 0}\frac{1}{t}[U(.,.-t) - I]\psi(.). \tag{5.26}$$

In the same manner we can define the multiplication operator $\mathcal{T}_0$ on the space E by replacing $U(t,s)$ with $U_0(t,s)$ in (5.25) and (5.26). Consequently, the Howland groups $e^{t\mathcal{G}_0}$ and $e^{t\mathcal{G}}$ can be defined by

$$e^{t\mathcal{G}_i}\psi(.) := U_i(.,.-t)\psi(.-t), \qquad \text{for} \quad \psi \in E \quad i = 0 \text{ or void}. \tag{5.27}$$

The following theorem characterizes $\mathcal{G}$, the generator of the evolution group $e^{t\mathcal{G}}$. Let us resume the notations of Sect. 1.13.

Theorem 5.4.2 *Let $h_0 \in D(T_0) \cap D(\widetilde{T_0}) \cap X^+$. With the notation above, $\mathbf{D} := D(\mathbf{d}) \cap D(\mathcal{T})$ is a core for $\mathcal{G}$ and for any $\psi \in \mathbf{D}$ we have*

$$\mathcal{G}\psi = \mathcal{T}\psi - \psi' = \mathcal{T}_0\psi + \mathcal{B}\psi. \tag{5.28}$$

Proof For $\psi \in E$ and $t > 0$, let us write

$$\delta(t) = \frac{1}{t}(e^{t\mathcal{G}}\psi - \psi) = \delta_1(t) + \delta_2(t), \tag{5.29}$$

where

$$\delta_1(t) = \frac{1}{t}U(.,.-t)(\psi(.-t) - \psi(.)) \quad \text{and} \quad \delta_2(t) = \frac{1}{t}(U(.,.-t)\psi(.) - \psi(.)).$$

For $\psi \in D(\mathbf{d}) \cap D(\mathcal{T})$, $t \to 0$, $\delta_1(t)$ converges to $-\psi'$ and $\delta_2(t)$ converges to $\mathcal{T}\psi$ in E. Hence $\delta(t)$ converges to $\mathcal{G}\psi$, which implies that $\psi \in D(\mathcal{G})$. Conversely, if $\psi \in D(\mathbf{d}) \cap D(\mathcal{G})$, then the same argument yields $\psi \in D(\mathcal{T})$. This means that

$$\mathbf{D} = D(\mathbf{d}) \cap D(\mathcal{T}) = D(\mathbf{d}) \cap D(\mathcal{G}).$$

Since $-\mathbf{d}$ and $\mathcal{G}$ are the generators of the C_0-semigroups $\mathbf{T}_0(t)$ and $e^{t\mathcal{G}}$, $D(\mathbf{d})$, $D(\mathcal{G})$ and consequently $\mathbf{D}$ are dense in E.

In order to prove that $\mathbf{D}$ is a core for $\mathcal{G}$, it suffices to show that it is invariant under the action of the semigroup $e^{t\mathcal{G}}$ (see Lemma 1.2.6). Since for $\psi \in D(\mathcal{G})$ one has $e^{t\mathcal{G}}\psi \in D(\mathcal{G})$, it suffices to show that $e^{t\mathcal{G}}\psi \in D(\mathbf{d})$ for $\psi \in \mathbf{D}$. In other words, for $\psi \in \mathbf{D}$ we shall prove that $e^{t\mathcal{G}}\psi \in E$ and $\left(e^{t\mathcal{G}}\psi\right)'(\cdot) \in E$. The fact that $e^{t\mathcal{G}}\psi \in E$ follows from (5.19). To prove $\left(e^{t\mathcal{G}}\psi\right)'(\cdot) \in E$, we write formally for $\psi \in D(\mathcal{G})$,

$$\begin{aligned}\left(e^{t\mathcal{G}}\psi\right)'(s) &= \frac{\partial}{\partial s}U(s, s-t)\psi(s-t)\\ &= \frac{\partial}{\partial \tau}U(\tau, s-t)\bigg|_{\tau=s}\psi(s-t) + \frac{\partial}{\partial \tau}U(s, \tau)\bigg|_{\tau=s-t}\psi(s-t)\\ &\quad + U(s, s-t)\psi'(s-t).\end{aligned} \tag{5.30}$$

Since

$$\begin{aligned}\left(\frac{\partial}{\partial t}e^{t\mathcal{G}}\psi\right)(s) &= \frac{\partial}{\partial t}U(s, s-t)\psi(s-t)\\ &= -\frac{\partial}{\partial \tau}U(s, \tau)\bigg|_{\tau=s-t}\psi(s-t) - U(s, s-t)\psi'(s-t),\end{aligned}$$

the last two terms of (5.30) are well-defined and belong to E, since $\frac{\partial}{\partial t}e^{t\mathcal{G}}\psi = \mathcal{G}e^{t\mathcal{G}}\psi$ and $e^{t\mathcal{G}}\psi \in D(\mathcal{G})$ when $\phi \in D(\mathcal{G})$. The first term is

$$\frac{\partial}{\partial \tau}U(\tau, s-t)\bigg|_{\tau=s}\psi(s-t) = T(s)U(s, s-t)\psi(s-t).$$

In order to prove that

$$\lim_{|s|\to\infty}\|T(s)U(s, s-t)\psi(s-t)\|_X = 0,$$

we will decompose $T(s)$ as $T(s) = T_0 + A(s) + K(s)$, and will use the uniform boundedness of $A(s)$ and $K(s)$ (see Lemma 5.2.1, (ii) and (v)), isometry of $U(s, s-t)$ for $t > 0$ and the fact that

$$\|T_0U(s, s-t)\psi(s-t)\|_X = \|U(s, s-t)T_0\psi(s-t)\|_X = \|T_0\psi(s-t)\|_X$$

goes to zero as $|s| \to \infty$, for all $\psi \in C_0(\mathbb{R}, D(T_0))$ which is dense in E. Finally (5.28) follows from (5.29) by taking $t \to 0$. □

As a consequence of this theorem we can consider the operator $\mathcal{G}$ as the closure of $\mathcal{G}$ defined by (5.28) on $\mathbf{D}$.

In Theorem 5.2.2 (i), we have shown that $T_0 + A(t)$ generates the following propagator:

$$[U_a(t,s)f](x,v) = f(x-(t-s)v, v)\exp\left\{-\int_s^t H(x-(t-r)v, r)dr\right\} \tag{5.31}$$

and that $T_0 + A(t) + K(t)$ generates a propagator $U(t,s)$ which satisfies Duhamel's formula

$$U(t,s) = U_a(t,s) + \int_s^t U_a(t,r)K(r)U(r,s)dr.$$

Hence, for $\psi \in E$,

$$U(s,s-t)\psi(s-t) = U_a(s,s-t)\psi(s-t) + \int_{s-t}^s U_0(s,r)K(r)U(r,s-t)\psi(s-t)dr. \tag{5.32}$$

Now, if we denote by $\{e^{t\mathcal{G}_a}\}_{t\in\mathbb{R}}$ the evolution group given by

$$[e^{t\mathcal{G}_a}\psi](.) = U_a(.,.-t)\psi(.-t), \quad \text{for all} \quad \psi \in E,$$

by putting $\tau = r - s + t$ in the integral of (5.32), we obtain Duhamel's formula for $e^{t\mathcal{G}}$; namely

$$e^{t\mathcal{G}} = e^{t\mathcal{G}_a} + \int_0^t e^{(t-\tau)\mathcal{G}_a}\mathcal{K}e^{\tau\mathcal{G}}\,d\tau, \tag{5.33}$$

where $\mathcal{K}$ is the bounded multiplication operator corresponding to the family $\{K(t)\}_{t\in\mathbb{R}}$. In the same way we can write

$$e^{t\mathcal{G}} = e^{t\mathcal{G}_0} + \int_0^t e^{(t-\tau)\mathcal{G}}\mathcal{B}e^{\tau\mathcal{G}_0}\,d\tau. \tag{5.34}$$

Since $U_0(t,s)$, $U_a(t,s)$ are positive for all $t, s \in \mathbb{R}$ and $U(t,s)$ is positive for $t \geq s$, we obtain the positivity of $e^{t\mathcal{G}_0}$, $e^{t\mathcal{G}_a}$ for all $t \in \mathbb{R}$ and positivity of $e^{t\mathcal{G}}$ for $t \geq 0$ on the Banach lattice E. Furthermore, since $\{K(t)\}_{t\in\mathbb{R}}$ is a family of positive operators in X, for any $\psi \geq 0$ in E the function $t \mapsto \int_0^t e^{(t-\tau)\mathcal{G}_a}\mathcal{K}e^{\tau\mathcal{G}}\psi\,d\tau$ is positively increasing in E, consequently

$$e^{t\mathcal{G}} \geq e^{t\mathcal{G}_a} \qquad \text{for all} \quad t \geq 0. \tag{5.35}$$

Lemma 5.4.3 *Let $h_0 \in X^+$, and $H(x,t) = \int_V h_0(x-tv, v)dv$, then*

$$\|e^{t\mathcal{G}_a}\psi\|_E \leq \begin{cases} \|\psi\|_E & \text{if } t \geq 0, \\ \sup_{s,x,v} \exp\left\{\int_0^{-t} H(x+\sigma v, s+\sigma)\, d\sigma\right\} \|\psi\|_E & \text{if } t \leq 0. \end{cases} \tag{5.36}$$

Proof For $t \geq 0$, $\|e^{t\mathcal{G}_a}\psi\|_E \leq \|\psi\|_E$ is direct consequence of (5.31). For $t \leq 0$, from (5.31)

$$[U_a(s, s-t)f](x, v) = f(x+tv, v)\exp\left\{-\int_{s-t}^{s} H(x+(t+r-s)v, r)dr\right\}$$

and by taking $r = s + \sigma$ we obtain

$$\begin{aligned} \|[e^{t\mathcal{G}_a}\psi](s)\|_X &= \|U_0(s, s-t)\psi(s-t)\|_X \\ &\leq \|\psi(s-t)\|_X \sup_{x,v} \exp\left\{\int_0^{-t} H(x+\sigma v, s+\sigma)d\sigma\right\}. \end{aligned}$$

This implies (5.36). □

5.5 Howland's Formalism

In the context of the charged particle transport equation, suppose that two bounded evolution groups defined by (5.27) are given. Since $U_0(t, s)$ is isometric in X for all $t, s \in R$, $e^{t\mathcal{G}_0}$ is also isometric in E, and $U(t, s)$ being isometric for $t \geq s$ means that $e^{t\mathcal{G}}$ is also isometric in E for $t \geq 0$. Hence one can view the existence of two wave operators

$$\mathcal{W}_+(\mathcal{G}_0, \mathcal{G}) \equiv s - \lim_{t\to+\infty} e^{-t\mathcal{G}_0}e^{t\mathcal{G}}, \tag{5.37}$$

$$\mathcal{W}_-(\mathcal{G}, \mathcal{G}_0) \equiv s - \lim_{t\to-\infty} e^{-t\mathcal{G}}e^{t\mathcal{G}_0} \tag{5.38}$$

and the related scattering operator $\mathcal{S}$ defined by

$$\mathcal{S} := \mathcal{W}_+(\mathcal{G}_0, \mathcal{G})\mathcal{W}_-(\mathcal{G}, \mathcal{G}_0), \tag{5.39}$$

in E.

There are two ways to prove an existence theorem for (5.39). The first way is to use Theorem 5.3.3 together with the fact that for $\psi \in E$,

$$[e^{-t\mathcal{G}_0}e^{t\mathcal{G}}\psi](.) = U_0(., .+t)U(.+t, .)\psi(.) \tag{5.40}$$

and

$$[e^{-t\mathcal{G}}e^{t\mathcal{G}_0}\psi](.) = U(., .+t)U_0(.+t, .)\psi(.) \tag{5.41}$$

by taking $t \to \pm\infty$ in (5.41). These considerations show that the wave operators $\mathcal{W}_\pm$ are the multiplication operators by $W_\pm(.)$ and the scattering operator $\mathcal{S}$ is therefore the multiplication operator by $S(.)$. If we denote by $S = S(0)$ the usual scattering operator which is referred to initial time $t = 0$, then $\mathcal{U}_0 \mathcal{S} \mathcal{U}_0^{-1}$ is a multiplication by S.

The second way is to apply directly the smooth perturbation theory for the existence of $\mathcal{W}_\pm$. The uniform sufficient conditions for the existence of $W_\pm(.)$ in X imply the existence of $\mathcal{W}_\pm$ in E. This method has the advantage of not using the regularity condition on the initial datum h_0.

Theorem 5.5.1 *Let* $h_0 \in X_+$, $H(x,t) = \int_V h_0(x - tv, v)dv$ *and*

$$\alpha_- = \sup_{(x,v,s)\in\mathbb{R}^n\times V\times\mathbb{R}} \int_{-\infty}^{0} H(x + tv, s)\, dt < \infty. \tag{5.42}$$

Then the wave operator $\mathcal{W}_-(\mathcal{G}, \mathcal{G}_0)$ *exists in* $\mathcal{L}(E)$.

Proof Suppose that for any ψ in some dense subspace of E we have

$$\int_{-\infty}^{0} \|\mathcal{B}e^{t\mathcal{G}_0}\psi\|_E dt < \infty. \tag{5.43}$$

By taking in Duhamel's formula (1.17), $C := \mathcal{G}$, $A := \mathcal{G}_0$ and replacing t by $-t$ for $t \le 0$, we get

$$e^{-t\mathcal{G}} = e^{-t\mathcal{G}_0} + \int_0^{-t} e^{r\mathcal{G}}\mathcal{B}e^{-(t+r)\mathcal{G}_0}dr.$$

Apply this formula to $e^{t\mathcal{G}_0}\psi$, we obtain

$$e^{-t\mathcal{G}}e^{t\mathcal{G}_0}\psi = \psi + \int_t^0 e^{-s\mathcal{G}}\mathcal{B}e^{s\mathcal{G}_0}\psi ds. \tag{5.44}$$

By virtue of Cook's Lemma, the existence of the limit (5.37) follows from (5.43), while $\|e^{t\mathcal{G}}\|_{\mathcal{L}(E)} = 1$ for $t \ge 0$.

In order to prove (5.43), we remark that

$$\begin{aligned} \|\mathcal{B}e^{t\mathcal{G}_0}\psi\|_E &= \sup_{s\in\mathbb{R}} \|B(s)U_0(s, s-t)\psi(s-t)\|_X \\ &= \sup_{s\in\mathbb{R}} \int_{\mathbb{R}^3\times V} |(A(s) + K(s))\psi(s-t, x - tv, v)|\, dxdv. \end{aligned}$$

Since for $h_0 \ge 0$ we have

$$\int_{\mathbb{R}^3\times V} |K(s)\psi(s-t,x-tv,v)|\,dxdv$$
$$\le \int_{\mathbb{R}^3\times V} h_0(x-sv,v)\int_V |\psi(s-t,x-tv',v')|\,dv'dxdv$$
$$= \int_{\mathbb{R}^3\times V} H(x,s)|\psi(s-t,x-tv',v')|\,dxdv'$$
$$= \int_{\mathbb{R}^3\times V} |A(s)\psi(s-t,x-tv,v)|\,dxdv,$$

the following estimate

$$\int_{-\infty}^0 \|\mathcal{B}e^{t\mathcal{G}_0}\psi\|_E dt$$
$$\le 2\sup_{s\in\mathbb{R}}\int_{-\infty}^0\int_{\mathbb{R}^3\times V} H(x,s)|\psi(s-t,x-tv,v)|\,dxdv$$
$$\le 2\left(\sup_{(x,v,s)\in\mathbb{R}^3\times V\times\mathbb{R}}\int_{-\infty}^0 H(x+tv,s)dt\right)\|\psi\|_E,$$
$$\le 2\alpha_-\|\psi\|_E \tag{5.45}$$

together with (5.42) implies the theorem. □

For the existence of $\mathcal{W}_+(\mathcal{G}_0,\mathcal{G})$ we need the following lemma.

Lemma 5.5.2 *Let $h_0\in X_+$, $H(x,t)=\int_V h_0(x-tv,v)dv$ and*

$$\alpha_+ = \sup_{(x,v,s)\in\mathbb{R}^3\times V\times\mathbb{R}}\int_0^\infty H(x+tv,s)\,dt<\infty. \tag{5.46}$$

Then the wave operator $\mathcal{W}_+(\mathcal{G}_0,\mathcal{G})$ exists iff $\mathcal{W}_+(\mathcal{G}_a,\mathcal{G})$ exists in $\mathcal{L}(E)$.

Proof Suppose that the wave operator $\mathcal{W}_+(\mathcal{G}_0,\mathcal{G}_a)$ exists. Since

$$e^{-t\mathcal{G}_0}e^{t\mathcal{G}}\psi = e^{-t\mathcal{G}_0}e^{t\mathcal{G}_a}e^{-t\mathcal{G}_a}e^{t\mathcal{G}}\psi,$$

the wave operator $\mathcal{W}_+(\mathcal{G}_0,\mathcal{G})$ exists iff $\mathcal{W}_+(\mathcal{G}_a,\mathcal{G})$ exists in $\mathcal{L}(E)$. In fact, for the existence of $\mathcal{W}_+(\mathcal{G}_0,\mathcal{G}_a)$ we write

$$\left[e^{-t\mathcal{G}_0}e^{t\mathcal{G}_a}\right]\psi(s) = U_0(s,s+t)U_0(s+t,s)\psi(s)$$
$$= \exp\left\{\int_0^t h(x+\tau v,\tau+s)\,d\tau\right\}\psi(s),$$

for all $\psi\in E$. Thus, Lebesgue's dominated convergence theorem, together with (5.46), shows the existence of $\mathcal{W}_+(\mathcal{G}_0,\mathcal{G}_a)$. □

Remark 5.5.3 As the existence of $\mathcal{W}_+(\mathcal{G}_0, \mathcal{G}_a)$ is implied by (5.46), it follows readily that this condition implies also the existence of $\mathcal{W}_-(\mathcal{G}_a, \mathcal{G}_0)$. Similarly, the existence of $\mathcal{W}_-(\mathcal{G}_0, \mathcal{G}_a)$ and $\mathcal{W}_+(\mathcal{G}_a, \mathcal{G}_0)$ follows from (5.42).

Theorem 5.5.4 *Under the hypothesis of Lemma 5.5.2, the wave operator $\mathcal{W}_+(\mathcal{G}_0, \mathcal{G})$ exists in $\mathcal{L}(E)$.*

Proof By virtue of Lemma 5.5.2, it suffices to show that $\mathcal{W}_+(\mathcal{G}_a, \mathcal{G})$ exists. The action of the semigroup $e^{-t\mathcal{G}_a}$ on $e^{t\mathcal{G}}$ given by (5.34) implies that

$$e^{-t\mathcal{G}_a}e^{t\mathcal{G}}\psi = \psi + \int_0^t e^{-s\mathcal{G}_a}\mathcal{K}e^{s\mathcal{G}}\psi ds, \quad \text{for all} \quad \psi \in E.$$

Lemma 5.4.3 yields

$$\begin{aligned}\int_0^\infty \|e^{-r\mathcal{G}_a}\mathcal{K}e^{r\mathcal{G}}\psi\|_E dr &\leq \int_0^\infty \exp\left(\int_0^r H(x+tv, t+r)dt\right)\|\mathcal{K}e^{r\mathcal{G}}\psi\|_E dr \\ &\leq e^{\alpha_+}\int_0^\infty \|\mathcal{K}e^{r\mathcal{G}}\psi\|_E dr,\end{aligned}$$

for all $\psi \in E$. On the other hand,

$$\begin{aligned}\int_0^\infty \|\mathcal{K}e^{r\mathcal{G}}\psi\|_E dr &\leq \sup_{s\in\mathbb{R}}\int_0^\infty \|K(s)U(s, s-r)\psi(s-r)\|_X dr \\ &= \sup_{s\in\mathbb{R}}\int_{-\infty}^s \|K(t+r)U(t+r, t)\psi(t)\|_X dt \\ &\leq 4\beta\|\psi\|_E,\end{aligned}$$

where $\beta = e^{\alpha_+} - 1$. Here the last inequality is shown in Remark 5.3.4. Hence the existence of the wave operator $\mathcal{W}_+(\mathcal{G}_0, \mathcal{G})$ follows from Cook's Lemma. □

5.6 Similarity Between Perturbed and Unperturbed Operators

In the previous section we have analyzed the existence of the wave operators $\mathcal{W}_+(\mathcal{G}_0, \mathcal{G})$ and $\mathcal{W}_-(\mathcal{G}, \mathcal{G}_0)$ given by (5.37) and (5.38). In the same way one can prove the existence of four wave operators $\mathcal{W}_\pm(\mathcal{G}_0, \mathcal{G})$ and $\mathcal{W}_\pm(\mathcal{G}, \mathcal{G}_0)$ given by

$$\mathcal{W}_\pm(\mathcal{G}_0, \mathcal{G}) \equiv s - \lim_{t\to\pm\infty} e^{-t\mathcal{G}_0}e^{t\mathcal{G}}, \tag{5.47}$$

$$\mathcal{W}_\pm(\mathcal{G}, \mathcal{G}_0) \equiv s - \lim_{t\to\pm\infty} e^{-t\mathcal{G}}e^{t\mathcal{G}_0}. \tag{5.48}$$

In Lemma 2.1.3 we have shown that the existence of each wave operator implies the correspondent intertwining identity and consequently the similarity of two operators $\mathcal{G}$ and $\mathcal{G}_0$ and the correspondent C_0-groups in the following sense:

$$\begin{aligned}
\mathcal{W}_{\pm}(\mathcal{G}_0, \mathcal{G})\mathcal{G}\mathcal{W}_{\pm}(\mathcal{G}, \mathcal{G}_0)f &= \mathcal{G}_0 f; \quad f \in D(\mathcal{G}_0),\\
\mathcal{W}_{\pm}(\mathcal{G}, \mathcal{G}_0)\mathcal{G}_0\mathcal{W}_{\pm}(\mathcal{G}_0, \mathcal{G})f &= \mathcal{G} f; \quad f \in D(\mathcal{G}),\\
\mathcal{W}_{\pm}(\mathcal{G}_0, \mathcal{G})e^{t\mathcal{G}}\mathcal{W}_{\pm}(\mathcal{G}, \mathcal{G}_0)f &= e^{t\mathcal{G}_0} f; \quad f \in E,\\
\mathcal{W}_{\pm}(\mathcal{G}, \mathcal{G}_0)e^{t\mathcal{G}_0}\mathcal{W}_{\pm}(\mathcal{G}_0, \mathcal{G})f &= e^{t\mathcal{G}} f; \quad f \in E.
\end{aligned}$$

In Sect. 2.3 we have shown how the existence of the four wave operators follows from the smooth pertubation. The following theorem ensures the existence of the wave operators (5.47) and (5.47).

Theorem 5.6.1 *Let* α_- *and* α_+ *be the real numbers defined in* (5.22) *and* (5.21). *If*

$$\alpha_- + \alpha_+ < 1, \tag{5.49}$$

then the four wave operators $\mathcal{W}_{\pm}(\mathcal{G}_0, \mathcal{G})$, $\mathcal{W}_{\pm}(\mathcal{G}, \mathcal{G}_0)$ *exist.*

Proof As (5.45), we can prove that

$$\int_0^{\infty} \|\mathcal{B}e^{t\mathcal{G}_0}\psi\|_E dt \le 2\alpha_+\|\psi\|_E.$$

Hence, in order to get $\alpha < 2$, where α satisfies

$$\int_{-\infty}^{\infty} \|\mathcal{B}e^{t\mathcal{G}_0}\psi\|_E dt \le \alpha\|\psi\|_E$$

and using Theorem 2.3.2, α_- and α_+ should satisfy (5.49). □

5.7 Notes and Comments

A thorough description of the derivation of the charged particles time evolution and the underlying physics can be found in [53]. The mathematical study of the corresponding Cauchy problem in the space $L^1(\mathbb{R}^3 \times V)$ and some remarks on the stability of this problem are given by Busoni [23]. The particularity of the Eq. (5.6) is that the equation is time dependent and the existence of the wave operators need some special technique. In [22], Busoni gave the conditions (5.21) and (5.22) for the existence of the time dependent wave operators and we retrieve the same conditions for the existence of the stationary wave operators. This chapter is a complete review of two papers [23] and [24]. Actually Sects. 5.2 and 5.3 are taken from [23] and Busoni and Emamirad [24] contributed the materials for Sects. 5.4, 5.5 and 5.6.

Chapter 6
Relationship Between the Albedo and Scattering Operators

6.1 Scattering Theory in Computerized Tomography

We formulate the problem of reconstructing objects from projection as an inverse problem of the following evolutionary equation:

$$\textbf{(P1)} \qquad \begin{cases} \dfrac{\partial u}{\partial t} = -\omega.\text{grad}u - \sigma(x)u \\ u(x,\omega,0) = f(x,\omega). \end{cases}$$

with some boundary condition, where x belongs to an open bounded subset Ω of $\mathbb{R}^2$, and ω to the unit circle S^1. This inverse problem is the reconstruction of an unknown function $\sigma(x)$ from its Radon transformation

$$[R\sigma](x,\omega) = \int_{-\infty}^{+\infty} \sigma(x+\omega t)dt. \tag{6.1}$$

The first mathematical solution to the problem of reconstructing a function from its projections was given by Radon [101] in 1917. Today, there are high-performance numerical methods which permit reversal of the formula (6.1) and calculation of the value of $\sigma(x)$ with high precision. In computerized tomography what we know experimentally is not directly the value of the Radon transformation but the measurement of the outgoing flux as a function of the incoming flux of the x-rays. In fact, if we denote by $\partial\Omega$ the boundary of Ω and by $f(x,\omega)$ the density of the particles at the point $x \in \partial\Omega$ of the source, a receiver placed at the point $x^* \in \partial\Omega$ at the other side of x and on the ray $\{x_0 + t\omega : t \in \mathbb{R}\}$, receives the intensity $g(x,\omega)$ from the point (x,ω) in phase space. The Radon transformation can be calculated from the following formula (Fig. 6.1):

$$[R\sigma](x,\omega) = \ln \frac{f(x_0,\omega)}{g(x,\omega)}. \tag{6.2}$$

H. Emamirad, *Scattering Theory for Transport Phenomena*, Mathematical Physics Studies, https://doi.org/10.1007/978-981-16-2373-8_6

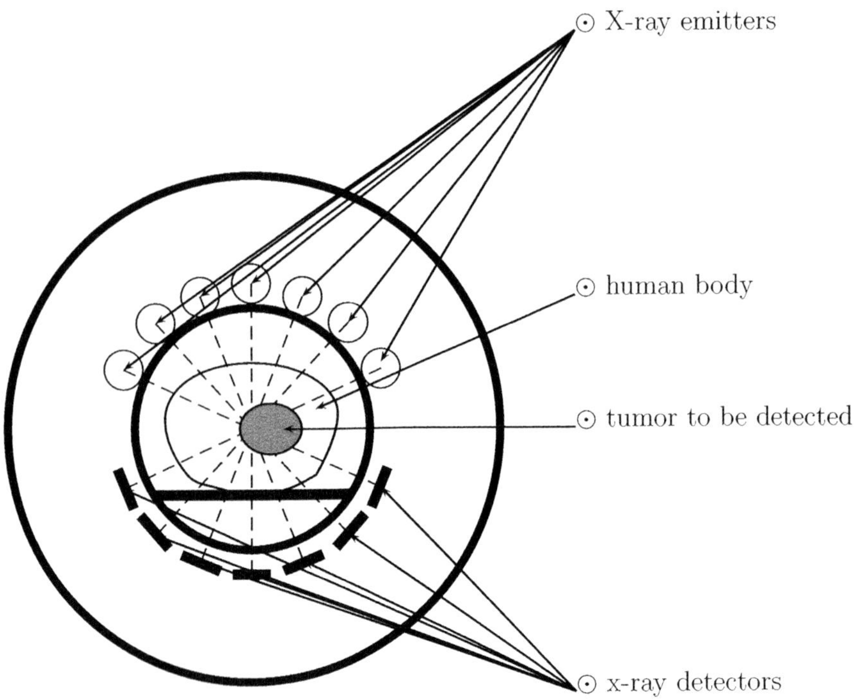

Fig. 6.1 Basic operation of a CT scan

In 1980, Hejtmanek [62] gave an interesting justification of the above formula by replacing g by Sf. The connection between computerized tomography and the problem (**P1**) can be derived as follows: Supposing that the particles (here the photons) do not meet any absorbing obstacle, the dynamic of the system is defined by

$$U_0(t) \;:\; f(x,\omega) \longrightarrow f(x-\omega t,\omega),$$

which is a strongly continuous group over $L^p(\mathbb{R}^2 \times S^1)$, $(0 \le p < \infty)$, and generates the mild solution of the following advection problem:

$$(\mathbf{P0}) \qquad \begin{cases} \dfrac{\partial u}{\partial t} = -\omega.\text{grad } u = T_0 u \\ u(x,\omega,0) = f(x,\omega). \end{cases}$$

Now, if the particles meet an absorbing obstacle (human body), the term $-\sigma(x)u$ must be added to the right-hand side of the above equation, where $\sigma(x)$ is the absorbing rate of obstacle. So the problem (**P0**) changes to (**P1**).

The solution of Problem (**P1**) is given by the following explicit formula:

$$\begin{aligned} u(x, \omega, t) &= \exp\{-\int_0^t \sigma(x - s\omega)ds\} f(x - t\omega, \omega) \\ &= [U_1(t) f](x, \omega). \end{aligned} \tag{6.3}$$

$U_1(t)$ is the dynamic of the perturbed system which is a strongly continuous group generated by $T_1 = T_0 + A_1$ where $A_1 u = -\sigma(x)u$.

Now if the Radon transformation of σ exists, then according to (6.3) the wave operators have also an explicit expression

$$\begin{aligned} [\mathcal{W}_+ f](x, \omega) &= [\mathcal{W}_+(T_0, T_1) f](x, \omega) \\ &= s - \lim_{t \to +\infty} [U_0(-t) U_1(t) f](x, \omega) \\ &= \lim_{t \to +\infty} \exp\{-\int_0^t \sigma(x - (s - t)\omega)ds\} f(x, \omega) \\ &= \exp\{-\int_{-\infty}^0 \sigma(x - \tau\omega)d\tau\} f(x, \omega) \end{aligned}$$

and

$$\begin{aligned} [\mathcal{W}_- f](x, \omega) &= [\mathcal{W}_-(T_1, T_0) f](x, \omega) \\ &= s - \lim_{t \to -\infty} [U_1(-t) U_0(t) f](x, \omega) \\ &= \lim_{t \to -\infty} \exp\{-\int_0^{-t} \sigma(x - \tau\omega)d\tau\} f(x, \omega) \\ &= \exp\{-\int_0^{\infty} \sigma(x - \tau\omega)d\tau\} f(x, \omega). \end{aligned}$$

Thus, the scattering operator is given by

$$\begin{aligned} [\mathcal{S} f](x, \omega) &= [W_+ W_- f](x, \omega) \\ &= \exp\{-\int_{-\infty}^{\infty} \sigma(x - \tau\omega)d\tau\} f(x, \omega) \\ &= \exp\{[R\sigma](x, \omega)\} f(x, \omega). \end{aligned}$$

From where we obtain

$$[R\sigma](x, \omega) = \ln \frac{f(x, \omega)}{[\mathcal{S} f](x, \omega)}, \tag{6.4}$$

which justifies the Hejtmanek interpretation.

The identification between (6.2) and (6.4) raises the following question:

(Q) *Why can one replace* $g(x, \omega)$ *by* $[\mathcal{S} f](x, \omega)$*?*

In fact, this interpretation is rather surprising, since the scattering operator is defined in the context of quantum mechanics theory and involves the space $L^2(\mathbb{R}^2 \times S^1)$, while our problem involves the albedo operator which links up the incoming flux to the outgoing flux. Our main aim in the forthcoming sections is to define the albedo operator

$$\mathcal{A} : \psi_{|\Gamma_-} \longrightarrow \psi_{|\Gamma_+}$$

which maps the incoming flux to outgoing flux of the stationary transport problem and define also the trace operators $\gamma_\pm : f \longrightarrow f_{|\Gamma_\pm}$, which links the function f in $D(T_0)$ the maximal domain of T_0, to the trace of f on the outgoing [incoming] surface Γ_+ [Γ_-]. These preliminaries permit us to give a mathematical formulation, in order to connect the scattering operator to the albedo operator by

$$\gamma_+ \mathcal{S} = \mathcal{A}\gamma_- \tag{6.5}$$

on the set of the solutions of the problem (P0), which clarify mathematically Hejtmanek's interpretation.

6.2 The Trace Theorems

Henceforth we consider the transport operator as in Chap. 3. We denote by Ω a convex bounded open subset of $\mathbb{R}^n$ with a smooth boundary $\partial\Omega$. This requirement about the smoothness of the boundary is needed in order to avoid pathological behavior of the particles hitting the boundary and to assure the validity of the Gauss integral theorem. We consider the velocity space V a spherical shell

$$V := \{v \in \mathbb{R}^n \quad ; \quad 0 < v_{min} \leq |v| \leq v_{max} \leq +\infty\}.$$

Let us denote by $\nu = \nu_x$ the outer normal vector at $x \in \partial\Omega$ and $d\gamma$ the surface measure of $\partial\Omega$ and let

$$\Gamma = \partial\Omega \times V, \quad \Gamma_\pm = \{(x, v) \in \Gamma \; ; \; \pm\nu \cdot v > 0\}, \quad \Gamma_0 = \{(x, v) \in \Gamma \; ; \; \nu \cdot v = 0\}.$$

We call Γ_+ the *outgoing surface*, Γ_- the *incoming surface* and Γ_0 the *tangent surface*. On Γ we will use the measure $d\mu := |\nu_x \cdot v| d\gamma dv$ and we write $L^1(\Gamma_\pm) := L^1(\Gamma_\pm, |\nu_x \cdot v| d\gamma dv)$, equipped with the norm

$$\|\gamma_\pm f\|_{L^1(\Gamma_\pm)} = \int_{\Gamma_\pm} |\gamma_\pm f| |\nu_x \cdot v| d\gamma dv.$$

It is clear that for this measure the tangent surface has the measure $\mu(\Gamma_0) = 0$. We have also $\Gamma_+ \cap \Gamma_- = \emptyset$.

Let $L^1(\Omega \times V) = L^1(\Omega \times V, dxdv)$. We write

$$\mathcal{W}^1(\Omega \times V) = \{u \in L^1(\Omega \times V)\, ; \quad v \cdot \nabla_x u \in L^1(\Omega \times V)\}.$$

where the differentiation is defined in the distribution sense. We endow this space with the norm

$$\|f\|_{\mathcal{W}^1(\Omega \times V)} = \int_{\Omega \times V} (|v \cdot \nabla_x f(x, v)| + |f(x, v)|)\, dxdv$$

and we write

$$\mathcal{W}^1_\pm(\Omega \times V) := \{f \in \mathcal{W}^1(\Omega \times V) \mid \gamma_\mp f = 0\}.$$

For each $(x, v) \in \overline{\Omega} \times V$, we define

$$t_i^\pm(x, v) = \inf\{s > 0\, ;\ x \pm sv \notin \Omega\} = \sup\{t > 0\, ;\ x \pm sv \in \Omega,\ \forall\, s \in (0, t)\}$$

(In t_i, i stands for interior and in t_e, e stands for exterior)
and for $(x, v) \in \Gamma_\mp$, we denote by $\tau_i^\pm(x, v) = t_i^\pm(x, v)$ the *sojourn time* of a trajectory in Ω. We remark that

$$t_i^-(x \pm rv, v) = t_i^-(x, v) \pm r, \qquad t_i^+(x \pm rv, v) = t_i^+(x, v) \mp r \tag{6.6}$$

and

$$\tau_i^+(x, v) = \tau_i^-(x, -v). \tag{6.7}$$

We define also

$$P_+ := \{(x, v) \in \Omega \times V \ \mid\ (x + t_i^+(x, v)v, v) \in \Gamma_+\}$$

and

$$P_- := \{(x, v) \in \Omega \times V \ \mid\ (x - t_i^-(x, v)v, v) \in \Gamma_-\}.$$

We remark that in $\Omega \times V$, the sets P_+ and P_- are equal.

Concerning the trace operators

$$\gamma_\pm\ :\ \mathcal{W}^1(\Omega \times V) \longrightarrow\ L^1(\Gamma_\pm, d\xi).$$

The following theorems are based strongly on the following *Gauss integral formula*:

$$\int_{\Omega \times V} v \cdot \nabla_x f(x, v) dxdv = \int_{\Gamma_+} (\gamma_+ f) v \cdot \nu_x d\gamma dv - \int_{\Gamma_-} (\gamma_- f) v \cdot \nu_x d\gamma dv. \tag{6.8}$$

Theorem 6.2.1 *The trace operators* $\gamma_\pm : u \longrightarrow u|_{\Gamma_\pm}$ *are continuous and surjective from* $\mathcal{W}^1(\Omega \times V)$ *to* $L^1(\Gamma_\pm, d\xi)$, *with continuous lifting.*

Proof For $0 \le f \in \mathcal{W}^1(\Omega \times V)$ the formula (6.8) implies that

$$\begin{aligned}\|\gamma_\pm f\|_{L^1(\Gamma_\pm)} = \|\gamma_\pm f\|_{L^1(\Gamma)} &= \left| \int_{\Gamma_+} f|_{\Gamma_+} v \cdot \nu_x d\gamma dv - \int_{\Gamma_-} f|_{\Gamma_-} v \cdot \nu_x d\gamma dv \right| \\ &= \left| \int_{\Omega \times V} v \cdot \nabla_x f(x, v) dx dv \right| \\ &\le \int_{\Omega \times V} |v \cdot \nabla_x f(x, v)| dx dv \le \|f\|_{\mathcal{W}^1(\Omega \times V)},\end{aligned}$$

which proves the continuity of the trace operators $\gamma_\pm$ on the positive cone $L^1_+(\Gamma_\pm)$. Now, since the positive cone in $L^1(\Gamma_\pm)$ is generating we can conclude the continuity in the whole space.

For the surjectivity, let $g_- \in L^1(\Gamma_-)$ and

$$u(x, v) = e^{\lambda t_i^-(x,v)} g_-(x - t_i^-(x, v)v, v) \qquad (\lambda < 0)$$

defined on $\Omega \times V$. Let us show that this function is the solution of the problem

$$(\mathbf{PL})_- \qquad \begin{cases} v \cdot \nabla_x u = \lambda u & in\ \Omega \times V \\ u_{|\Gamma_-} = g_- & on\ \Gamma_- \end{cases}$$

In fact we notice that for $y = x - t_i^-(x, v)v$, $g_-(y, v) \in L^1(\Gamma_-)$, hence g_- is constant with respect to x regarding in the direction v and its directional derivative is

$$\partial g_- / \partial v = v \cdot \nabla_x g_- = 0. \tag{6.9}$$

Furthermore, (6.6) implies that

$$v \cdot \nabla_x t_i^-(x, v) = \lim_{h \to 0} \frac{t_i^-(x + hv, v) - t_i^-(x, v)}{h} = 1. \tag{6.10}$$

Since

$$\begin{aligned}v \cdot \nabla_x u(x, v) = (v \cdot \nabla_x t_i^-) \lambda e^{\lambda t_i^-(x,v)} g_-(x - t_i^-(x, v)v, v) \\ + e^{\lambda t_i^-(x,v)} (v \cdot \nabla_x g_-(x - t_i^-(x, v)v, v))\end{aligned}$$

the relations (6.9) and (6.10) imply that u satisfies the problem (PL). To show that $u \in \mathcal{W}^1(\Omega \times V)$, it suffices to to prove $u \in L^1(\Omega \times V)$. We fix $v \in V$ such that $(x, v) \in \Gamma_-$ for all $x \in \overline{\Omega}$ and we write

$$\int_{\Omega} |e^{\lambda t_i^-(x,v)} g_-(x - t_i^-(x,v)v, v)| dx = \int_{\Gamma_-} \int_0^{\tau_i^+(x,v)} |e^{\lambda s} g_-(x,v)||v.\nu| d\gamma ds$$
$$\leq \int_{\Gamma_-} \frac{|e^{-|\lambda|\tau_i^+(x,v)} - 1|}{|\lambda|} |g_-(x,v)||v \cdot \nu_x| d\gamma$$
$$\leq \frac{1}{|\lambda|} \int_{\Gamma_-} |g_-(x,v)||v \cdot \nu_x| d\gamma.$$

By integrating over V we obtain

$$\|e^{\lambda t_i^-(x,v)} g_-(x - t_i^-(x,v)v, v)\|_{L^1(\Omega\times V)} \leq \frac{1}{|\lambda|} \|g_-\|_{L^1(\Gamma_-,|v\cdot\nu_x|d\gamma dv)}. \tag{6.11}$$

which implies the surjectivity and the fact that u is a continuous lifting on $L^1(\Gamma_-)$. The proof for $g_+ \in L^1(\Gamma_+)$ is analogous. Notice only that in this case the lifting

$$u(x,v) := e^{\lambda t_i^+(x,v)} g_+(x + t_i^+(x,v)v, v) \qquad (\lambda < 0)$$

satisfies the boundary problem

$$(\mathbf{PL})_+ \qquad \begin{cases} v \cdot \nabla_x u = \lambda u & \text{in } \Omega \times V, \\ u_{|\Gamma_+} = g_+ & \text{on } \Gamma_+. \end{cases}$$

□

Theorem 6.2.2 *Let*

$$\widetilde{W}^1(\Omega \times V) \ := \ \{u \in \mathcal{W}^1(\Omega \times V) \ ; \ u|_{\Gamma_\pm} \in L^1(\Gamma_\pm, |\nu \cdot v| d\gamma dv)\}.$$

We have

$$\widetilde{W}^1(\Omega \times V) = \{u \in \mathcal{W}^1(\Omega \times V); \quad u|_{\Gamma_-} \in L^1(\Gamma_-, |\nu \cdot v| d\gamma dv)\}$$
$$= \{u \in \mathcal{W}^1(\Omega \times V); \quad u|_{\Gamma_+} \in L^1(\Gamma_+, |\nu \cdot v| d\gamma dv)\}.$$

Proof From (6.8) it follows that if $f \in \mathcal{W}^1(\Omega \times V) \cap L^1(\Gamma_\pm)$ then $f \in L^1(\Gamma_\mp)$. □

Now let us denote by

$$R \ : \ L^1(\mathbb{R}^n \times V) \ \longrightarrow \ L^1(\Omega \times V) \tag{6.12}$$

the truncation operator on $L^1(\Omega \times V)$. It is a projection on $L^1(\Omega \times V)$. It is clear that

$$R \ : \ \mathcal{W}^1(\mathbb{R}^n \times V) \ \longrightarrow \ \mathcal{W}^1(\Omega \times V)$$

is continuous and for all $h \in \mathcal{W}^1(\mathbb{R}^n \times V)$

$$\gamma_{\pm}h = \gamma_{\pm}Rh. \tag{6.13}$$

In Sect. 3.4, we have defined the notions of *free forward and trapped backward, free backward and trapped forward point sets*. Similar notions will be defined in the sequel which suit more our discussion. For $(x, v) \in \overline{\Omega}^c \times V$, we define

$$t_e^{\pm} : \overline{\Omega}^c \times V \mapsto [0, \infty]$$

by

$$t_e^{\pm}(x, v) = \inf\{s > 0 \; ; \; x \pm sv \in \overline{\Omega}\}.$$

Let us denote by

$$Q_+ := \{(x, v) \in \overline{\Omega}^c \times V \mid t_e^+(x, v) < \infty\}$$

the points $(x, v) \in Q_+$ are called *reflected forward* and by

$$Q_- := \{(x, v) \in \overline{\Omega}^c \times V \mid t_e^-(x, v) < \infty\}$$

the points $(x, v) \in Q_-$ are called *reflected backward*. Finally,

$$Q_0 := \{(x, v) \in \overline{\Omega}^c \times V \mid t_e^+(x, v) \text{ and } t_e^-(x, v) = \infty\}$$

the points $(x, v) \in Q_0$ are called *nonreflected points*. The convexity of Ω implies that a reflected backward point is *free forward* and conversely a reflected forward point is *free backward*. We remark that any point stemming from Q_- ends on Γ_-. For any $(x, v) \in Q_-$ we denote by $s_-(x, v)$ the time when $(x - s_-(x, v)v, v) \in \Gamma_-$.

Theorem 6.2.3 *The trace operators $\gamma_{\pm} : g \longrightarrow g|_{\Gamma_{\pm}}$ are continuous and surjective from $\mathcal{W}^1(\mathbb{R}^n \times V)$ to $L^1(\Gamma_{\pm})$, with continuous lifting.*

Proof For $g \in \mathcal{W}^1(\mathbb{R}^n \times V)$, we have $Rg \in \mathcal{W}^1(\Omega \times V)$. Now, by (6.13) and Theorem 6.2.1, $g|_{\Gamma_{\pm}} = \gamma_{\pm}Rg \in L^1(\Gamma_{\pm}, |\nu \cdot v|d\gamma dv)$. The continuity of $\gamma_{\pm}$ from $\mathcal{W}^1(\mathbb{R}^n \times V)$ to $L^1(\Gamma_{\pm}, |\nu \cdot v|d\gamma dv)$ is a result of the continuity of $\gamma_{\pm}$ from $\mathcal{W}^1(\Omega \times V)$ to $L^1(\Gamma_{\pm}, |\nu \cdot v|d\gamma dv)$ and the continuity of R from $\mathcal{W}^1(\mathbb{R}^n \times V)$ to $\mathcal{W}^1(\Omega \times V)$.

Now given $g_- \in L^1(\Gamma_-, |\nu \cdot v|d\gamma dv)$, for $\lambda < 0$ we construct the following lifting in $\mathcal{W}^1(\Omega \times V)$:

$$\widetilde{g}_-(x, v) = \begin{cases} e^{\lambda s_-(x,v)} g_-(x - s_-(x, v)v, v) & \text{if } (x, v) \in Q_-, \\ e^{\lambda t_i^-(x,v)} g_-(x - t_i^-(x, v), v) & \text{if } (x, v) \in P_-, \\ e^{\lambda \tau_e^+(x,v)} g_-(x + \tau_e^+(x, v), v) & \text{if } (x, v) \in Q_+, \\ 0 & \text{if } (x, v) \in Q_0. \end{cases}$$

To show that $\widetilde{g}_- \in \mathcal{W}^1(\mathbb{R}^n \times V)$, we remark that

$$\int_{\mathbb{R}^n \times V} |\widetilde{g}_-(x, v)| dx dv = \|\widetilde{g}_-\|_{L^1(\Omega \times V)} + \|\widetilde{g}_-\|_{L^1(\Omega^c \times V)}.$$

The first integral is finite as is shown in (6.11) and the integral over $\overline{\Omega}^c \times V$ splits into two integrals,

$$\begin{aligned}
\int_{Q_-} |\widetilde{g}_-(x, v)| dx dv &\leq \int_{\Gamma_-} \int_{\tau_i^-(x,v)}^{\infty} |e^{\lambda s} g_-(x, v)||v.\nu| ds d\gamma dv \\
&\leq \int_{\Gamma_-} \frac{e^{-|\lambda|\tau_i^+(x,v)}}{|\lambda|} |g_-(x, v)||v \cdot \nu_x| d\gamma dv \\
&\leq \frac{1}{|\lambda|} \int_{\Gamma_-} |g_-(x, v)||v \cdot \nu_x| d\gamma dv,
\end{aligned}$$

so $\tilde{g}_- \in L^1(\mathbb{R}^3 \times V)$. It remains to show that $\lim_{t\to 0}(U_0(t)\tilde{g}_- - \tilde{g}_-)/t$ exists in the distribution sense and belongs to $L^1(\Omega \times V)$. For any $\varphi \in C_0^\infty(\mathbb{R}^3 \times V)$, consider

$$\begin{aligned}
I_t = \langle (U_0(t)\tilde{g}_- - \tilde{g}_-)/t, \varphi\rangle &= \int_{\mathbb{R}^3 \times V} \frac{\tilde{g}_-(x - tv, v) - \tilde{g}_-(x, v)}{t} \varphi(x, v) dx dv \\
&= \int_{\mathbb{R}^3 \times V} \tilde{g}_-(x', v) \frac{\varphi(x' + tv, v) - \varphi(x', v)}{t} dx dv.
\end{aligned}$$

Since $(U_0(-t)\varphi - \varphi)/t$ converges to $v \cdot \nabla_x \varphi$ uniformly in $\mathbb{R}^3 \times V$, the fact that $\tilde{g}_- \in L^1(\mathbb{R}^3 \times V)$ shows that

$$I_t \longrightarrow I = \int_{\mathbb{R}^3 \times V} \tilde{g}_-(x, v)[v \cdot \nabla_x \varphi](x, v) dx dv.$$

Hence, $[-v \cdot \nabla_x g_-]$ exists in the distribution sense and to end the proof it suffices to show that it belongs to $L^1(\Omega \times V)$. In fact

$$\begin{aligned}
I = \langle [-v \cdot \nabla_x g_-], \varphi\rangle &= \int_{\mathbb{R}^3 \times V} \tilde{g}_-(x, v)[v \cdot \nabla_x \varphi](x, v) dx dv \\
&= \int_{\Omega \times V} \hat{g}_-(x, v)[v \cdot \nabla_x \varphi](x, v) dx dv \qquad (6.14) \\
&\quad + \int_{\overline{\Omega}^c \times V} \hat{g}_-^c(x, v)[v \cdot \nabla_x \varphi](x, v) dx dv. \qquad (6.15)
\end{aligned}$$

Now, by using Green's formula the integral in (6.14) is equal to

$$\int_{\Omega \times V} [-v \cdot \nabla_x \hat{g}_-](x, v)\varphi(x, v) dx dv + \int_{\Gamma} \hat{g}_-(x, v)\varphi(x, v)(\nu \cdot v) d\gamma dv$$

and the integral appearing in (6.15) is equal to

$$\int_{\overline{\Omega}^c \times V} [-v \cdot \nabla_x \hat{g}_-^c](x, v)\varphi(x, v)dxdv + \int_{\Gamma} \hat{g}_-^c(x, v)\varphi(x, v)(\nu' \cdot v)d\gamma dv.$$

Here, $\nu' = -\nu$, and the fact that $\hat{g}_-$ and $\hat{g}_-^c$ are identical on Γ implies that

$$I = \int_{\Omega \times V} [-v \cdot \nabla_x \hat{g}_-](x, v)\varphi(x, v)dxdv + \int_{\overline{\Omega}^c \times V} [-v \cdot \nabla_x \hat{g}_-^c](x, v)\varphi(x, v)dxdv.$$

Hence

$$v \cdot \nabla_x \tilde{g}_- = v \cdot \nabla_x \hat{g}_- + v \cdot \nabla_x \hat{g}_-^c \tag{6.16}$$

almost everywhere and each function of the right-hand side of (6.16) is integrable. □

6.3 Relationship Between the Albedo and Scattering Operators for Transparent Boundary Condition

The aim of this section is to prove the formula (6.5). For this we begin to define to define the *albedo operator* $\mathcal{A}$. Let us consider the interior evolution problem

$$\textbf{(IP)} \qquad \begin{cases} \dfrac{\partial \psi}{\partial t} = \widetilde{T}\psi \\ \psi(x, v, 0) = g(x, v) \in \widetilde{W}^1(\Omega \times V) \\ \gamma_-\psi(., ., t) = \psi_-(t) \in L^1(\Gamma_-, d\xi), \ t \in (0, \infty), \end{cases}$$

where $\widetilde{T}$ is the restriction of the operator T given in (3.3) to $\mathcal{W}^1(\Omega \times V)$, $\psi_- \in C([0, \infty); L^1(\Gamma_-, d\xi))$, and $\psi_-(0) = \gamma_- g$. In [57] it is shown that this problem admits a unique solution in $C([0, \infty); \mathcal{W}^1(\Omega \times V))$.

Definition 6.3.1 Let ψ be the solution of (IP). The *albedo operator* $\mathcal{A}$ is an operator from $C([0, \infty); L^1(\Gamma_-, d\xi))$ into $C([0, \infty); L^1(\Gamma_+, d\xi))$ given by

$$\mathcal{A}: \quad \psi\,|_{\Gamma_-} \longrightarrow \psi\,|_{\Gamma_+},$$

which depends to the initial data $g(x, v) \in \widetilde{W}^1(\Omega \times V)$.

In Sect. 3.6 we have given the conditions under which the wave operators $W_+ = W_+(T_0, T)$ and $W_- = W_-(T, T_0)$ exist. The following corollary is an obvious consequence of Lemma 3.4.3.

Corollary 6.3.2 *Let* $g \in L^1(\mathbb{R}^n \times V)$*, then:*

(i) $[W_+ g](x, v) = g(x, v)$ *for all* $(x, v) \in Q_+$;
(ii) $[W_- g](x, v) = g(x, v)$ *for all* $(x, v) \in Q_-$;
(iii) $[W_\pm g](x, v) = g(x, v)$ *for all* $(x, v) \in Q_0$.

Proof Since $Q_+ \subseteq \Sigma_+$, applying $U_0(-t)$ to the expression of Lemma 3.4.3 (a) and taking $t \to \infty$, we obtain (i). Similarly, (ii) is the consequence of $Q_- \subseteq \Sigma_-$ and the action of $U_0(t)$ on the expression of Lemma 3.4.3 (b) as $t \to -\infty$ and finally, (iii) follows in the same manner from Lemma 3.4.3 (c). □

Remark 6.3.3 Since $\Gamma_\pm \subset Q_\pm$, it follows from above corollary that

$$\gamma_\pm W_\pm g = \gamma_\pm g \tag{6.17}$$

for all $g \in \mathcal{W}^1(\mathbb{R}^n \times V)$.

Let R be the truncation operator given in (6.12) and $\varphi \in C(\mathbb{R}; \mathcal{W}^1(\mathbb{R}^n \times V))$ the unique solution of the following advection problem:

$$(\mathbf{AP}) \qquad \begin{cases} \dfrac{\partial \varphi}{\partial t} = -v.\text{grad}\, \varphi = T_0 \varphi \\ \varphi(x, v, 0) = f(x, v) \in \widetilde{W}^1(\mathbb{R}^n \times V). \end{cases}$$

Lemma 6.3.4 (i) *If* $f \in W^1(\mathbb{R}^n \times V)$ *then* $RTf = \widetilde{T} Rf$.
(ii) *Let* φ *be the solution of* (**AP**)*. Then for any* $t > 0$, $\psi(t) = Rw_-\varphi(t)$ *is the solution of* (IP) *with the initial data* $g = RW_- f$ *and we have* $\psi(t) \mid_{\Gamma_-} = \varphi(t) \mid_{\Gamma_-}$.

Proof The assumption (i) is obvious from the definition of $\widetilde{T}$. To prove (ii), we remark that $D(T_0) = W^1(\mathbb{R}^n \times V)$ and according to (2.4) $W_- D(T_0) \subseteq D(T)$. This together with (i) implies that

$$RTW_-\varphi(t) = \widetilde{T} RW_-\varphi(t). \tag{6.18}$$

Now, we will use this relation together with intertwining identity given in Lemma 2.1.3 (b), which leads to

$$RTW_-\varphi(t) = RW_- T_0 \varphi(t) = RW_- \frac{\partial}{\partial t}\varphi(t) = \frac{\partial}{\partial t} RW_-\varphi(t),$$

so $\psi(t) = RW_-\varphi(t)$ is the solution of (IP) with the initial data $g = RW_- f$. Finally,

$$\begin{aligned} \gamma_-\psi(t) &= \gamma_- W_-\varphi(t) && \text{(according to (6.13))} \\ &= \gamma_- W_-\varphi(t) && \text{(according to (6.17)).} \end{aligned}$$

□

Corollary 6.3.5 *Let φ be the solution of (AP) and $\mathcal{A}$ the albedo operator, then*

$$\mathcal{A}\gamma_- RW_-\varphi(t) = \gamma_+ RW_-\varphi(t). \tag{6.19}$$

Proof It is an obvious consequence of above lemma and the definition of the albedo operator. □

Now we are able to demonstrate the formula (6.5) through the following theorem.

Theorem 6.3.6 *Suppose that the scattering operator S exists and the albedo operator is well-defined. Then* (6.5) *holds true on* $\widetilde{W}^1(\mathbb{R}^n \times V)$.

Proof Let $f(x, v) \in \widetilde{W}^1(\mathbb{R}^n \times V)$ and φ be the solution of (AP), with $\varphi(0) = f$, then according to Lemma 6.3.4, $\psi(t) = RW_-\varphi(t)$ is the solution of (IP) with the initial data $g = RW_- f$ and we have

$$\psi(t)\big|_{\Gamma_-} = \varphi(t)\big|_{\gamma_-}. \tag{6.20}$$

So,

$$\begin{aligned}
\gamma_+ S\varphi(t) &= \gamma_+ W_+ W_-\varphi(t) \\
&= \gamma_+ W_-\varphi(t) && \text{(according to (6.17))} \\
&= \gamma_+ RW_-\varphi(t) && \text{(according to (6.13))} \\
&= \mathcal{A}\gamma_- RW_-\varphi(t) && \text{(according to (6.19))} \\
&= \mathcal{A}\gamma_-\psi(t) \\
&= \mathcal{A}\gamma_-\varphi(t) && \text{(according to (6.20))}.
\end{aligned}$$

The theorem follows for $t = 0$. □

6.4 Relationship Between the Albedo and Scattering Operators for Semi-transparent Boundary Condition

In all the scattering situations analyzed in the previous sections the surface of the body $\partial\Omega$ was perfectly transparent for both incident and emergent distributions. The new aspect considered here is a partial reflectivity of the surface for emergent particles, the effect of which is a prolonged temporary trapping of the distribution before its eventual release into the outside vacuum space. This partial reflectivity is described by a boundary operator K which acts as a strict positive contraction between the incoming and outgoing boundary spaces. Assuming there are no collisions between the particles in the interior of the finite convex scattering body Ω, then the dynamic can be written as the following advection problem

$$\textbf{(APK)}\qquad \begin{cases} \dfrac{\partial u}{\partial t} = -v.\text{grad } u = T_0^K u, \\ u(x, v, 0) = f(x, v) \in \widetilde{W}^1(\mathbb{R}^n \times V), \\ \gamma_- u = K\gamma_+ u \quad \text{on } \Gamma_-. \end{cases}$$

To study the generator of this problem we have to define its domain

$$D(T_0^K) := \widetilde{W}_K^1(\mathbb{R}^n \times V) = \{f \in \widetilde{W}^1(\mathbb{R}^n \times V) \; : \; \gamma_- f = K\gamma_+ h_- \\ \text{for all } \; f \in \widetilde{W}_+^1(\mathbb{R}^n \times V) \; \text{ and some } \; h_- \in L^1(\Gamma_-)\}.$$

The space $\widetilde{W}_K^1(\mathbb{R}^n \times V)$ being dense in $L^1(\mathbb{R}^n \times V)$ we can define the operator T_0^K as the closure of $T_0|_{\widetilde{W}_K^1(\mathbb{R}^n \times V)}$ where $T_0 f(x, v) = -v.\nabla_x f(x, v)$.

Lemma 6.4.1 ([57], ch. XII, Theorem 2.2 or [88] Theorem 3.1) *Suppose the boundary operator $K : L^1(\Gamma_+) \mapsto L^1(\Gamma_-)$ is positive and strictly contractive. Then T_0^K defined on $D(T_0^K)$ generates a positive C_0-semigroup of contractions $\{U_0^K(t)\}_{t \geq 0}$ on $L^1(\mathbb{R}^n \times V)$.*

Remark 6.4.2 Note that contrary to $U_0(t)$, $U_0^K(t)$ is not a (C_0)-group, it is only a (C_0)-semigroup.

Lemma 6.4.3 *For each $f \in L^1(\mathbb{R}^n \times V)$ and each $\varepsilon > 0$, there exists $T_\varepsilon > 0$ such that for all $t > T_\varepsilon$,*

$$\int_{\Omega \times V} \left|[U_0^K(t)f](x, v)\right| dxdv < \varepsilon. \tag{6.21}$$

Proof Let $\widehat{C_0^\infty} := \{f \in C_0^\infty(\mathbb{R}^n \times V) \; : \; \text{supp } f \subset \mathbb{R}^n \times (V \setminus \{0\})\}$, which is dense in $L^1(\mathbb{R}^n \times V)$. Since $\widehat{C_0^\infty} \subset D(T_0^K)$, it is not hard to see that $\widehat{C_0^\infty}$ is invariant under the action of $U_0^K(t)$, hence according to Lemma 1.2.6, it is a core for T_0^K and for $0 \leq f \in \widehat{C_0^\infty}$, $u(x, v, t) = U_0^K(t)f(x, v)$ is a positive solution of (**APK**). Thus

$$\mathcal{I}(t) := \int_{\Omega \times V} [U_0^K(t)f](x, v)dxdv \geq 0$$

and the Green identity implies that

$$\begin{aligned} \frac{d}{dt}\int_{\Omega \times V} [U_0^K(t)f](x, v)dxdv &= \int_{\Omega \times V} -v.\nabla_x u(x, v, t)dxdv \\ &- \int_{\Gamma_+} u(x, v, t)\nu.vd\gamma dv + \int_{\Gamma_-} u(x, v, t)\nu.vd\gamma dv \\ &= -\|\gamma_+ u(t)\|_{L^1(\Gamma_+)} + \|K\gamma_+ u(t)\|_{L^1(\Gamma_+)} < 0, \end{aligned}$$

since the boundary operator K is strictly contractive. This yields that $\mathcal{I}(t)$ is strictly decreasing to zero. Therefore after a time T_ε it becomes less than ε. In addition, since $U_0^K(t)$ is positive and the positive cone is generating, (6.24) holds for all $f \in L^1(\mathbb{R}^n \times V)$. □

We can also consider somehow T_0^K as a perturbation of T_0 and show the existence of wave operators.

Theorem 6.4.4 *If the boundary operator K is positive and strictly contractive, then the wave operators*

$$W_+^{(0)} = s - \lim_{t\to+\infty} U_0(-t)U_0^K(t) \quad \textit{and} \quad W_-^{(0)} = s - \lim_{t\to+\infty} U_0^K(t)U_0(-t)$$

exist.

Proof We will prove that the family of operators $\{W_t\}_{t\geq 0}$ where $W_t := U_0(-t)U_0^K(t)$ forms a strongly Cauchy net in $L^1(\mathbb{R}^n \times V)$.

As before, since $U_0(-t)U_0^K(t)$ is a contraction, it is sufficient to prove the existence of the wave operators on $C_0^1(\mathbb{R}^n \times V)_+$, which is dense in $L^1(\mathbb{R}^n \times V)_+$. Let us define the cones

$$\mathscr{C}_\pm := \{(x \pm tv, v) \ : \ (x, v) \in \Omega \times V, t \geq 0\} \setminus \{\Omega \times V\}$$

corresponding to the past (–) and future (+) of the domain $\Omega \times V$ with respect to the free evolution $U_0(t)$ and also the free backward–forward points

$$\mathscr{C}_0 := \{(x, v) \in (\mathbb{R}^n \setminus \Omega) \times V \quad : x \pm tv \notin \Omega \ \text{ for all } t \geq 0\}.$$

For each function f supported in $\mathscr{C}_+$ or in $\mathscr{C}_0$ and $t \geq 0$ the semigroup $U_0^K(t)$ acts like $U_0(t)$. Hence by decomposing $\mathbb{R}^n \times V = \mathscr{C}_+ \cup \mathscr{C}_- \cup \mathscr{C}_0 \cup \{\Omega \times V\}$, for $t > T_\varepsilon$, where T_ε is defined as in Lemma 6.4.3, we get

$$\int_S W_t f(x, v)dxdv - \int_S W_{T_\varepsilon} f(x, v)dxdv = 0,$$

for the sets $S = \mathscr{C}_+, \mathscr{C}_0, \Omega \times V$. Finally, we can choose the point (x, v) and the time T_ε such that $(x - T_\varepsilon v, v) \in \Gamma_-$ and $(x - tv, v) \in \Omega \times V$. Since all the points in $\mathscr{C}_-$ are to be trapped in the region Ω, for some $t > T_\varepsilon$, we have

$$\int_{\mathscr{C}_-} W_t f(x, v)dxdv - \int_{\mathscr{C}_-} W_{T_\varepsilon} f(x, v)dxdv = \int_{\Omega\times V} U_0^K(t) f(x, v)dxdv.$$

So Lemma 6.4.3 implies that

$$0 \leq \int_{\Omega\times V} W_t f(x, v)dxdv - \int_{\Omega\times V} W_{T_\varepsilon} f(x, v)dxdv \leq \varepsilon.$$

The theorem follows from the fact that W_t is a positive operator, so it is generating on $L^1(\mathbb{R}^n \times V)_+$.

For the existence of $W_-^{(0)}$, one proceeds similarly by reversing the roles of $\mathscr{C}_-$ and $\mathscr{C}_+$. □

We are now in a position to establish the existence of the wave operators

$$W_+^K = s - \lim_{t\to+\infty} U_0(-t)U^K(t) \quad \text{and} \quad W_-^K = s - \lim_{t\to+\infty} U^K(t)U_0(-t), \quad (6.22)$$

where $U^K(t)f(x,v)$ is the solution of

(APK)

$$\begin{cases} \dfrac{\partial u}{\partial t}(x,v,t) = -v.\text{grad}\, u(x,v,t) + \chi_\Omega(x)[-\sigma_a(x,v)u(x,v,t) + \\ \qquad\qquad \displaystyle\int_V k(x,v',v)u(x,v',t)dv'] = (T_0 + A_a + A_p)u(x,v,t) \\ \qquad\qquad = T^K u(x,v,t) \\ u(x,v,0) = f(x,v) \in \widetilde{W}^1(\mathbb{R}^n \times V), \\ \gamma_- u = K\gamma_+ u \quad \text{on } \Gamma_-. \end{cases}$$

In the above problem $T_0 + A_a + A_p = T$, but we have written T_K; the reason is that T_0^K is defined on $D(T_0^K)$ and for $f \in D(T_0^K)$, $T_0 f = T_0^K f$ and similarly $D(T^K) = D(T_0^K)$ and for $f \in D(T^K)$, $Tf = T^K f$ and we have also the following lemma

Lemma 6.4.5 *If the pair (k, σ_a) is subcritical then the semigroup et $T^K$is also uniformly bounded.*

Proof It is an obvious consequence of the above lemma and the definition of the albedo operator. □

Before proving the existence of the wave operators for the subcritical system, we have to give a lemma which is an important consequence of the existence of the wave operator $W_+^{(0)}$.

Lemma 6.4.6 *Let (k, σ_a) be an admissible pair that obeys*

$$F(\sigma_i) := \sup_{x,v} \int_{-\infty}^{\infty} \sigma_i(x - tv, v)dt < \infty \quad \textit{for } i = a, p. \quad (6.23)$$

Then the existence of $W_+^{(0)}$ implies that there exists a constant C, such that

$$\int_0^{\infty} \|A_i U_0^K(t) f\| dt \le C\|f\| \quad \textit{for all } f \in L^1(\mathbb{R}^n \times V), \quad i = a, p. \quad (6.24)$$

Proof According to Lemma 3.1.3 (6.23) implies

$$\int_0^\infty \|A_i U_0(t) f\| dt \leq \int_{-\infty}^\infty \|A_i U_0(t) f\| dt$$
$$\leq F(\sigma_i)\|f\| \quad \text{for all } f \in L^1(\mathbb{R}^n \times V), \quad i = a, p.$$

Hence, for any $f \in L^1(\mathbb{R}^n \times V)$, let $g = W_+^{(0)} f$ we get

$$\int_0^\infty \|A_i U_0^K(t) f\| dt \leq \int_0^\infty \|A_i U_0(t)[U_0(-t)U_0^K(t)f - g]\| dt$$
$$+ \int_0^\infty \|A_i U_0(t) g\| dt$$
$$\leq F(\sigma_i)\Big[\sup_{t\geq 0} \|U_0(-t)U_0^K(t)f - g\| + \|g\|\Big] < \infty$$

and (6.24) follows from the uniform boundedness principle. □

Theorem 6.4.7 *Suppose that the pair* (k, σ_a) *is admissible such that* $U^K(t)$ *is subcritical and* (6.24) *holds. Then* W_-^K *exists.*

Proof Since $\sup_{t\geq 0} \|U^K(t)U_0(-t)\| = \sup_{t\geq 0} \|U^K(t)\| < M$, it is sufficient to prove the existence of W_-^K on the total set $C_0^1(\mathbb{R}^n \times V)_+$.

By applying the second Duhamel formula on $U_0(-t)f$ for any bounded operator we get

$$U^K(t)U_0(-t)f = U_0^K(t)U_0(-t)f + \int_0^t U^K(t-s)AU_0^K(s)U_0(-t)f ds.$$

By Theorem 6.4.4, the first term of the right-hand side has a strong limit as $t \to \infty$. Moreover,

$$\|U^K(t-s)AU_0^K(s)U_0(-t)f\| \leq M \int_{\mathbb{R}^n \times} |AU_0^K(s)f(x + tv, v) dx dv$$
$$= M\|AU_0^K(s)f\|.$$

Using Lemma 6.4.6, we get

$$\int_0^\infty \|U^K(t-s)AU_0^K(s)U_0(-t)f\| ds \leq MC\|f\|,$$

which proves the existence of W_-^K. □

Remark 6.4.8 In the above theorem we have assume the subcriticality of $U^K(t)$ and here we give two ways to obtain such a condition. First we can assume that $U(t)$ is subcritical and T^K coincides with T on $D(T^K)$. Thus the C_0-semigroup

$U^K(t)$ coincides also on $D(T^K)$ with $U(t)$, which is subcritical, so $\|U^K(t)\| \leq C$ since $D(T^K) = D(T_0^K)$ is dense in$L^1(\mathbb{R}^n \times V)$. Hence for any $f \in L^1(\mathbb{R}^n \times V)$ we can find a sequence $f_n \in D(T^K)$ such that $f_n \to f$ and $\|U^K(t)\| \leq \|U^K(t)(f - f_n)\| + \|U^K(t)f_n\| \leq e^{t\omega_K}\|f - f_n\| + C$, so we can choose n large enough such that $\|f - f_n\| \leq e^{-t\omega_K}$, and the subcriticality of $U^K(t)$ follows from the uniform boundedness principle. The second condition is to assume

$$\Sigma_p := \int_0^\infty \|A_p U_0^K(t)\| dt < 1, \tag{6.25}$$

which will be proved in the following theorem.

Once the existence of the wave operators W_-^K is proved we can define the scattering operator

$$\mathcal{S}^K := W_+^K W_-^K, \tag{6.26}$$

which will be used to express its relationship with the albedo operator for the semi-transparent boundary condition.

Theorem 6.4.9 *Suppose that the pair (k, σ_a) is admissible and satisfies* (6.25). *Then the corresponding semigroup $U^K(t)$ is subcritical and the wave operators W_+^K and W_-^K exist.*

Proof Let $\widehat{U}^K(t)$ denote the evolution operator corresponding to the pair $(k, 0)$. By the Dyson–Phillips expansion

$$\widehat{U}^K(t) = \sum_{j=0}^{\infty} \widehat{U}_j^K(t),$$

where $\widehat{U}_0^K(t) = U_0^K(t)$, and

$$\widehat{U}_j^K(t) = \int_0^t \widehat{U}_0^K(t-s) A_p \widehat{U}_{j-1}^K(s) ds. \tag{6.27}$$

For any $f \in L^1(\mathbb{R}^n \times V)$ and $t \geq 0$

$$\int_0^t \|A_p U_0^K(t) f\| dt \leq \Sigma_p \|f\|.$$

The semigroup $U_0^K(t)$ being of contraction, by induction we have

$$\begin{aligned} \|\widehat{U}_j^K(t) f\| &\leq \int_0^t \|A_p \widehat{U}_{j-1}^K(t) f\| dt \leq \int_0^\infty \int_0^t \|A_p \widehat{U}_0^K(t-s) A_p \widehat{U}_{j-2}^K(s) f\| ds dt \\ &\leq \Sigma_p \int_0^\infty \|A_p \widehat{U}_{j-2}^K(s) f\| \leq \cdots \leq (\Sigma_p)^j \|f\|. \end{aligned} \tag{6.28}$$

Consequently,

$$\|\widehat{U}^K(t)\| \leq (1-\Sigma_p)^{-1}. \tag{6.29}$$

On the other hand, the absorption operator A_a being negative, for any $f \in L^1(\mathbb{R}^n \times V)_+$ we have

$$\|U^K(t)f\| = \|e^{tA_a}\widehat{U}^K(t)f\| \leq \|\widehat{U}^K(t)f\|.$$

Since $L^1(\mathbb{R}^n \times V)_+$ is generating, the subcriticality of $U^K(t)$ follows from (6.29).

For the existence of W_+^K, we proceed as in Theorem 6.4.7 , by considering

$$\begin{aligned} U_0(-t)U^K(t) =& U_0(-t)U_0^K(t) + U_0(-t)\int_0^t U_0^K(t-s)A_aU^K(s)ds \\ &+ U_0(-t)\int_0^t U_0^K(t-s)A_pU^K(s)ds. \end{aligned} \tag{6.30}$$

The first term of the right-hand side converges to $W_+^{(0)}$ as $t \to \infty$. Convergence of the second term of the right-hand side follows from the fact that for $f \in L^1(\mathbb{R}^n \times V)_+$ the family $\{U_0(-t)\int_0^t U_0^K(t-s)A_aU^K(s)ds\ ;\ t \geq 0\}$ is bounded decreasing negative functions, consequently the monotone convergence theorem implies the convergence of $U_0(-t)\int_0^t U_0^K(t-s)A_aU^K(s)ds$ as $t \to \infty$ on $L^1(\mathbb{R}^n \times V)_+$, which is generating. Finally, for the last term of (6.30), by writing once more the Dyson–Phillips expansion of $\widehat{U}^K(t)$ we find

$$\int_0^\infty \|A_p\widehat{U}^K(t)f\|dt \leq \sum_{j=0}^{\infty}\int_0^\infty \|A_p\widehat{U}_j^K(t)f\|dt.$$

The boundedness of the last term follows from (6.30) and the fact that

$$\int_0^\infty \|A_pU^K(t)f\|dt \leq \int_0^\infty \|A_pU^K(t)f\|dt.$$

This proves the existence of W_+^K.

With the assumptions of this theorem the existence of W_-^K follows similarly, just by moving $U_0(-t)$ from the left-hand side to the right-hand side and using the existence of $W_-^{(0)}$. □

By modifying slightly the proof of Lemma 2.1.3 concerning the intertwining properties of the wave operators, we can retrieve the same results for W_-^K and W_+K. Note that in Lemma 2.1.3 $U(t)$ was a group, but here $U^K(t)$ is only a semigroup.

Lemma 6.4.10 *The existence of each wave operator $W_\pm^K$ implies the correspondent identity in the following list:*

(a)

$$U^K(t)W_-^K f = W_-^K U_0(t) f \quad \textit{for all} \quad f \in L^1(\mathbb{R}^n \times V);$$

(b)

$$U_0(t)W_+^K f = W_+^K U^K(t) f \quad \textit{for all} \quad f \in L^1(\mathbb{R}^n \times V);$$

(c)

$$T^K W_-^K f = W_-^K T_0 f \quad \textit{for all} \quad f \in D(T_0);$$

(d)

$$T_0 W_+^K f = W_+^K T^K f \quad \textit{for all} \quad f \in D(T_0^K).$$

Proof For (a), it suffices to observe that for $s > t \geq 0$ we have

$$U^K(t)U^K(s-t)U_0(t-s)f = U^K(s)U_0(-s)U_0(t)f \quad \text{for all} \quad f \in L^1(\mathbb{R}^n \times V).$$

As $s \to \infty$, we obtain (a). By taking its Laplace transform we get

$$\begin{aligned}\int_0^\infty e^{-\lambda t} U^K(t) W_-^K f dt &= \int_0^\infty W_-^K e^{-\lambda t} U_0(t) f dt \\ &= W_-^K \int_0^\infty e^{-\lambda t} U_0(t) f dt.\end{aligned}$$

Thus

$$(1 - T^K)^{-1} W_-^K f = W_-^K (1 - T_0)^{-1} f.$$

Since $D(T^K) = D(T_0^K)$, this relation implies that $W_-^K(D(T_0)) \subseteq D(T^K)$, which allows us to take the derivative of (a) with respect to t and obtain

$$T^K U^K(t) W_-^K f = W_-^K T_0 U_0(t) f \quad \text{for all} \quad f \in D(T_0),$$

which is (c) for $t = 0$.

The proof of (b) and (d) is similar by considering

$$U_0(t)U_0(-s)U^K(s) = U_0(t-s)U^K(s-t)U^K(t).$$

□

Maintaining the hypotheses of the Theorem 6.4.9 which ensure the existence of the scattering operator $\mathcal{S}^K$, we consider the interior evolution problem

$$\textbf{(IPK)} \qquad \begin{cases} \dfrac{\partial \psi}{\partial t}(x, v, t) = \widetilde{T}^K \psi(x, v, t) \quad \text{in} \quad \Omega \times V, \\ \psi(x, v, 0) = g(x, v) \in \widetilde{W}_K^1(\Omega \times V), \\ \gamma_- \psi = K \gamma_+ \psi \quad \text{on} \quad \Gamma_-, \end{cases}$$

where $\widetilde{T}^K := \chi_\Omega T^K$. In [57, Ch. XI Theorem 4.3] it is shown that this problem admits a unique solution in $\mathcal{C}^0((0,\infty); \widetilde{W}^1_K(\Omega \times V))$.

Definition 6.4.11 Let ψ be the solution of (IPK).The *semi-transparence albedo operator* $\mathcal{A}^K$ is an operator from $C([0,); L^1(\Gamma_-, d\xi))$ into $C([0,); L^1(\Gamma_+, d\xi))$ given by

$$\mathcal{A}^K: \ \psi\Big|_{\Gamma_-} \to \psi\Big|_{\Gamma_+}$$

which depends to the initial data $g(x, v) \in \widetilde{W}^1_K(\Omega \times V)$.

Theorem 6.3.6 of the previous section can be announced similarly for the semi-transparent albedo operator. In fact the effect of partial reflectivity for the emergent particle do not change the proof of this theorem, since all materials in the proof do never takes into account the negative values of t for the group $T(t)$. So we can announce the following Theorem.

Theorem 6.4.12 *Suppose that the scattering operator $\mathcal{S}^K$ exists and the albedo operator is well-defined. Then*

$$\gamma_+ \mathcal{S}^K = \mathcal{A}^K \gamma_-$$

on $\widetilde{W}^1(\mathbb{R}^n \times V)$.

6.5 Notes and Comments

The idea of finding a relationship between albedo and scattering operators was triggered by the outcome of J. Hejtmanek's paper [62], which, as we have said, tried to interpret Radon's formula expressed by scattering operator. In this period, two papers by Cessenat [26, 27] and Voigt's thesis [121] have helped us to carry out successfully this relationship and publish it in [6]. Since Cessenat's results are published in C.R. Acad. Sci. Paris and the proof is very succinct, so instead of citing them I prefer to spread them out in three theorems, 6.2.1, 6.2.2 and 6.2.3.

Later, Protopopescu [99] gave an generalization of this paper by introducing the semi-transparent boundary conditions which are systematically elaborated in [48]. The essential characterization is to consider the boundary value problem for the transport equation and assuming that the boundary trapped partially the outgoing particles by introducing a boundary operator K which acts as a strict positive contraction between the incoming and outgoing boundary spaces. The proofs are based on classical arguments of boundedness, contractivity and positivity and rely on the fact that all the distributions tend strongly to asymptotically free distributions as long as:

(i) there exists a reasonable balance between absorption and production;
(ii) the boundary operator K is a strict positive contraction.

The boundary value problem for transport equation is extensively studied in the literature, and also the generation of a C_0-semigroup (see [11, 16]. This problem has been proved not only for contraction boundary operators, but also for compact or even continuous operators (see [20]) and one of the most beautiful papers on this subject is [88] in which S. Matucci split the whole space into two components, inner distribution density and outer distribution density, and the boundary operator acts as a 2×2 matrix on these two components.

Chapter 7
Scattering Theory for Quantum Transport Equation

7.1 Wigner Transform: A Bridge Between Quantum and Statistical Mechanics

Suppose that in the theory of quantum mechanics the potential V is chosen in such a way that the Hamiltonian

$$H = \frac{-\hbar^2}{2m}\Delta + V$$

($\hbar$ being Planck's constant) is self-adjoint in the Hilbert space $\mathcal{H} = L^2(\mathbb{R}^n)$. Let $\varphi_\hbar$ be the solution of the corresponding Schrödinger equation, i.e.

$$i\hbar\frac{\partial\varphi}{\partial t} = H\varphi,$$

and $\varphi_\hbar^*$ its complex conjugate. We suppose that $\varphi_\hbar$ is a pure state, that is,

$$\int_{\mathbb{R}^n} |\varphi_\hbar(x)|^2 dx = 1. \tag{7.1}$$

The correspondent density matrix

$$\rho_\hbar(x, y) = \varphi_\hbar(x) \cdot \varphi_\hbar^*(y) \tag{7.2}$$

belongs to $L^2(\mathbb{R}^{2n})$, for which we denote again by $\boldsymbol{\rho}_\hbar$ the correspondent trace class operator:

$$\boldsymbol{\rho}_\hbar f = \int_{\mathbb{R}^n} \rho_\hbar(x, y) f(y)dy, \qquad f \in \mathcal{H}.$$

H. Emamirad, *Scattering Theory for Transport Phenomena*, Mathematical Physics Studies, https://doi.org/10.1007/978-981-16-2373-8_7

The above representation implies that the operator $\rho_\hbar$ belongs to Hilbert–Schmidt class $\mathscr{I}_2(\mathcal{H})$. Moreover, it is well-known that $\rho_\hbar$ governs the Liouville–von Neumann equation

$$\mathrm{i}\,\hbar\rho_t = [H,\rho], \qquad \rho(0) = \rho_0 \tag{7.3}$$

on which the quantum theory of open system is based. The solution of (7.3) can be expressed by the so-called *quantum dynamical semigroup* $\Phi_\hbar(t)$,

$$\rho_\hbar(t) = \Phi_\hbar(t)\rho_0 = e^{-i\frac{t}{\hbar}H}\rho_0\, e^{i\frac{t}{\hbar}H}, \tag{7.4}$$

which is a unitary group on $\mathscr{I}_1(\mathcal{H})$.

There is also another link between quantum theory of the open systems and the theory of statistical mechanics. This connection is given by the *Wigner transform* of the state $\varphi_\hbar$ given by

$$\begin{aligned} W_{\varphi_\hbar}(x,\xi,t) := W_\hbar(x,\xi,t) &= (2\pi)^{-n/2}\int_{\mathbb{R}^n} e^{-i\xi y}\rho_\hbar(x+\frac{\hbar y}{2}, x-\frac{\hbar y}{2}, t)\,dy \\ &= (2\pi)^{-n/2}\int_{\mathbb{R}^n} e^{-i\xi y}\varphi_\hbar(x+\frac{\hbar y}{2}, t)\varphi_\hbar^*(x-\frac{\hbar y}{2}, t)dy. \end{aligned}$$

For any $g \in L^2(\mathbb{R}^n \times \mathbb{R}^n)$, let us define $\widetilde{g}(x,y) = g(u,v)$, where

$$u = x + \frac{\hbar y}{2}, \qquad v = x - \frac{\hbar y}{2}.$$

If $\rho_\hbar \in \mathscr{I}_1(\mathcal{H})$, then $\widetilde{\rho}_\hbar$ belongs also to $\mathscr{I}_1(\mathcal{H})$, since

$$\iint_{\mathbb{R}^n_x\times\mathbb{R}^n_y} \widetilde{\rho}(x,y)dxdy = \iint_{\mathbb{R}^n_u\times\mathbb{R}^n_v} \rho(u,v)\left(\hbar^n\right)dudv, \tag{7.5}$$

where $\hbar^n$ is the Jacobian determinant of the mapping

$$C_\hbar:\ (x,y) \to (x+\frac{\hbar y}{2}, x-\frac{\hbar y}{2}). \tag{7.6}$$

Now, denote the Fourier transform of the function $f(x,\eta,t)$ with respect to the momentum variable η by

$$\mathscr{F}_\eta f(x,\xi,t) = (2\pi)^{-n/2}\int_{\mathbb{R}^n_\eta} f(x,\eta,t)e^{-i\xi.\eta}d\eta$$

and

$$\mathscr{F}_\xi^{-1} f(x,\eta,t) = \overline{\mathscr{F}_\xi} f(x,\eta,t) = (2\pi)^{-n/2}\int_{\mathbb{R}^n_\xi} e^{i\xi.\eta} f(x,\xi,t)d\xi.$$

Then, the Wigner function can be expressed as

$$W_\hbar(x,\xi,t) = [\mathscr{F}_y \widetilde{\rho}_\hbar(x,y,t)](\xi). \tag{7.7}$$

This function provides a phase-space view of the quantum state. If $\{\psi_i;\, i \in I\}$ is an ONB of $\mathcal{H}$ and $W_i(x,\xi) = (2\pi)^{-n/2} \int_{\mathbb{R}^n_\eta} \psi_i(x+\frac{\eta}{2})\psi_i^*(x-\frac{\eta}{2})e^{-i\xi.\eta}d\eta$ the Wigner transform of ψ_i, then $\{W_i;\, i \in I\}$ forms an ONB in the phase Hilbert space $\mathscr{H} := L^2(\mathbb{R}^n_x \times \mathbb{R}^n_\xi)$. In fact, since $W_i^*(x,\xi) = \overline{\mathscr{F}_y(\widetilde{\rho}_i(x,y))(\xi)}$, we get by the Plancherel identity

$$\begin{aligned}
\iint_{\mathbb{R}^n_x\times\mathbb{R}^n_\xi} W_i(x,\xi)W_j^*(x,\xi)dxd\xi &= \iint_{\mathbb{R}^n_x\times\mathbb{R}^n_\xi} \big[\mathscr{F}_y(\widetilde{\rho}_i(x,y)\overline{\mathscr{F}_y}(\widetilde{\rho}_j(x,y)\big](\xi)dxd\xi \\
&= \iint_{\mathbb{R}^n_x\times\mathbb{R}^n_y} \widetilde{\rho}_i(x,y)\widetilde{\rho}_j^*(x,y)dydx. \\
&= \iint_{\mathbb{R}^n_x\times\mathbb{R}^n_z} \psi_i(x+\frac{\hbar z}{2})\psi_i^*(x-\frac{\hbar y}{2})\psi_j^*(x+\frac{\hbar y}{2}) \\
&\qquad \psi_j(x-\frac{\hbar y}{2})dydx.
\end{aligned}$$

By taking $u = x + \frac{\hbar y}{2}$ and $v = x - \frac{\hbar y}{2}$ we have

$$\begin{aligned}
\int_{\mathbb{R}^n_x}\int_{\mathbb{R}^n_\xi} W_i(x,\xi)W_j^*(x,\xi)dxd\xi &= \left(\int_{\mathbb{R}^n_u} \psi_i(u)\psi_j^*(u)du\right)\left(\int_{\mathbb{R}^n_v} \psi_i^*(v)\psi_j(v)dv\right) \\
&= \left|\langle\psi_i,\psi_j\rangle\right|^2.
\end{aligned}$$

For $i = j$ we get $\|W_i\|_{\mathscr{H}} = 1$ which implies that the set $\{W_i;\, i \in I\}$ forms an ONB in $\mathscr{H}$.

By the same argument we can prove that if W_φ and W_ψ are the Wigner transform of the states φ and ψ, then

$$\int_{\mathbb{R}^n_x}\int_{\mathbb{R}^n_\xi} W_\varphi(x,\xi)W_\psi^*(x,\xi)dxd\xi = \left|\int_{\mathbb{R}^n_x} \varphi(x)\psi^*(x)dx\right|^2. \tag{7.8}$$

In the next section we see that if the potential $V \equiv 0$, then $W_\hbar$ satisfies the advection equation

$$\frac{\partial w}{\partial t} + \xi \cdot \nabla_x w = 0 \tag{7.9}$$

of the statistical mechanics (see also [56, Théorème 10.4.5.]) and we deduce this property by means of quantum dynamical semigroup in Sect. 7.8. If the potential

$V \neq 0$, the Wigner function (7.7) satisfies, so-called *Wigner equation* or *quantum Liouville equation*:

$$\textbf{(QLE)} \qquad \begin{cases} \dfrac{\partial W_\hbar}{\partial t} + \xi \cdot \nabla_x W_\hbar - P_\hbar(x, \nabla_\xi) W_\hbar = 0, \\ W_\hbar(x, \xi, 0) = w^0(x, \xi), \end{cases}$$

where $P_\hbar \equiv P(x, \nabla_\xi)$ is a pseudo-differential operator, which is symbolically given by

$$\frac{1}{i\hbar}[V(x + \frac{i\hbar}{2}\nabla_\xi) - V(x - \frac{i\hbar}{2}\nabla_\xi)] \tag{7.10}$$

or equivalently

$$\begin{aligned} P_\hbar(x, \nabla_\xi) f &= (2\pi)^{-n} \int_{\mathbb{R}^n{}_y \times \mathbb{R}^n{}_\eta} e^{i(\xi-\eta)y} \frac{i}{\hbar} \left[V(x + \frac{\hbar}{2}y) - V(x - \frac{\hbar}{2}y) \right] f(x, \eta) dy d\eta \\ &= \overline{\mathscr{F}_y} \left[\delta_\hbar V(x, y) \left[\mathscr{F}_\eta f(x, \eta) \right](y) \right](\xi) \\ &= [k_\hbar(x, \cdot) \underset{\xi}{*} f(x, \cdot)](\xi), \end{aligned}$$

where

$$\delta_\hbar V(x, y) = \frac{i}{\hbar} \left[V\left(x + \frac{\hbar}{2}y\right) - V\left(x - \frac{\hbar}{2}y\right) \right]. \tag{7.11}$$

Since this is an odd function,

$$k_\hbar(x, \xi) = (2\pi)^{-\frac{n}{2}} \mathscr{F}_y \big[\delta_\hbar V(x, y) \big](\xi). \tag{7.12}$$

Here we have to introduce two important physical notions: the density of charged particles ρ and the current density J, given by

$$\rho(x, t) = \int_{\mathbb{R}^n} w(x, \xi, t) d\xi \quad \text{and} \quad J(x, t) = \int_{\mathbb{R}^n} w(x, \xi, t) \xi d\xi. \tag{7.13}$$

We remark that J is a vector-valued function.

Proposition 7.1.1 *The density of charged particles ρ and the current density J verify the following equation:*

$$\rho_t(x, t) + \nabla_x J(x, t) = 0. \tag{7.14}$$

Proof It is enough to integrate (**QLE**) with respect to ξ and use the fact that $\int_{\mathbb{R}^n} k_\hbar(x, \xi) d\xi = 0$. □

Since any trace class operator is of Hilbert–Schmidt type, and its kernel belongs to $L^2(\mathbb{R}^n \times \mathbb{R}^n)$, so does $\widetilde{\rho}_\hbar$. $\mathscr{F}_y$ being an isometric operator in $L^2(\mathbb{R}^n \times \mathbb{R}^n)$, the

Wigner function belongs also to $L^2(\mathbb{R}^n \times \mathbb{R}^n)$. This property inspired many authors to make their analysis in $L^2(\mathbb{R}^n \times \mathbb{R}^n)$ space.

To sum up these properties we can say that if $\varphi_\hbar$ is the solution of the Schrödinger equation in $L^2(\mathbb{R}^n)$, then $W(\varphi_\hbar)$ belongs to $L^2(\mathbb{R}^n \times \mathbb{R}^n)$, so W makes a bridge between quantum and statistical mechanics. But, in statistical mechanics the unknown of the problems is the density of particles (probability or mass) and the physically right space invokes $L^1_+(\mathbb{R}^n \times \mathbb{R}^n)$, the positive cone of the Banach lattice $L^1(\mathbb{R}^n \times \mathbb{R}^n)$. Related to this, in the next section we will study the well-posedness of the Wigner equation in $L^1(\mathbb{R}^n \times \mathbb{R}^n)$ and by interpolation we retrieve the well-posedness in $L^p(\mathbb{R}^n \times \mathbb{R}^n)$, which will be helpful in the existence theorems of the wave operators.

7.2 Well-Posedness of the Wigner Problem

For the well-posedness of the Wigner problem (**QLE**) in the $L^2(\mathbb{R}^{2n})$ (then by interpolation in $L^p(\mathbb{R}^{2n})$) space we need the following lemma. Let H^s be the Sobolev space

$$H^s(\mathbb{R}^n) = \{u \in \mathcal{S}'; \int_{\mathbb{R}^n} (1 + |\xi|^2)^s |\widehat{u}(\xi)|^2 \, \mathrm{d}\xi < \infty\},$$

($\mathcal{S}'$ being the space of the temperate distributions) with the norm

$$\|u\|_{H^s} = \left(\int_{\mathbb{R}^n} (1 + |\xi|^2)^s |\widehat{u}(\xi)|^2 \, \mathrm{d}\xi \right)^{1/2}.$$

We recall that, for $u \in H^s(\mathbb{R}^n)$, when $s > n/2$, we have $\widehat{u} \in L^1(\mathbb{R}^n)$, and

$$\|\widehat{u}\|_1 \leq C \|u\|_{H^s}, \tag{7.15}$$

with $C = \int_{\mathbb{R}^n} (1 + |\xi|^2)^{-s} \, \mathrm{d}\xi$. Here $\|\cdot\|_1$ denote the norm of $L^1(\mathbb{R}^n_\xi)$, but in the following lemma we designate $\|\cdot\|_{1,1}$ to be the norm of $L^1(\mathbb{R}^n_x \times \mathbb{R}^n_\xi)$.

Lemma 7.2.1 *Assume that the potential $V \in H^s(\mathbb{R}^n)$ with $s > n/2$. Then*

$$\|P_\hbar(x, \nabla_\xi) f\|_{1,1} \leq m_\hbar \|f\|_1, \tag{7.16}$$

where $m_\hbar = \frac{2C}{\hbar (2\pi)^{n/2}} \|V\|_{H^s(\mathbb{R}^n)}$. Consequently the operator $L_\hbar$ generates a (C_0) group of type $(1, m_\hbar)$, satisfying

$$\|\mathrm{e}^{tL_\hbar}\|_{\mathscr{L}(L^1(\mathbb{R}^2))} \leq \mathrm{e}^{tm_\hbar}. \tag{7.17}$$

Proof Since $P_\hbar(x,\nabla_\xi)f = k_\hbar \underset{\xi}{*} f$ and

$$\left\|\left[\overline{\mathscr{F}}_y V\left(x \pm \frac{\hbar}{2}y\right)\right](\xi)\right\|_1 = \frac{2^n}{\hbar^n}\left\|\left[\overline{\mathscr{F}}_z V(x\pm z)\right]\left(\frac{2\xi}{\hbar}\right)\right\|_1$$
$$= \|\left[\overline{\mathscr{F}}_z V(x\pm z)\right](\eta)\|_1.$$

Writing

$$\|P_\hbar(x,\nabla_\xi)f\|_1 = \|k_\hbar \underset{\xi}{*} f\|_1$$
$$\leq \sup_x\left(\int_{\mathbb{R}^n}|k_\hbar(x,\xi)|\,\mathrm{d}\xi\right)\|f\|_1.$$

According to (7.12) and using (7.15), one gets

$$\|k_\hbar(x,\xi)\|_1 \leq \frac{2(2\pi)^{-n/2}}{\hbar}\|\mathscr{F}_z V\|_1$$
$$\leq \frac{2C(2\pi)^{-n/2}}{\hbar}\|V\|_{H^s},$$

and consequently the estimation (7.16). In any $L^p(\mathbb{R}^n_x\times\mathbb{R}^n_\xi)$, $(1\leq p<\infty)$ the operator

$$L_\hbar := -\xi\cdot\nabla_\xi + P_\hbar(x,\nabla_\xi) \tag{7.18}$$

with domain $D_p(L_\hbar) := \{f\in L^p(\mathbb{R}^n_x\times\mathbb{R}^n_\xi)\mid L_\hbar f\in L^p(\mathbb{R}^n_x\times\mathbb{R}^n_\xi)\}$ is a perturbation of the advection operator

$$L_0 := -\xi\cdot\nabla_\xi. \tag{7.19}$$

When $V\equiv 0$, the advection equation generated by the unitary (C_0) group is $S_0(t)w(x,\xi) = w(x-\xi t,\xi)$. Since $P_\hbar$ is bounded, from the perturbation theorem by the bounded operators (see Corollary 1.8.8), $L_\hbar$ generates a (C_0) group of type $(1,\delta/\hbar)$, satisfying (7.17). Consequently, the problem (**QLE**) is well-posed in $L^1(\mathbb{R}^n_x\times\mathbb{R}^n_\xi)$. □

In the Hilbert space $L^2(\mathbb{R}^{2n})$ we can announce the following important result.

Theorem 7.2.2 *In the Hilbert space $L^2(\mathbb{R}^{2n})$ the operator $L_\hbar$ generates a (C_0) unitary group.*

Proof In order to use Stone's Theorem 1.7.4, it suffices to prove that $L_\hbar$ is skew-symmetric. Let f and g be two functions in $D_2(L_\hbar)$, then

$$\int_{\mathbb{R}^n_\xi} P_\hbar(x,\nabla_\xi) f(x,\xi)\overline{g}(x,\xi)d\xi = (2\pi)^{-\frac{n}{2}} \int_{\mathbb{R}^n_\xi}\int_{\mathbb{R}^n_y} e^{iy\xi}\delta_\hbar V(x,y)\big[\mathscr{F}_\eta f(x,\cdot)\big](y)dy\overline{g}(x,\xi)d\xi$$
$$= \int_{\mathbb{R}^n_y}\Big[\mathscr{F}_\xi\overline{g}(x,\xi)\Big](-y)\delta_\hbar V(x,y)\big[\mathscr{F}_\eta f(x,\cdot)\big](y)dy.$$

Replacing $-y$ by y in the integral, and using the fact that $\delta_\hbar V(x,y)$ is an odd function with respect to y gives

$$\int_{\mathbb{R}^n_\xi} P_\hbar(x,\nabla_\xi) f(x,\xi)\overline{g}(x,\xi)d\xi = -(2\pi)^{-\frac{n}{2}} \int_{\mathbb{R}^n_\xi} f(x,\xi)\int_{\mathbb{R}^n_y} e^{iy\xi}\delta_\hbar V(x,y)$$
$$\big[\mathscr{F}_\eta\overline{g}(x,\eta)\big](y)dyd\xi$$
$$= -\int_{\mathbb{R}^n_y} f(x,\xi)\delta_\hbar V(x,y)\big[\mathscr{F}_\eta\overline{g}(x,\eta)\big](y)dy$$
$$= -\int_{\mathbb{R}^n_\xi} f(x,\xi)P_\hbar(x,\nabla_\xi)\overline{g}(x,\xi)d\xi. \tag{7.20}$$

By integrating both side of (7.20) on $\mathbb{R}^n_x$ we prove the skew-symmetry of $L_\hbar$ and thanks to Stone's Theorem we get

$$\|e^{tL_\hbar}\|_{\mathscr{L}(L^2(\mathbb{R}^{2n}))} = 1. \tag{7.21}$$

□

Now we can establish, by using the well-known Riesz–Thorin interpolation theorem (following lemma), an estimation of the group $e^{tL_\hbar}$ on $L^p(\mathbb{R}^{2n})$ spaces, for $1 \le p \le 2$.

Lemma 7.2.3 *Let (M,μ) and (N,ν) be measure spaces with σ-finite measures. Let $1 \le p_0, p_1, r_0, r_1 \le \infty$, and suppose that T is a linear transformation from $L^{p_0}(M,d\mu)\cap L^{p_1}(M,d\mu)$ to $L^{r_0}(N,d\nu)\cap L^{r_1}(N,d\nu)$ which satisfies*

$$\|Tf\|_{r_0} \le M_0\|f\|_{p_0} \quad \text{and} \quad \|Tf\|_{r_1} \le M_1\|f\|_{p_1}.$$

Then for each $f \in L^{p_0}\cap L^{p_1}$ and each $\theta \in (0,1)$, $Tf \in L^r$ and $\|Tf\|_r \le M_\theta\|f\|_p$, where $\frac{1}{p} = \frac{\theta}{p_1} + \frac{1-\theta}{p_0}$, $\frac{1}{r} = \frac{\theta}{r_1} + \frac{1-\theta}{r_0}$ and $M_\theta = M_0^{1-\theta}M_1^\theta$.

Theorem 7.2.4 *Under the assumptions of Lemma 7.2.1, the group $e^{tL_\hbar}$ is also well-defined in $L^p(\mathbb{R}^{2n})$ $1 \le p \le 2$, and we have*

$$\|e^{tL_\hbar}f\|_{L^p(\mathbb{R}^{2n})} \le e^{\theta M_\theta|t|}\|f\|_{L^p(\mathbb{R}^{2n})} \qquad \forall t \in \mathbb{R}, \tag{7.22}$$

where $\theta = 2 - \frac{2}{p}$.

Proof For $f \in L^1 \cap L^2(\mathbb{R}^{2n})$, according to Lemma 7.2.1 and Theorem 7.2.2 we have for all $t \in \mathbb{R}$

$$\|e^{tL_\hbar} f\|_{L^1(\mathbb{R}^{2n})} \leq e^{m_\hbar |t|} \|f\|_{L^1(\mathbb{R}^{2n})} \quad \text{and} \quad \|e^{tL_\hbar} f\|_{L^2(\mathbb{R}^{2n})} = \|f\|_{L^2(\mathbb{R}^{2n})}.$$

Taking $p_0 = r_0 = 1$, $p_1 = A = 2$, $M_0 = e^{m_\hbar}$, $M_1 = 1$ and $\theta = 2 - \frac{2}{p}$ in the Riesz–Thorin theorem (Lemma 7.2.3), we retrieve (7.22). □

7.3 Liouville–Vlasov Equation

Another interesting feature of the Wigner equation is its link with classical mechanic. In fact this equation is the semiclassical version of the Liouville–Vlasov system:

$$\textbf{(LVS)} \qquad \begin{cases} \dfrac{\partial v}{\partial t} + \xi \cdot \nabla_x v = \nabla_x V \cdot \nabla_\xi v, \\ v(x, \xi, 0) = w^0(x, \xi). \end{cases}$$

This link can be simply derived by *Koopman formalism* or by Weyl calculus, which will be subject of this section. The Koopman formalism offers great utility in many branches of mathematics. One of the most impressive features of this formalism is to transfer a finite-dimensional nonlinear dynamical system in an infinite-dimensional linear Cauchy problem.

Let

$$\textbf{(DS)} \qquad \begin{cases} \dot{\Phi} = F(\Phi) \\ \Phi(0) = \Phi_0 \end{cases}$$

be a dynamical system in $\mathbb{R}^N$. In our case $N = 2n$, $\Phi_0 = (x_0, \xi_0)$ and

$$X(x_0, \xi_0, t) = \Phi(-t)[(x_0, \xi_0)], \qquad U(t)f(x, \xi) = f(X(x, \xi, t)), \tag{7.23}$$

where $\Phi(t)$ is the flow of the correspondent ODE on the phase space. The operator $U(t)$ is called, the *Koopman operator*. The simplest example of such an operator is the transport operator for the following free force Hamiltonian system.

$$\textbf{(FFHS)} \qquad \begin{cases} \dot{x} = \xi, \quad x(0) = x_0 \\ \dot{\xi} = 0 \quad \ \xi(0) = \xi_0. \end{cases}$$

Here

$$\Phi(t)(x, \xi) = (x + t\xi, \xi)$$

and

$$X(x_0, \xi_0, t) = (x_0 - t\xi_0, \xi_0), \qquad U(t)f(x, \xi) = f(x - t\xi, \xi).$$

In this case $u(x, \xi, t) = U(t)f(x, \xi)$ satisfies the *advection equation*

$$\textbf{(AE)} \qquad \begin{cases} u_t = -\xi.\nabla_x u \\ u(x, \xi, 0) = f(x, \xi). \end{cases}$$

The next natural example is the Hamiltonian system; when the external force is not zero and derives from a potential, the Hamiltonian can be written as

$$\textbf{(HS)} \qquad \begin{cases} \dot{x} = \xi, & x(0) = x_0 \\ \dot{\xi} = -\nabla_x V & \xi(0) = \xi_0. \end{cases}$$

By applying the Koopman formalism to this system we get the *Liouville–Vlasov system* (**LVS**). Concerning the correspondent Koopman operator we have:

Lemma 7.3.1 *Let* $V \in W^{1,\infty}(\mathbb{R}^n)$ *and* $[L_0 u](x, \xi) = -\xi.\nabla_x u(x, \xi) + \nabla_x V.\nabla_\xi u(x, \xi)$, *with domain* $D(L_0) = \{f \in L^2(\mathbb{R}^n_x \times \mathbb{R}^n_\xi) \; : \; \xi.\nabla_x f + b(x).\nabla_\xi f \in L^2(\mathbb{R}^n_x \times \mathbb{R}^n_\xi)$, $\forall b \in L^\infty(\mathbb{R}^n)\}$, *then* L_0 *generates a unitary* (C_0) *group in* $L^2(\mathbb{R}^n_x \times \mathbb{R}^n_\xi)$.

Proof It is not hard to see that the Schwartz space $\mathcal{S}(\mathbb{R}^n_x \times \mathbb{R}^n_\xi) \subset D(L_0)$, hence $D(L_0)$ is dense in $L^2(\mathbb{R}^n_x \times \mathbb{R}^n_\xi)$. The Koopman operator $U(t)f(x, \xi) = f(x(t), \xi(t))$, where $x(t)$ and $\xi(t)$ satisfy the Hamiltonian system (HS), with $(x(0), \xi(0)) = (x, \xi)$ is unitary in $L^2(\mathbb{R}^n_x \times \mathbb{R}^n_\xi)$. For this let $f \in L^2(\mathbb{R}^n_x \times \mathbb{R}^n_\xi)$ and take $y = x(t)$, $v = \xi(t)$ in the following integral:

$$\|U(t)f\|^2 = \int_{\mathbb{R}^{2n}} |f(x(t), \xi(t))|^2 dx d\xi = \int_{\mathbb{R}^{2n}} |f(y, v)|^2 dy dv,$$

since the domain of integration is the whole of $\mathbb{R}^{2n}$. Thus

$$\|U(t)f\|^2 = \|f\|^2. \tag{7.24}$$

Now let us show the assumptions of Definition 1.2.1.

(i) The fact that $U(0)f(x, \xi) = f(x, \xi)$ follows from the definition of $U(t)$.
(ii) Since

$$U(t)U(s)f(x, \xi) = U(t)f(x(s), \xi(s)),$$

the second term is the solution of (**HS**) with the initial condition $(x(s), \xi(s))$, so this is the solution of (**HS**) with the initial condition (x, ξ) at the time $t + s$. Thus

$$U(t)U(s)f(x, \xi) = U(t + s)f(x, \xi).$$

(iii) For the continuity of $U(t)$ at origin, we write for $\varphi \in \mathcal{S}(\mathbb{R}^{2n})$

$$\begin{aligned}\|U(t)f - f\| &\le \|U(t)f - U(t)\varphi\| + \|U(t)\varphi - \varphi\| + \|\varphi - f\| \\ &= \left(\int_{\mathbb{R}^{2n}} |U(t)\varphi(x,\xi) - \varphi(x,\xi)|^2 dx d\xi \right)^{1/2} \\ &\quad + 2\left(\int_{\mathbb{R}^{2n}} |\varphi(x,\xi) - f(x,\xi)|^2 dx d\xi \right)^{1/2}.\end{aligned}$$

The first term of the right-hand side goes to zero since φ is continuous at origin and the second term goes to zero since $\mathcal{S}(\mathbb{R}^{2n})$ is dense in $L^2(\mathbb{R}^{2n})$. The fact that $U(t)$ is a (C_0) group follows from (7.24) and Stone's Theorem. □

7.4 Wigner Function in the Weyl Calculus

In Hamiltonian mechanics over a phase space $\mathbb{R}^n_x \times \mathbb{R}^n_\xi$, the *Poisson bracket* $\{u, v\}$ of two smooth observables $u, v : \mathbb{R}^n_x \times \mathbb{R}^n_\xi$ is a new observable w defined by

$$w = \{u, v\} = \sum_{j=1}^{n} \left(\frac{\partial u}{\partial x_j} \frac{\partial v}{\partial \xi_j} - \frac{\partial u}{\partial \xi_j} \frac{\partial v}{\partial x_j} \right).$$

For a state (x, ξ) of the phase space $\mathbb{R}^n_x \times \mathbb{R}^n_\xi$, the *position vector* is given by $x = (x_1, \cdots, x_n)$, while $\xi = (\xi_1, \cdots, \xi_n)$ is the *momentum vector*. The Poisson brackets of the coordinate functions x_j, ξ_k are given by

$$\{x_j, x_k\} = \{\xi_j, \xi_k\} = 0, \quad \{x_j, \xi_k\} = \delta_{jk}.$$

By comparison with quantum mechanics over $\mathbb{R}^n$, the position operator is $Q_j f := x_j f$ and the momentum operator is P_k given by

$$P_k f := \frac{\hbar}{\mathrm{i}} \frac{\partial f}{\partial x_k}.$$

If we replace the Poisson bracket $\{.,.\}$ by the Lie bracket $[.,.]$, we get relationships of the same type:

$$[Q_j, Q_k] = [P_j, P_k] = 0, \quad [P_j, Q_k] = \frac{\hbar}{\mathrm{i}} \delta_{jk} I.$$

Now let us consider the standard Hamiltonian $h(x, \xi) = \frac{|\xi|^2}{2} + V(x)$ and the Poisson bracket

$$\{h, w\} = \nabla_x V . \nabla_\xi w - \nabla_\xi h . \nabla_x w,$$

and if w is a function in $\mathbb{R}^{2n+1}$ of (x, ξ, t), then

$$\begin{aligned}\partial_t w &= \{h, w\} \\ &= \nabla_x V.\nabla_\xi w - \xi.\nabla_x w.\end{aligned}$$

This shows that if we replace w by the Wigner function and take $V \equiv 0$ then w satisfies the advection equation (7.9). And for $V \not\equiv 0$, w satisfies the the standard (**LVE**) system.

Here we are going to introduce the Weyl transform by means of a pseudo-differential operator. For this we need to present the class of the symbols, since the Weyl transform is a pseudo-differential operator for some particular symbol. Let m be a real number; we denote by S^m the set of all infinitely differentiable functions on $\mathbb{R}^{2n}$ such that for all multi-indices α and β, there is a positive constant $C_{\alpha,\beta}$, for which

$$|(\partial_x^\alpha \partial_\xi^\beta \sigma)(x, \xi)| \leq C_{\alpha,\beta}(1 + |\xi|)^{m-|\beta|}, \quad x, \xi \in \mathbb{R}^n.$$

Let $\sigma \in S^m$, the operator $Op(\sigma)$ defined on the Schwartz space $\mathcal{S}(\mathbb{R}^n)$ of C^∞ functions rapidly decreasing at infinity as

$$[Op(\sigma)]\varphi(x) = (2\pi)^{-n} \int_{\mathbb{R}^{2n}} e^{i(x-y)\xi} \sigma(x, \xi)\varphi(y) dy d\xi, \quad \varphi \in \mathcal{S}(\mathbb{R}^n) \tag{7.25}$$

is called the *pseudo-differential operator* with symbol σ.

Now if we replace $\sigma(x, \xi)$ in (7.25) by $\sigma((x + y)/2, \xi)$ we get the *Weyl transform* of σ which denotes by

$$[W_\sigma]\varphi(x) = (2\pi)^{-n} \int_{\mathbb{R}^{2n}} e^{i(x-y)\xi} \sigma\left(\frac{x+y}{2}, \xi\right) \varphi(y) dy d\xi, \quad \varphi \in \mathcal{S}(\mathbb{R}^n) \tag{7.26}$$

where $\sigma((x + y)/2, \xi)$ is called *Weyl symbol*.

In the first section we have defined the Wigner function for a single state $\varphi_\hbar$. Now in order to introduce the Wigner function in the Weyl calculus, most of the authors generalize this definition and introduce the mixed Wigner function for two states $\varphi_\hbar$ and $\psi_\hbar$ and define

$$W_{\varphi_\hbar,\psi_\hbar}(x, \xi) = (2\pi)^{-n/2} \int_{\mathbb{R}^n} e^{-i\xi y} \varphi_\hbar\left(x + \frac{\hbar y}{2}\right) \psi_\hbar^*\left(x - \frac{\hbar y}{2}\right) dy.$$

The above definition can be extended to any pair $(\varphi, \psi) \in \mathcal{S}(\mathbb{R}^n)^2$ by taking $\hbar = 1$ and writing

$$W_{\varphi,\psi}(x, \xi) = (2\pi)^{-n/2} \int_{\mathbb{R}^n} e^{-i\xi y} \varphi\left(x + \frac{y}{2}\right) \overline{\psi\left(x - \frac{y}{2}\right)} dy. \tag{7.27}$$

With this notation a Plancherel-type formula can be derived which is called the *Moyal identity* (see [93]),

$$(W_{\varphi_1,\psi_1}, W_{\varphi_2,\psi_2}) = (\varphi_1, \psi_1)(\varphi_2, \psi_2), \quad \text{for all } \varphi_1, \psi_1\varphi_2, \psi_2 \in \mathcal{S}(\mathbb{R}^n). \tag{7.28}$$

The following relationship between the Wigner distribution and the Weyl transform can be found in [125, Theorem 4.3]: For any Weyl symbol $\sigma \in S^m$,

$$(W_\sigma\varphi, \psi) = (2\pi)^{-n} \int_{\mathbb{R}^{2n}} \sigma(x, \xi) W_{\varphi,\psi}(x, \xi) dx d\xi, \quad \varphi, \psi \in \mathcal{S}(\mathbb{R}^n). \tag{7.29}$$

7.5 Semiclassical Limit of Wigner Equation

In this section we use the Lemma 1.16.3 to prove the ultraweak convergence of $e^{tL_\hbar}$ to e^{tL_0} as $\hbar \to 0$. This is in some manner foreseeable, since as $\hbar \to 0$, the expression (7.10) converges formally to the right-hand side of Liouville–Vlasov equation. More precisely we have:

Theorem 7.5.1 *Assume that the potential $V \in H^s(\mathbb{R}^n)$ with $s > n/2$. Then, the group $e^{tL_\hbar}$ is unitary and converges ultraweakly in $\mathcal{H} = L^2(\mathbb{R}^n_x \times \mathbb{R}^n_\xi)$ to e^{tL_0} as $\hbar \to 0$.*

Proof Since the group $e^{tL_\hbar}$ is unitary and the unit ball is weakly compact in $\mathcal{H}$, for any $w^0(x, \xi) \in \mathcal{H}$ we can extract a weak convergence subset of $\{e^{tL_\hbar} w^0(x, \xi)\}_\hbar$, since any Hilbert space is reflexive. According to the above lemma it suffices to show that $e^{tL_\hbar}$ converges weakly to e^{tL_0}. Since $e^{tL_\hbar}$ and e^{tL_0} are two groups, for all $t \in \mathbb{R}$, we define the space of test functions by $D = \mathcal{S}(\mathbb{R}^n_x \times \mathbb{R}^n_\xi \times \mathbb{R}_t)$, where $\mathcal{S}$ is the Schwarz space. $W_\hbar$ being the weak solution of (**QLE**), it satisfies

$$\int_{\mathbb{R}^{2n+1}} W_\hbar(x, \xi, t) \left(-\frac{\partial}{\partial t} - \xi \cdot \nabla_x + P_\hbar(x, \nabla_\xi) \right) \varphi(x, \xi, t) dx d\xi dt = 0,$$

for any $\varphi \in D$. Let w_0 be the weak limit of $W_\hbar$, thus

$$\lim_{\hbar \to 0} \int_{\mathbb{R}^{2n+1}} (W_\hbar - w_0) \left(\frac{\partial \varphi}{\partial t} + \xi \cdot \nabla_x \varphi \right) dx d\xi dt = 0.$$

Moreover, we have

$$\int_{\mathbb{R}^{2n+1}} W_\hbar (P_\hbar(x, \nabla_\xi \varphi) - w_0(\nabla_x V \cdot \nabla_\xi \varphi) dx d\xi dt = \int_{\mathbb{R}^{2n+1}} (W_\hbar - w_0) \nabla_x V \cdot \nabla_\xi \varphi \, dx d\xi dt$$
$$+ \int_{\mathbb{R}^{2n+1}} W_\hbar \left(P_\hbar(x, \nabla_\xi \varphi - \nabla_x V \cdot \nabla_\xi \varphi \right) dx d\xi dt. \tag{7.30}$$

We reformulate the first term on the right-hand side of (7.30) by using Plancherel's Theorem with respect to the ξ variable. That leads to

$$\int_{\mathbb{R}^{2n+1}} (W_\hbar - w_0)\nabla_x V \cdot \nabla_\xi \varphi \, dxd\xi dt = \mathrm{i} \int_{\mathbb{R}^{2n+1}} [\mathscr{F}_\xi (W_\hbar - w_0)](x, \eta, t)]$$
$$[\nabla_x V \cdot \eta \mathscr{F}_\xi \varphi](x, -\eta, t) \, dxd\eta dt.$$

Since $\mathscr{F}_\xi \varphi \in \mathcal{S}(\mathbb{R}^{2n})$ and $\mathscr{F}$ is an isometric isomorphism on $L^2((\mathbb{R}^{2n}), [\mathscr{F}_\xi (W_\hbar - w_0)](x, \eta, t)] \to 0$ weakly and $\nabla_x V \in L^\infty(\mathbb{R}^n)$, we have

$$\lim_{\hbar \to 0} \int_{\mathbb{R}^{2n+1}} (W_\hbar - w_0)\nabla_x V \cdot \nabla_\xi \varphi \, dxd\xi dt = 0.$$

Concerning the second term, let us write $\mathbb{H} := L^2((\mathbb{R}^{2n+1})$; then

$$\left| \int_{\mathbb{R}^{2n+1}} W_\hbar \left(P_\hbar(x, \nabla_\xi)\varphi - \nabla_x V \cdot \nabla_\xi \varphi \right) dxd\xi dt \right| \leq \|W_\hbar\|_{\mathbb{H}} \| P_\hbar(x, \nabla_\xi \varphi - \nabla_x V \cdot \nabla_\xi \varphi\|_{\mathbb{H}}.$$

We remark that $\|W_\hbar\|_{\mathbb{H}} \leq C$ is independent of $\hbar$ and

$$\begin{aligned} \|P_\hbar(x, \nabla_\xi \varphi) - \nabla_x V \cdot \nabla_\xi \varphi\|_{\mathbb{H}} &= \|\delta_\hbar V(x, y) \left[\mathscr{F}_\xi \varphi(x, \xi) \right](y)(\xi) - y.\nabla_x V \mathscr{F}_\xi \varphi(x, y)\|_{\mathbb{H}} \\ &\leq \|\delta_\hbar V(x, y) - y.\nabla_x V(x)\|_{L^\infty(\mathbb{R}^{2n})} \|\mathscr{F}_\xi \varphi(x, y)\|_{\mathbb{H}}, \end{aligned}$$

where $\delta_\hbar V(x, y)$ is defined in (7.11) and converges in $L^\infty(\mathbb{R}^{2n})$ to $y \cdot \nabla_x V$, which implies the theorem. □

7.6 Husimi Representation and Well-Posedness

It is well-known that the Wigner function $W_{\varphi_\hbar}(x, \xi)$ is not positive for all values (x, ξ) of the phase space, in spite of the fact that

$$\int_{\mathbb{R}_\xi} W_{\varphi_\hbar}(x, \xi) d\xi = \mathcal{F}^{-1}\left[\mathcal{F}_y[\varphi_\hbar(x + \frac{\hbar y}{2})\overline{\varphi}_\hbar(x - \frac{\hbar y}{2})]\right](0) = |\varphi_\hbar(x)|^2 > 0. \tag{7.31}$$

So, we cannot consider the Wigner function as a density function in the context of statistical mechanics. In [67], K. Husimi proposed the following procedure which leads to the definition of a new function which is called the *Husimi function*, which can be considered in some manner as a density function.

Define

$$\begin{aligned} H_\hbar(x, \xi, t) &= [W_{\varphi_\hbar} * G_\hbar](x, \xi, t) \\ &= \iint_{\mathbb{R}^n_x \times \mathbb{R}^n_\xi} W_{\varphi_\hbar}(y, \eta, t) G_\hbar(x - y, \xi - \eta) dy d\eta, \end{aligned}$$

where the Gaussian $G_\hbar$ is

$$G_\hbar(x,\xi) = (\pi\hbar)^{-n} e^{-(|x|^2+|\xi|^2)/\hbar}. \tag{7.32}$$

Let us write

$$C_\hbar : f \in L^1(\mathbb{R}^n_x \times \mathbb{R}^n_\xi) \mapsto G_\hbar * f = f * G_\hbar \in \mathcal{S}(\mathbb{R}^n_x \times \mathbb{R}^n_\xi), \tag{7.33}$$

then $H_\hbar(x,\xi) = C_\hbar W_\hbar$. The action of $C_\hbar$ on **(QLE)**, gives once more **(QLE)** with new initial data called *Husimi equation*.

$$\textbf{(HE)} \quad \begin{cases} \dfrac{\partial H_\hbar}{\partial t} + \xi \cdot \nabla_x H_\hbar = P_\hbar(x,\nabla_\xi) H_\hbar = k_\hbar(x,\cdot) \underset{\xi}{*} W_\hbar(x,.) * G_\hbar(x,\xi), \\ H_\hbar(x,\xi,0) = w^0 * G_\hbar(x,\xi). \end{cases}$$

By taking $\hbar = 4t$, it is well-known that $[T(t)f](x,\xi) = [G_{4t} * f](x,\xi)$ forms a Gaussian (or heat, or diffusion) semigroup and the strong continuity of this semigroup asserts that

$$\|G_{4t} * f - f\|_{L^p(\mathbb{R}^n_x \times \mathbb{R}^n_\xi)} \to 0 \quad \text{for any} \quad f \in L^p(\mathbb{R}^n_x \times \mathbb{R}^n_\xi), \quad (1 \le p < \infty) \tag{7.34}$$

as $t \to 0$.

Lemma 7.6.1 *For (x_0, ξ_0) in phase space, define the Gabor function*

$$\psi_{x_0,\xi_0}(x) = (\pi)^{-n/4} \mathrm{e}^{-\frac{|x-x_0|^2}{2\hbar}} \mathrm{e}^{\mathrm{i}\xi_0 \cdot x/\hbar};$$

then the Wigner transform of the Gabor function $W_{\psi_{x_0,\xi_0}}$ satisfies

$$W_{\psi_{x_0,\xi_0}}(x,\xi) = G_\hbar(x - x_0, \xi - \xi_0) = W_{\psi_{x,\xi}}(x_0,\xi_0). \tag{7.35}$$

Proof By using the expression

$$W_{\psi_{x_0,\xi_0}}(x,\xi) = (2\pi\hbar)^{-n} \int_{\mathbb{R}^n} \mathrm{e}^{-\mathrm{i}(\xi/\hbar).y} \psi_{x_0,\xi_0}(x + \frac{y}{2}) \overline{\psi}_{x_0,\xi_0}(x - \frac{y}{2}) dy$$

we get by using the parallelogram identity

$$\begin{aligned} W_{\psi_{x_0,\xi_0}}(x,\xi) &= (\pi\hbar)^{-n/2} (2\pi\hbar)^{-n} \int_{\mathbb{R}^n} \mathrm{e}^{-\mathrm{i}(\frac{\xi-\xi_0}{\hbar}.y)} \mathrm{e}^{-\left(\frac{|x-x_0|^2}{\hbar} + \frac{|y|^2}{4\hbar}\right)} dy \\ &= (\pi\hbar)^{-n} \mathrm{e}^{-(|x-x_0|^2+|\xi-\xi_0|^2)/\hbar}, \end{aligned} \tag{7.36}$$

since it is well-known that $\int_{\mathbb{R}^n} \mathrm{e}^{-\mathrm{i}\xi.y} \mathrm{e}^{-k|y|^2} dy = \left(\frac{\pi}{k}\right)^{\frac{n}{2}} \mathrm{e}^{-\frac{|\xi|^2}{4k}}$. □

Theorem 7.6.2 *The Husimi function $H_\hbar$ is positive and belongs to $L^1(\mathbb{R}^n_x \times \mathbb{R}^n_\xi)$. Further-more, $H_\hbar$ converges in this space to $w(x, \xi)$, the solution of the Liouville–Vlasov equation* **(LVE)**, *as $\hbar \to 0$.*

For the proof of this theorem we will use the following lemma.

Lemma 7.6.3 *Let us denote by $\chi_k(x, \xi, y, \eta)$ the characteristic function of $I_k = \{(x, \xi, y, \eta) \in \mathbb{R}^{4n} \ : \ (k-1)^2\hbar \le |x-y|^2 + |\xi - \eta|^2 < k^2\hbar\}$ and*

$$g_k(y, \eta) = \int_{\mathbb{R}^n_x \times \mathbb{R}^n_\xi} kG_\hbar(x-y, \xi-\eta)\chi_k(x, \xi, y, \eta)dxd\xi;$$

then

$$\sum_{k\ge 1} \|g_k\|^2_{L^2(\mathbb{R}^n_x \times \mathbb{R}^n_\xi)} < \infty. \tag{7.37}$$

Proof For the proof of this lemma we will use the Gaussian upper bound. Let $p(\hbar, (x, \xi), (y, \eta))$ be the kernel of the heat equation in $\mathbb{R}^n_x \times \mathbb{R}^n_\xi$. We remark that $G_{4\hbar}(x-y, \xi-\eta) = p(\hbar, (x, \xi), (y, \eta))$ is the Gaussian kernel of $\Delta_{x,\xi}$ in $\mathbb{R}^n_x \times \mathbb{R}^n_\xi$.

E.-M. Ouhabaz has obtained in his book [95, formula (7.5)] the following estimation:

$$|p(\hbar, (x, \xi), (y, \eta))| \le C(V((y, \eta), \sqrt{\hbar}))^{-1} \exp\left\{-c\left(\frac{|x-y|^2 + |\xi-\eta|^2}{\hbar}\right)\right\}, \tag{7.38}$$

where $V((y, \eta), r)$ is the volume of the ball $B(y, \eta), r)$ centered at (y, η) of radius r in $\mathbb{R}^n_x \times \mathbb{R}^n_\xi$. Using the facts that

$$\int_{\mathbb{R}^n_x \times \mathbb{R}^n_\xi} \chi_k(x, \xi, y, \eta)dxd\xi \le V((y, \eta), k\sqrt{\hbar})$$

and

$$I_k \subset \bigcup_{|(y,\eta)|\le k} \ \bigcup_{(x,\xi,y,\eta)\in I_k} B((y, \eta), k\sqrt{\hbar}),$$

it follows from (7.38) that

$$\sum_{k\ge 1} \|g_k\|^2_{L^2(\mathbb{R}^n_x \times \mathbb{R}^n_\xi)} = \sum_{k\ge 1} \int_{\mathbb{R}^n_x \times \mathbb{R}^n_\xi} \left| \int_{\mathbb{R}^n_x \times \mathbb{R}^n_\xi} kG_\hbar(x-y, \xi-\eta)\chi_k(x, \xi, y, \eta)dxd\xi \right|^2 dyd\eta$$

$$\le C^2 \sum_{k\ge 1} \int_{|(y,\eta)|\le k} \left(k e^{-8c(k-1)^2} \frac{V((y, \eta), k\sqrt{\hbar}))}{V\left((y, \eta), \sqrt{\hbar/2})\right)} \right)^2 dyd\eta.$$

Since in $\mathbb{R}^n_x \times \mathbb{R}^n_\xi$, $V((y,\eta),r)$ is proportional to r^{2n}, we get $\frac{V((y,\eta),k\sqrt{\hbar}))}{V\left((y,\eta),\sqrt{\hbar}/2)\right)} \le C'(2k)^{2n}$. This shows that

$$\sum_{k\ge 1} \|g_k\|^2_{L^2(\mathbb{R}^n_x\times\mathbb{R}^n_\xi)} \le C'' \sum_{k\ge 1} k^{6n+2} \mathrm{e}^{-8c(k-1)^2} < \infty,$$

where C'' is a constant independent of $\hbar$. □

Proof of Theorem 7.6.2. It follows from (7.36) that $W_{x_0,\xi_0} := W_{\psi_{x_0,\xi_0}} > 0$ and in $\left(L^1(\mathbb{R}^n_x \times \mathbb{R}^n_\xi), \|\cdot\|_{1,1}\right)$, we have $\|W_{x_0,\xi_0}\|_{1,1} = 1$ and from (7.8) and (7.35), that

$$H_\hbar(x,\xi) = (1/4\pi\hbar)^{-n} \left| \int_{\mathbb{R}^n} \varphi_\hbar(x)\overline{\psi_{x_0,\xi_0}(x)}dx \right|^2 > 0,$$

and the fact that $\|G_\hbar\|_{1,1} = 1$, implies that $\|H_\hbar\|_{1,1} = \|W_{x_0,\xi_0}\|_{1,}\|G_\hbar\|_{1,1} = 1$.

Now, in order to prove the L^1-convergence of $H_\hbar$ toward w we will use Theorem 1.16.1, together with some aspects of wavelet theory:

$$\|H_\hbar - w\|_{1,1} \le \|(W_\hbar - w) * G_\hbar\|_{1,1} + \|G_\hbar * w - w\|_{1,1}. \qquad (7.39)$$

The Koopman formalism (7.23), implies that if $w_0 \in L^p(\mathbb{R}^n_x \times \mathbb{R}^n_\xi)$, for any $p \in [1,\infty)$ then w belongs also to $L^p(\mathbb{R}^n_x \times \mathbb{R}^n_\xi)$. Hence (7.34) holds for $p = 1$, which yields the convergence of the last term of (7.39) to zero as $\hbar \to 0$. So, it remains to prove the convergence of $\|(W_\hbar - w) * G_\hbar\|_{1,1}$ to zero, as $\hbar \to 0$.

Let $I_k = \{(x,\xi) \in \mathbb{R}^n_x \times \mathbb{R}^n_\xi \ : \ k-1 \le \sqrt{|x|^2+|\xi|^2} < k\}$ and χ_I the characteristic function of I. We remark that $\sum_{k\ge1}\chi_{I_k} = 1$ for any $(x,\xi) \in \mathbb{R}^n_x \times \mathbb{R}^n_\xi$. Let $s(x,\xi) := \mathrm{e}^{-i\theta(x,\xi)}$, where $\theta(x,\xi) = \arg(W_\hbar - w)(x,\xi)$, so $[(W_\hbar - w)s](x,\xi) \ge 0$ and $s(x,\xi)s^*(x,\xi) = 1$. Finally, by writing $W_\hbar - w = (\mathrm{e}^{tL_\hbar} - \mathrm{e}^{tL})w_0$, we get

$$\|(W_\hbar - w) * G_\hbar\|_{1,1} = \int_{\mathbb{R}^n_x\times\mathbb{R}^n_\xi} \left| \int_{\mathbb{R}^n_y\times\mathbb{R}^n_\eta} (\mathrm{e}^{tL_\hbar} - \mathrm{e}^{tL})w_0(y,\eta)G_\hbar(x-y,\xi-\eta)dyd\eta \right| dxd\xi$$

$$\le \int_{\mathbb{R}^n_x\times\mathbb{R}^n_\xi} \int_{\mathbb{R}^n_y\times\mathbb{R}^n_\eta} (\mathrm{e}^{tL_\hbar} - \mathrm{e}^{tL})w_0(y,\eta)s(y,\eta)\left|s^*(y,\eta)G_\hbar(x-y,\xi-\eta)\right| dyd\eta dxd\xi$$

$$= \sum_{k\ge1} \int_{\mathbb{R}^n_y\times\mathbb{R}^n_\eta} (\mathrm{e}^{tL_\hbar} - \mathrm{e}^{tL}) f_k(y,\eta)g_k(y,\eta)dyd\eta,$$

where $f_k(y,\eta) = \frac{1}{k}w_0(y,\eta)s(y,\eta)$ and g_k is defined as in Lemma 7.6.3.

Now, since $\mathrm{e}^{tL_\hbar}$ converges ultraweakly to e^{tL}, it follows that $H_\hbar$ converges in $L^1(\mathbb{R}^n_x \times \mathbb{R}^n_\xi)$ toward w. □

7.7 Quantum Transport Equation

Let us consider the following unitary transform:

$$\mathcal{C} : L^2(\mathbb{R}^n_x \times \mathbb{R}^n_\eta) \ni g \longmapsto \mathcal{C}g \in L^2(\mathbb{R}^n_u \times \mathbb{R}^n_v),$$

where

$$\mathcal{C}g(u, v) = \hbar^{-n} g(x, \eta), \quad \text{with} \quad u = x + \frac{\hbar}{2}\eta \quad \text{and} \quad v = x - \frac{\hbar}{2}\eta. \tag{7.40}$$

Definition and Theorem 7.7.1 *Applying the Fourier transform on* (**QLE**) *leads back to a new Schrödinger type equation, called the* **quantum transport equation**[1]

$$\textbf{(QTE)} \qquad \begin{cases} \dfrac{\partial}{\partial t} z(u, v, t) - \dfrac{i}{\hbar}\left[\left(-\hbar^2 \dfrac{\Delta_u}{2} + V(u)\right)\right. \\ \qquad \left. - \left(-\hbar^2 \dfrac{\Delta_v}{2} + V(v)\right)\right] z(u, v, t) = 0, \\ z(u, v, 0) = z^0(u, v). \end{cases}$$

Proof Letting $W_\hbar(x, \xi, t)$ be the solution of (QLE) and $\widehat{w}_\hbar(x, \eta, t) = \mathscr{F}_\xi W_\hbar(x, \xi, t)(\eta)$, this function satisfies

$$\begin{cases} \dfrac{\partial}{\partial t} \widehat{w}_\hbar + \mathrm{i}\left(\mathrm{div}_\eta \nabla_x \widehat{w}_\hbar\right) - \delta_\hbar V(x, \eta) \widehat{w}_\hbar = 0, \\ \widehat{w}_\hbar(x, \eta, 0) = \mathscr{F}_\xi w^0(x, \xi)(\eta). \end{cases} \tag{7.41}$$

Since $x = (u + v)/2$ and $\eta = (u - v)/\hbar$, by applying the chain rule for derivatives of vector fields,

$$\mathrm{div}_\eta \nabla_x = \frac{1}{2\hbar}\left(\mathrm{div}_u \nabla_u + \mathrm{div}_u \nabla_v - \mathrm{div}_v \nabla_u - \mathrm{div}_v \nabla_u\right) = \frac{1}{2\hbar}\left(\Delta_u - \Delta_v\right).$$

On the other hand,

$$\delta_\hbar V(x, \eta) = \frac{i}{\hbar}(V(u) - V(v)).$$

By applying the unitary transformation $\mathcal{C}$ on $\widehat{w}_\hbar$, the function $z_\hbar := \mathcal{C}\widehat{w}_\hbar$ satisfies (QTE), where $z^0(u, v) = \mathcal{C}\widehat{w}^0_\hbar(x, \eta)$. □

---

[1]This equation is also called **Heisenberg–von Neumann equation** , which is written in terms of its kernel.

Now we will browse the opposite way: taking the solution of (**QTE**), $z_\hbar(u, v, t)$, and ending with the solution of (**QTE**), $W_\hbar(x, \xi, t)$. In order to do this, we shall first precise the meaning of the Hamiltonian which occurs in (**QTE**), which is $H = -\hbar^2 \frac{\Delta}{2} + V$. In the sequel we need this operator to be essentially self-adjoint with domain $D(H) = C_0^\infty(\mathbb{R}^n)$. There is wealth of literature on the potential V which ensures this property. The most popular one is Kato's condition:

$$\begin{cases} V = V_1 + V_2, \quad \text{where} \quad V_1 \in L^\infty(\mathbb{R}^n) \quad \text{and} \\ V_2 \in L^p(\mathbb{R}^n) \quad \text{with,} \quad p > \max\{\frac{n}{2}, 2\}. \end{cases} \tag{7.42}$$

Furthermore, we shall use also the notion of tensor products of unbounded operators which is developed in Chap. VIII of [102]. Let I_u and I_v be the identities of two Hilbert spaces $\mathcal{H}_u = L^2(\mathbb{R}^n_u)$ and $\mathcal{H}_v = L^2(\mathbb{R}^n_v)$ respectively. The tensor product of $\mathcal{H}_u$ and $\mathcal{H}_v$ can be identified with $\mathscr{H} = L^2(\mathbb{R}^n_u \times \mathbb{R}^n_v)$ by setting $(f \otimes g)(u, v) = f(u)g(v)$, which is denoted by $\mathcal{H}_u \otimes \mathcal{H}_v$. The following lemma is proved in [102, Chap. VIII].

Lemma 7.7.2 *Let A and B be two unbounded operators with densely defined domains $D(A) \subset \mathcal{H}_u$ and $D(B) \subset \mathcal{H}_v$. The operations $\otimes$ and $\oplus$ are given by:*

- $(A \otimes B)(f \otimes g)(u, v) = Af(u) \otimes Bg(v) \quad f \in D(A), g \in D(B)$
- $A \oplus B = A \otimes I_u + I_v \otimes B.$

(a) The domain of $A \otimes B$ is $D(A \otimes B) = D(A) \otimes D(B)$, the set of all finite linear combinations of elements of the form $\varphi \otimes \psi$, where $\varphi \in D(A)$ and $\psi \in D(B)$. In addition, $D(A \otimes B)$ is dense in $\mathscr{H}$.
(b) If A and B are closed then $A \otimes B$ is also closed.
(c) If A and B are bounded then $A \otimes B$ is also bounded and

$$\|A \otimes B\|_{\mathscr{L}(\mathscr{H})} = \|A\|_{\mathscr{L}(\mathcal{H})} \|B\|_{\mathscr{L}(\mathcal{H})}.$$

With these preliminaries we can define the operator

$$Q = -H_u \oplus H_v, \tag{7.43}$$

where H_u and H_v are the copies of H on $L^2(\mathbb{R}^n_u)$ and $L^2(\mathbb{R}^n_v)$, and for Hamitonian $H = -\frac{\hbar^2}{2}\Delta + V$ and

$$Q_0 = -H_u^0 \oplus H_v^0, \tag{7.44}$$

where H_u^0 and H_v^0 are the copies of the free Hamiltonian $H = -\frac{\hbar^2}{2}\Delta$ on $L^2(\mathbb{R}^n_u)$ and $L^2(\mathbb{R}^n_v)$.

Now, we can rewrite the problem (**QTE**) in the following Schrödinger tensorial form:

$$\textbf{(STF)} \qquad \begin{cases} i\hbar\dfrac{\partial}{\partial t}z(u,v,t) = Qz(u,v,t) \\ z(u,v,0) = z^0(u,v). \end{cases}$$

The relation between the operator Q and the quantum Liouville operator $L_\hbar$ appears from the relation

$$L_\hbar = \frac{\mathrm{i}}{\hbar}\mathscr{F}^{-1}C^{-1}QC\mathscr{F}. \tag{7.45}$$

We still denote by Q the closure of Q in $L^2(\mathbb{R}^n_u \times \mathbb{R}^n_v)$ and if we impose sufficient conditions on V in order that H to be self-adjoint (see (7.42)), then from (7.43), Q is also selfadjoint. This is another way to show that $L_\hbar$ is skew-adjoint. Hence, the operators $\mathrm{i}H$ and $\mathrm{i}Q$ generate unitary (C_0) groups $\{\mathrm{e}^{\mathrm{i}tH}\}_{t\in\mathbb{R}}$ on $L^2(\mathbb{R}^n)$ and $\{\mathrm{e}^{\mathrm{i}tQ}\}_{t\in\mathbb{R}}$ on $L^2(\mathbb{R}^n_u \times \mathbb{R}^n_v)$, linked by the relation

$$\mathrm{e}^{\mathrm{i}tQ} = \mathrm{e}^{-\mathrm{i}tH_u} \otimes \mathrm{e}^{\mathrm{i}tH_v} \qquad (t \in \mathbb{R}).$$

Similarly, the group generated by the quantum Liouville operator is given by

$$\exp(tL_\hbar) = \mathscr{F}^{-1}C^{-1}\exp\left(\frac{\mathrm{i}}{\hbar}tH_u\right) \otimes \exp\left(-\frac{\mathrm{i}}{\hbar}tH_v\right)C\mathscr{F}.$$

The above relation encourage us to study the scattering theory for the quantum Liouville equation via the scattering theory for the Schrödinger equation which is extensively studied by many authors. That is the central theme of the next section.

7.8 Scattering Theory for Quantum Liouville Equation

There are various conditions on the potential V which ensure the existence and completeness of the wave operators in the case of the Schrödinger problem. We have chosen a reference condition, so we worked with Agmon's potentials to illustrate our purpose (see [1]). Our goal, in fact, is not to establish a supplementary condition on V, but to justify the transmission of the completeness from a problem to the other.

Theorem 7.8.1 *Suppose that for an $\varepsilon > 0$ and a μ such that $0 < \mu < 4$, the potential V verifies*

$$\sup_{x\in\mathbb{R}^n}\left[(1+|x|)^{2+2\varepsilon}\int\limits_{|y-x|\le 1}|V(y)|^2|y-x|^{-n+\mu}\,\mathrm{d}y\right] < \infty.$$

Then the wave operators

$$W_{\pm}(H, H_0) = s - \lim_{t \to \pm\infty} \mathrm{e}^{\mathrm{i}tH} \mathrm{e}^{-\mathrm{i}tH_0}$$

exist, and are complete.

To prove the completeness of the wave operators for Wigner equation, we need some preliminary lemmas. These lemmas can be found in any standard manuscript (see e.g. [69, 103, 104]).

Lemma 7.8.2 *Suppose that $W_{\pm}(A, B)$ exist. The following conditions are equivalent:*

(a) *$W_{\pm}(A, B)$ are complete;*
(b) *$W_{\pm}(B, A)$ exist.*

Lemma 7.8.3 *Suppose that $W_{\pm}(A, B)$ exist and are complete. Then*

$$\sigma(B) = \sigma_{ac}(B) = \sigma_{ac}(A).$$

Consequently, if the wave operators $W_{\pm}(A, B)$ exist and are complete, then $W_{\pm}(B, A)$ are too, and we have

$$\sigma(B) = \sigma_{ac}(B) = \sigma_{ac}(A) = \sigma(A).$$

Now, let us define Q_0 as in (7.43):

$$Q_0 = -(H_0)_u \oplus (H_0)_v,$$

where $(H_0)_u$ and $(H_0)_v$ are the copies of H_0 on $L^2(\mathbb{R}^n_u)$ and $L^2(\mathbb{R}^n_v)$.

Let us define the operators $L_\hbar$ and L_0 by (7.18) and (7.19).

Proposition 7.8.4 *The operators Q_0 and L_0 satisfy:*

(a) *$\mathcal{H}_{ac}(Q_0) = L^2(\mathbb{R}^{2n})$;*
(b) *$\mathcal{H}_{ac}(L_0) = L^2(\mathbb{R}^{2n})$.*

To prove this proposition we will use the following result by Umeda [120].

Lemma 7.8.5 *Let A be a (pseudo)-differential operator on $L^2(\mathbb{R}^{2n})$, of symbol p, and let*

$$\Xi = \{(u, v) \in \mathbb{R}^{2n}; \nabla p(u, v) = 0\}. \tag{7.46}$$

If $p \in C^1(\mathbb{R}^{2n})$ and if $\mathrm{Mes}(\Xi) = 0$, then

$$\mathcal{H}_{ac}(A) = L^2(\mathbb{R}^{2n}).$$

Proof of Proposition 7.8.4.

(a) The symbol of $Q_0 = -H_{r,0} \oplus H_{s,0} = \frac{\hbar^2}{2}(\Delta_r - \Delta_s)$ being $\dfrac{\hbar^2}{2}(|r|^2 - |s|^2)$, so $\nabla p(r,s) = \hbar^2(r_1, \ldots, r_n, -s_1, \ldots, -s_n)$ and from (7.46), $\Xi = \{0\}$, by applying the Lemma 7.8.5, we get $\mathcal{H}_{ac}(Q_0) = L^2(\mathbb{R}^{2n})$.

(b) By taking $V \equiv 0$, the relation (7.45) becomes $\hbar L_0 = -\mathscr{F}^{-1}\mathcal{C}^{-1} i Q_0 \mathcal{C}\mathscr{F}$, which implies that $\mathcal{H}_{ac}(L_0) = \mathcal{H}_{ac}(Q_0)$ and (a) yields (b).

□

Theorem 7.8.6 *Suppose that the potential V has been chosen in such a way that $W_\pm(H, H_0)$ exist as unitary operators on $L^2(\mathbb{R}^n)$. Then the wave operators $W_\pm(L_\hbar, L_0)$ exist and are complete, as unitary operators on $L^2(\mathbb{R}^n_x \times \mathbb{R}^n_\xi)$, and are related by*

$$W_\pm(L_\hbar, L_0) = \mathscr{F}^{-1}\mathcal{C}^{-1} W_\pm(H, H_0) \otimes W_\mp(H, H_0)\mathcal{C}\mathscr{F}.$$

Proof By Proposition 7.8.4 (b), $P_{ac}(L_0) \equiv Id$. Hence

$$W_\pm(L_\hbar, L_0) = s - \lim_{t\to\pm\infty} \mathrm{e}^{tL_\hbar}\mathrm{e}^{-tL_0} P_{ac}(L_0) = s - \lim_{t\to\pm\infty} \mathrm{e}^{tL_\hbar}\mathrm{e}^{-tL_0}.$$

By hypothesis, we suppose that the operators $W_\pm(H, H_0)$ are complete. Then, by Lemma 7.8.2, $W_\pm(H_0, H)$ also exist and are complete. Moreover, by Lemma 7.8.3,

$$\sigma(H_0) = \sigma_{ac}(H_0) = \sigma_{ac}(H) = \sigma(H) = [0, +\infty).$$

This implies that $\mathcal{H}_{ac}(H) = L^2(\mathbb{R}^n)$. Thus we get

$$W_\pm(H_0, H) = s - \lim_{t\to\pm\infty} \mathrm{e}^{\mathrm{i}tH_0}\mathrm{e}^{-\mathrm{i}tH} P_{ac}(H) = s - \lim_{t\to\pm\infty} \mathrm{e}^{\mathrm{i}tH_0}\mathrm{e}^{-\mathrm{i}tH}.$$

Consequently,

$$s - \lim_{t\to\pm\infty} \mathscr{F}^{-1}\mathcal{C}^{-1} W_\pm(H_0, H) \otimes W_\mp(H_0, H)\mathcal{C}\mathscr{F}$$

is defined on $L^2(\mathbb{R}^n) \otimes L^2(\mathbb{R}^n)$, and is exactly

$$s - \lim_{t\to\pm\infty} \mathrm{e}^{tL_0}\mathrm{e}^{-tL_\hbar} = W_\pm(L_0, L_\hbar).$$

As $W_\pm(L_0, L_\hbar)$ and $W_\pm(L_\hbar, L_0)$ exist on $L^2(\mathbb{R}^n) \otimes L^2(\mathbb{R}^n)$, these wave operators are complete. □

7.9 Scattering Theory for Quantum Dynamical Semigroup

In this section we return back to the beginning of the chapter where we have defined the quantum dynamical semigroup (**QDS**) $\Phi_\hbar(t)$ in (7.4). Here we attempt to redefine the Wigner distribution by means of this semigroup. In fact instead writing the Wigner distribution as (7.7), we express it by

$$W_{\psi_\hbar}(x,\xi,t) = \mathscr{F}_y\left(\psi_\hbar(x+\frac{\hbar y}{2},t)\psi_\hbar^*(x-\frac{\hbar y}{2},t)\right), \tag{7.47}$$

where $\psi_\hbar(x,t) = e^{-\mathrm{i}tH_0}\psi(x)$ is the solution of free potential Schrödinger equation

$$\textbf{(FPS)} \qquad \begin{cases} \dfrac{\partial\psi_\hbar}{\partial t} = \frac{\mathrm{i}\hbar}{2}\Delta_x^0\psi_\hbar(x,t) = -\mathrm{i}H_0\psi_\hbar(x,t), \quad (m=1) \\ \psi_\hbar(x,0) = \psi^0(x). \end{cases}$$

Theorem 7.9.1 *If we confuse the notion of a quantum dynamical semigroup $\Phi_\hbar^0(t)\rho$ on $\mathscr{I}_1(\mathcal{H})$ with its kernel function $\rho(x,y)$, one has:*

(a) $W_{\psi_\hbar}(x,\xi,t) = \mathscr{F}_y\left(\Phi_\hbar^0(t)(\psi(x+\frac{\hbar y}{2})\psi^*(x-\frac{\hbar y}{2}))\right)(\xi)$;

(b)

$$\begin{aligned} \frac{d}{dt}W_\hbar(x,\xi,t) = -\frac{\mathrm{i}\hbar}{2}\Big[&\mathscr{F}_y\big(\psi(x+\frac{\hbar y}{2},t)\Delta_x^0\psi^*(x-\frac{\hbar y}{2},t) \\ &-\psi^*(x-\frac{\hbar y}{2},t)\Delta_x^0\psi(x+\frac{\hbar y}{2},t)\big)(\xi)\Big]. \end{aligned} \tag{7.48}$$

Proof (a) If $\psi\in\mathcal{H}$, $\rho(x,y)=\psi(x)\psi^*(y)$ and $\Phi_\hbar^0(t)\rho = e^{-\mathrm{i}tH_0}\rho e^{\mathrm{i}tH_0}$, then for the mapping $C_\hbar$ being defined in (7.6), the Wigner function is

$$\begin{aligned} W_{\psi_\hbar}(x,\xi,t) &= \mathscr{F}_y\left(\widetilde{\rho}_\hbar(x,y,t)\right)(\xi) \\ &= \mathscr{F}_y\left(C_\hbar(\psi(x,t)\psi^*(y,t))\right)(\xi) \\ &= \mathscr{F}_y\left(C_\hbar\Phi_\hbar^0(t)(\psi(x)\psi^*(y))\right)(\xi) \\ &= \mathscr{F}_y\left(\Phi_\hbar^0(t)C_\hbar(\psi(x)\psi^*(y))\right)(\xi). \end{aligned}$$

(b) First let us show that for the initial data $\psi\in\mathcal{S}(\mathbb{R}^n)$ of the Schrödinger equation $\widetilde{\rho}_\hbar(x,y)=\psi(x+\frac{\hbar y}{2})\psi^*(x-\frac{\hbar y}{2})\in D(ad_{H_0})$. Take $u\in D(ad_{H_0})$ and form $\widetilde{\rho}_\hbar u$:

$$\begin{aligned} \widetilde{\rho}_\hbar u &= \int_{\mathbb{R}_y^n}\psi(x+\frac{\hbar y}{2})\psi^*(x-\frac{\hbar y}{2})u(y)dy \\ &\le \int_{\mathbb{R}_y^n}\frac{1}{2}\left[\psi(x+\frac{\hbar y}{2})^2+\psi^*(x-\frac{\hbar y}{2})^2\right]u(y)dy. \end{aligned}$$

Now, it is enough to prove that $|\xi|$ times the Fourier transform of $x \mapsto \int_{\mathbb{R}^n_y} \frac{1}{2}(\psi(x + \frac{\hbar y}{2}))^2 u(y)dy$ belongs to $L^2(\mathbb{R}^n_\xi)$. That is,

$$\begin{aligned} G(\xi) &= \frac{1}{2}\left(\frac{1}{2\pi}\right)^{n/2} \int_{\mathbb{R}^n_x}\int_{\mathbb{R}^n_y} \psi(x+\frac{\hbar y}{2})^2 u(y)dy e^{-\mathrm{i}x\xi}dx \\ &= \frac{1}{2}\left(\frac{1}{2\pi}\right)^{n/2} \int_{\mathbb{R}^n_x}\int_{\mathbb{R}^n_z} \psi(z)^2 u\left(\frac{2}{\hbar}(z-x)\right)\left(\frac{2}{\hbar}\right)^n dz e^{-\mathrm{i}x\xi}dx \\ &= \frac{1}{2}\left(\frac{1}{2\pi}\right)^{n/2} \int_{\mathbb{R}^n_p}\int_{\mathbb{R}^n_z} \psi(z)^2 u(p)(-1)^n dz e^{-\mathrm{i}(z-\hbar p/2)\xi}dp. \end{aligned}$$

Hence $|G(\xi)| \leq C_n \|\psi\|_2^2 \left|\mathscr{F}^{-1}u(\hbar\xi/2)\right|$ and $|\xi| G(\xi) \in L^2(\mathbb{R}^n_\xi)$ since $|\xi|\mathscr{F}^{-1}u(\hbar\xi/2)) \in L^2(\mathbb{R}^n_\xi)$.

Now, since $\widetilde{\rho}_\hbar \in D(ad_{H_0})$, one can apply the formula (1.55) and calculate the derivative of $W_\hbar$,

$$\begin{aligned} \frac{d}{dt}W_{\psi_\hbar}(x,\xi,t) &= \mathscr{F}_y\left(C_\hbar \frac{d}{dt}\Phi^0_\hbar(t)(\psi(x)\psi^*(y))\right)(\xi). \\ &= \mathscr{F}_y\left(C_\hbar e^{-\mathrm{i}tH_0}\left[-\mathrm{i}(H_0\psi(x)\psi^*(y) - \psi(x)\psi^*(y)H)\right]e^{\mathrm{i}tH}\right)(\xi) \\ &= \mathscr{F}_y\Big(e^{-\mathrm{i}tH_0}\big[-\mathrm{i}(H_0\psi(x+\frac{\hbar y}{2})\psi^*(x-\frac{\hbar y}{2}) \\ &\qquad - \psi(x+\frac{\hbar y}{2})\psi^*(x-\frac{\hbar y}{2})H_0)\big]e^{\mathrm{i}tH_0}\Big)(\xi) \\ &= \mathscr{F}_y\Big(e^{-\mathrm{i}tH_0}\big[-\mathrm{i}(H_0\psi(x+\frac{\hbar y}{2})\psi^*(x-\frac{\hbar y}{2}) \\ &\qquad - \psi(x+\frac{\hbar y}{2})\psi^*(x-\frac{\hbar y}{2})H_0)\big]e^{\mathrm{i}tH_0}\Big)(\xi) \\ &= \mathscr{F}_y\Big(\big[-\mathrm{i}(H_0\psi(x+\frac{\hbar y}{2},t)\psi^*(x-\frac{\hbar y}{2},t) \\ &\qquad - \psi(x+\frac{\hbar y}{2},t)\psi^*(x-\frac{\hbar y}{2},t)H_0)\big]\Big)(\xi), \end{aligned}$$

which is identical to (7.48). □

The above theorem implies the following fundamental result.

Theorem 7.9.2 *The Wigner function satisfies the advection equation in the phase space $\mathscr{H}$, which means*

$$\begin{cases} \dfrac{d}{dt}W_{\psi_\hbar}(x,\xi,t) = -\xi.\nabla_x W_{\psi_\hbar}(x,\xi,t), \\ W_{\psi_\hbar}(x,\xi,0) = W_\psi(x,\xi). \end{cases}$$

Proof In the sequel we will use the formula

$$\operatorname{div}(f\nabla g) = \nabla f.\nabla g + f\Delta g, \tag{7.49}$$

for any $f, g \in H^2(\mathbb{R}^n)$, which applying to (7.48) gives

$$\frac{d}{dt}W_\hbar(x,\xi,t) = -\frac{i\hbar}{2}\Big[\mathscr{F}_y\big(\psi(x+\frac{\hbar y}{2},t)\Delta_x^0\psi^*(x-\frac{\hbar y}{2},t) - \psi^*(x-\frac{\hbar y}{2},t)$$

$$\Delta_x^0\psi(x+\frac{\hbar y}{2},t)\big)(\xi)\Big]$$

$$= -\frac{i\hbar}{2}\Big[\mathscr{F}_y\Big(\operatorname{div}\Big(\psi(x+\frac{\hbar y}{2},t)\nabla_x^0\psi^*(x-\frac{\hbar y}{2},t)$$

$$-\psi^*(x-\frac{\hbar y}{2},t)\nabla_x^0\psi^*(x+\frac{\hbar y}{2},t)\Big)\Big)(\xi)\Big].$$

Now by passing ∇_x to ∇_y, that is $\nabla_x\psi(x+\frac{\hbar y}{2}) = \frac{2}{\hbar}\nabla_y\psi(x+\frac{\hbar y}{2})$ and $\nabla_x\psi^*(x+\frac{\hbar y}{2}) = -\frac{2}{\hbar}\nabla_y\psi(x+\frac{\hbar y}{2})$ yields

$$\frac{d}{dt}W_\hbar(x,\xi,t) = i\Big[\mathscr{F}_y\Big(\operatorname{div}_x\nabla_y(\psi(x+\frac{\hbar y}{2},t)\psi^*(x-\frac{\hbar y}{2},t)\Big)(\xi)\Big]$$

$$= -\xi.\operatorname{div}_x[\mathscr{F}_y\Big(\psi(x+\frac{\hbar y}{2},t)\psi^*(x-\frac{\hbar y}{2},t)\Big)(\xi)\Big],$$

which proves the theorem. □

In Theorem 7.9.1 (a) we have seen that the Wigner function is connected to **QDS** $\Phi_\hbar^0(t)$. This connection inspired us to study the scattering formalism for **QDS**. The scattering theory for **QDS** is initiated by Alicki in [2]. In this paper he studied this theory for a simple model of heavy-ion nuclei collision of the quantum Fokker–Planck equation describing the relative motion of two nuclei. Here we restrict ourselves to the standard case where the free Hamiltonian is perturbed by a potential. That is, $H = -\hbar^2\frac{\Delta}{2} + V$, where V is Kato's potential (7.42). In this case H would be a self-adjoint and $e^{-\frac{i}{\hbar}H}$ a unitary group on $\mathcal{H}$. In this way, if, regarding (1.52), we write $\mathbb{L}\rho := (ad_H)\rho$ as the generator of **QDS**, $e^{t\mathbb{L}} := \Phi_\hbar(t)$ and $e^{t\mathbb{L}_0} := \Phi_\hbar^0(t)$, for the existence of the scattering operator

$$\mathbb{S} = \mathbb{W}_+(\mathbb{L}_0,\mathbb{L})\mathbb{W}_-(\mathbb{L},\mathbb{L}_0), \tag{7.50}$$

we have to prove the existence of the wave operators $\mathbb{W}_+(\mathbb{L}_0,\mathbb{L})$ and $\mathbb{W}_-(\mathbb{L},\mathbb{L}_0)$. For this purpose we need to ensure the existence of the wave operators $W_\pm(H_0,H)$ and $W_\pm(H,H_0)$ in the standard scattering theory for the Schrödinger equation which was amply discussed in Sect. 7.7.

Theorem 7.9.3 *Assume that the wave operators $W_-(H_0,H)$ and $W_-(H,H_0)$ exist and are complete. Then the wave operators $\mathbb{W}_+(\mathbb{L}_0,\mathbb{L})$ and $\mathbb{W}_-(\mathbb{L},\mathbb{L}_0)$ also exist.*

Proof Similarly to Definition 2.1.1,

$$\mathbb{W}_\pm(\mathbb{L}_0,\mathbb{L}) = \lim_{t\to\pm\infty} e^{-t\mathbb{L}_0}e^{t\mathbb{L}}$$

and

$$\mathbb{W}_{\pm}(\mathbb{L}, \mathbb{L}_0) = \lim_{t \to \pm\infty} e^{-t\mathbb{L}} e^{t\mathbb{L}_0},$$

but here the limit is ultraweak. We remark that the operator $\boldsymbol{\rho}$ is idempotent ($\boldsymbol{\rho}^2 = \boldsymbol{\rho}$),

$$e^{-t\mathbb{L}_0} e^{t\mathbb{L}} \boldsymbol{\rho} = e^{\mathrm{i}tH_0} e^{-\mathrm{i}tH} \boldsymbol{\rho}^2 e^{\mathrm{i}tH} e^{-\mathrm{i}tH_0}$$

and $e^{-\mathrm{i}tH_0} e^{\mathrm{i}tH}$ converges strongly to $W_-(H_0, H)$ as $t \to -\infty$ and $e^{-\mathrm{i}tH} e^{\mathrm{i}tH_0}$ converges strongly to $W_-(H, H_0)$ as $t \to -\infty$. Furthermore, the strong convergence is finer than the ultraweak convergence, thus

$$\lim_{t \to \infty} \mathrm{Tr}\left(e^{-t\mathbb{L}_0} e^{t\mathbb{L}} \boldsymbol{\rho}\right) = \mathrm{Tr}\left(W_-(H_0, H)\boldsymbol{\rho}\right) \mathrm{Tr}\left(\boldsymbol{\rho} W_-(H, H_0)\right),$$

which follows from Remark 1.16.5.
With the same argument one obtains the existence of $\mathbb{W}_-(\mathbb{L}_0, \mathbb{L})$. This prove that the wave operators $\mathbb{W}_+(\mathbb{L}_0, \mathbb{L})$ and $\mathbb{W}_-(\mathbb{L}, \mathbb{L}_0)$ exist and according to (7.50) we obtain the existence of the scattering operator $\mathbb{S}$. □

7.10 Notes and Comments

The Wigner function was introduced by Eugene Wigner in 1932 to study quantum corrections to classical statistical mechanics (see [124]). The goal was to link the wave-function that appears in Schrödinger's equation to a probability distribution in phase space. It was later performed by Jean Ville in 1948 as a quadratic in representation of the local time–frequency energy of an analytic signal. In 1949, José Enrique Moyal, recognized this function as the quantum moment-generating functional, and hence he take it as the basis in $L^2(\mathbb{R}^n_x \times \mathbb{R}^n_\xi)$. Since then many other applications have been discovered in statistical mechanics, quantum chemistry, quantum optics, classical optics and signal analysis in diverse fields such as electrical engineering, seismology, time–frequency analysis for music signals, spectrograms in biology and speech processing and the most cited book in semiconductor analysis is [87]. One of the defects of this function is that we cannot consider it as a density function in the context of statistical mechanics, since for some values of (x, ξ) in phase space $W(x, \xi)$ is not positive; that is why it is called a quasi-probabilistic function. The Wigner function built from a time dependent pure state satisfies the quantum Liouville equation and the solution of this equation converges formally to the solution of the Vlasov–Liouville equation as the Planck constant $\hbar$ goes to zero. In [85] it is shown that this convergence is in fact a weak convergence. In the first chapter we have defined the notion of ultraweak convergence and in Sect. 7.4 we show that it is not only a weak convergence but also a ultraweak convergence, which is a finer notion. To achieve this result we start in Sect. 7.2 with standard materials for well-posedness of the Wigner problem in $L^2(\mathbb{R}^n_x \times \mathbb{R}^n_\xi)$ by considering the Wigner function as the solution of the quantum Liouville equation. As mentioned in [85], the well-posedness of this equation in $L^2(\mathbb{R}^n_x \times \mathbb{R}^n_\xi)$ is a simple consequence of the bounded perturbation theorem. Using the Riesz–Thorin interpolation theorem we transfer it in any $L^p(\mathbb{R}^n_x \times \mathbb{R}^n_\xi)$, $1 \leq p \leq \infty$.

Concerning the Vlasov–Liouville equation in Sect. 7.3, we derived it by Koopman formalism. In [3] there is a nice exploratory article which indicate the great utility of this formalism. The original goal of Wigner in [124] was to find quantum correction to classical statistical mechanics, which is described in the phase space coordinates (x, ξ). In Sect. 7.4 we develop this point of view by introducing the Weyl calculus, which is expressed using the position vector x and momentum vector ξ. We compare the Poisson bracket with the Lie bracket. Since the Weyl transform is a pseudo-differential operator for some particular symbol, we define the class of symbols and give the Weyl symbol and the relationship of the Weyl transform with Wigner distribution. All these materials can be found in [125].

As we have mentioned, the Wigner function cannot be used for the probability density function since it is not positive for some points in the phase space. To remedy the lack of positivity, Husimi proposed in [67] a procedure to address this deficiency. This procedure consists of taking the convolution product with respect to ξ of the Wigner function with a Gaussian, and as a result the equation which satisfies the Husimi function is exactly the quantum Liouville equation with different initial data. This equation is well-posed in $L^1(\mathbb{R}^n_x \times \mathbb{R}^n_\xi)$ and as $\hbar \to 0$, thanks to the ultraweak convergence of the Wigner function to the solution of Liouville–Vlasov equation, we deduce the L^1 convergence of the Husimi function to the same solution. That was the principal goal of Sect. 7.7.

The scattering theory for the quantum Liouville equation was initiated by Markowich in [86]; this is undertaken in the materials of Sects. 7.6 and 7.7. Finally, since the Wigner function is intimately related to the quantum dynamical semigroup which was studied in Sect. 1.17, hence the Sect. 7.8 is devoted to scattering theory for **QDS**. The scattering theory for **QDS** was initiated by Alicki in [2], where he proved the existence of the scattering operator for the quantum mechanical Fokker–Planck equation using Cook's method. Here we have used directly the existence of wave operators for the Schrödinger equation.

List of Symbols

X	complex Banach space
$(\cdot, \cdot)$	inner product
$\|A\|$	norm of the operator A
$\mathcal{G}(A)$	graph of A
$\overline{A}$	closure of A
$\Phi(t)$	quantum dynamical semigroup
$\mathcal{J}(f)$	duality map
T_0	streaming free operator
$\langle f^*, f\rangle$	dual pair
$\sigma(A)$	spectrum of A
$\sigma_r(A)$	residual spectrum of A
$s(A)$	spectral bound of A
$\mathcal{F}(\mathcal{H})$	finite rank operators
$\mathscr{I}_2(\mathcal{H})$	Hilbert–Schmidt operators
$\mathrm{Tr}(A)$	trace of A
χ_S	characteristic function of S
$M(\sigma)$	mean free path
$\mathcal{H}$	complex Hilbert space
$\|x\| = \sqrt{(x, x)}$	norm of Hilbert space
$A \mapsto A^*$	involution mapping
$D(A)$	domain of A
$\rho(A)$	resolvent of A
ad_H	Lie bracket
(C_0)	strongly continuous
T_a	absorption operator
$\omega(A)$	growth bound
$\sigma_p(A)$	point spectrum of A

H. Emamirad, *Scattering Theory for Transport Phenomena*, Mathematical Physics Studies, https://doi.org/10.1007/978-981-16-2373-8

$\sigma_c(A)$	continuous spectrum of A
$\mathcal{M}$	multiplication operator
$\mathscr{I}_1(\mathcal{H})$	trace class operators
$\mathscr{I}_\infty(\mathcal{H})$	compact operators
$\widehat{f}$	Fourier transform of f
$F(\sigma)$	finite Radon transform
$F(\sigma)$	finite Radon transform

$A_a f(x,v) := -\sigma_a(x,v) f(x,v)$	$A_p f(x,v) := \int_V k(x,v',v) f(x,v') dv'$
$T_p := T_0 + A_p$	$T_a := T_0 + A_a$
$r_+(x,v) := \{x + tv \ : \ t > 0\}$	$r_-(x,v) := \{x + tv \ : \ t < 0\}$
$\Sigma_+ := \{(x,v) \in \mathbb{R}^n \times V \ : \ r_+ \cap \Omega = \emptyset$ and $r_- \cap \Omega \neq \emptyset\}$	$\Sigma_- := \{(x,v) \in \mathbb{R}^n \times V \ : \ r_+ \cap \Omega \neq \emptyset$ and $r_- \cap \Omega = \emptyset\}$
$\Sigma_0 := \{(x,v) \in \mathbb{R}^n \times V \ : \ r_+ \cap \Omega = \emptyset$ and $r_- \cap \Omega = \emptyset\}$	$\Sigma := \mathbb{R}^n \times V \setminus (\Sigma_+ \cup \Sigma_- \cup \Sigma_0)$
$\Sigma_\pm^0 = \Sigma_\pm \cup \Sigma_0$	$\Gamma_\pm f := \chi_{\Sigma_\pm^0} f$
$E_\rho^+ := \{(x,v) \mid x.v \geq \rho\lvert v\rvert\}$	$E_\rho^+ := \{(x,v) \mid x.v \leq -\rho\lvert v\rvert\}$
$D_\rho^+ := \{f(x,v)$ supp $f \subset E_\rho^+\}$	$D_\rho^- := \{f(x,v)$ supp $f \subset E_\rho^-\}$
$\chi_\rho^+ := \chi_{E_\rho^+}$	$\chi_\rho^- := \chi_{E_\rho^-}$
$P_\rho^+ f := (1 - \chi_\rho^+) f$	$P_\rho^- f := (1 - \chi_\rho^-) f$
$Q_\rho^+ f := \chi_\rho^+ f$	$Q_\rho^- f := \chi_\rho^- f$

Bibliography

1. S. Agmon, Spectral properties of Schrödinger operators and scattering theory. Annali della Scuola Norm. Sup. di Pisa **2**, 151–217 (1975)
2. R. Alicki, On the scattering theory for quantum dynamical semigroups. Annales de I.H.P. **35**, 97–103 (1981)
3. H. Arbabi. Introduction to Koopman operator theory of dynamical systems (2020). https://www.mit.edu/~arbabi/research/KoopmanIntro.pdf
4. W. Arendt, Resolvent positive operator. Proc. London Math. Soc. **54**, 321–349 (1987)
5. W. Arendt, C.J. Batty, M. Hieber, *Vector-Valued Laplace Transforms and Cauchy Problems*. Monographs in Mathematics, vol. 96 (Birkhäuser Verlag, Basel, 2001)
6. P. Arianfar, H. Emamirad, Relationship between scattering and albedo operators in linear transport theory. Transp. Theory Stat. Phys. **23**, 517–531 (1994)
7. A. Arnold, On absorbing boundary conditions for quantum transport equations. RAIRO Modél. Math. Anal. Numér. **28**, 853–872 (1994)
8. A. Arnold, C. Sparber, Quantum dynamical semigroups for diffusion models with Hartree interaction. Commun. Math. Phys. **251**, 179–207 (2004)
9. M. Balabane, H. Emamirad, Smooth distribution group and Schrödinger equation in $L^p(\mathbb{R}^n)$. J. Math. Anal. Appl. **292**, 61–71 (1985)
10. M. Balabane, H. Emamirad, L^p estimates for Schrödinger evolution equation. Trans. Am. Math. Soc. **70**, 357–373 (1979)
11. C. Bardos. Problèmes aux limites pour les équations aux derivées partielles du premier ordre a coefficients reels, Théorèmes d'approximation. Application à l'équation de transport. Ann. Sci. Ecole Norm. Sup. **4**, 185–233 (1970)
12. C. Bardos, F. Golse, P. Markowich, T. Paul, On the classical limit of the Schrödinger equation. Discret. Contin. Dyn. Syst. **35**, 5689–5709 (2015)
13. C. Bardos, F. Golse, P. Markowich, T. Paul, Hamiltonian evolution of monokinetic measures with rough momentum profile. Arch. Ration. Mech. Anal. **217**, 71–111 (2015)
14. C.J.K. Batty, D.W. Robinson, Positive one-parameter semigroup on ordered banach space. Acta Appl. Math. **1**, 221–296 (1984)
15. H. Baumgärtel, M. Wollenberg, *Mathematical Scattering Theory*. Operator Theory: Advances and Applications, vol. 19 (Birkhäuser-Verlag, Stuttgart, 1983)
16. R. Beals, V. Protopopescu, Abstract time-dependent transport equations. J. Math. Anal. Appl. **121**, 370–405 (1987)

H. Emamirad, *Scattering Theory for Transport Phenomena*, Mathematical Physics Studies, https://doi.org/10.1007/978-981-16-2373-8

17. T. Bermùdez, A. Bonilla, H. Emamirad, Chaotic tensor product semigroups. Semigroup Forum **71**, 252–264 (2005)
18. G.I. Bell, S. Glasstone, *Nuclear Reactor Theory* (Van Nostrand Reinhold, New York, 1970)
19. G.D. Birkhoff. Démonstration d'un théorème élémentaire sur les fonctions entières. C.R. Acad. Sci. Paris **189**, 473–475 (1929)
20. M. Boulanouar, Study of the streaming operator with general Maxwell's boundary conditions. J. Dyn. Control Syst. **7**, 299–326 (2001)
21. Ph Brenner, The Cauchy problem for symmetric hyperbolic system in L_p. Math. Scand. **19**, 27–37 (1966)
22. G. Busoni, Asymptotic behaviour and wave operators in charge exchange. J. Math. Anal. Appl. **212**, 257–269 (1988)
23. G. Busoni, Remarks on stability in charge exchange dynamics in a monoatomic ionized gas. Transp. Theory Stat. Phys. **17**, 190–208 (1997)
24. G. Busoni, H. Emamirad, Stationary scattering theory for a charged particles transport problem. J. Stat. Phys. **96**, 377–401 (1999)
25. F. Buttazzoni, A. Nocentini, On the dynamics of the perturbations in an ionized gas dominated by charge exchange. Meccanica **20**, 285–290 (1985)
26. M. Cessenat, Théorèmes de traces L^P pour des espaces de fonctions de la neutronique. C.R. Acad. Sci. Paris, Série I **299**, 831–834 (1984)
27. M. Cessenat, Théorèmes de trace pour des espaces de fonctions de la neutronique. C.R. Acad. Sci. Paris, Série I **300**, 89–92 (1985)
28. P. Chernoff, Note on product formulas for operator semigroups. J. Funct. Anal. **2**, 238–242 (1968)
29. C. Chicone and Y. Latushkin. *Evolution semigroups in dynamical systems and differential equations*, vol. 70 (American Mathematical Society MSM, Providence, R.I., 1999)
30. Ph Clément et al., *One-Parameter Semigroups* (North-Holland, Amsrterdam, New York, 1987)
31. J.M. Cook, Convergence to the Moller wave matrix. J. Math. Phys. **36**, 82–87 (1952)
32. G. DaPrato, Semigruppi regolarizzabili. Riserche Mat. **15**, 223–248 (1966)
33. E.B. Davies, *Quantum Theory of Open System* (Academic Press, London, New York, San Francisco, 1976)
34. E.B. Davies, *One-Parameter Semigroups* (Academic Press, New York, 1980)
35. R. deLaubenfels, *Existence Families, Functional Calculi and Evolution Equations* (Springer, Berlin, New York, Heidelberg, 1994)
36. R. deLaubenfels, H. Emamirad, K. Grosse-Erdmann, Chaotic semigroups of unbounded operators. Math. Nachr. **261–262**, 47–59 (2003)
37. W. Desch, W. Schappacher, G.F. Webb, Hypercyclic and chaotic semigroups of linear operators. Ergodic Theory Dyn. Syst. **17**, 793–819 (1997)
38. R.L. Devaney, *An Introduction to Chaotic Dynamical Systems* (Addison-Wesley, New York, 1989)
39. P.A.M. Dirac, A new notation for quantum mechanics. Math. Proc. Camb. Philos. Soc. **35**(3), 416–418 (1939)
40. J. Dixmier, *Les Algèbres d'Opérateurs dans l'Espace Hilbertien* (Gauthier-Villars, Paris, 1969)
41. J.D. Dollard, C.N. Friedmann, *Product Integration* (Mass, Addison-Wesley, Reading, 1979)
42. N. Dunford, J.T. Schwarz, *Linear Operators*. General Theory (Interscience, New York, 1958)
43. D.M. Eidus, The principle of limiting absorption. Transl. Amer. Math. Soc. **47**(2), 157–191 (1965)
44. H. Emamirad, On the Lax and Phillips scattering theory for transport equation. J. Funct. Anal. **62**, 276–303 (1985)
45. H. Emamirad, Scattering theory for linearized Boltzmann equation: survey and new results. Transp. Theory Stat. Phys. **16**, 503–528 (1987)
46. H. Emamirad, Hypercyclicity in the scattering theory for linear transport equation. Trans. Am. Math. Soc. **350**, 3707–3716 (1998)
47. H. Emamirad, On the theory of remediability. Banach Center Publ. **63**, 165–176 (2003)

48. H. Emamirad, V. Protopopescu, Relationship between the albedo and scattering operators for the Boltzmann equation with semi-transparent boundary conditions. Math. Meth. Appl. Sci. **19**, 1–13 (1996)
49. K.J. Engel, R. Nagel, *A Short Course on Operator Semigroups* (Springer, New York, Berlin, 2005)
50. D.E. Evans, Time dependent perturbations and scattering of strongly continuous groups on Banach spaces. Math. Ann. **221**, 275–290 (1976)
51. H.O. Fattorini, *The Abstract Cauchy Problem* (Mass, Addison-Wesley, Reading, 1983)
52. G. Godefroy, J.H. Shapiro, Operators with dense, invariant, cyclic vector manifolds. J. Funct. Anal. **98**, 229–269 (1991)
53. V.A. Golant, A.P. Zilinskii, I.E. Saharov, *Foundations of Plasma Physics* (Nauka, Moscow, 1983)
54. J.A. Goldstein, Abstract evolution equations. Trans. Am. Math. Soc. **141**, 159–186 (1969)
55. J.A. Goldstein, *Semigroups of Linear Operators and Applications* (Oxford University Press, New York, 1985)
56. F. Golse, Distributions, analyse de Fourier, équations aux dérivées partielles. Cours MAT432 de l'École Polytechnique (2018)
57. W. Greenberg, C.V.M. van der Mee, V. Protopopescu, *Boundary Value Problems in Abstract Kinetic Theory* (Birkhauser, Basel, 1987)
58. G. Greiner, Spectral properties and asymptotic behaviour of linear transport equation. Math. Z. **185**, 167–177 (1984)
59. K.G. Grosse-Erdmann, Universal families and hypercyclic operators. Bull. Am. Math. Soc. **36**, 345–381 (1999)
60. J. Hejtmanek, Scattering theory of the linear Boltzmann operator. Comm. Math. Phys. **43**, 109–120 (1975)
61. J. Hejtmanek, Dynamics and spectrum of the linear multiple scattering operator in the Banach lattice $L^1(\mathbb{R}^3 \times \mathbb{R}^3)$. Transp. Theory Stat. Phys. **8**, 29–44 (1979)
62. J. Hejtemanek, The problem of reconstructing objects from projection as an inverse problem in scattering theory of the linear transport operator, in *Mathematical Aspect of Computerized Tomography*, ed. by G.T. Herman, F. Natterer. Lecture Notes in Medical Informatics, vol. 8 (Springer, New York, 1980)
63. E. Hille, Une généralisation du problème de Cauchy. Ann. Inst. Fourier **4**, 31–48 (1952)
64. E. Hille, R.S. Phillips, *Functional Analysis and Semi-groups* (American Mathematical Society Colloquium. Publications, Providence, RI, 1957)
65. L. Hörmander, *The Analysis of Linear Partial Differential Operators*, vol. III (Springer, Berlin New York, 1985)
66. J.S. Howland, Stationary scattering theory for time-dependent Hamiltonians. Math. Ann. **207**, 315–335 (1974)
67. K. Husimi, Some formal properties of the density matrix. Proc. Phys. Math. Soc. Japan (III) **22**, 317–464 (1940)
68. H.G. Kaper, C.G. Lekkerkerker, J. Hejtmanek, *Spectral Methods in Linear Transport Theory* (Birkhäusen Verlag, Basel, 1982)
69. T. Kato, *Perturbation Theory for Linear Operators* (Springer, Berlin, New York, Heidelberg, 1966)
70. T. Kato, Wave operators and similarity for some non-selfadjoint operators. Math. Ann. **162**, 258–279 (1966)
71. T. Kato, Smooth operators and commutators. Studia Math. **31**, 535–546 (1968)
72. M.G. Krein, *Linear Differential Equations in Banach Spaces*. Translations of Mathematical Monographs (Amer. Math. Soc. Providence, RI, 1972)
73. G.E. Ladas, V. Lakshmikanthn, *Differential Equations in Abstract Spaces* (Academic Press, New York, 1972)
74. P. Lax, *Functional Analysis* (Wiley, Hoboken, New Jersey, 2002)
75. P. Lax, *Linear Algebra* (Wiley, Hoboken, New Jersey, 2007)
76. P. Lax, R. Phillips, *Scattering Theory* (Academic Press, New York, 1967)

77. P. Lax, R. Phillips, Scattering theory for transport phenomena, in *Proceedings Conference on Functional Analysis*, University of California, Irvine, Calif., 1967, pp. 119–130
78. E.H. Lieb, Y. Ostrover, Localization of multidimensional Wigner distributions. J. Math. Phys. **51**, 102101 (2010)
79. G. Lindbladt, On the generator of quantum dynamical semigroup. Commun. Math. Phys. **48**, 119–130 (1976)
80. J.L. Lions, Semi-groupes distributions. Portugal Math. **19**, 141–164 (1960)
81. P.-L. Lions, T. Paul, Sur les mesures de Wigner. Rev. Mat. Iberoamericana **9**, 553–618 (1993)
82. W. Littman, C. McCarthy, N. Rivière, The non-existence of L^p estimates for certain translation-invariant operators. Studia Math. **30**, 219–229 (1968)
83. C.R. MacCluer, Chaos in linear distributed systems. J. Dyn. Syst. Meas. Control. **114**, 322–324 (1992)
84. G.R. MacLane, Sequences of derivatives and normal families. J. Anal. Math. **2**, 72–87 (1952–53)
85. P. Markowich, C. Ringhofer, An analysis of the quantum Liouville equation. Z. Angew. Math. Mech. **69**, 121–127 (1989)
86. P.A. Markowich, On the equivalence of the Schrödinger and the quantum Liouville equations. Math. Methods Appl. Sci. **11**, 459–469 (1989)
87. P. Markowich, C. Ringhofer, C. Schmeiser, *Semiconductor Equations* (Springer, Berlin, New York, Heidelberg, 1990)
88. S. Matucci, A transport problem with semitransparent boundary conditions. Nonlinear Anal. **38**, 605–624 (1999)
89. P. Meyer-Nieberg, *Banach Lattices* (Springer, Berlin, New York, Heidelberg, 1991)
90. I. Miyadera, On the perturbation theory for semi-groups of operators. Tôhoku Math. **18**, 299–310 (1966)
91. M. Mokhtar-Kharroubi, Limiting absorption principles and wave operators on $L^1(\mu)$ spaces. Application to transport theory. J. Funct. Anal. **115**, 119–145 (1993)
92. M. Mokhtar-Kharroubi, *Mathematical Topics in Neutron Transport Theory: New Aspects* (World Scientific, 1997)
93. J.E. Moyal, Quantum mechanics as a statistical theory. Proc. Camb. Phil. Soc. **45**, 99–124 (1949)
94. R. Nagel (Ed.), *One-Parameter Semigroups of Positive Operators*. Lecture Notes in Mathematics, vol. 1184 (Springer, Berlin, New York, Heidelberg, 1986)
95. E.-M. Ouhabaz, *Analysis of Heat Equations on Domains* (Princeton Univ. Press, Princeton, 2005)
96. A. Pazy, *Semigroups of Linear Operators and Applications to Partial Differential Equations* (Springer, Berlin, New York, Heidelberg, 1983)
97. R. Phillips, Perturbation theory for semi-groups of linear operators. Trans. Am. Math. Soc. **74**, 199–221 (1953)
98. V. Protopopescu, On the scattering matrix for the linear Boltzmann equation. Rev. Roum. Phys. **21**, 991–994 (1976)
99. V. Protopopescu, Relation entre les opérateurs d'albédo et de scattering avec des conditions aux frontières non-transparentes. C.R. Acad. Sc. Paris Sér. I **318**, 83–86 (1994)
100. V. Protopopescu, Y.Y. Azmy, Topological chaos for a class of linear models. Math. Models Methods Appl. Sci. **2**, 79–90 (1992)
101. J. Radon, Ueber die Bestimmung von Funktionen durch ihr Integralwerte longs gewisser Mannigfaltigkeiten. Ber. Verh. Sachs. Acad. Wiss. Leipzig, Math. Nat. Kl. **69**, 262–277 (1917)
102. M. Reed, B. Simon, *Methods of Modern Mathematical Physics*. Functional Analysis, vol. I (Academic Press, New York, 1972)
103. M. Reed, B. Simon, *Methods of Modern Mathematical Physics*. Scattering Theory, vol. III (Academic Press, London, New York, San Francisco, 1979)
104. G.F. Roach, *An Introduction to Linear and Nonlinear Scattering Theory* (Longman, Harlow, Essex, 1995)

105. D. Robert. *Autour de l'Approximation Semi-Classique* (Bikhäuser (PM68), Basel, Boston, 1987)
106. H.H. Schaefer, *Banach Lattices and Positive Operators* (Springer, Berlin, New York, Heidelberg, 1974)
107. R. Schatten, *Norm Ideals of Completely Continuous Operators* (Springer, Berlin, New York, Heidelberg, 1960)
108. B. Simon, Existence of the scattering matrix for the linearized Boltzmann equation. Commun. Math. Phys. **41**, 99–108 (1975)
109. B. Simon, *Trace Ideals and Their Applications*. Mathematical Surveys and Monographs, vol. 120 (American Mathematical Society, Providence, RI, 2005)
110. P. Stefanov, Spectral and scattering theory for the linear Boltzmann equation in exterior domain. Math. Nachr. **137**, 63–77 (1988)
111. M.H. Stone, On one-parameter unitary groups in Hilbert space. Ann. Math. **33**(2), 643–648 (1932)
112. R.S. Strichartz, L^p estimates for integral transforms. Trans. Am. Math. Soc. **136**, 33–50 (1969)
113. H. Tanabe, *Equations of Evolution* (Pitman, London, 1979)
114. H. Tanabe, *Functional Analytic Methods for Partial Differential Equations* (Marcel Dekker, Amsterdam, 1997)
115. J.R. Taylor, *Scattering Theory* (Wiley, New York, 1972)
116. H.F. Trotter, On the product of semi-groups of operators. Proc. Am. Math. Soc. **10**, 545–551 (1959)
117. T. Umeda, Scattering and spectral theory for the linear Boltzmann equation. J. Math. Kyoto Univ. **24**, 205–218 (1984)
118. T. Umeda, Smooth perturbations in ordered Banach spaces and similarity for the linear transport operators. J. Math. Soc. Japan **38**, 617–625 (1986)
119. T. Umeda, Scattering theory for pseudo-differential operators I. The existence of wave operators. Osaka J. Math. **18**, 361–392 (1981)
120. T. Umeda. Scattering theory for pseudo-dfferential operators (II). The completness of wave operators. Osaka J. Math. **19**, 511–526 (1982)
121. J. Voigt, Functional analytic treatment of the initial boundary value problem for collisionless gases. Ph.D. Thesis, Universität München, 1985
122. J. Voigt, On the perturbation theory for strongly continuous semigroups. Math. Ann. **229**, 163–171 (1977)
123. J. Voigt, On the existence of the scattering operator for the linear Boltzmann equation. J. Math. Anal. Appl. **58**, 541–558 (1977)
124. E. Wigner, On the quantum correction for thermodynamic equilibrium. Phys. Rev. **40**, 749–759 (1932)
125. M.W. Wong, *Weyl Transforms* (Springer, New York, 1998)
126. D.R. Yafaev, *Mathematical Scattering Theory*. Transl. Math. Monographs, vol. 105 (Amer. Math. Soc., Providence, RI, 1992)

Index

H. Emamirad, *Scattering Theory for Transport Phenomena*, Mathematical Physics Studies, https://doi.org/10.1007/978-981-16-2373-8

Printed by Printforce, the Netherlands

Labelled Non-Classical Logics

by

Luca Viganò

Institut für Informatik,
Albert-Ludwigs-Universität Freiburg,
Freiburg im Breisgau, Germany

KLUWER ACADEMIC PUBLISHERS
BOSTON / DORDRECHT / LONDON

A C.I.P. Catalogue record for this book is available from the Library of Congress.

ISBN 0-7923-7749-4

Published by Kluwer Academic Publishers,
P.O. Box 17, 3300 AA Dordrecht, The Netherlands.

Sold and distributed in North, Central and South America
by Kluwer Academic Publishers,
101 Philip Drive, Norwell, MA 02061, U.S.A.

In all other countries, sold and distributed
by Kluwer Academic Publishers,
P.O. Box 322, 3300 AH Dordrecht, The Netherlands.

Printed on acid-free paper

Printed in the Netherlands.

Contents

List of Figures

List of Tables

Acknowledgments

I could not have written this book without the advice, collaboration and friendship of David Basin and Seán Matthews. Not only have they taught me a great deal and provided constant support and encouragement, but they have always found the time and patience to listen to my intellectual ramblings and never failed to ask the right question at the right time. So, David and Seán, thank you very much!

I am grateful to David also for inviting me to join the Lehrstuhl für Softwaretechnik in Freiburg, for whose pleasant and stimulating working atmosphere thanks are due to Abdel Ayari, Ina Eckert, Stefan Friedrich, Martin Preen, Georg Struth and Burkhart Wolff.

I greatly benefited from inspiring discussions with various colleagues and friends. To name a few: Krysia Broda, Shiva Chaudhuri, J. Michael Dunn, Harald Ganzinger and all the members of the Programming Logics group at the Max-Planck-Institut für Informatik, Fausto Giunchiglia and all the members of the MRG groups in Genova and Trento, Rajeev Goré, Manfred Jaeger, Thomas Kasper, Greg Restall, Renate Schmidt, Helmut Schwichtenberg, Roberto Sebastiani, Luciano Serafini, Amílcar Sernadas and his group in Lisboa, Alex Simpson, Lincoln Wallen, Christoph Weidenbach, Christos Zaroliagis, and all the participants of LD'98 and HyLo'99. My apologies to all those I have forgotten to mention.

Alessandro Armando, Matteo Baldoni, Patrick Blackburn, Marta Cialdea, Marcello D'Agostino, Stéphane Demri, Fabio Massacci, Andreas Nonnengart and Nicola Olivetti deserve a special mention for their fundamental comments on my work and on preliminary drafts of this book. So does Dov Gabbay, for the many discussions, for his contagious enthusiasm, and for writing the foreword to this book.

I thank James Finlay at Kluwer for his patience and support.

Thank you Susanne.

As usual, I owe Aldo and Claudia, sine quibus non.

Luca Viganò
Freiburg im Breisgau
October 1999

Foreword

I am very happy to have this opportunity to introduce Luca Viganò's book on Labelled Non-Classical Logics.

I put forward the methodology of labelled deductive systems to the participants of Logic Colloquium'90 (Labelled Deductive systems, a Position Paper, In J. Oikkonen and J. Vaananen, editors, *Logic Colloquium'90*, Volume 2 of Lecture Notes in Logic, pages 66–68, Springer, Berlin, 1993), in an attempt to bring labelling as a recognised and significant component of our logic culture. It was a response to earlier isolated uses of labels by various distinguished authors, as a means to achieve local proof-theoretic goals. Labelling was used in many different areas such as resource labelling in relevance logics, prefix tableaux in modal logics, annotated logic programs in logic programming, proof tracing in truth maintenance systems, and various side annotations in higher-order proof theory, arithmetic and analysis. This widespread local use of labels was an indication of an underlying logical pattern, namely the simultaneous side-by-side manipulation of several kinds of logical information. It was clear that there was a need to establish the labelled deductive systems methodology.

Modal logic is one major area where labelling can be developed quickly and systematically with a view of demonstrating its power and significant advantage. In modal logic the labels can play a double role. On the one hand they can bring the semantics into the syntax by naming possible worlds (as labels) and on the other hand the very same labels can act as proof theoretical resource labels.

This conceptual advantage in the case of modal logics is due to the natural correspondence between the possible world semantical interpretation and natural deduction. (Consider strict implication $\rightarrow$. The $\rightarrow$ introduction rule says that $\Delta \vdash A \rightarrow B$ iff $\Delta + A \vdash B$. If Δ is a theory of a world then $\Delta + A$ is a theory of an accessible world.) We can therefore expect a sharpening of the proof theory of modal logic through the use of labels, yielding a wealth of results, both on the algorithmic front (complexity) and on the conceptual front (Skolemisation, ε-calculus, quantifiers, nominals, and more).

The present book demonstrates admirably and skillfully the advantages of labelled deductive systems in non-classical logics, and pioneers, in my opinion, the way modal logic is going to be studied in the future.

I welcome this excellent book as well as the author, who will no doubt produce more excellent work in years to come.

Dov M. Gabbay
August De Morgan Professor of Logic
King's College, London

1 INTRODUCTION

1.1 BACKGROUND AND MOTIVATION

Non-classical logics such as modal, temporal, relevance or substructural logics are extensions or restrictions of classical logic that provide languages for formalizing and reasoning about knowledge, belief, time, space, resources, and other dynamic 'state-oriented' properties. As such, they are increasingly applied in various fields of computer science, artificial intelligence, engineering, cognitive science and computational linguistics, as well as in philosophy and mathematics, where most of them actually originated. For instance, non-classical logics are used for formalizing computability and provability [36, 37, 157], for representing knowledge, belief, common sense and contextual reasoning [82, 111, 123, 155, 208, 209], for planning and spatial reasoning [50, 56, 191], and for the formal specification and verification of distributed and concurrent systems, of programs, of circuit designs and of protocols for computer security or other safety critical applications [59, 69, 124, 156, 189, 218, 219].

In this book we do not consider applications, but rather investigate theoretical aspects of non-classical logics, so the question of which logic, among the many available ones, is best suited for a specific problem will not be an issue for us. Nevertheless, even without a particular application in mind, there are general questions that we must answer, such as: How are we going to present, reason with and about non-classical logics? And, in particular: can we give a uniform and modular presentation of related logics, so as to be able to abstract and reuse insights and results when switching from one logic to the other? In other words: is there a general methodology, a *framework*,

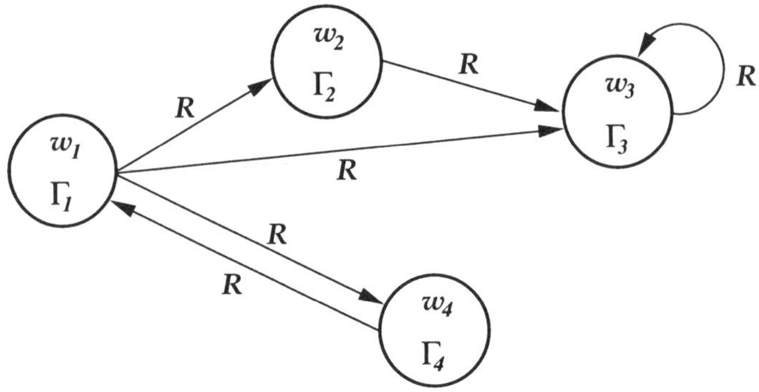

Figure 1.1. Evolution of state (example)

for presenting and working with non-classical logics, independent of the particular application and the logic we have chosen for it?

We build such a framework for presenting and using large families of non-classical logics, focussing in particular on modal and relevance logics, and suggest generalizations necessary to capture other logics. However, we also show that our framework is not a panacea for the above problems, and that in some cases there are tradeoffs or even theoretical and practical limitations that must be taken into account.

To motivate and illustrate the approach that we pursue, let us consider in more detail how we can reason about evolution of state in an arbitrary non-classical logic. For concreteness, consider Figure 1.1. We represent different states and actions on them by means of circles connected by arrows: states are characterized by the set of formulas holding at them (e.g. the state w_1 is characterized by the associated set of formulas Γ_1) and are connected by transition relations denoting actions (in this case, we have one binary relation R) telling us how to move from one state to the other (e.g. from state w_1 we can access the states w_2, w_3 and w_4, but w_3 only accesses itself).

This representation naturally suggests that we adopt for non-classical logics a semantics based on *possible worlds* as developed by Kripke [150, 152] and thus often referred to as *Kripke semantics* (but see also the precursory work of Carnap [47, 48], Hintikka [132] and Kanger [146]).[1] For example, we can identify the representation in Figure 1.1 with the standard Kripke semantics for propositional modal logics [58, 141], and consider a *Kripke frame* consisting of a set of possible worlds $\mathfrak{W}$ connected by a binary accessibility relation $\mathfrak{R}$. A *Kripke model* $\mathfrak{M}$ is the extension of a frame with a function $\mathfrak{V}$ mapping elements of $\mathfrak{W}$ and propositional variables to truth values, based on which we define a truth relation $\models^{\mathfrak{M}}$ to evaluate formulas of the logic at each possible world. That is, a propositional modal formula A is evaluated not globally in $\mathfrak{M}$ but locally at some particular $w \in \mathfrak{W}$; in symbols, $\models^{\mathfrak{M}} w{:}A$.[2] For example, we evaluate classical (material) implication $\supset$ by a straightforward extension of the

[1]Kripke semantics is also often referred to as *possible world semantics* or *relational semantics*.

[2]Other notation is possible, e.g. $\models^{\mathfrak{M}}_{w} A$ as in [58] or $(\mathfrak{M}, w) \models A$ as in [57].

standard clause for propositional classical logic,

$$\models^{\mathfrak{M}} w{:}A \supset B \text{ iff } \models^{\mathfrak{M}} w{:}A \text{ implies } \models^{\mathfrak{M}} w{:}B\,,$$

while the meaning of the modal operator $\Box$ at some world w is interpreted in terms of conditions holding at others: the formula $\Box A$ holds at $w \in \mathfrak{W}$ iff A holds at each world $w' \in \mathfrak{W}$ accessible from w according to $\mathfrak{R}$, i.e.

$$\models^{\mathfrak{M}} w{:}\Box A \text{ iff } \models^{\mathfrak{M}} w'{:}A \text{ for all } w' \text{ such that } (w, w') \in \mathfrak{R}\,.$$

Although propositional modal logics provide quite a powerful language for formalizing state-oriented properties, we can extend this language in various ways. For instance, we can introduce other operators of arbitrary arity, e.g. other unary modal operators, binary substructural implication, unary non-classical negation, and so on. We can also move from the propositional to the quantified case by introducing quantifiers, e.g. the universal quantifier $\forall$, which we treat in each world w as ranging only over the domain of quantification associated to w; we can then obtain different languages by modifying the properties of the domains of quantification (e.g. we can require that when we move from a world to another world accessible from it, objects persist or that no new objects are created). All these changes reflect themselves directly in modifications to our frames and models, e.g. selecting one specific $\mathfrak{o} \in \mathfrak{W}$ as the *actual world* and using the other worlds only to determine the truth of formulas at $\mathfrak{o}$; or connecting worlds with different binary relations (as is the case in multi-modal logics), or even introducing relations of different arities (e.g. to model implications in substructural logics).

The range of possibilities is enormous and different combinations of such changes yield not only different classes of non-classical logics and semantics for them (e.g. modal or relevance logics), but also different families of logics in a class (e.g. the normal modal logics K, T, S4, etc.; the relevance logics B, N, R, etc.). A large number of these logics have been studied and new ones are frequently proposed, prompted by theoretical or practical needs. The result is a multi-dimensional space of logics, each logic demanding, at a minimum, a semantics and a deduction system, and a set of metatheorems relating them together. This development is often non-trivial and, in many cases, has not been systematized; i.e. the characterization of a new logic may demand novel extensions of old techniques or even the invention of completely new ones. Thus a framework in which we can reason with and about non-classical logics in a uniform and modular way is called for.

1.2 CONTRIBUTION

1.2.1 A framework for non-classical logics

We are now in a position to state our contribution more clearly. The particular families of non-classical logics we consider here are extensions or restrictions of classical logic with *non-classical logical operators* that can be interpreted using a Kripke-style semantics consisting of a set of worlds between which relations have been defined. While the meaning of classical operators and quantifiers in some world w is defined only in terms of conditions holding at w, the meaning of a non-classical operator at some world is defined in terms of conditions at other worlds by associating each n–ary non-classical operator with an $n+1$–ary relation on worlds. Thus, e.g., the operators $\Box$ and $\Diamond$ of modal logic can be interpreted in terms of a binary relation (as we saw above), relevant implication can be interpreted using a ternary relation [77], and non-classical negation can be interpreted again using a binary relation [74, 78]. In each case a family of logics in a particular class is defined by variations of the behavior of the relations alone. We show how to exploit this view of non-classical logics as a basis for a framework, where we can

(i) exploit modularity in the semantics so that related logics (their deduction systems and their implementations) result from modifications just to the behavior of the relations, and

(ii) prove metatheoretical results in a modular fashion; i.e. the proofs are parameterized, along with the presentations themselves, over the properties of the relations.

We develop our framework by examining how (ideas from) two complementary proposals for dealing with the enormous range of non-classical logics combine together in practice.

The first is the use of a *generic theorem prover*, based on a *Logical Framework* [18, 125, 166, 181], which can be used to implement deduction systems for many logics in a uniform manner. These theorem provers are based on a metalogic in which the syntax and inference rules of object logics are encoded, so that theorems of the object logic are constructed by proving theorems in the metalogic.

The second is that of *labelling* (or *labelled deduction* [17, 87, 90]), a method for giving uniform presentations of logics typically associated with radically different deduction systems, e.g. modal, substructural, or non-monotonic logics. In the labelled deduction approach, instead of a *consequence relation* being defined over formulas $(\ldots A \vdash B \ldots)$, it is defined over pairs consisting of a label and a formula $(\ldots\ w{:}A \vdash w'{:}B \ldots)$. The labels then allow information needed to formalize the more subtle metatheoretical aspects of the relation to be tracked. For modal logic, for instance, we might want to distinguish between 'local' (with respect to some world) and 'global' (with respect to some frame or model) consequence, so the label might keep track of the possible world in which the formula lives. Or for a substructural logic, where the consequence relation should be sensitive to operations like weakening and contraction, the labels might track resources and their use.

We study this combination in the case of modal and relevance logics, and show how it can provide simple and usable presentations and implementations of large families of logics, including the modal logics K, D, T, B, K4, S4, S4.2, KD45, S5 and their quantified extensions, and the relevance logics B^+, B, N, T and R.

1.2.2 Why combine paradigms?

Why should the labelling and Logical Framework paradigms be combined when Logical Frameworks themselves should suffice to present and implement logics? We contend, and we hope our development illustrates, that the combination is sensible and advantageous since each paradigm can provide something that the other lacks. On one hand, labelling can help tailor the consequence relation of a logic to fit better that of the metalogic. On the other, a Logical Framework provides a means of directly implementing certain kinds of labelled presentations as natural deduction systems, and thus provides a concrete metalogic for reasoning about the correctness of the implementation, and may, as in the case of Isabelle [181], the generic theorem prover we employ, support structured theory development. Below we consider these points in more detail.

Non-classical logics are usually presented in terms of *Hilbert systems* (i.e. Hilbert-style axiomatizations), but these are notoriously difficult to use in practice. Thus alternative, more 'natural', deduction systems must be found.

Many of the Logical Framework logics that have been actively studied, e.g. the type theory of the Edinburgh LF [125], the higher-order logic of Isabelle, and even programming languages like λ-Prolog [85], lend themselves well to the representation of logics that can be presented as collections of inference rules for *proof under assumption*. An example of such a rule is the standard implication introduction rule

$$\frac{\Gamma, A \vdash B}{\Gamma \vdash A \supset B} \supset I$$

of *natural deduction* (ND) [106, 186].

Unfortunately, natural deduction, even though usually recognized as one of the more practical foundations for a deduction system, is often considered badly suited for non-classical logics. The problem is that proof under assumption typically requires a deduction theorem:

if assuming A true we can show B true, then $A \supset B$ is true.

But for implications weaker or substantially different from intuitionistic $\supset$, this fails (at least for the conventional reading of 'if-then' that we get in natural deduction). To illustrate, take the example of modal logic. The standard Kripke completeness theorem tells us that A is provable if and only if A is true at every world in every suitable Kripke model $\mathfrak{M}$, in symbols: $\vdash A$ iff $\models^{\mathfrak{M}} w{:}A$ for all $w \in \mathfrak{W}$. The deduction theorem, as formulated above, then corresponds to

$$(\forall w \in \mathfrak{W}\,(\models^{\mathfrak{M}} w{:}A) \Rightarrow \forall w \in \mathfrak{W}\,(\models^{\mathfrak{M}} w{:}B)) \Rightarrow \forall w \in \mathfrak{W}\,(\models^{\mathfrak{M}} w{:}A \supset B)\,,$$

where $\Rightarrow$ is implication in the meta-language and $\supset$ is implication in the object language. But this is false; the semantics of $\supset$ in a Kripke model is just the weaker:

$$\forall w \in \mathfrak{W}\,((\models^{\mathfrak{M}} w{:}A \Rightarrow \models^{\mathfrak{M}} w{:}B) \Rightarrow \models^{\mathfrak{M}} w{:}A \supset B)\,. \tag{1.1}$$

Thus a naïve attempt to embed modal logic in a ND system will fail.[3]

Approaches to building ND presentations of non-classical logics other than intuitionistic logic have introduced various technical devices to get around the problem.

[3]Note that more complex versions of the deduction theorem do hold for some non-classical logics, see, e.g., [77, 87, 182].

For instance, Dunn [77], for some relevance logics, considers 'relevant' natural deduction, where rules have side conditions on discharged assumptions. In the case of modal logics, Bull and Segerberg [46], Fitting [87] and Prawitz [186], among others, have shown that some, albeit *not* all, logics based on K can be given a natural deduction presentation in which we have $\supset$I together with the rule

$$\frac{\Gamma \vdash A}{\Box\Gamma \vdash \Box A}\ \Box\text{I}$$

where $\Box\Gamma$ indicates that each assumption in Γ has $\Box$ as its main logical operator.[4] The problem with this rule is that it carries a non-local side condition, i.e. a condition on the complete set of assumptions. While Logical Frameworks work well in encoding certain kinds of natural deduction systems, namely those with rules that are *ordinary* (insensitive to thinning or contraction of assumptions), *pure* (have no non-local side conditions), and *single-conclusioned*, encodings of systems that do not meet these criteria, such as the systems based on $\Box$I above, can require considerable ingenuity.[5]

Of course, there may be other sets of inference rules, which are pure, that present the same logic. For example, a pure presentation of S4 for the Edinburgh LF can be found in [9, § 4.4], where two judgements (*true* and *valid*) are used which, in essence, factor the deduction system into two parts, in one of which only propositional reasoning is possible. Although it may be possible to develop other presentations in this fashion, there does not appear to be a systematic way for doing this; each new modal system requires insight and its own justification of soundness and completeness with respect to the corresponding Kripke semantics. Further, even when given such presentations, we have no reason to expect them to have the same combinational properties as their corresponding Hilbert systems; e.g. given pure presentations of the modal logics K4 and KT (i.e. T), we do not know if their combination corresponds to KT4 (i.e. S4).

To summarize, in the case of propositional modal logics (but analogous problems apply for the quantified case and for other non-classical logics), ingenuity is required not only for inventing natural deduction systems in the first place (in fact, Bull and Segerberg [46, p.30] point out that "only exceptional systems ... seem to be characterizable in terms of reasonably simple rules"), but also for the Logical Framework encoding of the rules we have invented. The continuing primacy of Hilbert presentations in non-classical logics, despite the difficulty in actually using them, is evidence that these inventions have not been completely successful.

---

[4]Different approaches to proof under assumption in modal logics, based on modified deduction systems (ND, sequent or tableaux systems), have been proposed, e.g. [10, 73, 87, 158, 233, 235]; see §7 for a detailed comparison and discussion. Note also that there are approaches to the presentation of non-classical logics that are based on neither Hilbert-style axiomatizations nor natural deduction, e.g. the semantics-based approach [126, 169, 170, 171], in which a non-classical logic is translated into a 'suitable' first-order theory. We return to this below.

[5]We use the vocabulary of [6], which should be consulted for a technical discussion of consequence relations and degrees of impurity of natural deduction rules, and which notes (§ 5.5) that "every ordinary, pure single-conclusioned natural deduction system can, e.g., quite easily be implemented on the Edinburgh LF." Note, however, that with sufficient effort a Logical Framework can implement *any* (recursively enumerable) deduction system, but the resulting encoding does not necessarily 'fit' well; see [103], where a concept of a *natural representation* in a Logical Framework is formalized and investigated.

1.2.3 Our framework: labelled deduction systems for non-classical logics

In our framework we present non-classical logics as natural deduction systems, namely as *labelled natural deduction systems*. We show that the combination of the labelling and Logical Framework paradigms provides systems that, unlike those considered above, fit well in a standard Logical Framework, in that our inference rules are ordinary, pure and single-conclusioned.

To illustrate, consider again the deduction theorem and suppose that we extend natural deduction to be over pairs drawn from the language of modal logic *and labels*; i.e. instead of $\vdash A$ we consider $\vdash w{:}A$, where the label w represents a possible world, and intuitively we have: $\vdash A$ iff $\vdash w{:}A$ for all $w \in \mathfrak{W}$. This provides a language in which we can formulate a deduction theorem corresponding to (1.1), namely

if assuming $w{:}A$ true we can show $w{:}B$ true, then $w{:}A \supset B$ is true,

and thus provides a basis for a ND system of the sort we need. Moreover, we can use the same notation to express the general behavior of non-classical operators like $\Box$ in a way that is independent of the (relational) details of the Kripke models providing the semantics, i.e. $\vdash w{:}\Box A$ iff $\vdash w'{:}A$ for all $w' \in \mathfrak{W}$ accessible from w. We formalize this by considering two kinds of formulas, namely *labelled formulas* of the form $w{:}A$, intuitively expressing that the propositional modal formula A holds at world w, and *relational formulas* of the form wRw', capturing the modal accessibility relation (i.e. expressing that w' is accessible from w). This allows us to give ND introduction and elimination rules for $\Box$ and other modal operators that are fixed for all the logics we consider, e.g. for $\Box$ we have the rules

$$\dfrac{\begin{matrix}[wRw'] \\ \vdots \\ w'{:}A\end{matrix}}{w{:}\Box A}\;\Box\mathrm{I} \qquad \text{and} \qquad \dfrac{w{:}\Box A \quad wRw'}{w'{:}A}\;\Box\mathrm{E}$$

where $\Box$I has the side condition that w' is different from w and does not occur in any assumption on which $w'{:}A$ depends other than wRw'.[6] Then we can produce ND systems for particular modal logics simply by formalizing the details of particular accessibility relations, i.e. by specifying how we can infer relational formulas.

Similar insights and intuitions apply for other propositional non-classical logics. We present a propositional non-classical logic in terms of a ND system consisting of two parts: a *base system* for manipulating labelled formulas, and a separate *labelling algebra* for reasoning about the labels, i.e. for manipulating relational formulas. (The term 'labelling algebra' is adopted from Gabbay's *Labelled Deductive Systems* (LDS) [90].)

The base system is a labelled ND presentation of classical logic extended (or restricted) with introduction and elimination rules for the non-classical operators; the base system thus presents the base logic of a family of propositional non-classical logics (e.g. the base ND system N(K) presents the propositional modal logic K).

---

[6] Note that this is a side condition that can be directly encoded in a Logical Framework; see [6] and §5, §6 and §7.

The labelling algebras we consider are *relational theories* comprised of Horn clause axioms formalizing the relations between worlds in Kripke models (i.e. for reasoning about relational formulas of appropriate arities).

These two parts are separate and communicate through an *interface* provided by the rules for the non-classical operators; the intuition behind all this is that for a family or class of related logics the base system stays fixed and we obtain a presentation of the particular logic we want by 'plugging in' the appropriate relational theory.

1.2.4 Finding a 'good' presentation

In order to provide a labelled presentation of a propositional non-classical logic we thus need two things: a base deduction system and a general notion of a labelling algebra. However, for each of these there may be more than one possible candidate. For instance in this book we consider ND and sequent systems but not the closely related tableaux systems. Also, we focus mainly on labelling algebras corresponding to Horn relational theories, which is one possibility out of many, and perhaps not even the most obvious one. Why restrict ourselves to Horn clause logic, instead of first or even higher-order logic?

What we need are criteria for assessing the relative merits of the range of possibilities. We can, of course, consider the basic metatheoretical properties that deduction systems are expected to satisfy, such as *normalization of derivations* and the *subformula property*, but we can extend this list. There are also pragmatic considerations, such as 'is it easy to use?'. For example, Horn relational theories can be directly encoded in the Horn fragment of the metalogics we use for our implementations (it is not necessary first to embed first-order logic or formalize additional judgements; see [102, 125]). But there are other theoretical considerations; for instance, when introducing their Labelled Deductive Systems for substructural logics, D'Agostino and Gabbay [63, p. 244] write:

> The labelling algebra represents this metalevel information as a *separate* component of a standard derivation system and can be treated as an independent parameter. In the LDS approach, logical systems are not studied statically, in isolation, but dynamically, observing the process of their generation and their interaction (via modifications of the labelling algebras) on the basis of a fixed proof-theoretical hard core (the underlying system of deduction). [their emphasis]

In other words, a good presentation should correspond not just to some logic, but to a space of logics, which vary in a well-behaved way according to the details of the labelling algebra; e.g. we would expect that a labelled presentation of K4 combined with one of T does result in S4. By this standard, for instance, while the presentation of S4 in [9] could be seen as a labelled deduction system where the two judgements correspond to labels, it would not be a good one since there is no labelling algebra to vary.[7]

[7]This is not meant as a criticism of that presentation, which was not motivated by such concerns.

1.2.5 Properties and limitations of our labelled deduction framework

The labelled deduction framework we propose does well by the above measures since it cleanly separates the labelling algebras (our Horn relational theories) from the base systems. We show that our framework has good, modular, compositional properties, behaving in the way we would expect when we combine labelling algebras together, and providing a natural hierarchy of systems that inherit theorems and derived rules. We have implemented our approach in Isabelle, which supports management of separate theories and their structured combination, and the result is a parameterized proof development environment where (although this is not a formally quantifiable property) proof construction is natural and intuitive. Moreover, we use the parameterized relational theories to prove, in a parameterized way, metatheoretical properties such as correctness of our encodings in Isabelle, and soundness and completeness of our labelled ND systems with respect to the corresponding Kripke semantics. (We provide a modified canonical model construction that accounts for the explicit formalization of labels and of the relations between them). These theorems show that our implementation not only properly captures non-classical provability within our hierarchies of logics, but also a satisfactory notion of proof under assumption, i.e. consequence.

In our framework we are able to interpret the 'separate' in the previous quotation of D'Agostino and Gabbay in a strong way: not only do we have a separation between the base systems and the labelling algebras, but that separation is maintained even when building derivations: in the relational theory we reason only about relational formulas, while in the base system we exploit labelled and relational formulas to infer only labelled formulas, so that a derivation in the base system may depend on a derivation in the relational theory but not vice versa. That is, derivations of labelled formulas consist of a tree built from the base system, which is decorated with a fringe of derivations in the relational theory alone. Moreover, we exploit this separation to show that derivations in our labelled ND systems *normalize*, i.e. they reduce to a normal form, which we can then characterize further by identifying particular sequences of labelled formulas, and showing that in these sequences there is an ordering on inferences. By exploiting this ordering, we can then show that derivations in normal form satisfy a *subformula property*, in the sense that the formulas appearing in a normal derivation Π of $w{:}A$ must be subformulas of A or of the assumptions of Π.

Thus, our natural deduction systems have 'good' structural properties, in that they possess a well-defined structure, which we can exploit also to investigate advantages and limitations of our framework.

To illustrate some of these advantages and limitations, consider again propositional modal logics. Our treatment has obvious similarities to traditional *semantic embeddings* (i.e. translations of modal logics into predicate logic [126, 169, 170, 171]), but it offers advantages in comparison: our formalization does not require all of first-order logic and it yields structured labelled ND systems where the separation between the base system and the relational theories gives us better normalization results, in that the normal form of a derivation in our approach preserves more structure than the normal form of a derivation in the translation approach. (As we point out below, we can use this extra structure to investigate the complexity of the decision problem for the logics we present).

This extra structure depends on the choices we have made, namely the use of a 'partial' translation (i.e. the introduction of labels and relational formulas) and the extension of fixed base systems with separate Horn relational theories. In fact, it turns out that the structural properties of our systems are directly related to the behavior of *falsum* ($\perp$) in the base system: falsum is able to propagate between different worlds, i.e. from $w{:}\perp$ we can derive $w'{:}\perp$ in any world w', a property we call *global falsum*. We show that global falsum is enough to present, among others, many of the non-classical logics we are likely to encounter in practice, but not enough to present all non-classical logics with first or higher-order definable frames (in contrast to traditional semantic embeddings).

Having identified this property of falsum, we can vary it to produce different candidate 'hard cores'. We investigate the other two obvious possibilities. The first of these, an extension we call *universal falsum*, allows $\perp$ to propagate not only from one world to another, but also between worlds and the labelling algebra (assuming that the labelling algebra is also extended with falsum). The second, a restriction where $\perp$ is no longer able to propagate even between different worlds, we call *local falsum*.

A system with universal falsum is strictly more general than one with global falsum. In fact, we can show that it is essentially equivalent to a traditional semantic embedding in first-order logic, and therefore able to treat not just the logics we can capture with global falsum and Horn relational theories, but any first-order (or even higher-order) axiomatizable non-classical logic. However, in exchange for this greater scope we lose the better behaved proof theory of a system with global falsum, and the result does not seem to offer any advantages over semantic embedding in first-order logic (where there is no separation at all). If we restrict ourselves to a local falsum on the other hand, we obtain deduction systems that possess interesting paraconsistency properties but are in general not suitable for presenting the usual non-classical logics. Thus a base system with global falsum seems to be the weakest base system that we can extend to a useful range of labelled ND systems for non-classical logics.

1.2.6 Introducing quantifiers

The development and results are similar when we move to the quantified case (in this book we consider only quantified modal logics [89, 104, 141] as a significant case study). Here difficulties not present in the propositional case arise, since quantifiers introduce additional complexity to the range of possible semantics that might be appropriate for the logic of an intended application: we must choose not only properties of the accessibility relation in the Kripke model, as in the propositional case, but also how the domains of individuals change between worlds (e.g. varying or constant domains). Since these two choices can be made independently, the result is a two-dimensional space of possible quantified modal logics.[8] We give a labelled presentation of quantified modal logics that is modular in two dimensions, reflecting these two degrees of freedom. As before, it is based on a fixed base ND system (now $\mathrm{N(QK)}$, for

[8]Other dimensions are possible, e.g. non-rigid designators [89, 104], but here we consider only the rigid case.

quantified K) where extensions are made by independently instantiating two separate theories: a relational theory (as before), and a *domain theory*, which formalizes the behavior of the domains of quantification. This second theory requires the introduction of *labelled terms*, $w{:}t$, expressing the existence of the term (variable or constant) t at world w. Thus $\vdash w{:}\forall x(A)$ iff $\vdash w{:}A[t/x]$ for all t such that $\vdash w{:}t$. This formulation naturally suggests that we adopt quantifier rules similar to those of *free logic* [28], i.e. N(QK) contains the following introduction and elimination rules for $\forall$:

$$\dfrac{\begin{matrix}[w{:}t]\\ \vdots\\ w{:}A[t/x]\end{matrix}}{w{:}\forall x(A)}\ \forall\mathrm{I} \qquad \text{and} \qquad \dfrac{w{:}\forall x(A) \quad w{:}t}{w{:}A[t/x]}\ \forall\mathrm{E}$$

where $\forall$I has the side condition that t does not occur in any assumption on which $w{:}A[t/x]$ depends other than $w{:}t$.

We give predicate extensions (with varying, increasing, decreasing or constant domains) of propositional modal logics by appropriately instantiating the relational and domain theories extending the base system. As in the propositional case, the metatheory of our quantified systems is also modular, in that we can prove, e.g., soundness, completeness and normalization in a parameterized way. Furthermore, we show that there are tradeoffs in formalizations of the base system and the theories extending it: we show not only that we can extend the results (and the arguments about advantages and limitations) from the propositional to the quantified case, but also that, in the latter, new tradeoffs must be considered.

1.2.7 Substructural and complexity analysis

Normalization of derivations and the subformula property allow us not only to assess the merits and limits of our approach, they are also pragmatically useful. In particular, we can exploit them to show that we can use our framework to establish complexity results: we develop a proof-theoretical method for establishing the decidability of (some of) the logics we present and for bounding the computational complexity of decision procedures for these logics.

These kinds of analyses are more easily carried out when logics are presented as sequent systems (although ND systems could be used as well). We therefore introduce *cut-free labelled sequent systems*, which we then show equivalent to our normalizing labelled ND systems. As before, we separate fixed base systems from the theories extending them. For example, the base sequent system S(K), presenting the propositional modal logic K, contains the following left and right rules for $\Box$, corresponding to the elimination and introduction rules in the ND system N(K):

$$\dfrac{\Delta \vdash wRw' \quad w'{:}A, \Gamma, \Delta \vdash \Gamma'}{w{:}\Box A, \Gamma, \Delta \vdash \Gamma'}\ \Box\mathrm{L} \qquad \text{and} \qquad \dfrac{\Gamma, \Delta, wRw' \vdash \Gamma', w'{:}A}{\Gamma, \Delta \vdash \Gamma', w{:}\Box A}\ \Box\mathrm{R}$$

where $\vdash$ is the sequent symbol, Γ and Γ' are multisets of labelled formulas, Δ is a multiset of relational formulas, and $\Box$R has the side condition that w' does not occur in $\Gamma, \Delta \vdash \Gamma', w{:}\Box A$.

S(K) also contains *axioms* (i.e. initial sequents), left and right rules for the other logical operators (e.g. implication), and *structural rules*, weakenings and contractions, which tell us how we can alter the structure of sequents by deleting or duplicating formulas (when reading the rules backwards, i.e. upwards, as is usually done); for example:

$$\frac{\Gamma, \Delta \vdash \Gamma'}{w{:}A, \Gamma, \Delta \vdash \Gamma'}\ \text{WlL} \qquad \frac{\Gamma, \Delta \vdash \Gamma'}{\Gamma, \Delta \vdash \Gamma', w{:}A}\ \text{WlR}$$

$$\frac{w{:}A, w{:}A, \Gamma, \Delta \vdash \Gamma'}{w{:}A, \Gamma, \Delta \vdash \Gamma'}\ \text{ClL} \qquad \frac{\Gamma, \Delta \vdash \Gamma', w{:}A, w{:}A}{\Gamma, \Delta \vdash \Gamma', w{:}A}\ \text{ClR} \quad .$$

Our sequent systems satisfy a subformula property, in that proofs of a sequent S may contain only subformulas of the labelled formulas in S. In other words, the subformula property tells us which formulas are allowed to appear in proofs. However, it does not limit the number of times these formulas may appear in a sequent. In fact, we can immediately see that, when reasoning backwards, the contraction rules are always applicable. In order to show that proof search terminates and establish decidability and complexity results, we must therefore find a way of bounding applications of these rules, and so bound the number of times we can make use of formulas in proofs. This is achieved by a substructural analysis of our sequent systems, i.e. by a proof-theoretical analysis of applications of the structural rules (in particular, contractions).

Although contraction-elimination is in general impossible in non-classical logics (analogously to the necessity of contracting universally quantified formulas in first-order logic), we are able to show that bounds on applications of contraction do exist for particular logics. Specifically, our substructural analysis shows that contraction can be bounded in our sequent systems for the propositional modal logics K, T, K4 and S4. (And we discuss extensions to other logics.) This, combined with an analysis of the accessibility relation of the corresponding Kripke frames, yields decision procedures with bounded space requirements, specifically, Polynomial Space. Moreover, as a by-product of our substructural analysis, we are able to give proof-theoretical justifications and partial refinements of the rules of standard (unlabelled) sequent systems for these logics.

1.3 MAIN RESULTS

We summarize our main results as follows.

Presentation. We give modular presentations of non-classical logics in terms of labelled natural deduction systems. A family of related logics is presented by extending a fixed base system with separate Horn relational theories formalizing the properties of the relations in the corresponding Kripke semantics. In the case of quantified modal logics, we extend the base system also with domain theories formalizing the behavior of the domains of quantification (varying, increasing, decreasing or constant domains).

Soundness and completeness. We uniformly show that our labelled natural deduction systems are sound and complete with respect to the corresponding Kripke semantics. The use of explicit labels leads to a modular proof of soundness and completeness for all the logics we consider, which differs from the standard one for unlabelled deduction systems: we provide a modified canonical model construction that accounts for the explicit formalization of labels, of the relations, and of the properties of the domains of quantification.

Proof theory. We exploit proof-theoretical results to explore tradeoffs in the formulation of the base system and the theories extending it. For example, in the propositional case, we show that when the relational theory can be formulated as a set of Horn clauses (as opposed to a set of first or higher-order axioms), then derivations normalize and satisfy the subformula property, and there is a strong separation between the base system and the relational theory; i.e. derivations in the base system may depend on derivations in the relational theory, but not vice versa. Analogous properties hold for quantified modal logics. We then exploit these structural properties to delineate advantages and limitations of our labelled deduction framework.

Implementation. We show that our labelled natural deduction systems can be encoded in a Logical Framework based on a minimal metalogic with higher-order quantification, e.g. the metalogic of Isabelle [181] or the Edinburgh LF [125]. We implement our approach in Isabelle and the result is a simple and natural environment for interactive proof development in which it is possible to structure non-classical logics hierarchically, extending a logic with new properties to generate a new one, and having theorems inherited by these extensions.

Substructural analysis. We exploit our normalization results for labelled natural deduction systems to give a new proof-theoretical method for bounding the complexity of the decision problem for non-classical logics. We present logics in a uniform way as cut-free labelled sequent systems, which we show equivalent to normalizing labelled natural deduction systems, and then restrict the structural rules for particular sequent systems; we consider the systems for K, T, K4 and S4 as case studies. This substructural analysis, combined with an analysis of the accessibility relation of the corresponding Kripke frames, yields decision procedures with bounded space requirements, specifically Polynomial Space. Moreover, it also yields justifications (and, in some cases, refinements) of the rules of standard sequent systems.

1.4 SYNOPSIS

The rest of this book consists of twelve chapters, §2 – §13, divided into two parts, and of a final chapter, §14, in which we draw conclusions. In Part I, §2 – §7, we introduce labelled deduction systems for non-classical logics. In Part II, §8 – §13, we perform a substructural and complexity analysis of (some of) our modal sequent systems. The contents of these chapters are as follows.

Part I

In §2 we formalize labelled ND deduction systems for propositional modal logics and analyze their metatheoretical properties; specifically, soundness and completeness with respect to the corresponding Kripke semantics, normalization of derivations, and the subformula property. We then exploit these structural properties to delineate advantages and limitations of our systems.

In §3 we generalize the development of §2 to formalize and analyze labelled ND systems for propositional non-classical logics.

In §4 we extend the development of §2 to formalize and analyze labelled ND systems for quantified modal logics.

In §5 we encode our labelled ND systems in Isabelle, prove our encodings correct, and show how our approach supports modular (interactive) proof construction.

In §6 we exploit the normalizing labelled ND systems of the previous chapters to present non-classical logics in terms of cut-free labelled sequent systems.

In §7 we summarize the results of Part I and compare with related work.

Part II

In §8 we prove preliminary results that we exploit to perform a substructural and complexity analysis of our labelled sequent systems for propositional modal logics.

In §9, §10 and §11 we investigate applications of structural rules in our sequent systems for K, T, and K4 and S4, respectively. We show that in each of these systems applications of the contraction rules can be bounded, and that our analyses can be exploited to justify and refine the corresponding standard sequent systems.

In §12, based on the results in §9 – §11, we give decision procedures with bounded space requirements for the logics we considered.

In §13 we summarize the results of Part II and compare with related work.

Notation

Most of the notation and terminology we use is standard. Nevertheless, in order to keep this book as self-contained as possible, we have tried to be systematic about providing explicit definitions (and listing all definitions, symbols and topics in the index). We do assume some familiarity on the part of readers with the basic ideas underlying standard deduction systems and semantics for modal, relevance, and other non-classical logics.

I Labelled deduction for non-classical logics

2 LABELLED NATURAL DEDUCTION SYSTEMS FOR PROPOSITIONAL MODAL LOGICS

We give a framework for presenting families of propositional modal logics (including the logics K, D, T, B, S4, S4.2, KD45 and S5) in a uniform and modular way as labelled natural deduction (ND) systems. Our approach is based on a separation between a base ND system and a labelling algebra, which interact through a fixed interface. While the base system stays fixed, ND systems for different modal logics are generated by 'plugging in' appropriate algebras, i.e. by adding rules formalizing the properties of the corresponding accessibility relations. This leads to a hierarchical structuring of systems with inheritance of theorems. Moreover, it allows modular proofs of metatheoretical properties, in that these proofs, along with the presentations themselves, are parameterized over the properties of the relations.

The outline of this chapter is as follows. In §2.1 we give a uniform and modular presentation of propositional modal logics in terms of a base ND system N(K) extended with separate Horn relational theories. In §2.2 we prove the soundness and completeness of our systems with respect to the corresponding Kripke semantics, by providing a modified canonical model construction that accounts for the explicit formalization of labels and of the relations between them. In §2.3 we consider proof-theoretical properties of our systems; in particular, we show that derivations normalize, and that derivations in normal form possess a well-defined structure and satisfy a subformula property. We then exploit these properties to contrast our approach with related formalizations and investigate the tradeoffs in possible presentations.

In §5 we will then present our encodings of modal and other non-classical logics in the generic theorem prover Isabelle, prove their correctness (i.e. that they are faithful and adequate with respect to the ND systems they implement), and, corresponding to our example ND derivations, give proof scripts from Isabelle sessions that demonstrate modular (interactive) proof construction with our implementations.

2.1 A MODULAR PRESENTATION OF PROPOSITIONAL MODAL LOGICS

2.1.1 *The base system* N(K)

We begin by reviewing the standard syntax of propositional modal logics.

Definition 2.1.1 *The language of propositional modal logics consists of a denumerable infinite set of* propositional variables, *the brackets '(' and ')', and the following primitive* logical operators*:*

- *the classical connectives* $\perp$ (falsum) *and* $\supset$ (implication, 'implies'), *and*
- *the modal operator* $\Box$ (necessity, 'box').

The propositional variables and $\perp$ *stand for the indecomposable propositions, which we also call* atomic formulas.

The set of propositional well-formed modal formulas (hereafter simply called propositional modal formulas) *is the smallest set that contains the atomic formulas and is closed under the following formation rules:*

1. *if A and B are formulas, then so is* $(A \supset B)$*;*
2. *if A is a formula, then so is* $(\Box A)$*; and*
3. *all formulas are given by the above clauses.*

Other connectives and modal operators, e.g. $\sim$ (negation, 'not'), $\wedge$ (conjunction, 'and'), $\vee$ (disjunction, 'or'), *and* $\Diamond$ (possibility, 'diamond'), *can be defined in the usual manner, e.g.* $(\sim A) =_{def} (A \supset \perp)$ *and* $(\Diamond A) =_{def} (\sim(\Box(\sim A)))$.

We call boxed formula *a formula of the form* $\Box A$*, i.e. a formula that has a* $\Box$ *as its main logical operator, and we call* box-free formula *a formula that does not contain any* $\Box$ *(or, in general, any modal operator, as we define* $\Diamond$ *in terms of* $\Box$*).* ■

Notation 2.1.2 Let A, B, C and D be formulas. In order to simplify our notation, we will omit brackets whenever no confusion can arise. We will always discard the outermost brackets and we will discard brackets in the case of negation; for example, we will write $A \supset B$ for $(A \supset B)$ and $\sim\Box\sim A$ for $(\sim(\Box(\sim A)))$. Furthermore, we adopt the convention that $\sim$, $\Box$ and $\Diamond$ are of equal binding strength and bind tighter than $\wedge$, which binds tighter than $\vee$, which binds tighter than $\supset$. So, for example, $\Diamond\Box A \wedge \sim B \supset \Box C \vee \Diamond D$ stands for $((\Diamond(\Box A)) \wedge (\sim B)) \supset ((\Box C) \vee (\Diamond D))$. ■

Based on these definitions, we introduce the language of our labelled presentations of propositional modal logics.

Definition 2.1.3 *Let W be a set of* labels *and R a binary relation over W. If x and y are labels and A is a propositional modal formula, then xRy is a* relational well-formed formula *(hereafter simply called* relational formula *or* rwff *for short) and $x{:}A$ is a* labelled well-formed formula *(hereafter simply called* labelled formula *or* lwff *for short).* ■

Definition 2.1.4 *The* grade *of an lwff $x{:}A$, in symbols $grade(x{:}A)$, is the number of times $\supset$ and $\Box$ occur in A.* ■

Notation 2.1.5 For the rest of this chapter, we assume that the variables x, y, z, w, ... range over labels, the variables A, B, ... range over propositional modal formulas, φ is an arbitrary rwff or lwff, and

$$\Gamma = \{x_1{:}A_1, \ldots, x_n{:}A_n\} \quad \text{and} \quad \Delta = \{x_1 R y_1, \ldots, x_m R y_m\}$$

are arbitrary sets of lwffs and rwffs. All variables may be annotated with subscripts or superscripts. ■

The rules given in Figure 2.1 determine N(K), the base ND system presenting the modal logic K. That N(K) presents K is proved in Theorem 2.2.5 below. Note also that we do not enforce Prawitz's side condition on $\bot$E that $A \neq \bot$; see the discussion of the different types of falsum in §2.3, in particular the rule for 'global falsum' in Fact 2.3.1 and Footnote 4. For brevity, in the following we also use the derived rules of N(K) given in Figure 2.2; their derivations are given in Example 2.1.14 below.

Formally, a *ND system* (or *ND calculus*) is a collection of *rules* formalizing proof under assumption. The system N(K) consists of an *introduction rule*, $\bullet$I, and an *elimination rule*, $\bullet$E, for each logical operator $\bullet$ except *falsum*, i.e. $\bot$, for which only an elimination rule is given. In other words, the rules define the behavior of logical operators: they introduce or eliminate instances of the operators of the logic. We call the formula below the line in a rule the *conclusion* of the rule, and the formulas above the line the *premises* of the rule. In an application of an elimination rule, we call the premise in which the eliminated operator is exhibited the *major premise* of the rule and the other premises, if any, the *minor premises*. At some rule applications, e.g. $\Box$I, the conclusion becomes independent of some or all assumptions, e.g. xRy. When this is the case, we say that we *discharge* the assumptions in question, and display this by enclosing them in square brackets; the remaining assumptions, if any, we call *open* assumptions. Further notation and terminology for ND systems is introduced below, e.g. Definition 2.1.11; for a full account see [186] or [106, 221, 230].

Note that there is a close correspondence between our rules for $\Box$ and $\supset$; this holds since we express $x{:}\Box A$ as the metalevel implication $xRy \Rightarrow y{:}A$ for an arbitrary world y accessible from x. Furthermore, as we anticipated in §1, the introduction and elimination rules for $\Box$ are independent of the properties of R; the derived rules for $\Diamond$ enjoy the same independence since $\Diamond$ is defined in terms of $\Box$ and $\sim$.

$$\dfrac{\begin{array}{c}[x{:}A \supset \bot]\\ \vdots \\ y{:}\bot\end{array}}{x{:}A}\ \bot\mathrm{E} \qquad \dfrac{\begin{array}{c}[x{:}A]\\ \vdots \\ x{:}B\end{array}}{x{:}A \supset B}\ \supset\mathrm{I} \qquad \dfrac{x{:}A \supset B \quad x{:}A}{x{:}B}\ \supset\mathrm{E}$$

$$\dfrac{\begin{array}{c}[x R y]\\ \vdots \\ y{:}A\end{array}}{x{:}\Box A}\ \Box\mathrm{I} \qquad \dfrac{x{:}\Box A \quad x R y}{y{:}A}\ \Box\mathrm{E}$$

In $\Box$I, y is different from x and does not occur in any assumption on which $y{:}A$ depends other than $x R y$.

Figure 2.1. The rules of N(K)

$$\dfrac{\begin{array}{c}[x{:}A]\\ \vdots \\ x{:}\bot\end{array}}{x{:}\sim A}\ \sim\mathrm{I} \qquad \dfrac{x{:}\sim A \quad x{:}A}{x{:}\bot}\ \sim\mathrm{E} \qquad \dfrac{y{:}A \quad x R y}{x{:}\Diamond A}\ \Diamond\mathrm{I} \qquad \dfrac{x{:}\Diamond A \quad \begin{array}{c}[y{:}A]\ [x R y]\\ \vdots \\ z{:}B\end{array}}{z{:}B}\ \Diamond\mathrm{E}$$

In $\Diamond$E, y is different from x and z and does not occur in any assumption on which the upper occurrence of $z{:}B$ depends other than $y{:}A$ and $x R y$.

Figure 2.2. Some derived rules of N(K)

2.1.2 Relational theories

Modal logics are traditionally [57, 58, 141] presented by extending a Hilbert system for propositional classical logic with a collection of axiom schemas and inference rules.[1] For example, the axiom schemas and rules given in Figure 2.3 determine a Hilbert system H(K) for K. Note that we consider axiom schemas instead of axioms: any substitutional instance of an axiom schema or theorem of a Hilbert system is also a theorem of that system (and of its extensions).

Of particular interest to us are *normal* modal logics. A normal modal logic is any set of formulas which contains the theorems of H(K), and which is closed under modus ponens and necessitation. Hilbert systems for other normal modal logics are obtained by extending H(K) with axiom schemas formalizing the behavior of the modal operator $\Box$. Examples of such axiom schemas are given in Table 2.1, where

[1] In fact, modal and other non-classical logics were investigated long before a semantics for them had been devised. For a brief but detailed history of modal logics see the overview in [32].

Axiom schemas:

1. All axiom schemas for propositional classical logic.
2. The axiom schema K: $\Box(A \supset B) \supset (\Box A \supset \Box B)$.

Inference rules:

1. *Modus ponens*: if $A \supset B$ and A are theorems, then so is B.
2. *Necessitation*: If A is a theorem, then so is $\Box A$.

Figure 2.3. H(K), a Hilbert system for K

Table 2.1. Some modal axiom schemas

Name	*Axiom schema*	*Name*	*Axiom schema*
K	$\Box(A \supset B) \supset (\Box A \supset \Box B)$	3	$\Box(\Box A \supset B) \vee \Box(\Box B \supset A)$
D	$\Box A \supset \Diamond A$	R	$\Diamond\Box A \supset (A \supset \Box A)$
T	$\Box A \supset A$	MV	$\Diamond\Box A \vee \Box A$
B	$A \supset \Box\Diamond A$	Löb	$\Box(\Box A \supset A) \supset \Box A$
4	$\Box A \supset \Box\Box A$	Grz	$\Box(\Box(A \supset \Box A) \supset A) \supset A$
5	$\Diamond A \supset \Box\Diamond A$	Go	$\Box(\Box(A \supset \Box A) \supset A) \supset \Box A$
2	$\Diamond\Box A \supset \Box\Diamond A$	M	$\Box\Diamond A \supset \Diamond\Box A$
Cxt	$\Diamond\Box A \supset \Box\Box A$	Z	$\Box(\Box A \supset A) \supset (\Diamond\Box A \supset \Box A)$
X	$\Box\Box A \supset \Box A$	Zem	$\Box\Diamond\Box A \supset (A \supset \Box A)$

for simplicity we employ also the defined operators; extensive lists of modal axiom schemas can be found in, e.g., [57, 58, 141].

We here present particular normal propositional modal logics by extending the labelled ND system N(K) with *relational theories*, which axiomatize properties of R formalizing the accessibility relation $\mathfrak{R}$ in Kripke frames.

Correspondence theory [164, 204, 227, 228] provides a tool for telling us which modal axiom schemas correspond to which axioms for R. For example, the T axiom $\Box A \supset A$ corresponds to the first-order axiom $\forall x(x R\, x)$, and the 4 axiom $\Box A \supset \Box\Box A$ corresponds to the first-order axiom $\forall x \forall y \forall z((x R\, y \wedge y R\, z) \supset x R\, z)$.

Not all modal axiom schemas can be captured in a first-order setting, e.g. the McKinsey axiom M, and the Löb axiom of provability logic GL [36]. So there is an important decision that we must make: Should our relational theories be axiomatized in higher-order logic (and thus allow the formalization of all normal propositional modal logics), first-order logic, or some subset thereof?

This decision is important. We show in §2.3 that different choices of *interface* between N(K) and the relational theory (labelling algebra) result in essentially different systems. Our choice is based on our intention to 'encode' these theories as sets of rules using a metalogic corresponding to minimal implicational predicate logic (see also §5.1.1). Thus, we choose to admit precisely those theories of R that can be directly formulated in the Horn-fragment of this metalogic without requiring additional axioms (e.g. for auxiliary predicates) or judgements (e.g. for equality). We further justify this choice by showing that it captures a large family of propositional modal logics including most of the common ones; in the next chapters, we then show that the Horn-fragment is sufficient to formalize large families of propositional and quantified non-classical logics. Moreover, and most importantly, the use of the Horn-fragment results in ND systems where derivations have good normalization properties, in contrast with what we get from systems where relational properties are axiomatized in first or higher-order logic.

2.1.3 Horn relational theories

Definition 2.1.6 *A* Horn relational formula *is a closed formula of the form*

$$\forall x_1 \ldots \forall x_n((s_1 R\, t_1 \wedge \ldots \wedge s_m R\, t_m) \supset s_0 R\, t_0)\,,$$

where $m \geq 0$, and the s_i and t_i are terms built from the labels $x_1, \ldots, x_n$ and constant function symbols (i.e. Skolem function constants; see Proposition 2.1.8 below). Corresponding to each such formula is a Horn relational rule

$$\frac{s_1 R\, t_1 \quad \ldots \quad s_m R\, t_m}{s_0 R\, t_0}\,,$$

which has no premises when $m = 0$. A Horn relational theory $\mathrm{N}(\mathcal{T})$ *is a theory generated by a set of such rules.* ■

In first-order logic the addition of a Horn formula to a theory is equivalent to adding the corresponding rule; thus, in the context of our metatheories we shall talk about additions based on either formulas or rules as is convenient.

We now illustrate that restricting our attention to Horn theories is often sufficient in practice. Let i, j, m and n be natural numbers, and let $\Box^n$ $[\Diamond^n]$ stand for a sequence of n consecutive $\Box$'s [$\Diamond$'s]; for example $\Diamond^2\Box^3\Diamond^0 A$ is $\Diamond\Diamond\Box\Box\Box A$. A large and important class of propositional modal logics falls under the *generalized Geach axiom schema*

$$\Diamond^i\Box^m A \supset \Box^j\Diamond^n A\,,$$

which corresponds to the semantic notion of (i, j, m, n) *convergency* (or 'incestuality' in the terminology of [58]),

$$\forall x \forall y \forall z(x\, R^i\, y \wedge x\, R^j\, z \supset \exists u(y\, R^m\, u \wedge z\, R^n\, u))\,,$$

where $x\, R^0\, y$ means $x = y$ and $x\, R^{i+1}\, y$ means $\exists v(x R\, v \wedge v\, R^i\, y)$.

There are instances of (i, j, m, n) convergency that explicitly require the equality predicate $=$, e.g. $(1, 0, 0, 0)$ yields *vacuity*, $\forall x \forall y(x R\, y \supset x = y)$. For simplicity, here we do not consider theories with equality, and we introduce the subclass of *restricted* (i, j, m, n) *convergency axioms* as the class of properties of the accessibility relation that can be expressed as Horn rules in the theory of one binary predicate R.[2] These theories yield, among others, labelled ND systems for most of the propositional modal logics usually of interest, e.g. K, D, T, B, K4, S4, S4.2, KD45, S5, etc.

Definition 2.1.7 Restricted (i, j, m, n) convergency axioms *are closed formulas of the form*

$$\forall x \forall y \forall z ((x\, R^i\, y \wedge x\, R^j\, z) \supset \exists u (y\, R^m\, u \wedge z\, R^n\, u))\,,$$

where $m = n = 0$ *implies* $i = j = 0$. ■

Proposition 2.1.8 *If* $\mathcal{T}_G$ *is a theory corresponding to a collection of restricted* (i, j, m, n) *convergency axioms, then there is a Horn relational theory* $\mathrm{N}(\mathcal{T})$ *conservatively extending it.*

Proof As noted in [216], the restriction that $m = n = 0$ implies $i = j = 0$ is a necessary and sufficient condition for equality to be inessential (the necessity can be checked semantically). Now, for each convergency axiom α^k in $\mathcal{T}_G$, let β^k be formed by prenexing quantifiers followed by skolemizing remaining existential quantifiers. β^k must be of the form

$$\forall x_1 \ldots \forall x_l ((s_1 R\, t_1 \wedge \ldots \wedge s_p R\, t_p) \supset (s'_1 R\, t'_1 \wedge \ldots \wedge s'_q R\, t'_q))\,,$$

where $q = m + n \neq 0$, and where Skolem functions occur only in the consequent. We can translate β^k into q Horn relational formulas, β^k_r for $r \in \{1, \ldots, q\}$, of the form

$$\forall x_1 \ldots \forall x_l ((s_1 R\, t_1 \wedge \ldots \wedge s_p R\, t_p) \supset s'_r R\, t'_r)\,.$$

Let $\mathrm{N}(\mathcal{T})$ be the theory generated by the union of the β^k_r rules. The conservativity of $\mathrm{N}(\mathcal{T})$ follows by the theorem on functional extensions [213, p. 55], and the observation that Skolem constants occur only positively in the β^k_r. An alternative proof of the conservativity of $\mathrm{N}(\mathcal{T})$ can be obtained by adapting Theorem 3.4.4.(i) in [230, p. 137]. ■

Some properties corresponding to instances of restricted (i, j, m, n) convergency are given in Table 2.2. We also present there the Horn relational rules that result from applying the above translation to these axioms, together with the corresponding *characteristic* modal axiom schemas. (An axiom schema is said to *characterize* the class of frames (and thus of modal logics) in which it is valid [227, 228].) Horn relational rules for some properties of R that are not instances of restricted (i, j, m, n) convergency are given in Table 2.3.

Various combinations of Horn relational rules define labelled ND systems for common propositional modal logics:

[2]Equality can however be easily introduced, provided that we add also rules to characterize its properties.

Table 2.2. Some (i, j, m, n) convergency properties of R, corresponding characteristic axiom schemas and Horn relational rules

Property	(i, j, m, n)	*Axiom schema*	*Horn relational rule*
Seriality	$(0, 0, 1, 1)$	D: $\Box A \supset \Diamond A$	$\dfrac{}{xR\,f(x)}\,ser$
Reflexivity	$(0, 0, 1, 0)$	T: $\Box A \supset A$	$\dfrac{}{xR\,x}\,refl$
Symmetry	$(0, 1, 0, 1)$	B: $A \supset \Box\Diamond A$	$\dfrac{xR\,y}{yR\,x}\,symm$
Transitivity	$(0, 2, 1, 0)$	4: $\Box A \supset \Box\Box A$	$\dfrac{xR\,y \quad yR\,z}{xR\,z}\,trans$
Euclideaness	$(1, 1, 0, 1)$	5: $\Diamond A \supset \Box\Diamond A$	$\dfrac{xR\,y \quad xR\,z}{zR\,y}\,eucl$
Convergency	$(1, 1, 1, 1)$	2: $\Diamond\Box A \supset \Box\Diamond A$	$\dfrac{xR\,y \quad xR\,z}{yR\,g(x, y, z)}\,conv1$ $\dfrac{xR\,y \quad xR\,z}{zR\,g(x, y, z)}\,conv2$
Contextuality	$(1, 2, 1, 0)$	Cxt: $\Diamond\Box A \supset \Box\Box A$	$\dfrac{xR\,y \; xR\,z \; zR\,w}{yR\,w}\,cxt$
Density	$(0, 1, 2, 0)$	X: $\Box\Box A \supset \Box A$	$\dfrac{xR\,y}{xR\,h(x, y)}\,dens1$ $\dfrac{xR\,y}{h(x, y)R\,y}\,dens2$

Where f, g and h are (Skolem) function constants.

Definition 2.1.9 *The labelled ND system* $\mathrm{N}(\mathcal{L}) = \mathrm{N}(\mathrm{K}) + \mathrm{N}(\mathcal{T})$ *for the propositional modal logic* $\mathcal{L}$ *is obtained by extending* $\mathrm{N}(\mathrm{K})$ *with a given Horn relational theory* $\mathrm{N}(\mathcal{T})$. ■

Notation 2.1.10 We refer to the system $\mathrm{N}(\mathcal{L}) = \mathrm{N}(\mathrm{K}) + \mathrm{N}(\mathcal{T})$ also as $\mathrm{N}(\mathrm{K}Ax)$, where Ax is a string consisting of the standard names of the characteristic axioms corresponding to the relational rules contained in $\mathrm{N}(\mathcal{T})$. ■

Then, for example, the systems N(KD), N(KT), N(KTB), N(KT4) and N(KT5), and their synonyms N(D), N(T), N(B), N(S4) and N(S5), present the modal logics D, T, B, S4 and S5.

Table 2.3. Further properties of R, corresponding characteristic axiom schemas and Horn relational rules

Property	*Axiom schema*	*Horn relational rule*
Weak reflexivity	$\Box(\Box A \supset A)$	$\dfrac{wRx}{xRx}$ *wrefl*
Weak symmetry	$\Box(A \supset \Box\Diamond A)$	$\dfrac{wRx \quad xRy}{yRx}$ *wsymm*
Weak transitivity	$\Box(\Box A \supset \Box\Box A)$	$\dfrac{wRx \quad xRy \quad yRz}{xRz}$ *wtrans*
Weak euclideaness	$\Box(\Diamond A \supset \Box\Diamond A)$	$\dfrac{wRx \quad xRy \quad xRz}{zRy}$ *weucl*

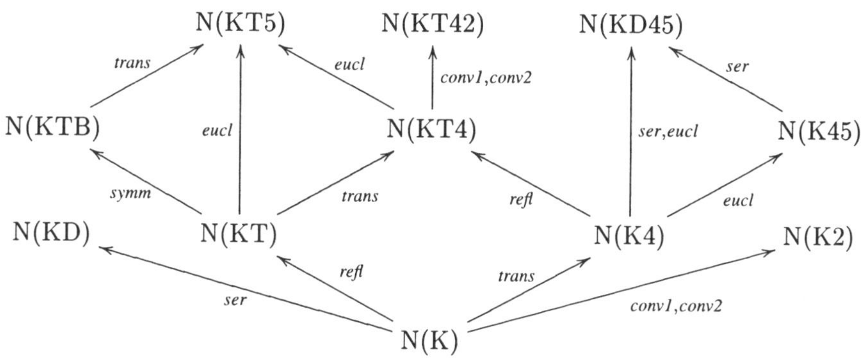

where N(D) = N(KD), N(T) = N(KT), N(B) = N(KTB), N(S4) = N(KT4), N(S4.2) = N(KT42) and N(S5) = N(KT5) = N(KTB4) = N(KT45).

Figure 2.4. A hierarchy of propositional modal systems (fragment)

Figure 2.4 shows a fragment of the resulting hierarchical dependency. For example, N(KT4), i.e. N(S4), is obtained by extending N(K) with the rules *refl* and *trans*, or alternatively by extending N(KT) with *trans* or N(K4) with *refl*.

Our approach of extending N(K) with a relational theory N($\mathcal{T}$) provides a general method for presenting logics in a modular and transparent way. The relational theory can be viewed as an independent parameter: the base system N(K) stays fixed for a given family of related logics and we generate (a labelled ND system for) the logic we want in the family by combining N(K) with the appropriate relational theory. In §2.3 we return to the question of extensions to full first or higher-order theories: we show there that it is possible to generalize our presentation, but, perhaps surprisingly, for some extensions the 'interface' between N(K) and the relational theory must be changed if completeness with respect to the corresponding Kripke semantics is to be preserved, and the metatheoretical properties of the system change.

Note that when the relational theory $N(\mathcal{T})$ contains Skolem function constants, e.g. in the case of N(D), then our language must be extended accordingly. More specifically, we extend the definitions of lwffs and rwffs to $t_i{:}A$ and $t_i R t_j$, where the labels t_i's are now terms built from labels and Skolem functions. We can then distinguish *atomic labels* x_i, i.e. 'variable-labels', and *composite labels* t_i, i.e. 'term-labels' built from the application of Skolem functions, e.g. $f(f(w))$. Rules must also be changed accordingly, and □I in particular should then read:

$$\frac{\begin{array}{c}[uRv]\\ \vdots\\ v{:}A\end{array}}{u{:}\Box A}\ \Box I$$

where the atomic label (i.e. variable) v is different from the possibly composite label u and does not occur in any assumption on which $v{:}A$ depends other than uRv. For simplicity, we will continue using the variables x, y, z, etc., to range over labels although these may now be built using Skolem function constants.

2.1.4 Derivations

We adapt the standard definition of Prawitz [186] to define derivations of lwffs and rwffs relative to a given relational theory $N(\mathcal{T})$ used to extend N(K).

Definition 2.1.11 *A* derivation of an lwff or rwff φ from a set of lwffs Γ and a set of rwffs Δ in a ND system $N(\mathcal{L}) = N(K) + N(\mathcal{T})$ *is a tree formed using the rules in* $N(\mathcal{L})$, *ending with* φ *and depending only on* $\Gamma \cup \Delta$. *We write* $\Gamma, \Delta \vdash_{N(\mathcal{L})} \varphi$ *when* φ *can be so derived. A derivation of* φ *in* $N(\mathcal{L})$ *depending on the empty set,* $\vdash_{N(\mathcal{L})} \varphi$, *is a* proof of φ in $N(\mathcal{L})$, *and we then say that* φ *is a* $N(\mathcal{L})$-theorem. ■

We also call a derivation [proof] in $N(\mathcal{L})$ a $N(\mathcal{L})$*-derivation* [$N(\mathcal{L})$*-proof*], and we will omit the '$N(\mathcal{L})$' when the details of the particular logic are not relevant or are clear from context.

When φ is an rwff xRy we have:

Fact 2.1.12

(i) $\Gamma, \Delta \vdash_{N(K)} xRy$ *iff* $xRy \in \Delta$.

(ii) $\Gamma, \Delta \vdash_{N(K)+N(\mathcal{T})} xRy$ *iff* $\Delta \vdash_{N(\mathcal{T})} xRy$. ■

Notation 2.1.13 We systematically use Π, possibly annotated, to range over derivations; when we do not need to refer to a derivation, we simply display it with vertical dots. We write $\begin{array}{c}\Pi\\ \varphi\end{array}$ to specify that the formula φ is the conclusion of the derivation Π. Similarly, we distinguish a possibly empty set of occurrences of the open or discharged assumption φ in Π by writing $\begin{matrix}\varphi\\ \Pi\end{matrix}$ or $\begin{matrix}[\varphi]\\ \Pi\end{matrix}$. Further, we sometimes combine derivations graphically; for example, we can combine the derivations

$$\begin{matrix}\Gamma_1 \quad \Delta_1\\ \Pi_1\\ \varphi_1\end{matrix} \quad \text{and} \quad \begin{matrix}\varphi_1 \quad \Gamma_2 \quad \Delta_2\\ \Pi_2\\ \varphi_2\end{matrix} \quad \text{to} \quad \begin{matrix}\begin{matrix}\Gamma_1 \quad \Delta_1\\ \Pi_1\\ \varphi_1\end{matrix} \quad \Gamma_2 \quad \Delta_2\\ \Pi_2\\ \varphi_2\end{matrix} .$$

Finally, in longer derivations we use superscripts to associate discharged assumptions with rule applications. ■

We conclude this section with a few examples of derivations; Isabelle proofs corresponding to some of these derivations are given in §5.1.3.

Example 2.1.14 We begin with three examples of relational reasoning. To illustrate that N(S5) $=$ N(KT5) $=$ N(KTB4) $=$ N(KT45) we give N(KT5)-derivations of the rules *symm* and *trans*, corresponding to the axiom schemas B and 4, respectively:

$$\dfrac{\begin{matrix}\Pi\\ xRy\end{matrix}}{yRx}\,\textit{symm} \quad \rightsquigarrow \quad \dfrac{\begin{matrix}\Pi\\ xRy\end{matrix} \quad \dfrac{}{xRx}\,\textit{refl}}{yRx}\,\textit{eucl}$$

$$\dfrac{\begin{matrix}\Pi_1\\ xRy\end{matrix} \quad \begin{matrix}\Pi_2\\ yRz\end{matrix}}{xRz}\,\textit{trans} \quad \rightsquigarrow \quad \dfrac{\dfrac{\begin{matrix}\Pi_1\\ xRy\end{matrix} \quad \dfrac{}{xRx}\,\textit{refl}}{yRx}\,\textit{eucl} \quad \begin{matrix}\Pi_2\\ yRz\end{matrix}}{xRz}\,\textit{eucl} .$$

Similarly, we can derive in N(KTB4) the rule *eucl*, corresponding to the axiom schema 5, as follows:

$$\dfrac{\begin{matrix}\Pi_1\\ xRy\end{matrix} \quad \begin{matrix}\Pi_2\\ xRz\end{matrix}}{yRz}\,\textit{eucl} \quad \rightsquigarrow \quad \dfrac{\dfrac{\begin{matrix}\Pi_1\\ xRy\end{matrix}}{yRx}\,\textit{symm} \quad \begin{matrix}\Pi_2\\ xRz\end{matrix}}{yRz}\,\textit{trans} .$$

We now derive the rules for $\sim$ and $\Diamond$ using the rules of N(K). The rules for $\sim$ follow immediately from the ones for $\supset$, i.e.

$$\dfrac{\begin{matrix}[x{:}A]\\ \Pi\\ x{:}\bot\end{matrix}}{x{:}\sim A}\,{\sim}\mathrm{I} \quad \rightsquigarrow \quad \dfrac{\begin{matrix}[x{:}A]\\ \Pi\\ x{:}\bot\end{matrix}}{x{:}A \supset \bot}\,{\supset}\mathrm{I}$$

$$\dfrac{x{:}\sim A \quad x{:}A}{x{:}\bot}\,{\sim}\mathrm{E} \quad \rightsquigarrow \quad \dfrac{x{:}A \supset \bot \quad x{:}A}{x{:}\bot}\,{\supset}\mathrm{E} .$$

Then we can use these rules to derive the ones for $\Diamond$, where the side condition on the application of $\Diamond$E follows from the condition on the application of $\Box$I:

$$\dfrac{y{:}A \quad xRy}{x{:}\Diamond A}\,\Diamond\mathrm{I} \quad\rightsquigarrow\quad \dfrac{\dfrac{\dfrac{\dfrac{[x{:}\Box\sim A]^1 \quad xRy}{y{:}\sim A}\,\Box\mathrm{E} \qquad y{:}A}{y{:}\bot}\,\sim\mathrm{E}}{x{:}\bot}\,\bot\mathrm{E}}{x{:}\sim\Box\sim A}\,\sim\mathrm{I}^1 \tag{2.1}$$

$$\dfrac{x{:}\Diamond A \qquad \begin{matrix}[y{:}A]\ [xRy]\\ \Pi\\ z{:}B\end{matrix}}{z{:}B}\,\Diamond\mathrm{E} \quad\rightsquigarrow\quad \dfrac{\dfrac{x{:}\sim\Box\sim A \qquad \dfrac{\dfrac{\dfrac{\dfrac{[z{:}B\supset\bot]^3 \qquad \begin{matrix}[y{:}A]^1\ [xRy]^2\\ \Pi\\ z{:}B\end{matrix}}{z{:}\bot}\,\supset\mathrm{E}}{y{:}\bot}\,\bot\mathrm{E}}{y{:}\sim A}\,\sim\mathrm{I}^1}{x{:}\Box\sim A}\,\Box\mathrm{I}^2}{x{:}\bot}\,\sim\mathrm{E}}{z{:}B}\,\bot\mathrm{E}^3\ . \tag{2.2}$$

Note that, dually, we could take $\Diamond$ as primitive and derive the rules for $\Box$ where $\Box A =_{def} \sim\Diamond\sim A$, e.g.

$$\dfrac{\begin{matrix}[xRy]\\ \Pi\\ y{:}A\end{matrix}}{x{:}\Box A}\,\Box\mathrm{I} \quad\rightsquigarrow\quad \dfrac{\dfrac{[x{:}\Diamond\sim A]^2 \qquad \dfrac{\dfrac{[y{:}\sim A]^1 \qquad \begin{matrix}[xRy]^1\\ \Pi\\ y{:}A\end{matrix}}{y{:}\bot}\,\sim\mathrm{E}}{x{:}\bot}\,\bot\mathrm{E}}{x{:}\bot}\,\Diamond\mathrm{E}^1}{x{:}\sim\Diamond\sim A}\,\sim\mathrm{I}^2\ . \tag{2.3}$$

In other words, $\Box$ and $\Diamond$, and the corresponding rules, are interdefinable in $\mathrm{N(K)}$.

Using the rules for $\Diamond$ we can give the following $\mathrm{N(K2)}$-proof of the characteristic axiom corresponding to convergency, i.e. $\vdash_{\mathrm{N(K2)}} x{:}\Diamond\Box A\supset\Box\Diamond A$.

$$\dfrac{\dfrac{\dfrac{[x{:}\Diamond\Box A]^3 \qquad \dfrac{\dfrac{[y{:}\Box A]^1 \qquad \dfrac{[xRy]^1 \quad [xRz]^2}{yRg(x,y,z)}\,\mathit{conv1}}{g(x,y,z){:}A}\,\Box\mathrm{E} \qquad \dfrac{[xRy]^1 \quad [xRz]^2}{zRg(x,y,z)}\,\mathit{conv2}}{z{:}\Diamond A}\,\Diamond\mathrm{I}}{z{:}\Diamond A}\,\Diamond\mathrm{E}^1}{x{:}\Box\Diamond A}\,\Box\mathrm{I}^2}{x{:}\Diamond\Box A\supset\Box\Diamond A}\,\supset\mathrm{I}^3\ . \tag{2.4}$$

As a final example, taken from [90, p. 48], we give a N(K)-derivation of $x{:}\Diamond\Diamond B$ from the assumptions $x{:}\Box\Box A$, $y{:}\Diamond(A \supset B)$ and $x R y$.

$$
\dfrac{\dfrac{y{:}\Diamond(A \supset B) \qquad \dfrac{\dfrac{[z{:}A \supset B]^1 \qquad \dfrac{\dfrac{x{:}\Box\Box A \quad xRy}{y{:}\Box A}\,\Box\mathrm{E} \qquad [yRz]^1}{z{:}A}\,\Box\mathrm{E}}{z{:}B}\,{\supset}\mathrm{E} \qquad [yRz]^1}{y{:}\Diamond B}\,\Diamond\mathrm{I}}{y{:}\Diamond B}\,\Diamond\mathrm{E}^1 \qquad xRy}{x{:}\Diamond\Diamond B}\,\Diamond\mathrm{I}\ .
$$

■

2.2 SOUNDNESS AND COMPLETENESS

We now introduce a Kripke semantics for our ND systems and prove that any system $\mathrm{N}(\mathcal{L})$ obtained by extending N(K) with a Horn relational theory $\mathrm{N}(\mathcal{T})$ is sound and complete with respect to its semantics.

Definition 2.2.1 *A* (Kripke) frame *for* $\mathrm{N}(\mathcal{L})$ *is a pair* $(\mathfrak{W}, \mathfrak{R})$, *where* $\mathfrak{W}$ *is a non-empty set of* worlds *and* $\mathfrak{R} \subseteq \mathfrak{W} \times \mathfrak{W}$. *A* (Kripke) model *for* $\mathrm{N}(\mathcal{L})$ *is a triple* $\mathfrak{M} = (\mathfrak{W}, \mathfrak{R}, \mathfrak{V})$, *where* $(\mathfrak{W}, \mathfrak{R})$ *is a frame for* $\mathrm{N}(\mathcal{L})$, *and the valuation* $\mathfrak{V}$ *maps an element of* $\mathfrak{W}$ *and a propositional variable to a truth value (0 or 1). We say that a frame* $(\mathfrak{W}, \mathfrak{R})$ *and a model* $(\mathfrak{W}, \mathfrak{R}, \mathfrak{V})$ *have some property of binary relations (e.g. transitivity) iff* $\mathfrak{R}$ *has that property.* ■

Note that our models do not contain functions corresponding to possible Skolem functions in the signature. When such constants are present the appropriate Skolem expansion of the model [230, p. 137] is required.

Definition 2.2.2 *Given a set of lwffs* Γ *and a set of rwffs* Δ, *we call the ordered pair* (Γ, Δ) *a* proof context. *When* $\Gamma_1 \subseteq \Gamma_2$ *and* $\Delta_1 \subseteq \Delta_2$, *we write* $(\Gamma_1, \Delta_1) \subseteq (\Gamma_2, \Delta_2)$ *and say that* (Γ_1, Δ_1) is included in (Γ_2, Δ_2). *When* $w{:}A \in \Gamma$, *we write* $w{:}A \in (\Gamma, \Delta)$ *irrespective of* Δ, *and when* $xRy \in \Delta$, *we write* $xRy \in (\Gamma, \Delta)$ *irrespective of* Γ. *Finally, we say that a label* x occurs in (Γ, Δ), *in symbols* $x \Subset (\Gamma, \Delta)$, *if there exists an* A *such that* $x{:}A \in \Gamma$ *or there exists a* y *such that* $xRy \in \Delta$ *or* $yRx \in \Delta$. ■

Definition 2.2.3 Truth for an rwff or lwff φ in a model $\mathfrak{M}$, $\models^{\mathfrak{M}} \varphi$, *is the smallest relation* $\models^{\mathfrak{M}}$ *satisfying:*

$$
\begin{array}{lll}
\models^{\mathfrak{M}} xRy & \textit{iff} & (x,y) \in \mathfrak{R}; \\
\models^{\mathfrak{M}} x{:}p & \textit{iff} & \mathfrak{V}(x,p) = 1; \\
\models^{\mathfrak{M}} x{:}A \supset B & \textit{iff} & \models^{\mathfrak{M}} x{:}A \textit{ implies} \models^{\mathfrak{M}} x{:}B; \\
\models^{\mathfrak{M}} x{:}\Box A & \textit{iff} & \textit{for all } y, \models^{\mathfrak{M}} xRy \textit{ implies} \models^{\mathfrak{M}} y{:}A.
\end{array}
$$

When $\models^{\mathfrak{M}} \varphi$, *we say that* φ *is* true *in* $\mathfrak{M}$. *By extension:*

$\models^{\mathfrak{M}} \Gamma$	*means that*	$\models^{\mathfrak{M}} x{:}A$ *for all* $x{:}A \in \Gamma$;
$\models^{\mathfrak{M}} \Delta$	*means that*	$\models^{\mathfrak{M}} xRy$ *for all* $xRy \in \Delta$;
$\models^{\mathfrak{M}} (\Gamma, \Delta)$	*means that*	$\models^{\mathfrak{M}} \Gamma$ *and* $\models^{\mathfrak{M}} \Delta$;
$\Delta \models^{\mathfrak{M}} xRy$	*means that*	$\models^{\mathfrak{M}} \Delta$ *implies* $\models^{\mathfrak{M}} xRy$;
$\Delta \models xRy$	*means that*	$\Delta \models^{\mathfrak{M}} xRy$ *for all* $\mathfrak{M}$;
$\Gamma, \Delta \models^{\mathfrak{M}} x{:}A$	*means that*	$\models^{\mathfrak{M}} (\Gamma, \Delta)$ *implies* $\models^{\mathfrak{M}} x{:}A$;
$\Gamma, \Delta \models x{:}A$	*means that*	$\Gamma, \Delta \models^{\mathfrak{M}} x{:}A$ *for all* $\mathfrak{M}$.

■

Truth for lwffs built using other operators can be defined in the usual manner, where $\not\models^{\mathfrak{M}} x{:}\bot$ for every x by Definition 2.2.3. For example:

$$\begin{array}{lll} \models^{\mathfrak{M}} x{:}\sim A & \text{iff} & \not\models^{\mathfrak{M}} x{:}A \\ & \text{iff} & \models^{\mathfrak{M}} x{:}A \text{ implies } \models^{\mathfrak{M}} x{:}\bot\,; \\ \models^{\mathfrak{M}} x{:}\Diamond A & \text{iff} & \not\models^{\mathfrak{M}} x{:}\Box \sim A \\ & \text{iff} & \text{for some } y\,,\ \models^{\mathfrak{M}} xRy \text{ and } \models^{\mathfrak{M}} y{:}A\,. \end{array}$$

Note also that, as a further simplification, we do not define an interpretation function mapping labels into worlds in $\mathfrak{W}$. To reduce notational overhead, we instead directly identify the label x with the world $x \in \mathfrak{W}$, i.e. we identify W with $\mathfrak{W}$; similarly we identify a Skolem function f of arity n with an identically named n–ary total function over $\mathfrak{W}$ in the corresponding Skolem expansion of the model. Furthermore, truth for lwffs is related to the standard truth relation for unlabelled modal logics, e.g. [58], by observing that $\models^{\mathfrak{M}} x{:}A$ iff $\models^{\mathfrak{M}}_x A$.

The explicit embedding of properties of the models and the capability of explicitly reasoning about them, via rwffs and relational rules, require us to consider soundness and completeness also for rwffs, where we show that $\Delta \vdash_{\mathrm{N}(\mathcal{L})} xRy$ iff $\Delta \models xRy$.

Definition 2.2.4 *The system* $\mathrm{N}(\mathcal{L}) = \mathrm{N(K)} + \mathrm{N}(\mathcal{T})$ *is* sound *iff*

(i) $\Delta \vdash_{\mathrm{N}(\mathcal{L})} xRy$ *implies* $\Delta \models xRy$, *and*
(ii) $\Gamma, \Delta \vdash_{\mathrm{N}(\mathcal{L})} x{:}A$ *implies* $\Gamma, \Delta \models x{:}A$.

$\mathrm{N}(\mathcal{L})$ *is* complete *iff the converses hold, i.e. iff*

(i) $\Delta \models xRy$ *implies* $\Delta \vdash_{\mathrm{N}(\mathcal{L})} xRy$, *and*
(ii) $\Gamma, \Delta \models x{:}A$ *implies* $\Gamma, \Delta \vdash_{\mathrm{N}(\mathcal{L})} x{:}A$. ■

By Lemma 2.2.6 and Lemma 2.2.16 below, we have:

Theorem 2.2.5 $\mathrm{N}(\mathcal{L}) = \mathrm{N(K)} + \mathrm{N}(\mathcal{T})$ *is sound and complete.* ■

2.2.1 Soundness

Lemma 2.2.6 $\mathrm{N}(\mathcal{L}) = \mathrm{N(K)} + \mathrm{N}(\mathcal{T})$ *is sound.*

Proof Throughout the proof let $\mathfrak{M} = (\mathfrak{W}, \mathfrak{R}, \mathfrak{V})$ be an arbitrary model for $\mathrm{N}(\mathcal{L})$. We prove (i) of Definition 2.2.4 by induction on the structure of the derivation of xRy from

Δ. The base case, where $xRy \in \Delta$, is trivial. There is one step case for each Horn relational rule of $\mathrm{N}(\mathcal{T})$, and we treat only transitivity and convergency as examples; the cases for the other rules follow similarly.

For transitivity, assume that $\mathfrak{R}$ is transitive and consider an application of the rule *trans*,

$$\dfrac{\begin{matrix}\Pi_1 & \Pi_2 \\ xRy & yRz\end{matrix}}{xRz}\ trans \quad ,$$

where Π_1 and Π_2 are the derivations $\Delta_1 \vdash_{\mathrm{N}(\mathcal{L})} xRy$ and $\Delta_2 \vdash_{\mathrm{N}(\mathcal{L})} yRz$, with $\Delta = \Delta_1 \cup \Delta_2$. By the induction hypotheses, $\Delta_1 \vdash_{\mathrm{N}(\mathcal{L})} xRy$ implies $\Delta_1 \models xRy$, and $\Delta_2 \vdash_{\mathrm{N}(\mathcal{L})} yRz$ implies $\Delta_2 \models yRz$. Assume $\models^{\mathfrak{M}} \Delta$. Then, from the induction hypotheses we obtain $\models^{\mathfrak{M}} xRy$ and $\models^{\mathfrak{M}} yRz$, i.e. $(x, y) \in \mathfrak{R}$ and $(y, z) \in \mathfrak{R}$. Since $\mathfrak{R}$ is transitive, we conclude $\models^{\mathfrak{M}} xRz$ by Definition 2.2.3.

When Skolem constants are present, $\mathfrak{M}$ is a Skolem expansion; for example, for convergency we assume that $\mathfrak{R}$ is convergent and consider applications of the rules *conv1* and *conv2*,

$$\dfrac{\begin{matrix}\Pi_1 & \Pi_2 \\ xRy & xRz\end{matrix}}{yRg(x,y,z)}\ conv1 \qquad \text{and} \qquad \dfrac{\begin{matrix}\Pi_1 & \Pi_2 \\ xRy & xRz\end{matrix}}{zRg(x,y,z)}\ conv2 \quad ,$$

where Π_1 and Π_2 are the derivations $\Delta_1 \vdash_{\mathrm{N}(\mathcal{L})} xRy$ and $\Delta_2 \vdash_{\mathrm{N}(\mathcal{L})} xRz$, with $\Delta = \Delta_1 \cup \Delta_2$. By Proposition 2.1.8, the theory $\mathrm{N}(\mathcal{T})$ generated by *conv1* and *conv2* is a conservative extension of the first-order theory $\mathcal{T}_G$ corresponding to the convergency axiom. By Theorem 3.4.4.(ii) in [230, p. 137], each model of the theory $\mathcal{T}_G$ has a Skolem expansion, contained in $\mathfrak{M}$, which is a model of $\mathrm{N}(\mathcal{T})$. Assume $\models^{\mathfrak{M}} \Delta$. Then, from the induction hypotheses we obtain $\models^{\mathfrak{M}} xRy$ and $\models^{\mathfrak{M}} xRz$, i.e. $(x, y) \in \mathfrak{R}$ and $(x, z) \in \mathfrak{R}$. Since $\mathfrak{R}$ is convergent, we conclude $\models^{\mathfrak{M}} yRg(x, y, z)$ and $\models^{\mathfrak{M}} zRg(x, y, z)$ by Definition 2.2.3.

We prove (ii) of Definition 2.2.4 by induction on the structure of the derivation of $x{:}A$ from Γ and Δ. The base case, where $x{:}A \in \Gamma$, is trivial. There is one step for each inference rule of N(K), and we treat only applications of the rules for $\bot$ and $\Box$ as examples; the steps for applications of $\supset$ I and $\supset$ E follow by a straightforward adaptation of the standard proofs for propositional logic.

Consider an application of the rule $\bot$E,

$$\dfrac{\begin{matrix}[x{:}A \supset \bot] \\ \Pi \\ y{:}\bot\end{matrix}}{x{:}A}\ \bot\mathrm{E} \quad ,$$

where Π is the derivation $\Gamma_1, \Delta \vdash_{\mathrm{N}(\mathcal{L})} y{:}\bot$, with $\Gamma_1 = \Gamma \cup \{x{:}A \supset \bot\}$. By the induction hypothesis, $\Gamma_1, \Delta \vdash_{\mathrm{N}(\mathcal{L})} y{:}\bot$ implies $\Gamma_1, \Delta \models y{:}\bot$. We assume $\models^{\mathfrak{M}} (\Gamma, \Delta)$ and prove $\models^{\mathfrak{M}} x{:}A$. Since $\not\models^{\mathfrak{M}} y{:}\bot$ for any y, from the induction hypothesis we obtain $\not\models^{\mathfrak{M}} \Gamma_1$, and therefore $\not\models^{\mathfrak{M}} x{:}A \supset \bot$, i.e. $\models^{\mathfrak{M}} x{:}A$ and $\not\models^{\mathfrak{M}} x{:}\bot$ by Definition 2.2.3.

Consider an application of the rule $\Box$I,

$$\dfrac{\begin{array}{c}[xRy]\\ \Pi \\ y{:}A\end{array}}{x{:}\Box A}\,\Box\text{I}\quad ,$$

where Π is the derivation $\Gamma, \Delta_1 \vdash_{\mathrm{N}(\mathcal{L})} y{:}A$, with $\Delta_1 = \Delta \cup \{xRy\}$. By the induction hypothesis, $\Gamma, \Delta_1 \vdash_{\mathrm{N}(\mathcal{L})} y{:}A$ implies $\Gamma, \Delta_1 \models y{:}A$. Assume $\models^{\mathfrak{M}} (\Gamma, \Delta)$. Considering the restriction on the application of $\Box$I, we can extend Δ to $\Delta' = \Delta \cup \{xRz\}$ for an arbitrary $z \notin (\Gamma, \Delta)$, and assume $\models^{\mathfrak{M}} \Delta'$. Since $\models^{\mathfrak{M}} \Delta'$ implies $\models^{\mathfrak{M}} \Delta_1$, from the induction hypothesis we obtain $\models^{\mathfrak{M}} y{:}A$, that is $\models^{\mathfrak{M}} z{:}A$ for an arbitrary $z \notin (\Gamma, \Delta)$ such that $\models^{\mathfrak{M}} xRz$. We conclude $\models^{\mathfrak{M}} x{:}\Box A$ by Definition 2.2.3.

Consider an application of the rule $\Box$E,

$$\dfrac{\begin{array}{cc}\Pi_1 & \Pi_2\\ x{:}\Box A & xRy\end{array}}{y{:}A}\,\Box\text{E}\quad ,$$

where Π_1 and Π_2 are the derivations $\Gamma, \Delta_1 \vdash_{\mathrm{N}(\mathcal{L})} x{:}\Box A$ and $\Delta_2 \vdash_{\mathrm{N}(\mathcal{L})} xRy$, with $\Delta = \Delta_1 \cup \Delta_2$. Assume $\models^{\mathfrak{M}} (\Gamma, \Delta)$. Then, from the induction hypotheses we obtain $\models^{\mathfrak{M}} x{:}\Box A$ and $\models^{\mathfrak{M}} xRy$, and thus $\models^{\mathfrak{M}} y{:}A$ by Definition 2.2.3. ■

2.2.2 Completeness

We begin by giving some preliminary definitions and results.

Definition 2.2.7 *Let* $\mathrm{N}(\mathcal{L}) = \mathrm{N}(\mathrm{K}) + \mathrm{N}(\mathcal{T})$ *be a* consistent system, *i.e.* $\nvdash_{\mathrm{N}(\mathcal{L})} x{:}\bot$ *for every* x. *A proof context* (Γ, Δ) *is* $\mathrm{N}(\mathcal{L})$-consistent *iff* $\Gamma, \Delta \nvdash_{\mathrm{N}(\mathcal{L})} x{:}\bot$ *for every* x. (Γ, Δ) *is* $\mathrm{N}(\mathcal{L})$-inconsistent *iff it is not* $\mathrm{N}(\mathcal{L})$*-consistent.* ■

When speaking in general terms, we will omit the '$\mathrm{N}(\mathcal{L})$' and simply speak of consistent and inconsistent proof contexts.

Fact 2.2.8 *If* (Γ, Δ) *is consistent, then for every* x *and every* A, *either* $(\Gamma \cup \{x{:}A\}, \Delta)$ *is consistent or* $(\Gamma \cup \{x{:}\sim A\}, \Delta)$ *is consistent.* ■

Definition 2.2.9 *For any system* $\mathrm{N}(\mathcal{L}) = \mathrm{N}(\mathrm{K}) + \mathrm{N}(\mathcal{T})$, *let* $\Delta_{\mathrm{N}(\mathcal{L})}$ *be the* deductive closure of Δ under $\mathrm{N}(\mathcal{L})$, *i.e.*

$$\Delta_{\mathrm{N}(\mathcal{L})} =_{def} \{xRy \mid \Delta \vdash_{\mathrm{N}(\mathcal{L})} xRy\}\,.$$

■

Note that, by Fact 2.1.12,

$$\Delta_{\mathrm{N}(\mathcal{L})} = \{xRy \mid \Delta \vdash_{\mathrm{N}(\mathcal{T})} xRy\}\,,$$

$$\Gamma, \Delta \vdash_{\mathrm{N}(\mathcal{L})} \varphi \quad \text{iff} \quad \Gamma, \Delta_{\mathrm{N}(\mathcal{L})} \vdash_{\mathrm{N}(\mathcal{L})} \varphi \,,$$

and that $\Delta_{\mathrm{N}(\mathcal{L})}$ might be empty when Δ is empty and, e.g., $\mathrm{N}(\mathcal{L})$ is $\mathrm{N}(\mathrm{K})$ or $\mathrm{N}(\mathrm{K4})$.

Definition 2.2.10 *A proof context* (Γ, Δ) *is* maximally consistent *iff*

(i) *it is consistent,*
(ii) $\Delta = \Delta_{\mathrm{N}(\mathcal{L})}$, *and*
(iii) *for every* x *and every* A, *either* $x{:}A \in \Gamma$ *or* $x{:} \sim A \in \Gamma$. ■

Completeness follows by a Henkin-style proof, where a canonical model

$$\mathfrak{M}^C = (\mathfrak{W}^C, \mathfrak{R}^C, \mathfrak{V}^C)$$

is built to show the contrapositives of the conditions in Definition 2.2.4, i.e.

$$\Delta \nvdash_{\mathrm{N}(\mathcal{L})} xRy \text{ implies } \Delta \nvDash^{\mathfrak{M}^C} xRy \,,$$

and

$$\Gamma, \Delta \nvdash_{\mathrm{N}(\mathcal{L})} x{:}A \text{ implies } \Gamma, \Delta \nvDash^{\mathfrak{M}^C} x{:}A \,.$$

In standard proofs for unlabelled modal logics, e.g. [58], the set $\mathfrak{W}^C$ is obtained by progressively building maximally consistent sets of formulas, where consistency is locally checked within each set. In our case, given the presence of labelled formulas and explicit assumptions on the relations between the labels, i.e. Δ, we modify the Lindenbaum lemma (Lemma 2.2.11 below) to extend (Γ, Δ) to one single maximally consistent proof context (Γ^*, Δ^*), where consistency is 'globally' checked also against the additional assumptions in Δ.[3] The elements of $\mathfrak{W}^C$ are then built by partitioning Γ^* and Δ^* with respect to the labels, and accessibility is defined by exploiting the information in Δ^*. Moreover, in standard proofs the way in which $\mathfrak{W}^C$ is built depends on the particular propositional modal logic $\mathcal{L}$, in particular on the accessibility conditions holding for $\mathcal{L}$. In our case, the proof is independent of the details of $\mathcal{L}$, since the same procedure applies for any logic.

In the Lindenbaum lemma for predicate logic a maximally consistent and ω-complete set of formulas is inductively built by adding for every formula $\sim \forall x(A)$ a *witness* to its truth, namely a formula $\sim A[c/x]$ for some new individual constant c. This ensures that the resulting set is ω-complete, i.e. that if, for every closed term t, $A[t/x]$ is contained in the set, then so is $\forall x(A)$. A similar procedure applies here in the case of lwffs of the form $x{:} \sim \Box A$. That is, together with $x{:} \sim \Box A$ we consistently add $y{:} \sim A$ and xRy for some new y, which acts as a *witness world* to the truth of $x{:} \sim \Box A$. This ensures that the maximally consistent proof context (Γ^*, Δ^*) is such

[3] We consider only consistent proof contexts. If (Γ, Δ) is inconsistent, then $\Gamma, \Delta \vdash_{\mathrm{N}(\mathcal{L})} x{:}A$ for all $x{:}A$, and thus completeness immediately holds for lwffs. Our labelling algebra does not allow us to define inconsistency for a set of rwffs, but, if (Γ, Δ) is inconsistent, the canonical model built in the following is nonetheless a counter-model to non-derivable rwffs.

that if $xRz \in (\Gamma^*, \Delta^*)$ implies $z{:}B \in (\Gamma^*, \Delta^*)$ for every z, then $x{:}\Box B \in (\Gamma^*, \Delta^*)$, as shown in Lemma 2.2.12 below. Note that in the standard completeness proof for unlabelled modal logics, one shows instead that if $w \in \mathfrak{W}^C$ and $\models^{\mathfrak{M}^C} w{:} \sim\Box A$, then $\mathfrak{W}^C$ also contains a world w' accessible from w that serves as a witness world to the truth of $w{:} \sim\Box A$, i.e. $\models^{\mathfrak{M}^C} w'{:} \sim A$.

Lemma 2.2.11 *Every consistent proof context* (Γ, Δ) *can be extended to a maximally consistent proof context* (Γ^*, Δ^*).

Proof We first extend the language of $\mathrm{N}(\mathcal{L})$ with infinitely many new constants for witness worlds. Systematically let w range over labels, v range over the new constants for witness worlds, and u range over both. All these may be subscripted. Let $l_1, l_2, \ldots$ be an enumeration of all lwffs in the extended language; when l_i is $u{:}A$, we write $\sim l_i$ for $u{:} \sim A$. Starting from $(\Gamma_0, \Delta_0) = (\Gamma, \Delta)$, we inductively build a sequence of consistent proof contexts by defining $(\Gamma_{i+1}, \Delta_{i+1})$ to be:

- (Γ_i, Δ_i), if $(\Gamma_i \cup \{l_{i+1}\}, \Delta_i)$ is inconsistent; else
- $(\Gamma_i \cup \{l_{i+1}\}, \Delta_i)$, if l_{i+1} is not $u{:} \sim\Box A$; else
- $(\Gamma_i \cup \{u{:} \sim\Box A, v{:} \sim A\}, \Delta_i \cup \{uRv\})$ for a $v \notin (\Gamma_i \cup \{u{:} \sim\Box A\}, \Delta_i)$, if l_{i+1} is $u{:} \sim\Box A$.

Every (Γ_i, Δ_i) is consistent. To show this we show that if $(\Gamma_i \cup \{u{:} \sim\Box A\}, \Delta_i)$ is consistent, then so is $(\Gamma_i \cup \{u{:} \sim\Box A, v{:} \sim A\}, \Delta_i \cup \{uRv\})$ for a $v \notin (\Gamma_i \cup \{u{:} \sim \Box A\}, \Delta_i)$; the other cases follow by construction. We proceed by contraposition. Suppose that

$$\Gamma_i \cup \{u{:} \sim\Box A, v{:} \sim A\}, \Delta_i \cup \{uRv\} \vdash_{\mathrm{N}(\mathcal{L})} u_j{:}\bot$$

where $v \notin (\Gamma_i \cup \{u{:} \sim\Box A\}, \Delta_i)$. Then, by $\bot$E,

$$\Gamma_i \cup \{u{:} \sim\Box A\}, \Delta_i \cup \{uRv\} \vdash_{\mathrm{N}(\mathcal{L})} v{:}A\,,$$

and $\Box$I yields

$$\Gamma_i \cup \{u{:} \sim\Box A\}, \Delta_i \vdash_{\mathrm{N}(\mathcal{L})} u{:}\Box A\,.$$

Since also

$$\Gamma_i \cup \{u{:} \sim\Box A\}, \Delta_i \vdash_{\mathrm{N}(\mathcal{L})} u{:} \sim\Box A\,,$$

by $\sim$E we have

$$\Gamma_i \cup \{u{:} \sim\Box A\}, \Delta_i \vdash_{\mathrm{N}(\mathcal{L})} u{:}\bot\,,$$

i.e. $(\Gamma_i \cup \{u{:} \sim\Box A\}, \Delta_i)$ is inconsistent. Contradiction.

Now define

$$\Gamma^* = \bigcup_{i \geq 0} \Gamma_i \quad \text{and} \quad \Delta^* = \bigcup_{i \geq 0} (\Delta_i)_{\mathrm{N}(\mathcal{L})}\,.$$

We show that (Γ^*, Δ^*) is maximally consistent by proving that it satisfies the conditions in Definition 2.2.10. For (i), note that

$$\text{if } (\bigcup_{i\geq 0} \Gamma_i, \bigcup_{i\geq 0} \Delta_i) \text{ is consistent, then so is } (\bigcup_{i\geq 0} \Gamma_i, \bigcup_{i\geq 0} (\Delta_i)_{\mathrm{N}(\mathcal{L})}) .$$

Now suppose that (Γ^*, Δ^*) is inconsistent. Then for some finite (Γ', Δ') included in (Γ^*, Δ^*) there exists a u such that $\Gamma', \Delta' \vdash_{\mathrm{N}(\mathcal{L})} u{:}\bot$. Every lwff $l \in (\Gamma', \Delta')$ is in some (Γ_j, Δ_j). For each $l \in (\Gamma', \Delta')$, let i_l be the least j such that $l \in (\Gamma_j, \Delta_j)$, and let $i = \max\{i_l \mid l \in (\Gamma', \Delta')\}$. Then $(\Gamma', \Delta') \subseteq (\Gamma_i, \Delta_i)$, and (Γ_i, Δ_i) is inconsistent, which is not the case. Condition (ii) is satisfied by definition of Δ^*. For (iii), suppose that $l_{i+1} \notin (\Gamma^*, \Delta^*)$. Then $l_{i+1} \notin (\Gamma_{i+1}, \Delta_{i+1})$ and $(\Gamma_i \cup \{l_{i+1}\}, \Delta_i)$ is inconsistent. Thus, by Fact 2.2.8, $(\Gamma_i \cup \{\sim l_{i+1}\}, \Delta_i)$ is consistent, and $\sim l_{i+1}$ is consistently added to some (Γ_j, Δ_j) during the construction, and therefore $\sim l_{i+1} \in (\Gamma^*, \Delta^*)$. ∎

The following lemma states some properties of maximally consistent proof contexts.

Lemma 2.2.12 *Let* (Γ^*, Δ^*) *be a maximally consistent proof context. Then*

(i) $\Delta^* \vdash_{\mathrm{N}(\mathcal{L})} u_i R u_j$ *iff* $u_i R u_j \in \Delta^*$.
(ii) $\Gamma^*, \Delta^* \vdash_{\mathrm{N}(\mathcal{L})} u{:}A$ *iff* $u{:}A \in \Gamma^*$.
(iii) $u{:}B \supset C \in \Gamma^*$ *iff* $u{:}B \in \Gamma^*$ *implies* $u{:}C \in \Gamma^*$.
(iv) $u_i{:}\Box B \in \Gamma^*$ *iff* $u_i R u_j \in \Delta^*$ *implies* $u_j{:}B \in \Gamma^*$ *for all* u_j.

Proof (i) and (ii) follow immediately by definition and Fact 2.1.12. We only treat (iv); (iii) follows analogously. For the left-to-right direction, suppose that $u_i{:}\Box B \in \Gamma^*$. Then, by (ii), $\Gamma^*, \Delta^* \vdash_{\mathrm{N}(\mathcal{L})} u_i{:}\Box B$, and, by $\Box$E, we have $\Delta^* \vdash_{\mathrm{N}(\mathcal{L})} u_i R u_j$ implies $\Gamma^*, \Delta^* \vdash_{\mathrm{N}(\mathcal{L})} u_j{:}B$ for all u_j. By (i) and (ii), conclude $u_i R u_j \in \Delta^*$ implies $u_j{:}B \in \Gamma^*$ for all u_j. For the converse, suppose that $u_i{:}\Box B \notin \Gamma^*$. Then $u_i{:} \sim\Box B \in \Gamma^*$, and, by the construction of (Γ^*, Δ^*), there exists a u_j such that $u_i R u_j \in \Delta^*$ and $u_j{:}B \notin \Gamma^*$. ∎

We can now define the canonical model $\mathfrak{M}^C = (\mathfrak{W}^C, \mathfrak{R}^C, \mathfrak{V}^C)$.

Definition 2.2.13 *Given a maximal consistent proof context* (Γ^*, Δ^*), *we define the* canonical model $\mathfrak{M}^C = (\mathfrak{W}^C, \mathfrak{R}^C, \mathfrak{V}^C)$ *for the system* $\mathrm{N}(\mathcal{L})$ *as follows:*

- $\mathfrak{W}^C = \{u \mid u \Subset (\Gamma^*, \Delta^*)\}$;
- $(u_i, u_j) \in \mathfrak{R}^C$ *iff* $u_i R u_j \in \Delta^*$;
- $\mathfrak{V}^C(u, p) = 1$ *iff* $u{:}p \in \Gamma^*$. ∎

Note that the standard definition of $\mathfrak{R}^C$, i.e.

$$(u_i, u_j) \in \mathfrak{R}^C \text{ iff } \{A \mid \Box A \in u_i\} \subseteq u_j ,$$

is not applicable in our setting, since $\{A \mid \Box A \in u_i\} \subseteq u_j$ does *not* imply $\vdash_{\mathrm{N}(\mathcal{L})} u_i R u_j$. We would therefore be unable to prove completeness for rwffs, since there would be cases, e.g. when $\mathrm{N}(\mathcal{L}) = \mathrm{N}(\mathrm{K})$ and $\Delta = \{\}$, where $\nvdash_{\mathrm{N}(\mathcal{L})} u_i R u_j$ but $(u_i, u_j) \in \mathfrak{R}^C$ and thus $\models^{\mathfrak{M}^C} u_i R u_j$. Hence, we instead define $(u_i, u_j) \in \mathfrak{R}^C$ iff $u_i R u_j \in \Delta^*$; note that therefore $u_i R u_j \in \Delta^*$ implies $\{A \mid \Box A \in u_i\} \subseteq u_j$. As a further comparison with the standard definition, note that in the canonical model the label u can be identified with the set of formulas $\{A \mid u{:}A \in \Gamma^*\}$. Moreover, we immediately have:

Fact 2.2.14 $u_i R u_j \in \Delta^*$ *iff* $\Delta^* \models^{\mathfrak{M}^C} u_i R u_j$. ■

The deductive closure of Δ^* ensures not only completeness for rwffs, as shown in Lemma 2.2.16 below, but also that the conditions on $\mathfrak{R}^C$ are satisfied, so that $\mathfrak{M}^C$ is really a model for $\mathrm{N}(\mathcal{L})$. As an example, we show that if $\mathrm{N}(\mathcal{L})$ contains *conv*1 and *conv*2, then $\mathfrak{R}^C$ is convergent. Consider an arbitrary proof context (Γ, Δ) from which we build $\mathfrak{M}^C$. Assume $(u_i, u_j) \in \mathfrak{R}^C$ and $(u_i, u_k) \in \mathfrak{R}^C$. Then $u_i R u_j \in \Delta^*$ and $u_i R u_k \in \Delta^*$. Since Δ^* is deductively closed, by (i) in Lemma 2.2.12, we have $u_j R\, g(u_i, u_j, u_k) \in \Delta^*$ and $u_k R\, g(u_i, u_j, u_k) \in \Delta^*$. Thus, there exists a u_l such that $(u_j, u_l) \in \mathfrak{R}^C$ and $(u_k, u_l) \in \mathfrak{R}^C$, and $\mathfrak{R}^C$ is indeed convergent.

By Lemma 2.2.12 and Fact 2.2.14, it follows that:

Lemma 2.2.15 $u{:}A \in (\Gamma^*, \Delta^*)$ *iff* $\Gamma^*, \Delta^* \models^{\mathfrak{M}^C} u{:}A$.

Proof We proceed by induction on the grade of $u{:}A$, and we treat only the step case where $u{:}A$ is $u_i{:}\Box B$; the other cases follow analogously. For the left-to-right direction, assume $u_i{:}\Box B \in \Gamma^*$. Then, by Lemma 2.2.12, $u_i R u_j \in \Delta^*$ implies $u_j{:}B \in \Gamma^*$, for all u_j. Fact 2.2.14 and the induction hypothesis yield $\Gamma^*, \Delta^* \models^{\mathfrak{M}^C} u_j{:}B$ for all u_j such that $\Delta^* \models^{\mathfrak{M}^C} u_i R u_j$, i.e. $\Gamma^*, \Delta^* \models^{\mathfrak{M}^C} u_i{:}\Box B$ by Definition 2.2.3. For the converse, assume $u_i{:} \sim\Box B \in \Gamma^*$. Then, by Lemma 2.2.12, $u_i R u_j \in \Delta^*$ and $u_j{:} \sim B \in \Gamma^*$, for some u_j. Fact 2.2.14 and the induction hypothesis yield $\Delta^* \models^{\mathfrak{M}^C} u_i R u_j$ and $\Gamma^*, \Delta^* \models^{\mathfrak{M}^C} u_j{:} \sim B$, i.e. $\Gamma^*, \Delta^* \models^{\mathfrak{M}^C} u_i{:} \sim\Box B$ by Definition 2.2.3. ■

We can now finally show that:

Lemma 2.2.16 $\mathrm{N}(\mathcal{L}) = \mathrm{N}(\mathrm{K}) + \mathrm{N}(\mathcal{T})$ *is complete.*

Proof (i) If $\Delta \nvdash_{\mathrm{N}(\mathcal{L})} w_i R w_j$, then $w_i R w_j \notin \Delta^*$, and thus $\Delta^* \not\models^{\mathfrak{M}^C} w_i R w_j$ by Fact 2.2.14. (ii) If $\Gamma, \Delta \nvdash_{\mathrm{N}(\mathcal{L})} w{:}A$, then $(\Gamma \cup \{w{:} \sim A\}, \Delta)$ is consistent; otherwise there exists a w_i such that $\Gamma \cup \{w{:} \sim A\}, \Delta \vdash_{\mathrm{N}(\mathcal{L})} w_i{:}\bot$, and then $\Gamma, \Delta \vdash_{\mathrm{N}(\mathcal{L})} w{:}A$. Therefore, by Lemma 2.2.11, $(\Gamma \cup \{w{:} \sim A\}, \Delta)$ is included in a maximally consistent proof context $((\Gamma \cup \{w{:} \sim A\})^*, \Delta^*)$. Then, by Lemma 2.2.15, $(\Gamma \cup \{w{:} \sim A\})^*, \Delta^* \models^{\mathfrak{M}^C} w{:} \sim A$, i.e. $(\Gamma \cup \{w{:} \sim A\})^*, \Delta^* \not\models^{\mathfrak{M}^C} w{:}A$, and thus $\Gamma, \Delta \not\models^{\mathfrak{M}^C} w{:}A$. ■

2.3 NORMALIZATION AND ITS CONSEQUENCES

We have given a modular presentation of propositional modal logics as labelled ND systems based on two separate parts: a base system N(K) and Horn relational theories extending it. In this section we consider alternatives for defining hierarchies of logics and systems, and classify them based on their metatheoretical properties. We organize this investigation around the interface between the two parts: since the rules for $\Box$ cannot be sensibly changed, this amounts to studying how falsum, i.e. $\bot$, propagates between worlds. We show that this question directly relates to which kinds of relational theories we can formalize while retaining completeness.

We start in §2.3.1 with the base system N(K) we have developed above, where we have what we call global falsum: $\bot$ can propagate from one world to another (Fact 2.3.1). We prove that in this system derivations have good normalization properties (Theorem 2.3.5, Corollary 2.3.6, Lemma 2.3.11 and Lemma 2.3.13) in comparison with what we get from semantic embedding (Theorem 2.3.14). Then, in §2.3.2, we show that in exchange for these structural properties, we cannot use N(K) as a base to present all modal logics with first-order axiomatizable frames (Theorem 2.3.17).

In §2.3.3 we consider what happens if we allow $\bot$ to propagate between the base system and the labelling algebra in either direction, i.e. if we introduce what we call a universal falsum. By doing this, we lose the good normalization properties of N(K) (Fact 2.3.18) in exchange for a system $N(K^{uf})$, i.e. N(K) with universal falsum, that is essentially equivalent to semantic embedding in first-order logic (Theorem 2.3.20).

Finally, in §2.3.4 we investigate the properties of $N(K^{lf})$, i.e N(K) with local falsum, the base system we get by restricting $\bot$E in N(K) so that all references are local to one world. Here, unlike in N(K), we cannot propagate $\bot$ freely from one world to another (Proposition 2.3.22). We argue that though certain modal logics can be formalized in extensions of $N(K^{lf})$, the system lacks basic properties, such as duality between $\Box$ and $\Diamond$ (Proposition 2.3.24) or normal form derivations (Proposition 2.3.26), which we might look for in a 'good' presentation.

2.3.1 Global falsum and normalization

We begin by observing that in N(K), and therefore in N(K) + N($\mathcal{T}$) as well, $\bot$ propagates 'globally' between all worlds. We call this property *global falsum*, and as an immediate consequence of $\bot$E (where no assumptions are discharged) we have:

Fact 2.3.1 *The rule* $\dfrac{x{:}\bot}{y{:}\bot}\ gf$ *is derivable in* N(K). ■

We can exploit *gf* to exhibit the propagation of falsum in the derivations of the rules for $\Diamond$. That is, we can replace the (undischarging) applications of $\bot$E in (2.1) and (2.2) with applications of *gf*. For example, we can transform (2.2) as follows.

$$\dfrac{x{:}\Diamond A \qquad \begin{array}{c}[y{:}A]\ [x R y]\\ \Pi \\ z{:}B\end{array}}{z{:}B}\,\Diamond\mathrm{E} \quad\rightsquigarrow\quad \dfrac{\dfrac{x{:}\sim\Box\sim A \qquad \dfrac{\dfrac{\dfrac{\dfrac{[z{:}B\supset\bot]^3 \qquad \begin{array}{c}[y{:}A]^1\ [x R y]^2\\ \Pi\\ z{:}B\end{array}}{z{:}\bot}\supset\mathrm{E}}{y{:}\bot}\,gf}{y{:}\sim A}\sim\mathrm{I}^1}{x{:}\Box\sim A}\,\Box\mathrm{I}^2}{x{:}\bot}\sim\mathrm{E}}{z{:}B}\,\bot\mathrm{E}^3 \;.$$

Dually, we can use *gf* to give an alternative version of the derivation (2.3) of □I from ◇E, i.e.

$$\dfrac{\begin{array}{c}[x R y]\\ \Pi \\ y{:}A\end{array}}{x{:}\Box A}\,\Box\mathrm{I} \quad\rightsquigarrow\quad \dfrac{\dfrac{[x{:}\Diamond\sim A]^2 \qquad \dfrac{\dfrac{[y{:}\sim A]^1 \qquad \begin{array}{c}[x R y]^1\\ \Pi\\ y{:}A\end{array}}{y{:}\bot}\sim\mathrm{E}}{x{:}\bot}\,gf}{x{:}\bot}\,\Diamond\mathrm{E}^1}{x{:}\sim\Diamond\sim A}\sim\mathrm{I}^2 \;.$$

To show normalization, we follow, where possible, Prawitz [186], and like him, we introduce some restrictions to simplify the development; in particular, we restrict applications of ⊥E to the case where the conclusion $x{:}A$ is atomic (i.e. A is atomic).[4]

Lemma 2.3.2 *If* $\Gamma, \Delta \vdash_{\mathrm{N(K)}} x{:}A$*, then there is a* N(K)*-derivation of* $x{:}A$ *from* Γ, Δ *where the conclusions of applications of* ⊥E *are atomic.*

Proof We show that any application of ⊥E with a non-atomic conclusion can be replaced with a derivation in which ⊥E is applied only to lwffs of smaller grade. There are two possible cases, depending on whether the conclusion is $x{:}B \supset C$ or $x{:}\Box B$.

---

[4] When presenting classical first-order logic, Prawitz [186] first introduces a natural deduction system consisting of an elimination rule for ⊥ and introduction and elimination rules for all the other connectives, and then, to show normalization, restricts his attention to the functionally complete ⊥, ∧, ⊃, ∀ fragment, where ⊥E is restricted to atomic conclusions (that are also different from ⊥). In this way he avoids having to treat the rules for ∨ and ∃, which behave 'badly' for normalization. Here, since we have already focused on the functionally complete ⊥, ⊃, □ system, we do not need further restrictions than the one on ⊥E (where however we allow the atomic conclusion A to be falsum itself, albeit labelled differently as in *gf*). In §3.3, where we discuss normalization for a ND system for an arbitrary non-classical logic with a 'classical' negation, we follow Prawitz's development more closely.

(*Case 1*)

$$\dfrac{\begin{array}{c}[x{:}(B\supset C)\supset\bot]\\ \Pi\\ y{:}\bot\end{array}}{x{:}B\supset C}\,\bot\mathrm{E} \quad\rightsquigarrow\quad \dfrac{\dfrac{\begin{array}{c}\dfrac{\dfrac{[x{:}C\supset\bot]^2 \quad \dfrac{[x{:}B\supset C]^1\quad [x{:}B]^3}{x{:}C}\,\supset\mathrm{E}}{x{:}\bot}\,\supset\mathrm{E}}{x{:}(B\supset C)\supset\bot}\,\supset\mathrm{I}^1\\ \Pi \\ y{:}\bot\end{array}}{x{:}C}\,\bot\mathrm{E}^2}{x{:}B\supset C}\,\supset\mathrm{I}^3 .$$

(*Case 2*)

$$\dfrac{\begin{array}{c}[x{:}\Box B\supset\bot]\\ \Pi\\ y{:}\bot\end{array}}{x{:}\Box B}\,\bot\mathrm{E} \quad\rightsquigarrow\quad \dfrac{\dfrac{\begin{array}{c}\dfrac{\dfrac{\dfrac{[y{:}B\supset\bot]^2 \quad \dfrac{[x{:}\Box B]^1\quad [xRy]^3}{y{:}B}\,\Box\mathrm{E}}{y{:}\bot}\,\supset\mathrm{E}}{x{:}\bot}\,\bot\mathrm{E}}{x{:}\Box B\supset\bot}\,\supset\mathrm{I}^1\\ \Pi \\ y{:}\bot\end{array}}{y{:}B}\,\bot\mathrm{E}^2}{x{:}\Box B}\,\Box\mathrm{I}^3 .$$

By iterating these transformations, we transform an arbitrary derivation $\Gamma,\Delta\vdash_{\mathrm{N(K)}} x{:}A$ into a N(K)-derivation of $x{:}A$ from Γ,Δ where the conclusions of applications of $\bot$E are atomic. ■

An immediate consequence of this lemma is the equivalence of the restricted and the unrestricted ND systems. In the rest of this section we will therefore assume applications of $\bot$E to be restricted in this way.

We now show that the derivation of an lwff can be reduced to a normal form that does not contain unnecessary detours and satisfies a subformula property. In our ND systems for propositional modal logics there is only one possible form of detour, the application of an elimination rule immediately below the application of the corresponding introduction rule, which we remove by the reduction operations defined below; the intuition for this is that if an lwff is introduced and then immediately eliminated, then we can avoid introducing it in the first place.[5]

Definition 2.3.3 *Any lwff $x{:}A$ in a derivation is the root of a tree of rule applications leading back to assumptions. The lwffs in this tree other than $x{:}A$ we call* side lwffs *of $x{:}A$. A* maximal lwff *in a derivation is an lwff that is both the conclusion of an introduction rule and the major premise of an elimination rule.* ■

Maximal lwffs are removed from a derivation by (finitely many applications of) *proper reductions*. Two possible configurations, for $\supset$ and $\Box$, result in a maximal lwff in a

[5]ND systems for other non-classical logics allow additional forms of detour; see, e.g., §3.3.

derivation; they, and their corresponding proper reductions in N(K) are

$$\dfrac{\dfrac{\begin{array}{c}[x{:}A]\\ \Pi_1\\ x{:}B\end{array}}{x{:}A\supset B}\supset\mathrm{I}\qquad \begin{array}{c}\Pi_2\\ x{:}A\end{array}}{x{:}B}\supset\mathrm{E}\qquad\rightsquigarrow\qquad \begin{array}{c}\Pi_2\\ x{:}A\\ \Pi_1\\ x{:}B\end{array}\tag{2.5}$$

$$\dfrac{\dfrac{\begin{array}{c}[xR\,y]\\ \Pi\\ y{:}A\end{array}}{x{:}\Box A}\Box\mathrm{I}\qquad xR\,z}{z{:}A}\Box\mathrm{E}\qquad\rightsquigarrow\qquad \begin{array}{c}xR\,z\\ \Pi[z/y]\\ z{:}A\end{array}\tag{2.6}$$

where $\Pi[z/y]$ is obtained from Π by systematically substituting z for y, with a suitable renaming of the variables to avoid possible (variable) clashes. Note that we only show the part of the derivation where the reduction, denoted by $\rightsquigarrow$, actually takes place; the missing parts remain unchanged.

Definition 2.3.4 *A derivation is in* normal form (*is a* normal derivation) *iff it contains no maximal lwffs.* ■

Theorem 2.3.5 *Every derivation of $x{:}A$ from Γ, Δ in* N(K) *reduces to a derivation in normal form.*

Proof If Π is a derivation of $x{:}A$ from Γ, Δ in N(K), then from the set of maximal lwffs of Π pick some $y{:}B$ which has the highest grade and has maximal lwffs only of lower grade as side lwffs. Let Π' be the proper reduction of Π at $y{:}B$. Π' is also a derivation of $x{:}A$ from Γ, Δ in N(K) and no new maximal lwff as large, or larger than $y{:}B$ has been introduced. Hence, by a finite number of similar reductions we obtain a derivation of $x{:}A$ from Γ, Δ in N(K) containing no maximal lwffs. ■

Note that we do not define normal derivations of rwffs; no 'maximal rwffs' can be introduced in $\mathrm{N}(\mathcal{T})$, since Horn relational theories contain only rwffs of the form $xR\,y$. In fact, since all rwffs are atomic, maximal rwffs cannot exist in the first place. Thus, in the following we will sometimes speak simply of normal derivations, meaning derivations of lwffs that are in normal form. Since derivations in a Horn relational theory $\mathrm{N}(\mathcal{T})$ cannot introduce maximal lwffs (and all the rwffs are of the form $xR\,y$), by replacing $xR\,z$ with $\begin{array}{c}\Pi_2\\ xR\,z\end{array}$ in (2.6), we have:

Corollary 2.3.6 *Every derivation of $x{:}A$ from Γ, Δ in* $\mathrm{N(K)} + \mathrm{N}(\mathcal{T})$ *reduces to a derivation in normal form.* ■

2.3.1.1 The form of normal derivations. By inspection of the rules of $\mathrm{N(K)} + \mathrm{N}(\mathcal{T})$ and by Fact 2.1.12, it follows that we have a separation between the base system and the relational theories extending it, i.e.

Fact 2.3.7 *The two parts of the deduction system* $N(K)+N(\mathcal{T})$ *are strictly separated: derivations of lwffs can depend on derivations of rwffs, but not vice versa.* ■

Thus, a derivation of an lwff consists of a central subderivation Π in which only rules of the base system $N(K)$ are applied, 'decorated' with subderivations of rwffs in the relational theory, which attach onto Π through instances of $\Box$E.

We now show that normal derivations possess a well-defined structure that has several desirable properties. Specifically, by analyzing the structure of a normal derivation, we can characterize the form of the central subderivations in the base system: we can identify particular sequences of formulas, and show that in these sequences there is an ordering on inferences. By exploiting this ordering, we can then show a subformula property for our modal ND systems.[6]

To analyze the structure of normal derivations, we adapt standard terminology and results [186, 187, 221, 230].

Definition 2.3.8 *A* thread *in a derivation* Π *in* $N(K)+N(\mathcal{T})$ *is a sequence of formulas* $\varphi_1, \dots, \varphi_n$ *such that (i)* φ_1 *is an assumption of* Π*, (ii)* φ_i *stands immediately above* φ_{i+1}*, for* $1 \le i < n$*, and (iii)* φ_n *is the conclusion of* Π*.*

We further characterize a thread in terms of the formulas occurring in it: an lwff-thread *is a thread where* $\varphi_1, \dots, \varphi_n$ *are all lwffs, and an* rwff-thread *is a thread where* $\varphi_1, \dots, \varphi_n$ *are all rwffs.*

A track *in a derivation* Π *is an initial part of an lwff-thread in* Π *which stops either at the first minor premise of an elimination rule in the lwff-thread or at the conclusion of the lwff-thread. In other words, a track can only pass through the major premises of elimination rules of* $N(K)$*, and it ends either at the first minor premise of an application of* $\supset$E *(the only 'primitive' elimination rule of* $N(K)$ *that has an lwff as minor premise) or at the conclusion of* Π*. We call* main track *a track that is also an lwff-thread and ends at the conclusion of the derivation.* ■

Example 2.3.9 As a first example, consider the following normal $N(K4)$-derivation Π of the characteristic axiom 4, together with its underlying tree (with numbers substituted for the formulas): the rwff-threads of Π, displayed with dashed lines, are $(1,4)$ and

[6] In §6.3 we exploit the soundness and completeness of our normalizing ND systems with respect to the corresponding Kripke semantics to show that we can give cut-free sequent systems that are sound and complete with respect to the same semantics; in Part II we then use these sequent systems to bound the complexity of the decision problem for some of the propositional modal logics we present. Note also that there are other applications of normalization, such as interpolation, that we do not explicitly consider here.

$(2, 4)$; the main (and only) track of Π, displayed with dotted lines, is $(3, 5, 6, 7, 8)$.

$$\dfrac{\dfrac{\dfrac{\dfrac{x{:}\Box A \qquad \dfrac{xRy \quad yRz}{xRz}\,\textit{trans}}{z{:}A}\,\Box\mathrm{E}}{x{:}\Box A}\,\Box\mathrm{I}^{3}}{x{:}\Box\Box A}\,\Box\mathrm{I}^{2}}{x{:}\Box A \supset \Box\Box A}\,\supset\mathrm{I}^{3}$$

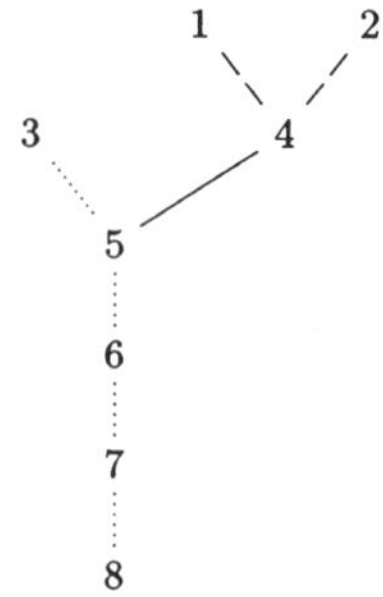

As a second example, consider the following normal N(K)-derivation Π of the characteristic axiom K, together with its underlying tree: the lwff-threads of Π are $(1, 5, 7, 8, 9, 10)$ and $(3, 6, 7, 8, 9, 10)$; the tracks of Π, displayed with dotted lines, are $(1, 5, 7, 8, 9, 10)$, which is also the main track of Π, and $(3, 6)$.

$$\dfrac{\dfrac{\dfrac{\dfrac{\dfrac{[x{:}\Box(A \supset B)]^{3} \quad [xRy]^{1}}{y{:}A \supset B}\,\Box\mathrm{E} \qquad \dfrac{[x{:}\Box A]^{2} \quad [xRy]^{1}}{y{:}A}\,\Box\mathrm{E}}{y{:}B}\,\supset\mathrm{E}}{x{:}\Box B}\,\Box\mathrm{I}^{1}}{x{:}\Box A \supset \Box B}\,\supset\mathrm{I}^{2}}{x{:}\Box(A \supset B) \supset (\Box A \supset \Box B)}\,\supset\mathrm{I}^{3}$$

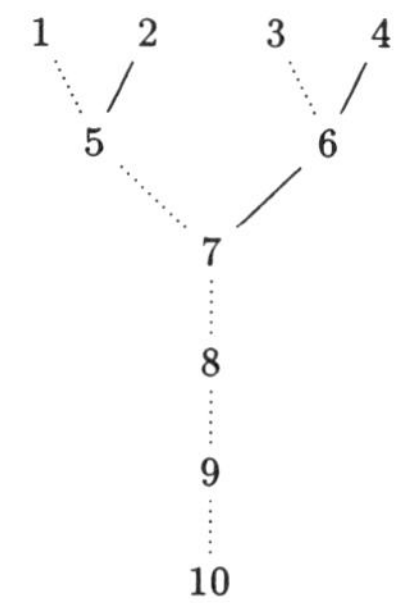

■

We adapt the standard definition of subformula as follows.

Definition 2.3.10 *B is a* subformula *of A iff (i) A is B; or (ii) A is $A_1 \supset A_2$ and B is a subformula of A_1 or A_2; or (iii) A is $\Box A_1$ and B is a subformula of A_1. We say that $y{:}B$ is a (labelled) subformula of $x{:}A$ iff B is a subformula of A.* ■

Lemma 2.3.11 *Let Π be a normal derivation, and let t be a track $x_1{:}A_1, x_2{:}A_2, \ldots, x_n{:}A_n$ in Π. Then t contains an lwff $x_i{:}A_i$, called the* minimal lwff, *which separates two possibly empty parts of t, called the* elimination part *and the* introduction part *of t, where:*

(i) *each $x_j{:}A_j$ in the elimination part, i.e. $j < i$, is a major premise of an elimination rule and contains $x_{j+1}{:}A_{j+1}$ as a subformula;*

(ii) *$x_i{:}A_i$, provided that $i \neq n$, is premise of an introduction rule or of $\bot$E;*

(iii) *each $x_j{:}A_j$ in the introduction part except the last one, i.e. $i < j < n$, is a premise of an introduction rule and is a subformula of $x_{j+1}{:}A_{j+1}$.* ■

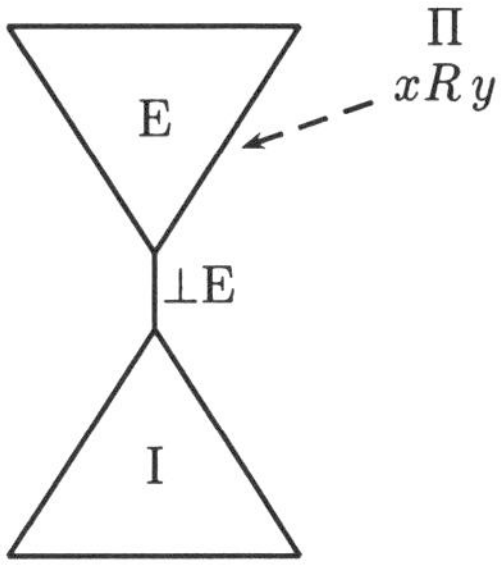

Figure 2.5. The form of tracks in a normal derivation of an lwff in $\mathrm{N(K)} + \mathrm{N}(\mathcal{T})$

In other words, a track in normal derivation is divided into at most three parts, each of which might be empty: an elimination part containing only major premises of elimination rules, a $\bot$-part in which $\bot$E is applied, and an introduction part containing only premises of introduction rules.

The lemma follows immediately by observing that in a track t in a normal derivation no introduction rule application can precede an application of an elimination rule. In other words, the lwffs in t that are major premises of elimination rules precede all lwffs in t that are premises of introduction rules or of $\bot$E, as shown in Figure 2.5.

Mirroring [186], we can exploit the form of tracks in normal derivations to show that our ND systems are consistent, i.e. $\nvdash x{:}\bot$ in $\mathrm{N(K)} + \mathrm{N}(\mathcal{T})$. It also follows that a normal derivation Π in $\mathrm{N(K)} + \mathrm{N}(\mathcal{T})$ consists of a central normal subderivation Π in $\mathrm{N(K)}$ 'decorated' with subderivations of rwffs in $\mathrm{N}(\mathcal{T})$, which attach onto Π through instances of $\Box$E, and where Π consists of a sequence of tracks, all of which have the form described in Lemma 2.3.11 and in Figure 2.5. Note that the observation about the subderivations in $\mathrm{N}(\mathcal{T})$ (the way they are connected to the central $\mathrm{N(K)}$-subderivation) holds for all, normal or 'non-normal', derivations in $\mathrm{N(K)} + \mathrm{N}(\mathcal{T})$. Moreover, if $\Diamond$I is added explicitly, together with its rules, then $\mathrm{N}(\mathcal{T})$-derivations appear also at the fringes of the introduction part through instances of $\Diamond$I (and the lemma can be suitably modified). As a final remark, note that, like in [187], the definition of normal form in our systems can be further refined, e.g. by requiring minimal lwffs to be atomic, to define derivations in *fully normal form* or in *expanded normal form*.[7]

The above results allow us to show that normal derivations in $\mathrm{N(K)} + \mathrm{N}(\mathcal{T})$ satisfy the following subformula property.

Definition 2.3.12 *Given a derivation $\Gamma, \Delta \vdash x{:}A$, let $\mathcal{S}$ be the set of subformulas of the formulas in $\{C \mid z{:}C \in \Gamma \cup \{x{:}A\}$ for some $z\}$, i.e. $\mathcal{S}$ is the set consisting of the subformulas of the assumptions Γ and of the conclusion $x{:}A$. We say that*

[7] Like for the definition of normal form, the definitions of fully and expanded normal form depend on the rules of the particular system we employ. Note also that in the proof of Lemma 5.1.3 we exploit the fact that derivations in the metalogic of Isabelle reduce to expanded normal form [179], and that in the proof of Theorem 2.3.20 we directly exploit Prawitz's results for the ND system for first-order logic.

$\Gamma, \Delta \vdash x{:}A$ *satisfies the* subformula property *iff for all lwffs* $y{:}B$ *used in the derivation (i)* $B \in \mathcal{S}$*; or (ii)* B *is an assumption* $D \supset \bot$ *discharged by an application of* $\bot$E*, where* $D \in \mathcal{S}$*; or (iii)* B *is an occurrence of* $\bot$ *obtained by* $\supset$E *from an assumption* $D \supset \bot$ *discharged by an application of* $\bot$E*, where* $D \in \mathcal{S}$*; or (iv)* B *is an occurrence of* $\bot$ *obtained by an application of* $\bot$E *that does not discharge any assumption (i.e. an occurrence of* $\bot$ *obtained by an application of gf).* ■

In other words, we define $\Gamma, \Delta \vdash x{:}A$ to have the subformula property iff for all $y{:}B$ in the derivation, either B is a subformula of the assumptions or of the conclusion of the derivation, or B is the negation of such a subformula and is discharged by $\bot$E, or B is an occurrence of $\bot$ immediately below the negation of a subformula, or, by *gf*, B is an occurrence of $\bot$ immediately below another occurrence of $\bot$ that is labelled differently.

Lemma 2.3.13 *Every normal derivation of* $x{:}A$ *from* Γ, Δ *in* $\mathrm{N(K)} + \mathrm{N}(\mathcal{T})$ *satisfies the subformula property.* ■

This follows immediately from the standard proof, which is based on the introduction of an ordering of the tracks in a normal derivation depending on their distance from the main track.

To summarize, our labelled ND systems have the following properties.

Theorem 2.3.14

(i) *The deduction machinery is minimal: labelled ND systems formalize a minimum fragment of first-order logic required by the semantics of propositional modal logics with Horn axiomatizable properties of the relations.*

(ii) *Derivations are strictly separated: derivations of lwffs may depend, via the rules for* $\Box$*, on derivations of rwffs, but not vice versa.*

(iii) *Derivations normalize: derivations of lwffs have a well-structured normal form that satisfies the subformula property.* ■

For comparison, consider the semantic embedding approach, e.g. [126, 169, 171], in which a propositional modal logic is encoded as a first-order theory by axiomatizing an appropriate definition of truth: (i) a propositional modal logic with Horn axiomatizable properties of the relations constitutes a theory of full first-order logic, as opposed to an extension of labelled propositional logic with Horn-clauses; (ii) all structure is lost as propositions and relations are flattened into first-order formulas; (iii) there are normal forms, those of ND for first-order logic, but derivations of lwffs are mingled with derivations of rwffs, as opposed to the separation between the base system and the relational theory that we have enforced.

This separation is in the philosophical spirit of labelled deduction and it also provides extra structure that is pragmatically useful: since derivations of rwffs use only the resources of the relational theory, we may be able to employ system-specific reasoners to automate proof construction. We can also exploit the existence of normal forms to design equivalent cut-free labelled sequent systems and automate proof search (see §6 and Part II). However, in exchange for this extra structure there are limits to the generality of the formulation: the properties in Theorem 2.3.14 depend on design decisions we have made, in particular, the use of Horn relational theories. This, of course, places stronger limitations on what we can formalize than a semantic embedding in first-order logic, as we show in the next sections.

2.3.2 Global falsum and first-order relational theories

So far, we have considered extensions of N(K) with Horn relational theories. There is, however, no reason why we should not have relational theories that make use of an arbitrary logic. We just have to extend the language and add appropriate rules and axioms. However, *irrespective* of which logic we allow in the labelling algebra, the rules of N(K) dictate that the only way that derivations there can contribute to lwff derivations is via propositions of the form $x R y$, thus our normalization results in fact extend to N(K) extended with an arbitrary relational theory $\mathrm{N}(\mathcal{T})$: any normal derivation of an lwff in $\mathrm{N}(\mathrm{K}) + \mathrm{N}(\mathcal{T})$ satisfies the subformula property and is structured as a central normal derivation Π in the base system N(K) 'decorated' with subderivations in the relational theory $\mathrm{N}(\mathcal{T})$, which attach onto Π through instances of $\Box$E. That is, by the above results, we have:[8]

Lemma 2.3.15 *Let* $\mathrm{N}(\mathcal{T})$ *be an* arbitrary *relational theory. The two parts of the labelled ND system* $\mathrm{N}(\mathrm{K}) + \mathrm{N}(\mathcal{T})$ *are strictly separated: derivations of lwffs can depend on derivations of rwffs, but not vice versa. Derivations of lwffs in* $\mathrm{N}(\mathrm{K}) + \mathrm{N}(\mathcal{T})$ *normalize, and any normal derivation in* $\mathrm{N}(\mathrm{K}) + \mathrm{N}(\mathcal{T})$ *satisfies the subformula property.* ∎

For concreteness, consider now an extension of the labelling algebra to a first-order theory.

Notation 2.3.16 To keep distinct the syntax of the base system from the labelling algebra, we build formulas in the labelling algebra using the connectives $\emptyset$ ('falsum') and $\sqsupset$ ('implies'), and the quantifier $\bigsqcap$ ('for all'). We henceforth assume that the possibly subscripted variable ρ ranges over such formulas. ∎

In Figure 2.6 we give the rules of N_R, the first-order ND system of R; formulas over other connectives and quantifiers, e.g. $-$ ('not'), $\sqcap$ ('and'), $\sqcup$ ('or'), $\bigsqcup$ ('for some'), and corresponding rules, are defined as usual, e.g.

$$\dfrac{\rho_1 \sqcup \rho_2 \qquad \begin{matrix}[\rho_1] \\ \vdots \\ \rho\end{matrix} \qquad \begin{matrix}[\rho_2] \\ \vdots \\ \rho\end{matrix}}{\rho}\ \sqcup\mathrm{E} \quad .$$

First-order properties of R are now added as axioms (or rules) directly in their full form, and a *first-order relational theory* $\mathrm{N}(\mathcal{T}_\mathrm{F}) = \mathrm{N}_R + \mathcal{C}_R$ is obtained by extending

[8] Note that we are talking about *labelled* subformulas, and not about subformulas of arbitrary, non-atomic, relational formulas built according to Notation 2.3.16.

$$\frac{\begin{array}{c}[\rho \sqsupset \emptyset] \\ \vdots \\ \emptyset\end{array}}{\rho}\ \emptyset\mathrm{E} \qquad \frac{\begin{array}{c}[\rho_1] \\ \vdots \\ \rho_2\end{array}}{\rho_1 \sqsupset \rho_2}\ \sqsupset\mathrm{I} \qquad \frac{\rho_1 \sqsupset \rho_2 \quad \rho_1}{\rho_2}\ \sqsupset\mathrm{E}$$

$$\frac{\rho}{\bigsqcap x(\rho)}\ \bigsqcap\mathrm{I} \qquad \frac{\bigsqcap x(\rho)}{\rho[t/x]}\ \bigsqcap\mathrm{E}$$

In $\bigsqcap$I, the variable x must not occur free in any open assumption on which ρ depends.

Figure 2.6. The rules of N_R

Table 2.4. Some first-order properties of R and corresponding rules

Property	*Rule*
Irreflexivity	$\dfrac{}{\bigsqcap x(-\ (x R\, x))}\ \mathit{irrefl}$
Intransitivity	$\dfrac{}{\bigsqcap x \bigsqcap y \bigsqcap z((x R\, y \sqcap y R\, z) \sqsupset -\ (x R\, z))}\ \mathit{intrans}$
Antisymmetry	$\dfrac{}{\bigsqcap x \bigsqcap y((x R\, y \sqcap y R\, x) \sqsupset x = y)}\ \mathit{antisymm}$
Asymmetry	$\dfrac{}{\bigsqcap x \bigsqcap y(x R\, y \sqsupset -\ (y R\, z))}\ \mathit{asymm}$
Connectedness	$\dfrac{}{\bigsqcap x \bigsqcap y \bigsqcap z(x R\, y \sqcup x = y \sqcup y R\, x)}\ \mathit{conn}$

Note that none of these properties corresponds to a modal axiom schema.

N_R with a collection $\mathcal{C}_R$ of such axioms. For example, for instances of restricted (i, j, m, n) convergency we now add the corresponding instances of the (schematic) rule

$$\frac{}{\bigsqcap x \bigsqcap y \bigsqcap z((x\, R^i\, y \sqcap x\, R^j\, z) \sqsupset \bigsqcup u(y\, R^m\, u \sqcap z\, R^n\, u))}\ \mathit{rconv} \quad ,$$

and for irreflexivity we add

$$\frac{}{\bigsqcap x(-\ (x R\, x))}\ \mathit{irrefl} \quad .$$

In Table 2.4 we give other first-order properties of R, which, like irreflexivity, cannot be expressed as Horn relational rules. Note that some of them explicitly require equality. Most importantly, none of these properties can be axiomatized by means of modal axioms [118, 140]; thus our language allows us to capture logics that cannot be presented by means of Hilbert-style axiomatizations. However, simply adding a first-order relational theory to N(K) may result in an incomplete system. Namely, we have:

Theorem 2.3.17 *There are systems* $\mathrm{N}(\mathrm{K}) + \mathrm{N}(\mathcal{T}_{\mathrm{F}})$ *with* $\mathrm{N}(\mathcal{T}_{\mathrm{F}}) = \mathrm{N}_R + \mathcal{C}_R$ *that are incomplete with respect to the corresponding Kripke models with accessibility relation defined by a collection* $\mathcal{C}_R$ *of first-order axioms.*

Proof We give an example of incompleteness. According to [227, p. 173], the Kripke model with accessibility relation defined by

$$\mathcal{C}_R = \{ \textstyle\prod x \prod y \prod z((x R y \sqcap x R z) \sqsupset (y R z \sqcup z R y)) \}$$

corresponds to the modal logic with axiom schema 3:

$$\sim\Box(\Box A \supset B) \supset \Box(\Box B \supset A)\,,$$

or, equivalently, $\Box(\Box A \supset B) \vee \Box(\Box B \supset A)$. If we assume that A and B are different propositional variables, then, reasoning backwards from the conclusion to the assumptions, a normal proof of this in $\mathrm{N}(\mathrm{K}) + \mathrm{N}_R + \mathcal{C}_R$ must have the form

$$\dfrac{\dfrac{\dfrac{\begin{array}{c}[x{:}\sim\Box(\Box A \supset B)]^1 \; [x R y]^2 \; [y{:}\Box B]^3 \\ \Pi \\ y{:}A\end{array}}{y{:}\Box B \supset A}\supset\mathrm{I}^3}{x{:}\Box(\Box B \supset A)}\Box\mathrm{I}^2}{x{:}\sim\Box(\Box A \supset B) \supset \Box(\Box B \supset A)}\supset\mathrm{I}^1 \;.$$

What might Π be? We can use Lemma 2.3.15 to explore all the possibilities.[9] Since A is a propositional variable, Π must end in an application of an elimination rule; by examining the possibilities we see that it must be an application of $\bot$E, since it is not possible to derive $y{:}A$ directly from the available hypotheses using other elimination rules. Thus the only candidate for Π is

$$\dfrac{\dfrac{[x{:}\sim\Box(\Box A \supset B)]^1 \qquad \dfrac{\dfrac{\dfrac{[y{:}\Box B]^3 \qquad \begin{array}{c}[x R y]^2 \; [y{:}A \supset \bot]^4 \; [x R z]^5 \; [z{:}\Box A]^6 \\ \Pi_1 \\ y R z\end{array}}{z{:}B}\Box\mathrm{E}}{z{:}\Box A \supset B}\supset\mathrm{I}^6}{x{:}\Box(\Box A \supset B)}\Box\mathrm{I}^5}{x{:}\bot}\sim\mathrm{E}}{y{:}A}\bot\mathrm{E}^4 \tag{2.7}$$

[9] Alternatively, we can exploit the fact that derivations in $\mathrm{N}(\mathrm{K}) + \mathrm{N}(\mathcal{T}_{\mathrm{F}})$ reduce to an expanded normal form in which all minimal lwffs are atomic, and thus dispose of the assumption that A and B are different propositional variables.

where Π_1 is a derivation purely in the relational theory $\mathrm{N}_R + \mathcal{C}_R$. But

$$xRy, xRz \nvdash yRz \text{ in } \mathrm{N}_R + \mathcal{C}_R,$$

so $\mathrm{N(K)} + \mathrm{N}_R + \mathcal{C}_R$ cannot prove the characteristic axiom for the models with accessibility relation defined by $\mathcal{C}_R$, i.e. $\mathrm{N(K)} + \mathrm{N}_R + \mathcal{C}_R$ is not complete with respect to its corresponding semantics. ■

Note that $xRy, xRz \vdash yRz$ in $\mathrm{N}_R + \mathcal{C}_R + \{\sqcap x \sqcap y(xRy \supset yRx\}$. Therefore, this particular counter-example to completeness does not hold for extensions of systems where R is symmetric, e.g. N(KB), for which, however, other counter-examples can be found. Note also that incompleteness can be shown by means of other modal formulas, but the provability of the corresponding modal axiom is 'philosophically' the first requirement to be fulfilled by the addition of a relational rule (or axiom). For instance, by similar reasoning, we can show that $x{:}\Box A \supset \Diamond A$ is not provable in $\mathrm{N(K)} + \mathrm{N}_R + \{\sqcap x \sqcup y(xRy)\}$.

2.3.3 Universal falsum

The reason for the incompleteness of $\mathrm{N(K)} + \mathrm{N}(\mathcal{T}_\mathrm{F})$ in the proof of Theorem 2.3.17 is easy to identify: we could imagine replacing Π_1 in (2.7) above with

$$\dfrac{\dfrac{\dfrac{[y{:}A \supset \bot]^4 \qquad \dfrac{[z{:}\Box A]^6 \qquad \begin{matrix}[xRy]^2\ [xRz]^5\ [yRz \sqsupset \emptyset]^7 \\ \vdots \\ zRy\end{matrix}}{y{:}A}\ \Box\mathrm{E}}{y{:}\bot}\ \supset\mathrm{E}}{\emptyset}\ (r)}{yRz}\ \emptyset\mathrm{E}^7$$

since we can show that

$$xRy, xRz, yRz \sqsupset \emptyset \vdash zRy \text{ in } \mathrm{N}_R + \mathcal{C}_R.$$

What we need is some rule (r) to allow us to propagate falsum not only between worlds, like *gf*, but also between the base system and the relational theory; i.e. collapsing $x{:}\bot$ and $\emptyset$ together. We can achieve this by adding the rules

$$\dfrac{x{:}\bot}{\emptyset}\ uf_1 \qquad \text{and} \qquad \dfrac{\emptyset}{x{:}\bot}\ uf_2$$

to N(K) to obtain the system $\mathrm{N(K}^{uf})$ which has what we call a *universal falsum*. However, it immediately follows that with universal falsum we lose the separation between the two parts of the deduction system described above. In other words, we have:

Fact 2.3.18 *In* $\mathrm{N(K}^{uf})$, *and, a fortiori, in* $\mathrm{N(K}^{uf}) + \mathrm{N}(\mathcal{T}_\mathrm{F})$, *the two parts of the deduction system are* not *separated: derivations of lwffs can depend on derivations of rwffs,* and *vice versa.* ■

In fact, we can show that $\mathrm{N}(\mathrm{K}^{uf}) + \mathrm{N}(\mathcal{T}_{\mathrm{F}})$, unlike $\mathrm{N}(\mathrm{K}) + \mathrm{N}(\mathcal{T})$, is essentially equivalent to the usual semantic embedding of propositional modal logics in first-order logic.

Definition 2.3.19 *We define a* translation $[\![\cdot]\!]$ *of formulas of* $\mathrm{N}(\mathrm{K}^{uf}) + \mathrm{N}(\mathcal{T}_{\mathrm{F}})$ *into formulas of first-order logic as follows.*

$$\begin{array}{rclrcl}
[\![\emptyset]\!] & \leadsto & \bot\,; & [\![x{:}\bot]\!] & \leadsto & \bot\,; \\
[\![x R\, y]\!] & \leadsto & R(x,y)\,; & [\![x{:}p]\!] & \leadsto & P(x)\,; \\
[\![\rho_1 \sqsupset \rho_2]\!] & \leadsto & [\![\rho_1]\!] \supset [\![\rho_2]\!]\,; & [\![x{:}A \supset B]\!] & \leadsto & [\![x{:}A]\!] \supset [\![x{:}B]\!]\,; \\
[\![\textstyle\sqcap x(\rho)]\!] & \leadsto & \forall x([\![\rho]\!])\,; & [\![x{:}\Box A]\!] & \leadsto & \forall y(R(x,y) \supset [\![y{:}A]\!])\,; \\
[\![\Delta]\!] & \leadsto & \{[\![\rho]\!] \mid \rho \in \Delta\}\,; & [\![\Gamma]\!] & \leadsto & \{[\![x{:}A]\!] \mid x{:}A \in \Gamma\}\,.
\end{array}$$

■

Theorem 2.3.20 *Let $\mathcal{C}_R$ be an arbitrary collection of first-order axioms about R, and φ an arbitrary lwff or rwff. The following are equivalent:*

(i) $\Gamma, \Delta \vdash \varphi$ *in* $\mathrm{N}(\mathrm{K}^{uf}) + \mathrm{N}_R + \mathcal{C}_R$.

(ii) $\mathcal{C}_R, [\![\Gamma]\!], [\![\Delta]\!] \vdash [\![\varphi]\!]$ *in (the ND system for) first-order logic.*

Proof Since reasoning about labels is directly translated, we only treat the case when φ in an lwff. Transforming a derivation in $\mathrm{N}(\mathrm{K}^{uf}) + \mathrm{N}_R + \mathcal{C}_R$ into a derivation in the standard [186] ND system for first-order logic is simple, since we can find derived rules in first-order logic corresponding to each rule of $\mathrm{N}(\mathrm{K}^{uf})$. For example

$$\dfrac{\begin{matrix}[x R\, y]^1 \\ \vdots \\ y{:}A\end{matrix}}{x{:}\Box A}\,\Box\mathrm{I}^1 \quad \leadsto \quad \dfrac{\dfrac{\begin{matrix}[R(x,y)]^1 \\ \vdots \\ [\![y{:}A]\!]\end{matrix}}{R(x,y) \supset [\![y{:}A]\!]}\,\supset\mathrm{I}^1}{\forall y(R(x,y) \supset [\![y{:}A]\!])}\,\forall\mathrm{I} \quad [=_{def}\ [\![x{:}\Box A]\!]] \tag{2.8}$$

where the side condition on $\forall$I, that y is different from x and does not occur free in the assumptions on which $R(x,y) \supset [\![y{:}A]\!]$ depends, is satisfied since, by the condition on $\Box$I, y is different from x and does not occur in any assumption on which $y{:}A$ depends other than $x R\, y$. The other rules are dealt with similarly.

The other direction is trickier. However, we know that derivations in the ND system for first-order logic have expanded normal forms [187], thus we can assume that Π is an expanded normal derivation of $\mathcal{C}_R, [\![\Gamma]\!], [\![\Delta]\!] \vdash [\![x{:}A]\!]$, and observe that it is possible to translate this derivation directly into $\mathrm{N}(\mathrm{K}^{uf}) + \mathrm{N}_R + \mathcal{C}_R$; e.g. if we reverse $\leadsto$ in (2.8), we can see that since a normal derivation of $[\![x{:}\Box A]\!]$ must have exactly the form (the sequence of introduction rules) given there, and, by induction, the same translation can be performed on the subderivation of $[\![y{:}A]\!]$ from $[\![x R\, y]\!]$, it is possible to translate this into a derivation in $\mathrm{N}(\mathrm{K}^{uf}) + \mathrm{N}_R + \mathcal{C}_R$. We can do the same with the other rules. All we have to do is, occasionally, insert additional rules translating between falsum for rwffs and falsum for lwffs. ■

Since semantic embedding of a propositional modal logic in first-order logic is sound and complete with respect to the appropriate Kripke semantics [169], it then follows that:

Corollary 2.3.21 $\mathrm{N}(\mathrm{K}^{uf}) + \mathrm{N}(\mathcal{T}_{\mathrm{F}})$ *is sound and complete.* ■

2.3.4 Local falsum

In $\mathrm{N}(\mathrm{K})$, rwffs interact with lwffs through the $\Box$E rule and this changes the label of the major premise into that of the conclusion. This is however not the only rule that changes worlds: $\bot$E, as we have discussed, also has this property. To complete our investigation of alternative formulations, we consider the other end of the spectrum from universal falsum where, by restricting $\bot$E, falsum is *local* and cannot move arbitrarily between worlds:

$$\dfrac{\begin{array}{c}[x{:}A \supset \bot]\\ \vdots \\ x{:}\bot\end{array}}{x{:}A}\ lf \quad .$$

Call $\mathrm{N}(\mathrm{K}^{lf})$ the system obtained from $\mathrm{N}(\mathrm{K})$ by replacing $\bot$E with its restricted form *lf*. In $\mathrm{N}(\mathrm{K}^{lf})$ we can propagate $\bot$ forwards indirectly: given $x{:}\bot$ we have $x{:}\Box\bot$, and thus $y{:}\bot$ when xRy; i.e.

$$\dfrac{\dfrac{x{:}\bot}{x{:}\Box\bot}\ lf \qquad xRy}{y{:}\bot}\ \Box\mathrm{E} \quad . \tag{2.9}$$

But we cannot propagate $\bot$ to an arbitrary world, i.e.

Proposition 2.3.22 *There is no derivation of $y{:}\bot$ from $x{:}\bot$ in $\mathrm{N}(\mathrm{K}^{lf})$ where y is an arbitrary label.*

Proof Since $\mathrm{N}(\mathrm{K}^{lf})$ is a fragment of $\mathrm{N}(\mathrm{K})$, a derivation Π of $y{:}\bot$ from $x{:}\bot$ in $\mathrm{N}(\mathrm{K}^{lf})$ would have a normal derivation Π' in $\mathrm{N}(\mathrm{K})$. Since any such derivation needs to make use of $\bot$E, which, by Lemma 2.3.23 below, must already be present in the un-normalized form of Π, no such derivation can exist in $\mathrm{N}(\mathrm{K}^{lf})$. ■

Lemma 2.3.23 *If there are no applications of $\bot$E in a derivation in $\mathrm{N}(\mathrm{K})$, then normalization of the derivation cannot introduce one.*

Proof By examining the transformations involved in reducing a derivation to normal form. ■

In the same way, we can prove that, since *gf* is not derivable, the duality of $\Box$ and $\Diamond$ fails for $\mathrm{N}(\mathrm{K}^{lf})$, i.e.

Proposition 2.3.24 *The connectives $\Box$ and $\Diamond$ are not interdefinable in $\mathrm{N}(\mathrm{K}^{lf})$.*

Proof Consider the derivation (2.1) in Example 2.1.14, and assume that Π is a suitable derivation of $\Diamond$I in $\mathrm{N}(\mathrm{K}^{lf})$. Then, since Π is also a derivation in $\mathrm{N}(\mathrm{K})$, it has a normal form Π' in $\mathrm{N}(\mathrm{K})$. However, by Lemma 2.3.23 and Lemma 2.3.25 below, such a derivation in $\mathrm{N}(\mathrm{K}^{lf})$ does not exist, since Π', and thus Π, must contain unrestricted applications of $\bot$E. ■

Lemma 2.3.25 *A normal form derived rule in* $\mathrm{N}(\mathrm{K})$ *suitable for the substitution* (2.1) *in Example 2.1.14 involves a step application*

$$\begin{array}{c} [x{:}A \supset \bot] \\ \vdots \\ \dfrac{y{:}\bot}{x{:}A}\ \bot\mathrm{E} \end{array}$$

where we are not able to assume that $y\,R\,x$.

Proof By examination of the possible normal derivations. ■

Note that $\Box$ and $\Diamond$ are not even 'intuitionistically' related in $\mathrm{N}(\mathrm{K}^{lf})$, in the sense that $\Diamond \sim A$ does not imply $\sim\Box A$, and $\Box \sim A$ does not imply $\sim\Diamond A$.

Proposition 2.3.24 shows that $\mathrm{N}(\mathrm{K}^{lf})$ is not in general suitable for formalizing modal logics, since we are not able to propagate falsum to inaccessible worlds. However, it is easy to show that in fact we only ever have to deal with worlds accessible in some way from each other. Given, as we have observed, that we can propagate $\bot$ forwards in $\mathrm{N}(\mathrm{K}^{lf})$, if R is symmetrical we also have a backwards propagation:

$$\dfrac{\dfrac{x{:}\bot}{x{:}\Box\bot}\ lf \qquad \dfrac{y\,R\,x}{x\,R\,y}\ symm}{y{:}\bot}\ \Box\mathrm{E} \quad .$$

Thus $\mathrm{N}(\mathrm{K}^{lf})$ can be used to formalize certain logics after a fashion (if the relational theory $\mathrm{N}(\mathcal{T}_{\mathrm{F}})$ is inconsistent or if R is *universal*, so that $x\,R\,y$ for all x and y, then we get this much more simply).[10] However, the resulting formalization is unsatisfactory, since it lacks important metatheoretical properties that we get in $\mathrm{N}(\mathrm{K})$; namely, we have:

Proposition 2.3.26 *Derivations in* $\mathrm{N}(\mathrm{K}^{lf})$ *do not have normal forms satisfying the subformula property.*

Proof As we observed in (2.9), there is a derivation of $y{:}\bot$ from $x\,R\,y$ and $x{:}\bot$ in $\mathrm{N}(\mathrm{K}^{lf})$. However, there cannot be a normal one satisfying the subformula property, i.e. $x{:}\Box\bot$ is not a subformula in (2.9). ■

[10]Given that $\mathrm{N}(\mathrm{KT5})$, i.e. $\mathrm{N}(\mathrm{S5})$, is sound and complete with respect to the class of universal frames [58, p. 178], it is possible to prove that $\Gamma, \Delta \vdash x{:}A$ in $\mathrm{N}(\mathrm{KT5})$ iff $\Gamma, \Delta \vdash x{:}A$ in $\mathrm{N}(\mathrm{K}^{lf}\mathrm{T5})$, since, when R is universal, $\Box$ and $\Diamond$ are interdefinable, and $\bot$E and *lf* are interderivable (but the derivations are not normal).

While this could of course be obviated by suitably extending the definition of subformula, we leave further investigation of systems based on local falsum as future work.

3 LABELLED NATURAL DEDUCTION SYSTEMS FOR PROPOSITIONAL NON-CLASSICAL LOGICS

In the previous chapter we investigated labelled natural deduction presentations of propositional modal logics. Here we explore the generalizations needed to build ND systems for large families of propositional non-classical logics, including *relevance logics* (and, more generally, *substructural logics* [75, 76, 196]), where we can treat non-classical negation as a modal operator and also consider explicitly positive fragments. (The metatheory of positive logics is different from that for 'full' logics; see, e.g., Dunn's semantic analysis of positive modal logics in [80].) We generalize our framework to provide a uniform treatment of a wide range of non-classical operators ($\Box$, $\Diamond$, relevant and intuitionistic implication, non-classical negation, etc.), where we base our presentations on an abstract classification of non-classical operators as 'universal' or 'existential', and associated general metatheorems. We proceed as follows.

In §3.1 we formalize modular presentations of propositional non-classical logics as extensions of fixed base systems with Horn relational theories; we provide several examples of relevance logics, including in particular a labelled presentation of the relevance logic R, which also shows the advantages of our approach over Hilbert-style axiomatizations. In §3.2 we give parameterized proofs of soundness and completeness of our systems with respect to the corresponding Kripke-style semantics, and discuss the possible incompleteness of unrestricted positive fragments. In §3.3 we consider the proof-theoretical properties of our systems, including normalization and the subformula property. In §5.2 we will then present the Isabelle encodings of our systems, demonstrate their correctness, and give example proofs.

3.1 MODULAR PRESENTATIONS OF PROPOSITIONAL NON-CLASSICAL LOGICS

In this section we formalize our presentations. In §3.1.1 we introduce the fundamentals of how a labelled ND presentation relates to a Kripke semantics. In §3.1.2 and §3.1.3 we define the base ND systems and the associated class of relational theories over which it is parameterized. In §3.1.4 we give examples of labelled ND presentations for non-classical logics.

3.1.1 Labels and Kripke models

We generalize Definition 2.1.3 to arbitrary propositional non-classical logics. Let W be a set of *labels* ranging over worlds in a Kripke model, and R an n+1–ary relation over W. If $a, a_1, \ldots, a_n$ are labels and A is a formula, then we call $R\,a\,a_1 \ldots a_n$ a *relational formula* (*rwff*) and $a{:}A$ a *labelled formula* (*lwff*). Formulas are built from logical operators, which are partitioned into two families: 'local' and 'non-local'.

If a formula A is built from a *local operator* $\mathcal{O}$ of arity n, i.e. $A = \mathcal{O}A_1 \ldots A_n$, then the truth of the lwff $a{:}A$ depends only on the (local) truth of $a{:}A_1, \ldots, a{:}A_n$. Typical local operators are conjunction ($\wedge$), disjunction ($\vee$), material (classical) implication ($\supset$), and local (classical) negation ($\sim$); for notational simplicity, we omit brackets where possible and write binary operators in infix notation.

Where $\models^{\mathfrak{M}}$ is the truth relation for lwffs in the model $\mathfrak{M}$, we have:[1]

$$\models^{\mathfrak{M}} a{:}A \wedge B \text{ iff } \models^{\mathfrak{M}} a{:}A \text{ and } \models^{\mathfrak{M}} a{:}B; \tag{3.1}$$

$$\models^{\mathfrak{M}} a{:}A \vee B \text{ iff } \models^{\mathfrak{M}} a{:}A \text{ or } \models^{\mathfrak{M}} a{:}B; \tag{3.2}$$

$$\models^{\mathfrak{M}} a{:}A \supset B \text{ iff } \models^{\mathfrak{M}} a{:}A \text{ implies } \models^{\mathfrak{M}} a{:}B; \tag{3.3}$$

$$\models^{\mathfrak{M}} a{:} \sim A \text{ iff } \not\models^{\mathfrak{M}} a{:}A. \tag{3.4}$$

A *non-local operator* $\mathcal{M}$ of arity n is associated with an n+1–ary relation R on worlds, and the truth of $a{:}\mathcal{M}A_1 \ldots A_n$ is evaluated non-locally at the worlds R-accessible from a, i.e. in terms of the truth of $a_1{:}A_1, \ldots, a_n{:}A_n$ where $R\,a\,a_1 \ldots a_n$. Examples of non-local operators and associated relations are the unary modal operator $\Box$ and the binary accessibility relation on possible worlds, or *relevant implication* $\rightarrow$ and the ternary *compossibility relation*.

We extend $\models^{\mathfrak{M}}$ to express truths for rwffs in a Kripke model $\mathfrak{M}$ with an n+1–ary relation $\mathfrak{R}$ as

$$\models^{\mathfrak{M}} R\,a\,a_1 \ldots a_n \text{ iff } (a, a_1, \ldots, a_n) \in \mathfrak{R}\,, \tag{3.5}$$

[1]Note that $\sim$ can be defined in terms of $\supset$ and falsum ($\bot$) like we did in §2. When this is the case, we can compare, like for modal logics in §2.3, the systems/logics obtained when (i) $\models^{\mathfrak{M}} a{:} \sim A$ iff $\models^{\mathfrak{M}} a{:}A$ implies $\models^{\mathfrak{M}} b{:}\bot$, and (ii) $\bot$ is a global falsum, i.e. $\models^{\mathfrak{M}} a{:}\bot$ implies $\models^{\mathfrak{M}} b{:}A$, with the (paraconsistent) systems/logics where (i′) $\models^{\mathfrak{M}} a{:} \sim A$ iff $\models^{\mathfrak{M}} a{:}A$ implies $\models^{\mathfrak{M}} a{:}\bot$, and (ii′) $\bot$ is a local falsum, i.e. $\models^{\mathfrak{M}} a{:}\bot$ implies $\models^{\mathfrak{M}} a{:}A$.

and we call $\mathcal{M}$ a *universal non-local operator* when the metalevel quantification in the evaluation clause of $\mathcal{M}$ is universal (and the body is an implication), i.e.

$$\models^{\mathfrak{M}} a{:}\mathcal{M}A_1 \ldots A_n \text{ iff for all } a_1, \ldots, a_n \, ((\models^{\mathfrak{M}} R\, a\, a_1 \ldots a_n \text{ and } \models^{\mathfrak{M}} a_1{:}A_1 \text{ and} \ldots \text{and } \models^{\mathfrak{M}} a_{n-1}{:}A_{n-1}) \text{ imply } \models^{\mathfrak{M}} a_n{:}A_n) \,. \quad (3.6)$$

Similarly, $\mathcal{M}$ is an *existential non-local operator* when the metalevel quantification is existential (and the body is a conjunction), i.e.

$$\models^{\mathfrak{M}} a{:}\mathcal{M}A_1 \ldots A_n \text{ iff there exist } a_1, \ldots, a_n \, (\models^{\mathfrak{M}} R\, a\, a_1 \ldots a_n \text{ and } \models^{\mathfrak{M}} a_1{:}A_1 \text{ and} \ldots \text{and } \models^{\mathfrak{M}} a_{n-1}{:}A_{n-1} \text{ and } \models^{\mathfrak{M}} a_n{:}A_n) \,. \quad (3.7)$$

In these terms, $\Box$ and relevant $\rightarrow$ are universal non-local operators, $\Diamond$ is existential, and their evaluation clauses are special cases of (3.6) and (3.7), e.g.

$$\models^{\mathfrak{M}} a{:}A_1 \rightarrow A_2 \text{ iff for all } a_1, a_2 \, ((\models^{\mathfrak{M}} R\, a\, a_1\, a_2 \text{ and } \models^{\mathfrak{M}} a_1{:}A_1) \text{ imply } \models^{\mathfrak{M}} a_2{:}A_2) \,. \quad (3.8)$$

Note that many but not all non-local operators fall under this classification. For example, to capture the binary *until* operator of temporal logics, whose first argument has a universal character while its second argument has an existential one, we would need to extend the language of our framework appropriately.

A uniform treatment of negation plays a central role in our framework. However, in the Kripke semantics for relevance (and other) logics, both a formula and its 'negation' may be true at a world, which cannot be the case with $\sim$. Thus a new operator is introduced, a *non-local negation* $\neg$, formalized by a unary function $*$ on worlds [77]:

$$\models^{\mathfrak{M}} a{:}\neg A \text{ iff } \not\models^{\mathfrak{M}} a^*{:}A \,. \quad (3.9)$$

Informally, a^* is the world that does not deny what a asserts, i.e. a and a^* are *compatible* worlds. We generalize this by introducing the constant $\perp\!\!\!\perp$ that expresses *incoherence* of compatible worlds,

$$\models^{\mathfrak{M}} b{:}\perp\!\!\!\perp \text{ iff for some } a \, (\models^{\mathfrak{M}} a{:}\neg A \text{ and } \models^{\mathfrak{M}} a^*{:}A) \,,$$

and replace (3.9) with

$$\models^{\mathfrak{M}} a{:}\neg A \text{ iff for all } b \, (\models^{\mathfrak{M}} a^*{:}A \text{ implies } \models^{\mathfrak{M}} b{:}\perp\!\!\!\perp) \quad (3.10)$$

where $\not\models^{\mathfrak{M}} b{:}\perp\!\!\!\perp$ for every world b.

Some remarks. First, when relevant implication is present, we can define $\neg A$ as $A \rightarrow \perp\!\!\!\perp$ and postulate $R\, a\, a^*\, b$ for every b, so that (3.10) is just a special case of (3.8). (That a and a^* are 'compossible' according to every b, as stated by $R\, a\, a^*\, b$, is justified by the meaning of $*$.) Second, when $a = a^*$, e.g. for modal or classical logic, $\perp\!\!\!\perp$ reduces to $\perp$, $\neg$ to $\sim$, and (3.10) to (3.4).[2] Finally, we remark that there

[2]Note that, similarly, relevant implication $\rightarrow$ reduces to classical implication $\supset$ when we postulate that $\models^{\mathfrak{M}} R\, a\, a_1\, a_2$ iff $a = a_1 = a_2$ and $\models^{\mathfrak{M}} R\, a\, a\, a$. Note also that when both $\perp\!\!\!\perp$ and $\perp$ are present we may be required to find a semantic means of distinguishing them.

is an alternative approach to non-local negation, e.g. for relevance, linear and ortho-logic [74, 78, 99, 109, 117, 197], which uses a binary *incompatibility relation* N between worlds:

$$\models^{\mathfrak{M}} a{:}\neg A \text{ iff for all } b\ (\models^{\mathfrak{M}} b{:}A \text{ implies } bNa)\,. \tag{3.11}$$

Then a^* is the 'strongest' world b for which bNa does *not* hold. This can be shown to be equivalent to our approach; for a detailed discussion and comparison of (3.11) with (3.9) see [78].

We define the language of a propositional non-classical logic $\mathcal{L}$ and of the corresponding ND system $\mathrm{N}(\mathcal{L})$ as follows.

Definition 3.1.1 *Let $\mathcal{H}$ and $\mathcal{I}$ be two finite sets of indices. The language of a propositional non-classical logic $\mathcal{L}$ and of the corresponding ND system $\mathrm{N}(\mathcal{L})$ is a tuple $(W, *, S, O, F)$. W is a set of labels closed under $*$ of type $W \supset W$. S is a denumerably infinite set of propositional variables. O is the set whose members are*

(i) *the constant $\perp\!\!\!\perp$ (and/or $\perp$);*
(ii) *local and/or non-local negation (or neither for positive logics);*
(iii) *a set of local operators $\{\mathcal{O}_h \mid h \in \mathcal{H}\}$; and*
(iv) *a set of non-local operators $\{\mathcal{M}_i \mid i \in \mathcal{I}\}$ with an associated set $\overline{R} = \{R_i \mid i \in \mathcal{I}\}$ of relations of the appropriate arities.*

F is the set of rwffs and lwffs: if $a, a_1, \ldots, a_n$ are labels, R_i has arity $n{+}1$, and A is a formula built up from members of S and O, then $R_i\, a\, a_1 \ldots a_n$ is an rwff and $a{:}A$ is an lwff. ■

Note that by associating different relations to universal and existential non-local operators we make no a priori assumptions about their interrelationships; we show in Theorem 3.2.12 in §3.2.3 below that when the relations are not independent, incompleteness may arise.

We now generalize Notation 2.1.2 and Definitions 2.1.4, 2.2.1, 2.2.2 and 2.2.3.

Notation 3.1.2 In order to simplify our notation, we will omit brackets whenever no confusion can arise, and write binary local and non-local operators in infix notation. Furthermore, we adopt the convention that $\sim$, $\neg$ and $\mathcal{M}$ are of equal binding strength and bind tighter than $\wedge$, which binds tighter than $\vee$, which binds tighter than $\supset$. ■

Definition 3.1.3 *The* grade *of an lwff $a{:}A$, in symbols: $grade(a{:}A)$, is the number of local and non-local operators that occur in A.* ■

Definition 3.1.4 *Given a set of lwffs Γ and a set of rwffs Δ, we call the ordered pair (Γ, Δ) a* proof context*. When $\Gamma_1 \subseteq \Gamma_2$ and $\Delta_1 \subseteq \Delta_2$, we write $(\Gamma_1, \Delta_1) \subseteq (\Gamma_2, \Delta_2)$. When $a{:}A \in \Gamma$, we write $a{:}A \in (\Gamma, \Delta)$ irrespective of Δ, and when $R_i\, a\, a_1 \ldots a_n \in \Delta$, we write $R_i\, a\, a_1 \ldots a_n \in (\Gamma, \Delta)$ irrespective of Γ. Finally, we say that a label a* occurs in *(Γ, Δ), in symbols $a \Subset (\Gamma, \Delta)$, if there exists an A such that $a{:}A \in \Gamma$ or if a is an argument of an rwff in Δ.* ■

Definition 3.1.5 *A* (Kripke) frame *for* $\mathrm{N}(\mathcal{L})$ *is a tuple* $(\mathfrak{W}, o, \overline{\mathfrak{R}}, *)$*, where* $\mathfrak{W}$ *is a non-empty set of* worlds, $o \in \mathfrak{W}$ *is the* actual world, $\overline{\mathfrak{R}} = \{\mathfrak{R}_i \mid i \in I\}$ *is the set of relations over* $\mathfrak{W}$ *corresponding to* $\overline{R}$*, and* $*$ *is a function of type* $\mathfrak{W} \supset \mathfrak{W}$*. A* (Kripke) model $\mathfrak{M} = (\mathfrak{W}, o, \overline{\mathfrak{R}}, *, \mathfrak{V})$ *for* $\mathrm{N}(\mathcal{L})$ *consists of a frame and a function* $\mathfrak{V}$ *mapping elements of* $\mathfrak{W}$ *and propositional variables to truth values* $(0$ *or* $1)$*, where*

$$\models^{\mathfrak{M}} a{:}p \qquad \textit{iff} \qquad \mathfrak{V}(a,p) = 1\,. \tag{3.12}$$

$\models^{\mathfrak{M}}$ *is extended to lwffs with local and non-local operators and to rwffs as above, and when* $\models^{\mathfrak{M}} \varphi$*, for* φ *an lwff or an rwff, we say that* φ *is* true *in* $\mathfrak{M}$*. By extension:*

$\models^{\mathfrak{M}} \Gamma$	*means that*	$\models^{\mathfrak{M}} a{:}A$ *for all* $a{:}A \in \Gamma$;
$\models^{\mathfrak{M}} \Delta$	*means that*	$\models^{\mathfrak{M}} R_i\, a\, a_1 \ldots a_n$ *for all* $R_i\, a\, a_1 \ldots a_n \in \Delta$;
$\models^{\mathfrak{M}} (\Gamma, \Delta)$	*means that*	$\models^{\mathfrak{M}} \Gamma$ *and* $\models^{\mathfrak{M}} \Delta$;
$\Delta \models^{\mathfrak{M}} R_i\, a\, a_1 \ldots a_n$	*means that*	$\models^{\mathfrak{M}} \Delta$ *implies* $\models^{\mathfrak{M}} R_i\, a\, a_1 \ldots a_n$;
$\Delta \models R_i\, a\, a_1 \ldots a_n$	*means that*	$\Delta \models^{\mathfrak{M}} R_i\, a\, a_1 \ldots a_n$ *for all* $\mathfrak{M}$;
$\Gamma, \Delta \models^{\mathfrak{M}} a{:}A$	*means that*	$\models^{\mathfrak{M}} (\Gamma, \Delta)$ *implies* $\models^{\mathfrak{M}} a{:}A$;
$\Gamma, \Delta \models a{:}A$	*means that*	$\Gamma, \Delta \models^{\mathfrak{M}} a{:}A$ *for all* $\mathfrak{M}$.

■

A propositional non-classical logic $\mathcal{L}$ is characterized by its language and by its models, i.e. the conditions independently imposed on each $\mathfrak{R}_i$, on $*$, etc. Moreover, some logics, e.g. intuitionistic and relevance logics, require truth to be monotonic. To express this, we define a *partial order* $\sqsubseteq$ on worlds, where for intuitionistic logic $\sqsubseteq$ coincides with the accessibility relation, while for relevance logics it can be defined in terms of R, i.e. $a \sqsubseteq b$ iff $R\,0\,a\,b$, where 0 is the label denoting the actual world; for modal logic $\sqsubseteq$ reduces to equality. Then we require that $\mathfrak{V}$ satisfy the *atomic monotony condition*, i.e. for any a_i and a_j and for any propositional variable p,

$$\text{if } \models^{\mathfrak{M}} a_i{:}p \text{ and } \models^{\mathfrak{M}} a_i \sqsubseteq a_j\text{, then } \models^{\mathfrak{M}} a_j{:}p\,. \tag{3.13}$$

One might be tempted to generalize this immediately to arbitrary formulas. In fact, this generalization holds for the 'usual' non-classical logics, such as intuitionistic and relevance logics, where we can prove by induction on the structure of A that

$$\text{if } \models^{\mathfrak{M}} a_i{:}A \text{ and } \models^{\mathfrak{M}} a_i \sqsubseteq a_j\text{, then } \models^{\mathfrak{M}} a_j{:}A\,. \tag{3.14}$$

But there are logics for which (3.14) does not hold for every formula. For example, [84, 142] combine intuitionistic implication $\rightarrow$ with classical implication $\supset$, and show that (3.14) holds for $A \rightarrow B$ (in fact it holds, as one would expect, for every intuitionistic formula) but it fails for $A \supset B$. This problem is solved there by restricting (3.14) to *persistent* formulas. Formally, a formula A of the 'intuitionistic/classical' logic in [84] is persistent iff

(i) it is atomic, or
(ii) it is of the form $B \rightarrow C$ or $\neg B$, where $\neg$ is intuitionistic (and thus non-local) negation, or

(iii) it is of the form $B \wedge C$ or $B \vee C$, and B and C are both persistent.

Similar definitions of persistency can be given for other non-classical logics, depending on the particular language we are considering, and we can then restrict (3.14) to the following general property, the *monotony condition*, which is provable from (3.13) by induction on the structure of A.

Property 3.1.6 *For any a_i and a_j, and for any persistent formula A,*

$$\textit{if } \models^{\mathfrak{M}} a_i{:}A \textit{ and } \models^{\mathfrak{M}} a_i \sqsubseteq a_j\,, \textit{ then } \models^{\mathfrak{M}} a_j{:}A\,.$$

■

Monotony is defined also for rwffs: for an n+1–ary relation R_i we require that

$$\text{for all } j < n, \text{ if } \models^{\mathfrak{M}} R_i\, a_0 \ldots a_{j-1}\, a_j\, a_{j+1} \ldots a_n \text{ and } \models^{\mathfrak{M}} a \sqsubseteq a_j\,, \quad \text{then } \models^{\mathfrak{M}} R_i\, a_0 \ldots a_{j-1}\, a\, a_{j+1} \ldots a_n \tag{3.15}$$

and

$$\text{if } \models^{\mathfrak{M}} R_i\, a_0 \ldots a_{n-1}\, a_n \text{ and } \models^{\mathfrak{M}} a_n \sqsubseteq a, \text{ then } \models^{\mathfrak{M}} R_i\, a_0 \ldots a_{n-1}\, a\,. \tag{3.16}$$

In the following we assume formulas of the form $a \sqsubseteq b$ to be special cases of relational formulas, i.e. $R\,0\,a\,b$, but we note that we could introduce them explicitly as a third kind of formulas, independent of lwffs and rwffs (proof-theory and semantics are then extended accordingly). This assumption allows us to treat the properties of the partial order, reflexivity and transitivity, as instances of (3.15) and (3.16).

Definition 3.1.7 *As a notational simplification, we will often restrict our attention to (ND systems for) propositional non-classical logics with a restricted language containing the local operators $\wedge$, $\vee$ and $\supset$, one universal non-local operator $\mathcal{M}^u$ of arity u associated with a u+1–ary relation R^u, one existential non-local operator $\mathcal{M}^e$ of arity e associated with an e+1–ary relation R^e, non-local negation $\neg$, and the constant $\perp\!\!\!\perp$. (Since the language might not contain (non-local) $\rightarrow$, we take $\neg$ as a primitive operator as opposed to defined in terms of $\perp\!\!\!\perp$ and $\rightarrow$.)*

From Definition 3.1.5, a model for an ND system built from such a language is the tuple $\mathfrak{M} = (\mathfrak{W}, \text{o}, \mathfrak{R}^u, \mathfrak{R}^e, *, \mathfrak{V})$, and truth for an rwff or lwff φ in $\mathfrak{M}$, $\models^{\mathfrak{M}} \varphi$, is the smallest relation $\models^{\mathfrak{M}}$ satisfying (3.1), (3.2) *and* (3.3) *for local operators,* (3.5) *and* (3.6) *for R^u and $\mathcal{M}^u$,* (3.5) *and* (3.7) *for R^e and $\mathcal{M}^e$,* (3.10), (3.12), (3.13), (3.15) *and* (3.16) *for R^u and R^e.* ■

Finally note that we do not, here, consider logics like the relevance logic E for which models with more than one actual world are needed [192, 201]. These logics can be presented by employing a set $\mathcal{A}$ of actual worlds and modifying the postulates of the relational theory with a precondition testing membership in $\mathcal{A}$; for example the identity postulate $R\,0\,a\,a$ (see below) is replaced with '$x \in \mathcal{A}$ implies $R\,x\,a\,a$'.

3.1.2 The base system $\mathrm{N}(\mathcal{B})$

We now introduce the base system $\mathrm{N}(\mathcal{B})$ formalizing a ND presentation of the base propositional non-classical logic $\mathcal{B}$. $\mathrm{N}(\mathcal{B})$ provides the rules we need to reason about lwffs. Our formalization is motivated by pragmatic concerns: the base system should

(i) make no assumptions about the relational theories extending it,
(ii) be adequate for the logics we are interested in, and
(iii) have good proof-theoretical properties.

In §2 we provided a base system for a large family of propositional modal logics that satisfies all these criteria. Unfortunately, in the more general case considered here, things are not so clear-cut: the base system (and the base logic) depends on the particular family of non-classical logics we consider, and thus to achieve (ii) and (iii) we have to replace (i) with

(i') make as few assumptions as possible about the relational theories extending it, ideally none at all.

(See §3.1.3, where we discuss 'complementary rules', and §3.3, where we discuss extensions with first-order relational theories.)

The labelled ND system $\mathrm{N}(\mathcal{B})$ consists of an *introduction rule*, $\bullet$I, and an *elimination rule*, $\bullet$E, for each logical operator $\bullet$ except $\perp\!\!\!\perp$ (and, if present, $\perp$), for which only an elimination rule is given. We begin by considering the simplest logical operators, the local ones, for which, like for modal logics in §2, we adapt the traditional ND rules [186] by adding labels. For example, for classical (local) implication $\supset$ we give the rules

$$\frac{\begin{matrix}[a{:}A]\\ \vdots\\ a{:}B\end{matrix}}{a{:}A \supset B}\;\supset\mathrm{I} \qquad \text{and} \qquad \frac{a{:}A \supset B \quad a{:}A}{a{:}B}\;\supset\mathrm{E}\;. \tag{3.17}$$

Rules for $\wedge$, $\vee$ and other local operators are adapted similarly, e.g.

$$\frac{a{:}A \quad a{:}B}{a{:}A \wedge B}\;\wedge\mathrm{I} \qquad \text{and} \qquad \frac{a{:}A \vee B \quad \begin{matrix}[a{:}A]\\ \vdots\\ c{:}C\end{matrix} \quad \begin{matrix}[a{:}B]\\ \vdots\\ c{:}C\end{matrix}}{c{:}C}\;\vee\mathrm{E}\;.$$

For the non-local operators $\mathcal{M}^u$ and $\mathcal{M}^e$ we give the rules

$$\dfrac{\begin{matrix}[a_1{:}A_1] \cdots [a_{u-1}{:}A_{u-1}]\,[R^u\, a\, a_1 \ldots a_u] \\ \vdots \\ a_u{:}A_u\end{matrix}}{a{:}\mathcal{M}^u A_1 \ldots A_u}\ \mathcal{M}^u\mathrm{I}$$

$$\dfrac{a{:}\mathcal{M}^u A_1 \ldots A_u \quad a_1{:}A_1 \ \cdots\ a_{u-1}{:}A_{u-1} \quad R^u\, a\, a_1 \ldots a_u}{a_u{:}A_u}\ \mathcal{M}^u\mathrm{E}$$

(3.18)

$$\dfrac{a_1{:}A_1 \cdots a_e{:}A_e \quad R^e\, a\, a_1 \ldots a_e}{a{:}\mathcal{M}^e A_1 \ldots A_e}\ \mathcal{M}^e\mathrm{I}$$

$$\dfrac{a{:}\mathcal{M}^e A_1 \ldots A_e \qquad \begin{matrix}[a_1{:}A_1] \cdots [a_e{:}A_e]\,[R^e\, a\, a_1 \ldots a_e] \\ \vdots \\ b{:}B\end{matrix}}{b{:}B}\ \mathcal{M}^e\mathrm{E}$$

where, in $\mathcal{M}^u$I and $\mathcal{M}^e$E, each a_k and each a_l, for $1 \leq k \leq u$ and $1 \leq l \leq e$, is *fresh*. That is, in $\mathcal{M}^u$I, the labels $a_1, \ldots, a_u$ are all different from a and each other, and do not occur in any assumption on which $a_u{:}A_u$ depends other than those listed; in $\mathcal{M}^e$E, the labels $a_1, \ldots, a_e$ are all different from a, b and each other, and do not occur in any assumption on which the upper occurrence of $b{:}B$ depends other than those listed. Note that the introduction and elimination rules for $\mathcal{M}^u$ and $\mathcal{M}^e$ are independent of the properties of R^u and R^e.[3]

Comparing these rules with (3.1), (3.2), (3.3), (3.6) and (3.7), we see that they reflect the semantic definitions. When we treat negation, however, the correspondence between the rules and the semantics is more subtle, and we must choose which kind of negation we want to formalize. We begin by providing the rules

$$\dfrac{\begin{matrix}[a^*{:}A] \\ \vdots \\ b{:}\perp\!\!\!\perp\end{matrix}}{a{:}\neg A}\ \neg\mathrm{I} \qquad \text{and} \qquad \dfrac{a{:}\neg A \quad a^*{:}A}{b{:}\perp\!\!\!\perp}\ \neg\mathrm{E}\ , \tag{3.19}$$

which reflect (3.10). These rules capture only a *minimal* non-local negation. If we want a base system capable of formalizing *intuitionistic* or *classical* non-local negation we respectively need the additional rules

[3]Moreover, analogously to $\Box$ and $\supset$, note the close correspondence between our rules for $\mathcal{M}^u$ and $\supset$, which holds because we express $a{:}\mathcal{M}^u A_1 \ldots A_u$ in terms of a metalevel implication.

$$\frac{b{:}\perp\!\!\perp}{a{:}A}\perp\!\!\perp\mathrm{Ei} \qquad \text{and} \qquad \begin{array}{c}[a{:}\neg A]\\ \vdots\\ \dfrac{b{:}\perp\!\!\perp}{a^{*}{:}A}\perp\!\!\perp\mathrm{Ec}\end{array} \quad . \tag{3.20}$$

Finally, we express monotony at the level of lwffs using the rule

$$\frac{a_i{:}A \quad a_i \sqsubseteq a_j}{a_j{:}A}\ monl \tag{3.21}$$

where A is a persistent formula.[4] Since *monl* reflects Property 3.1.6, the finer details of its definition, including the side condition on its application, depend on the logic we are considering.

3.1.3 Relational theories

We present particular non-classical logics by extending (the appropriate) base system $\mathrm{N}(\mathcal{B})$ with a *relational theory* axiomatizing the properties of $*$ and of the relations $\mathfrak{R}_i$ in a Kripke model. Correspondence theory [227, 228] and known correspondence results, e.g. [164, 192, 201], allow us to determine which possible axiom schemas correspond to which semantic properties. As we observed in §2.1.2, some of these properties can only be expressed using higher-order logic, but for other properties first-order logic, or even fragments of it, is enough. Following our choice in §2.1.3, we restrict our attention to properties axiomatizable using *Horn relational rules*, i.e. rules of the form

$$\frac{R_i\, t_0^1 \ldots t_n^1 \quad \cdots \quad R_i\, t_0^m \ldots t_n^m}{R_i\, t_0^0 \ldots t_n^0}$$

where the t_k^j are terms built from labels and function symbols. (Some properties of $\mathfrak{R}_i$, e.g. *assoc1* and *assoc2* below, can be expressed as Horn relational rules only after the introduction of Skolem function constants; like for modal logics, cf. Proposition 2.1.8, we can show that the introduction of such constants constitutes a conservative extension.) A *Horn relational theory* $\mathrm{N}(\mathcal{T})$ is a theory generated by a set of such rules.

Even with such a restriction, we are able to capture many families of common propositional non-classical logics, including logics in the modal Geach hierarchy (K, T, S4, S4.2 and S5; cf. §2), and various relevance logics (e.g. B, N, T and R; cf. §3.1.4). For example, the modal axiom schema $\Box A \supset \Box\Box A$ corresponds to the transitivity of the accessibility relation, formalized by the Horn relational rule

$$\frac{aRb \quad bRc}{aRc}\ trans \quad .$$

[4] If this restriction is not imposed, then the result is not sound for some logics, e.g. an (attempted) encoding of intuitionistic implication collapses to classical implication, similar to what is shown to happen for Hilbert systems in [84, 142], as remarked above.

The axiom schemas of relevance logic $A \to A$ and $(A \to B) \to ((C \to A) \to (C \to B))$ correspond to identity ($\forall a(R\,0\,a\,a)$, formalized by *iden*) and associativity ($\forall a\,\forall b\,\forall c\,\forall d\,\forall x(R\,a\,b\,x \wedge R\,x\,c\,d \supset \exists y(R\,b\,c\,y \wedge R\,a\,y\,d))$, formalized by *assoc1* and *assoc2*) for the compossibility relation:

$$\frac{}{R\,0\,a\,a}\ iden \qquad \frac{R\,a\,b\,x \quad R\,x\,c\,d}{R\,b\,c\,f(a,b,c,d,x)}\ assoc1 \qquad \frac{R\,a\,b\,x \quad R\,x\,c\,d}{R\,a\,f(a,b,c,d,x)\,d}\ assoc2 \qquad (3.22)$$

where f is a 5–ary Skolem function constant.

For negation we give Horn rules that impose different behaviors of the $*$ function. For example, we can add the rules

$$\frac{}{a \sqsubseteq a^{**}}\ {**i} \qquad \frac{}{a^{**} \sqsubseteq a}\ {**c} \qquad \frac{}{a \sqsubseteq a^{*}}\ ortho1 \qquad \frac{}{a^{*} \sqsubseteq a}\ ortho2$$

to encode intuitionistic ($**$i), classical ($**$i and $**$c), or ortho (*ortho1* and *ortho2*) negation.

Tables 3.2 and 3.3 in §3.1.4 below list further axiom schemas of relevance logics, together with the corresponding properties of the compossibility relation R and of the $*$ function, as well as the corresponding Horn rules.

Finally, corresponding to (3.15) and (3.16), for each R_i we have n+1 rules for the monotony properties of rwffs:

$$\frac{R_i\,a_0 \ldots a_{j-1}\,a_j\,a_{j+1} \ldots a_n \quad a \sqsubseteq a_j}{R_i\,a_0 \ldots a_{j-1}\,a\,a_{j+1} \ldots a_n}\ monR_i(j)$$

$$\frac{R_i\,a_0 \ldots a_{n-1}\,a_n \quad a_n \sqsubseteq a}{R_i\,a_0 \ldots a_{n-1}\,a}\ monR_i(n)$$

where $0 \leq j < n$ in the schematic rule $monR_i(j)$.

Negation and monotony again raise the question of what exactly a base system (and a base logic) should be. The rules we have just given can be seen as rwff complements of the lwff rules given earlier. For instance, for an intuitionistic negation, i.e. where the base system contains $\bot\!\!\bot$Ei, we need also the rule $**$i, while for a classical negation, i.e. with $\bot\!\!\bot$Ec, we need also $**$i and $**$c; similarly, the $monR_i$ rules complement *monl*. Only by requiring these complementary rules can we establish desired proof-theoretical results (see, e.g., the proof of Theorem 3.3.9). Thus it is convenient, on pragmatic grounds, to assume that a base system $\mathrm{N}(\mathcal{B})$ is extended with a theory $\mathrm{N}(\mathcal{T})$ that includes these minimal relational rules (a characterization of the systems, and logics, in which this complementarity is not satisfied, e.g. $\bot\!\!\bot$Ei without $**$i, or $\bot\!\!\bot$Ec with only $**$c, is out of the scope of this book).

Definition 3.1.8 *The labelled ND system* $\mathrm{N}(\mathcal{L}) = \mathrm{N}(\mathcal{B}) + \mathrm{N}(\mathcal{T})$ *for the propositional non-classical logic* $\mathcal{L}$ *is the extension of an appropriate base system* $\mathrm{N}(\mathcal{B})$ *with a given Horn relational theory* $\mathrm{N}(\mathcal{T})$. ■

Note that we employ the same notational convention as in §2: when the relational theory $\mathrm{N}(\mathcal{T})$ contains Skolem function constants, then our language and rules must

Table 3.1. The systems $N(\mathcal{ML})$, $N(\mathcal{JL})$ and $N(\mathcal{CL})$

$N(\mathcal{L})$	$N(\mathcal{B})$	$N(\mathcal{T})$ *(includes at least)*
$N(\mathcal{ML})$	rules for $\wedge, \vee, \supset, \mathcal{M}^u, \mathcal{M}^e, \neg$	
	monl	$monR_i$ rules (for R^u and R^e)
$N(\mathcal{JL})$	rules for $\wedge, \vee, \supset, \mathcal{M}^u, \mathcal{M}^e, \neg$	
	monl	$monR_i$ rules (for R^u and R^e)
	$\perp\!\!\!\perp$Ei	$**$i
$N(\mathcal{CL})$	rules for $\wedge, \vee, \supset, \mathcal{M}^u, \mathcal{M}^e, \neg$	
	monl	$monR_i$ rules (for R^u and R^e)
	$\perp\!\!\!\perp$Ec	$**$i, $**$c

be extended to distinguish atomic and composite labels. As before, we will continue using the meta-variables *a*, *b*, *c*, etc., although labels may now be built using Skolem function constants.

Consider now the restricted language of Definition 3.1.7 (with the operators $\wedge$, $\vee$, $\supset$, $\mathcal{M}^u$, $\mathcal{M}^e$, $\neg$ and $\perp\!\!\!\perp$). Mirroring Prawitz [186, 187], in Table 3.1 we distinguish three families of ND systems according to their treatment of (non-local) negation: minimal, intuitionistic or classical; we make the distinction by considering the rules for $\perp\!\!\!\perp$, while Prawitz considers the rules for $\perp$.

The *minimal system* $N(\mathcal{ML})$ is determined by a base system including *monl* (with the appropriate restrictions) and introduction and elimination rules for local, e.g. (3.17), and non-local operators, e.g. (3.18) and (3.19), and by a relational theory including, at least, the $monR_i$ rules, to complement *monl*.[5] The *intuitionistic system* $N(\mathcal{JL})$ is obtained by extending $N(\mathcal{ML})$ with $\perp\!\!\!\perp$Ei and the complementary rule $**$i, and the *classical system*$N(\mathcal{CL})$ is obtained by extending $N(\mathcal{ML})$ with $\perp\!\!\!\perp$Ec and the complementary rules $**$i and $**$c. Alternatively, we can obtain $N(\mathcal{CL})$ by extending $N(\mathcal{JL})$ with $\perp\!\!\!\perp$Ec and $**$c (the derivation of $\perp\!\!\!\perp$Ei from these rules is straightforward). Furthermore, we can extend $N(\mathcal{CL})$ with the rules *ortho1* and *ortho2* to formalize ortho negation, where $a = a^* = a^{**}$. For each of these systems, we can then further extend the theory $N(\mathcal{T})$ with rules expressing properties of $*$ and the relations to obtain presentations of particular logics.

We conclude this section by generalizing Definition 2.1.11 and Fact 2.1.12; Notation 2.1.13 for derivations in modal logics also generalizes straightforwardly to derivations in $N(\mathcal{B}) + N(\mathcal{T})$.

Definition 3.1.9 *A derivation of an* lwff *or* rwff φ *from a set of* lwffs Γ *and a set of* rwffs Δ *in a ND system* $N(\mathcal{L}) = N(\mathcal{B}) + N(\mathcal{T})$ *is a tree formed using the rules in* $N(\mathcal{L})$, *ending with* φ *and depending only on* $\Gamma \cup \Delta$. *We write* $\Gamma, \Delta \vdash_{N(\mathcal{L})} \varphi$ *when* φ

[5]Note that, unlike Prawitz's, our minimal system does not satisfy the *inversion principle*, since it contains *monl* which is neither an introduction nor an elimination rule. Also, we do not enforce the restriction that the conclusions of the elimination rules for $\perp\!\!\!\perp$ or $\perp$ are different from $\perp\!\!\!\perp$ and $\perp$.

Axiom schemas:

A1: $A \to A$.

A2: $A \wedge B \to A$.

A3: $A \wedge B \to B$.

A4: $A \to A \vee B$.

A5: $B \to A \vee B$.

A6: $A \wedge (B \vee C) \to (A \wedge B) \vee (A \wedge C)$ [or $A \wedge (B \vee C) \to (A \wedge B) \vee C$].

A7: $(A \to B) \wedge (A \to C) \to (A \to B \wedge C)$.

A8: $(A \to C) \wedge (B \to C) \to (A \vee B \to C)$.

Inference rules:

R1: $\dfrac{A \to B \quad A}{B}$ *modus ponens*,

R2: $\dfrac{A \quad B}{A \wedge B}$ *adjunction*,

R3: $\dfrac{A \to B \quad C \to D}{(B \to C) \to (A \to D)}$ *affixing*,

along with their disjunctive forms, where if $\dfrac{A_1 \; \cdots \; A_n}{B}$ is a rule, then its *disjunctive form* is the rule $\dfrac{C \vee A_1 \; \cdots \; C \vee A_n}{C \vee B}$.

Figure 3.1. H(B$^+$), a Hilbert system for B$^+$

can be so derived. A derivation of φ *in* N($\mathcal{L}$) *depending on the empty set,* $\vdash_{\mathrm{N}(\mathcal{L})} \varphi$, *is a* proof of φ in N($\mathcal{L}$), *and we then say that* φ *is a* N($\mathcal{L}$)-theorem. ∎

Fact 3.1.10 $\Gamma, \Delta \vdash_{\mathrm{N}(\mathcal{B})+\mathrm{N}(\mathcal{T})} R_i\, a\, a_1 \ldots a_n$ *iff* $\Delta \vdash_{\mathrm{N}(\mathcal{T})} R_i\, a\, a_1 \ldots a_n$. ∎

3.1.4 Examples of propositional non-classical logics

Our framework can be specialized to present large families of (fragments of and full) non-classical logics. Among others, propositional modal and relevance logics. The important, though relatively simple, case of modal logics is discussed at length in §2, albeit for a slightly different notation. Here we consider examples of relevance logics, which, like modal logics, are traditionally [1, 2, 201] presented by using Hilbert systems. For example, the axiom schemas and inference rules given in Figure 3.1 determine a Hilbert system H(B$^+$) for the basic positive relevance logic B$^+$, e.g. [188, 192, 199].

$$\frac{a{:}A \quad a{:}B}{a{:}A \wedge B}\,\wedge\mathrm{I} \qquad \frac{a{:}A \wedge B}{a{:}A}\,\wedge\mathrm{E1} \qquad \frac{a{:}A \wedge B}{a{:}B}\,\wedge\mathrm{E2}$$

$$\frac{a{:}A}{a{:}A \vee B}\,\vee\mathrm{I1} \qquad \frac{a{:}B}{a{:}A \vee B}\,\vee\mathrm{I2} \qquad \frac{a{:}A \vee B \quad \begin{array}{c}[a{:}A]\\ \vdots \\ c{:}C\end{array} \quad \begin{array}{c}[a{:}B]\\ \vdots \\ c{:}C\end{array}}{c{:}C}\,\vee\mathrm{E}$$

$$\frac{\begin{array}{c}[b{:}A]\ [R\,a\,b\,c]\\ \vdots \\ c{:}B\end{array}}{a{:}A \rightarrow B}\,\rightarrow\mathrm{I} \qquad \frac{a{:}A \rightarrow B \quad b{:}A \quad R\,a\,b\,c}{c{:}B}\,\rightarrow\mathrm{E} \qquad \frac{a{:}A \quad R\,0\,a\,b}{b{:}A}\,\mathit{monl}$$

$$\frac{R\,a\,b\,c \quad R\,0\,x\,a}{R\,x\,b\,c}\,\mathit{monR}(1) \qquad \frac{R\,a\,b\,c \quad R\,0\,x\,b}{R\,a\,x\,c}\,\mathit{monR}(2)$$

$$\frac{R\,a\,b\,c \quad R\,0\,c\,x}{R\,a\,b\,x}\,\mathit{monR}(3) \qquad \frac{}{R\,0\,a\,a}\,\mathit{iden}$$

In $\rightarrow$I, b and c are different from a and each other, and do not occur in any assumption on which $a{:}A \rightarrow B$ depends other than those listed.

Figure 3.2. The natural deduction system $\mathrm{N}(\mathrm{B}^+)$

Hilbert systems for other propositional relevance logics are obtained by extending $\mathrm{H}(\mathrm{B}^+)$ with axiom schemas and rules formalizing the behavior of the non-local operators $\rightarrow$ and $\neg$. Examples of such axiom schemas and rules are given in Tables 3.2 and 3.3, where we also indicate the corresponding properties of the compossibility relation R and of the $*$ function. Note that most of these correspondences hold only under the assumption of the postulates for $\mathrm{H}(\mathrm{B}^+)$. For example, A9 corresponds to $\forall a \forall b (R\,0\,a\,b \supset R\,a\,a\,b)$, which simplifies to $\forall a (R\,a\,a\,a)$ under the assumption of identity $\forall a (R\,0\,a\,a)$. Extensive lists of such correspondence results can be found in [192, 201].

In Figure 3.2 we give a labelled ND system $\mathrm{N}(\mathrm{B}^+)$ for the logic B^+. $\mathrm{N}(\mathrm{B}^+)$ is an instance of a minimal base system, where $A \rightarrow B$ is defined as the binary universal modal operator $\mathcal{M}^u AB$ associated with the ternary relation R, and which we can extend with Horn rules formalizing properties of R and $*$ to present other propositional relevance logics. In Table 3.4 we give examples of labelled ND systems for some common propositional relevance logics, together with the corresponding Hilbert systems.

Table 3.2. Some axiom schemas and inference rules of relevance logics and corresponding properties of R and $*$

Name	*Axiom schema/Inference rule*	*Property*
A9	$A \wedge (A \to B) \to B$	$R\,a\,a\,a$ or $R\,0\,a\,b \supset R\,a\,a\,b$ (*idempotence*)
A10	$(A \to B) \wedge (B \to C) \to (A \to C)$	$R\,a\,b\,c \supset R^2\,a\,(ab)\,c$ (*transitivity*)
A11	$(A \to B) \to ((B \to C) \to (A \to C))$	$R^2\,a\,b\,c\,d \supset R^2\,b\,(ac)\,d$ (*suffixing*)
A12	$(A \to B) \to ((C \to A) \to (C \to B))$	$R^2\,a\,b\,c\,d \supset R^2\,a\,(bc)\,d$ (*associativity* or *prefixing*)
A13	$(A \to (A \to B)) \to (A \to B)$	$R\,a\,b\,c \supset R^2\,a\,b\,b\,c$ (*contraction*)
A14	$((A \to A) \to B) \to B$	$R\,a\,0\,a$ (*specialized assertion*)
A15	$A \to ((A \to B) \to B)$	$R\,a\,b\,c \supset R\,b\,a\,c$ (*commutativity* or *assertion*)
A16	$A \to (A \to A)$	$R\,a\,b\,c \supset (R\,0\,a\,c \vee R\,0\,b\,c)$ or $R\,0\,0\,a \vee R\,0\,0\,a^*$ (*mingle*)
A17	$A \to (B \to B)$	$R\,0\,0\,a$ or $R\,a\,b\,c \supset R\,0\,b\,c$ (*thinning*)
A18	$A \to (B \to A)$	$R\,a\,b\,c \supset R\,0\,a\,c$ (*positive paradox*)
R4	$\dfrac{A \to \neg B}{B \to \neg A}$ *contraposition*	$R\,0\,a\,b \supset R\,0\,b^*\,a^*$ (*antitonicity*)
A19	$(A \to \neg B) \to (B \to \neg A)$	$R\,a\,b\,c \supset R\,a\,c^*\,b^*$ (*inversion*)
A20	$\neg\neg A \to A$	$a^{**} = a$ (*period two*)
A21	$A \vee \neg A$	$R\,0\,0^*0$ (*excluded middle*)

- $R^2\,a\,b\,c\,d =_{def} \exists x(R\,a\,b\,x \wedge R\,x\,c\,d)$ and $R^2\,a\,(bc)\,d =_{def} \exists x(R\,b\,c\,x \wedge R\,a\,x\,d)$.
- All the properties of R are outermost universally quantified, e.g. the property corresponding to A11 is in fact $\forall a \forall b \forall c \forall d(R^2\,a\,b\,c\,d \supset R^2\,b\,(ac)\,d)$, which by the above definitions is equivalent to $\forall a \forall b \forall c \forall d(\exists x(R\,a\,b\,x \wedge R\,x\,c\,d) \supset (\exists y(R\,a\,c\,y \wedge R\,b\,y\,d)))$, or to $\forall a \forall b \forall c \forall d \forall x((R\,a\,b\,x \wedge R\,x\,c\,d) \supset (\exists y(R\,a\,c\,y \wedge R\,b\,y\,d)))$ by prenexing quantifiers.
- Using the definition of the partial order we could write $a \sqsubseteq b$ for $R\,0\,a\,b$.

Table 3.3. Some axiom schemas and inference rules of relevance logics and corresponding Horn rules

Name	*Horn relational rules*
A9	$\dfrac{}{R\,a\,a\,a}$ *idem* or $\dfrac{R\,0\,a\,b}{R\,a\,a\,b}$ *idem*
A10	$\dfrac{R\,a\,b\,c}{R\,a\,b\,f_1(a,b,c)}$ *trans1* and $\dfrac{R\,a\,b\,c}{R\,a\,f_1(a,b,c)\,c}$ *trans2*
A11	$\dfrac{R\,a\,b\,x \quad R\,x\,c\,d}{R\,a\,c\,f_2(a,b,c,d,x)}$ *suff1* and $\dfrac{R\,a\,b\,x \quad R\,x\,c\,d}{R\,b\,f_2(a,b,c,d,x)\,d}$ *suff2*
A12	$\dfrac{R\,a\,b\,x \quad R\,x\,c\,d}{R\,b\,c\,f_3(a,b,c,d,x)}$ *assoc1* and $\dfrac{R\,a\,b\,x \quad R\,x\,c\,d}{R\,a\,f_3(a,b,c,d,x)\,d}$ *assoc2*
A13	$\dfrac{R\,a\,b\,c}{R\,a\,b\,f_4(a,b,c)}$ *cont1* and $\dfrac{R\,a\,b\,c}{R\,f_4(a,b,c)\,b\,c}$ *cont2*
A14	$\dfrac{}{R\,a\,0\,a}$ *specassert*
A15	$\dfrac{R\,a\,b\,c}{R\,b\,a\,c}$ *comm*
A17	$\dfrac{}{R\,0\,0\,a}$ *thin* or $\dfrac{R\,a\,b\,c}{R\,0\,b\,c}$ *thin*
A18	$\dfrac{R\,a\,b\,c}{R\,0\,a\,c}$ *pospar*
R4	$\dfrac{R\,0\,a\,b}{R\,0\,b^*\,a^*}$ *anti*
A19	$\dfrac{R\,a\,b\,c}{R\,a\,c^*\,b^*}$ *inv*
A20	$\dfrac{}{R\,0\,a\,a^{**}}$ **i and $\dfrac{}{R\,0\,a^{**}\,a}$ **c
A21	$\dfrac{}{R\,0\,0^*0}$ *exmid*

Where the f_i's are Skolem function constants. Note that the property of R corresponding to the 'mingle' axiom A16 cannot be expressed by a Horn relational rule but by a first-order rule; we return to this at the end of this chapter when we discuss the logic RM.

Table 3.4. Extensions of B^+: Hilbert systems and labelled ND systems for some propositional relevance logics

Logic $\mathcal{L}$	*Hilbert system* $H(\mathcal{L})$	*Labelled ND system* $N(\mathcal{L})$
N^+	$H(B^+) + \{A11, A12\}$	$N(B^+) + \{\mathit{suff1}, \mathit{suff2}, \mathit{assoc1}, \mathit{assoc2}\}$
T^+	$H(N^+) + \{A13\}$	$N(N^+) + \{\mathit{cont1}, \mathit{cont2}\}$
E^+	$H(T^+) + \{A14\}$	$N(T^+) + \{\mathit{specassert}\}$
R^+	$H(E^+) + \{A15\}$	$N(E^+) + \{\mathit{comm}\}$
$S4^+$	$H(E^+) + \{A17\}$	$N(E^+) + \{\mathit{thin}\}$
J^+	$H(R^+) + \{A17\}$ $= H(S4^+) + \{A15\}$	$N(R^+) + \{\mathit{thin}\}$ $= N(S4^+) + \{\mathit{comm}\}$
B	$H(B^+) + \{A20, R4\}$	$N(B^+) + \{\neg I, \neg E, \perp\!\!\perp Ec, {*}{*}i, {*}{*}c, \mathit{anti}\}$
R	$H(B) + \{A11, A13, A15, A19\}$ $= H(B^+) + \{A11, A13, A15, A19, A20\}$	$N(B) + \{\mathit{suff1}, \mathit{suff2}, \mathit{cont1}, \mathit{cont2}, \mathit{comm}, \mathit{inv}\}$ $= N(B^+) + \{\neg I, \neg E, \perp\!\!\perp Ec, {*}{*}i, {*}{*}c, \mathit{suff1}, \mathit{suff2}, \mathit{cont1}, \mathit{cont2}, \mathit{comm}, \mathit{inv}\}$
G	$H(B) + \{A21\}$	$N(B) + \{\mathit{exmid}\}$
C	$H(R) + \{A17\}$	$N(R) + \{\mathit{thin}\}$

Note that we have chosen the 'economical' system $H(R)$ given by [2, 200], where, e.g., R4 is redundant as it can be derived using A19 and R1; similarly, in $N(R)$ we can trivially derive the rule *anti* using *inv*, and, albeit less trivially, the rule *idem* using identity and contraction, e.g.

$$\dfrac{\dfrac{\dfrac{}{R\,0\,a\,a}\,\mathit{iden}}{R\,f_4(0\,,a\,,a)\,a\,a}\,\mathit{cont2} \qquad \dfrac{\dfrac{}{R\,0\,a\,a}\,\mathit{iden}}{R\,0\,a\,f_4(0\,,a\,,a)}\,\mathit{cont1}}{R\,a\,a\,a}\,\mathit{mon}R(1)\ .$$

Alternative, equivalent, axiomatizations are possible, for R and other logics [1, 2, 188, 192, 199]. Note also that J^+ is positive intuitionistic logic, G is 'basic' classical logic and C is 'full' classical logic.

We postpone proofs that our systems are what we claim they are, i.e. equivalent to the corresponding Hilbert systems, until §3.2, where we show the soundness and completeness of our presentations with respect to the corresponding Kripke semantics (proofs of soundness and completeness of Hilbert systems can be found in various textbooks, e.g. [201]).

Here we are interested rather in comparing the modularity of the two presentations. As an example, we compare our system $N(R)$ with the Hilbert system $H(R)$ and show the advantages of our approach in the modular way we present the relevance logic R so that it can be extended to obtain (*positive* and *full*) *intuitionistic* and *classical* logic.

Routley and Meyer [200] show that there is a problem with Hilbert systems for relevance logics: $\mathrm{H}(\mathrm{R}^+)$ is a subsystem of the system $\mathrm{H}(\mathrm{J}^+)$ for positive intuitionistic logic J^+, but H(R) is a subsystem *only* of the system H(C) for 'full' classical logic C. That is, the Hilbert system H(J) for 'full' intuitionistic logic J cannot be *modularly* obtained by simply adding new axioms to H(R); H(J) can be obtained from H(R), but only in a specialized fashion, if relevant negation is rejected in favor of an intuitionistic one [200, p. 227].

Now consider our systems, and note that since $\neg\neg A \to A$ is an axiom of H(R), we have 'based' N(R) on the classical version of $\mathrm{N}(\mathcal{B})$ with $\perp\!\!\!\perp$Ec. As shown in Table 3.4, a ND system $\mathrm{N}(\mathrm{J}^+)$ for positive intuitionistic logic J^+ is obtained from $\mathrm{N}(\mathrm{R}^+)$ by adding the rule

$$\frac{}{R\,0\,0\,a}\ thin$$

corresponding to the (intuitionistically valid) 'thinning' axiom schema $A \to (B \to B)$, so that the ternary R reduces to a binary partial order (in fact to the usual accessibility relation of Kripke models for intuitionistic logic), and $\to$ reduces to intuitionistic implication. However, extending N(R) with the rule *thin* yields classical logic: in Example 3.1.12 below we show that we are then able to prove $R\,0\,a\,0$, so that, essentially, all the worlds collapse; i.e. $a = a^* = a^{**}$, $\to$ reduces to $\supset$, and $\neg$ to $\sim$. This should not come as a surprise, since N(R) contains, like H(R), a classical treatment of negation (because of the axiom schema A20 and the corresponding rules $\perp\!\!\!\perp$Ec, $**$i and $**$c). That is, with reference to Table 3.1, we have $\mathrm{N}(\mathrm{R}) = \mathrm{N}(\mathcal{C}\mathrm{R})$. But Table 3.1 also tells us that this problem can be naturally overcome in our setting. To restore the *modularity* of the extensions and obtain a ND system N(J) for full intuitionistic logic J, we just need to consider the system $\mathrm{N}(\mathcal{J}\mathrm{R})$, *intuitionistic* N(R), i.e. N(R) with an intuitionistic treatment of negation, which we obtain from N(R) by substituting $\perp\!\!\!\perp$Ec with $\perp\!\!\!\perp$Ei and deleting $**$c. Indeed, $\mathrm{N}(\mathcal{J}\mathrm{R})$ is an intermediate system between $\mathrm{N}(\mathrm{R}^+)$ and N(R), i.e. $\mathrm{N}(\mathrm{R}^+) \subset \mathrm{N}(\mathcal{J}\mathrm{R}) \subset \mathrm{N}(\mathrm{R})$, and we can extend it with *thin* to obtain full intuitionistic logic in a modular way.

Proposition 3.1.11 *Adding the rule thin to* $\mathrm{N}(\mathcal{J}\mathrm{R})$ *results in* N(J). ■

This follows immediately by showing that R reduces to a partial order, and that relevant $\to$, $\neg$, $\perp\!\!\!\perp$ and the corresponding relevance rules reduce to intuitionistic $\to$, $\neg$, $\perp$ and the corresponding intuitionistic rules.

We conclude this section with some examples of derivations, the Isabelle formalizations of which are given in §5.2.1.

Example 3.1.12 We begin by showing that using the antitonicity rule we can derive a labelled equivalent of a contraposition rule 'weaker' than R4, i.e. 'if $A \to B$ then

$\neg B \to \neg A$':[6]

$$\dfrac{\dfrac{[a{:}\neg B]^2 \qquad \dfrac{0{:}A \to B \quad [b^*{:}A]^1 \quad \dfrac{[R\,0\,a\,b]^2}{R\,0\,b^*\,a^*}\,\textit{anti}}{a^*{:}B}\,{\to}\mathrm{E}}{\dfrac{c{:}\perp\!\!\!\perp}{b{:}\neg A}\,\neg\mathrm{I}^1}\,\neg\mathrm{E}}{0{:}\neg B \to \neg A}\,{\to}\mathrm{I}^2 \quad . \tag{3.23}$$

Using the inversion rule *inv* and the rule $**$i, we can prove the axiom schema for contraposition A19, and, similarly, derive a labelled equivalent of the contraposition rule R4:

$$\dfrac{\dfrac{\dfrac{\dfrac{\dfrac{[a{:}A \to \neg B]^3 \quad [d^*{:}A]^1 \quad \dfrac{\dfrac{[R\,b\,c\,d]^2 \quad [R\,0\,a\,b]^3}{R\,a\,c\,d}\,\textit{monR}(1)}{R\,a\,d^*\,c^*}\,\textit{inv}}{c^*{:}\neg B}\,{\to}\mathrm{E} \qquad \dfrac{[c{:}B]^2 \quad \dfrac{}{R\,0\,c\,c^{**}}\,{**}\mathrm{i}}{c^{**}{:}B}\,\textit{monl}}{e{:}\perp\!\!\!\perp}\,\neg\mathrm{E}}{d{:}\neg A}\,\neg\mathrm{I}^1}{b{:}B \to \neg A}\,{\to}\mathrm{I}^2}{0{:}(A \to \neg B) \to (B \to \neg A)}\,{\to}\mathrm{I}^3 \quad .$$

We can prove A20 using the 'classical' rules $\perp\!\!\!\perp$Ec and $**$c, i.e.

$$\dfrac{\dfrac{\dfrac{\dfrac{\dfrac{[a{:}\neg\neg A]^2 \quad [R\,0\,a\,b]^2}{b{:}\neg\neg A}\,\textit{monl} \quad [b^*{:}\neg A]^1}{c{:}\perp\!\!\!\perp}\,\neg\mathrm{E}}{b^{**}{:}A}\,\perp\!\!\!\perp\mathrm{Ec}^1 \qquad \dfrac{}{R\,0\,b^{**}\,b}\,{**}\mathrm{c}}{b{:}A}\,\textit{monl}}{0{:}\neg\neg A \to A}\,{\to}\mathrm{I}^2 \quad . \tag{3.24}$$

Finally, we show that when we extend N(R) with *thin* we can prove $R\,0\,a\,0$:

$$\dfrac{\dfrac{\Pi}{R\,0\,a\,0^*} \qquad \dfrac{\dfrac{\dfrac{\dfrac{\dfrac{\dfrac{}{R\,0^*\,0^*\,0^*}\,\textit{idem}}{R\,0^*\,0^{**}\,0^{**}}\,\textit{inv} \quad \dfrac{}{R\,0\,0^{**}\,0}\,{**}\mathrm{c}}{R\,0^*\,0^{**}\,0}\,\textit{monR}(3) \qquad \dfrac{}{R\,0\,0\,0^{**}}\,{**}\mathrm{i}}{R\,0^*\,0\,0}\,\textit{monR}(2)}{R\,0\,0^*\,0}\,\textit{comm}}{}}{R\,0\,a\,0}\,\textit{monR}(3)$$

where Π is

$$\dfrac{\dfrac{\dfrac{}{R\,0\,0\,a^*}\,\textit{thin}}{R\,0\,a^{**}\,0^*}\,\textit{anti} \qquad \dfrac{}{R\,0\,a\,a^{**}}\,{**}\mathrm{i}}{R\,0\,a\,0^*}\,\textit{monR}(2) \quad .$$

■

---

[6]As discussed in [188, 192], adding this rule together with axiom schemas for the De Morgan laws, $\neg(A \vee B) \leftrightarrow (\neg A \wedge \neg B)$ and $\neg(A \wedge B) \leftrightarrow (\neg A \vee \neg B)$, yields H(BM), an axiomatization of BM, which is an intermediate logic between B^+ and B. H(B) can then be obtained from H(BM) by adding the axiom schema $A \leftrightarrow \neg\neg A$, making the De Morgan laws redundant.

3.2 SOUNDNESS AND COMPLETENESS

In this section we show that every non-classical system $\mathrm{N}(\mathcal{L})$ obtained by extending $\mathrm{N}(\mathcal{B})$ with a Horn relational theory $\mathrm{N}(\mathcal{T})$ is sound and complete with respect to the corresponding Kripke semantics. For notational simplicity, we consider again the restricted language of Definition 3.1.7 (with the operators $\wedge$, $\vee$, $\supset$, $\mathcal{M}^u$, $\mathcal{M}^e$, $\neg$ and $\perp\!\!\!\perp$); the results generalize straightforwardly to unrestricted languages.

Like for modal logics, the explicit embedding of properties of the models and the capability of explicitly reasoning about them, via rwffs and relational rules, require us to consider soundness and completeness also for rwffs, where we show that $\Delta \vdash_{\mathrm{N}(\mathcal{L})} R_i\, a\, a_1 \ldots a_n$ iff $\Delta \models R_i\, a\, a_1 \ldots a_n$.

Definition 3.2.1 *The system* $\mathrm{N}(\mathcal{L}) = \mathrm{N}(\mathcal{B}) + \mathrm{N}(\mathcal{T})$ *is* sound *iff*

(i) $\Delta \vdash_{\mathrm{N}(\mathcal{L})} R_i\, a\, a_1 \ldots a_n$ *implies* $\Delta \models R_i\, a\, a_1 \ldots a_n$, *and*
(ii) $\Gamma, \Delta \vdash_{\mathrm{N}(\mathcal{L})} a{:}A$ *implies* $\Gamma, \Delta \models a{:}A$.

$\mathrm{N}(\mathcal{L})$ *is* complete *iff the converses hold, i.e. iff*

(i) $\Delta \models R_i\, a\, a_1 \ldots a_n$ *implies* $\Delta \vdash_{\mathrm{N}(\mathcal{L})} R_i\, a\, a_1 \ldots a_n$, *and*
(ii) $\Gamma, \Delta \models a{:}A$ *implies* $\Gamma, \Delta \vdash_{\mathrm{N}(\mathcal{L})} a{:}A$. ■

By Lemma 3.2.3 and Lemma 3.2.11 below, we have:

Theorem 3.2.2 $\mathrm{N}(\mathcal{L}) = \mathrm{N}(\mathcal{B}) + \mathrm{N}(\mathcal{T})$ *is sound and complete.* ■

3.2.1 Soundness

We generalize Lemma 2.2.6.

Lemma 3.2.3 $\mathrm{N}(\mathcal{L}) = \mathrm{N}(\mathcal{B}) + \mathrm{N}(\mathcal{T})$ *is sound.*

Proof Throughout the proof let $\mathfrak{M} = (\mathfrak{W}, \circ, \mathfrak{R}^u, \mathfrak{R}^e, *, \mathfrak{V})$ be an arbitrary model for $\mathrm{N}(\mathcal{L})$. We prove (i) by induction on the structure of the derivation of the rwff $R_i\, a\, a_1 \ldots a_n$ from Δ. The base case, $R_i\, a\, a_1 \ldots a_n \in \Delta$, is trivial, and there is one step case for each Horn relational rule of $\mathrm{N}(\mathcal{T})$. We treat only one example, which involves Skolem functions; soundness of the other rules follows similarly.[7] Consider applications of the rules *assoc1* and *assoc2* for a ternary relation R^u,

$$\dfrac{\begin{array}{cc}\Pi_1 & \Pi_2 \\ R^u\, a\, b\, x & R^u\, x\, c\, d\end{array}}{R^u\, b\, c\, f(a,b,c,d,x)}\; assoc1 \qquad \text{and} \qquad \dfrac{\begin{array}{cc}\Pi_1 & \Pi_2 \\ R^u\, a\, b\, x & R^u\, x\, c\, d\end{array}}{R^u\, a\, f(a,b,c,d,x)\, d}\; assoc2 \quad ,$$

[7] Recall that our models do not contain functions corresponding to possible Skolem functions in the signature. When such constants are present the appropriate Skolem expansion of the model is required [230, p. 137]; for example, for associativity the signature of the relational theory is conservatively extended with a 5–ary Skolem function constant f, and f is also added to the model. As for propositional modal logics, we do not interpret labels, but simply identify them with the identically named worlds.

where Π_1 is the derivation $\Delta_1 \vdash_{\mathrm{N}(\mathcal{L})} R^u\, a\, b\, x$, and Π_2 is the derivation $\Delta_2 \vdash_{\mathrm{N}(\mathcal{L})} R^u\, x\, c\, d$, with $\Delta = \Delta_1 \cup \Delta_2$. Assume that $\mathfrak{R}^u$ is associative and that $\models^{\mathfrak{M}} \Delta$. Then from the induction hypotheses we obtain $\models^{\mathfrak{M}} R^u\, a\, b\, x$ and $\models^{\mathfrak{M}} R^u\, x\, c\, d$, and we conclude $\models^{\mathfrak{M}} R^u\, b\, c\, f(a,b,c,d,x)$ and $\models^{\mathfrak{M}} R^u\, a\, f(a,b,c,d,x)\, d$.

We prove (ii) by induction on the structure of the derivation of $a{:}A$ from Γ and Δ. The base case, $a{:}A \in \Gamma$, is trivial, and there is one step case for each inference rule of $\mathrm{N}(\mathcal{B})$. We treat only applications of $\mathcal{M}^u$I, $\mathcal{M}^u$E, $\mathcal{M}^e$I, $\mathcal{M}^e$E, $\neg$I, $\neg$E, $\perp\!\!\!\perp$Ei and $\perp\!\!\!\perp$Ec; soundness of the rules for local operators is straightforward, and soundness of *monl* with respect to Property 3.1.6 is immediate by the restriction on its application.

Consider an application of the rule $\mathcal{M}^u$I,

$$\dfrac{\begin{array}{c}[a_1{:}A_1]\ \cdots\ [a_{u-1}{:}A_{u-1}]\ [R^u\, a\, a_1 \ldots a_u]\\ \Pi_1\\ a_u{:}A_u\end{array}}{a{:}\mathcal{M}^u A_1 \ldots A_u}\ \mathcal{M}^u\mathrm{I}\ ,$$

where Π_1 is the derivation $\Gamma_1, \Delta_1 \vdash_{\mathrm{N}(\mathcal{L})} a_u{:}A_u$, with $\Gamma_1 = \Gamma \cup \{a_1{:}A_1, \ldots, a_{u-1}{:}A_{u-1}\}$ and $\Delta_1 = \Delta \cup \{R^u\, a\, a_1 \ldots a_u\}$. The induction hypothesis is $\Gamma_1, \Delta_1 \vdash_{\mathrm{N}(\mathcal{L})} a_u{:}A_u$ implies $\Gamma_1, \Delta_1 \models a_u{:}A_u$. Assume $\models^{\mathfrak{M}} (\Gamma, \Delta)$. Considering the restriction on the application of $\mathcal{M}^u$I, we can extend Γ and Δ to $\Gamma' = \Gamma \cup \{a'_1{:}A_1, \ldots, a'_{u-1}{:}A_{u-1}\}$ and $\Delta' = \Delta \cup \{R^u\, a\, a'_1 \ldots a'_u\}$ for arbitrary $a'_1, \ldots, a'_u \notin (\Gamma, \Delta)$, and assume $\models^{\mathfrak{M}} \Gamma'$ and $\models^{\mathfrak{M}} \Delta'$. Since $\models^{\mathfrak{M}} \Gamma'$ implies $\models^{\mathfrak{M}} \Gamma_1$ and $\models^{\mathfrak{M}} \Delta'$ implies $\models^{\mathfrak{M}} \Delta_1$, from the induction hypothesis we obtain $\models^{\mathfrak{M}} a'_u{:}A_u$ for arbitrary $a'_1, \ldots, a'_u \notin (\Gamma, \Delta)$ such that $\models^{\mathfrak{M}} R^u\, a\, a'_1 \ldots a'_u$ and $\models^{\mathfrak{M}} a'_1{:}A_1, \ldots, \models^{\mathfrak{M}} a'_{u-1}{:}A_{u-1}$. We conclude $\models^{\mathfrak{M}} a{:}\mathcal{M}^u A_1 \ldots A_u$ from the definition of $\models^{\mathfrak{M}}$.

Consider an application of the rule $\mathcal{M}^u$E,

$$\dfrac{\begin{array}{c}\Pi_0\\ a{:}\mathcal{M}^u A_1 \ldots A_u\end{array}\quad \begin{array}{c}\Pi_1\\ a_1{:}A_1\end{array}\quad \cdots \quad \begin{array}{c}\Pi_{u-1}\\ a_{u-1}{:}A_{u-1}\end{array}\quad \begin{array}{c}\Pi_u\\ R^u\, a\, a_1 \ldots a_u\end{array}}{a_u{:}A_u}\ \mathcal{M}^u\mathrm{E}\ ,$$

where Π_0 is the derivation $\Gamma_0, \Delta_0 \vdash_{\mathrm{N}(\mathcal{L})} a{:}\mathcal{M}^u A_1 \ldots A_u$; Π_i is the derivation $\Gamma_i, \Delta_i \vdash_{\mathrm{N}(\mathcal{L})} a_i{:}A_i$, for $1 \leq i \leq u-1$; Π_u is the derivation $\Delta_u \vdash_{\mathrm{N}(\mathcal{L})} R^u\, a\, a_1 \ldots a_u$; $\Gamma = \bigcup_{0 \leq i \leq u-1} \Gamma_i$ and $\Delta = \bigcup_{0 \leq i \leq u} \Delta_i$. Assume $\models^{\mathfrak{M}} (\Gamma, \Delta)$. Then, from the induction hypotheses we obtain $\models^{\mathfrak{M}} a{:}\mathcal{M}^u A_1 \ldots A_u$, $\models^{\mathfrak{M}} a_1{:}A_1, \ldots, \models^{\mathfrak{M}} a_{u-1}{:}A_{u-1}$, and $\models^{\mathfrak{M}} R^u\, a\, a_1 \ldots a_u$, and thus $\models^{\mathfrak{M}} a_u{:}A_u$ from the definition of $\models^{\mathfrak{M}}$.

Consider an application of the rule $\mathcal{M}^e$I,

$$\dfrac{\begin{array}{c}\Pi_1\\ a_1{:}A_1\end{array}\quad \cdots \quad \begin{array}{c}\Pi_e\\ a_e{:}A_e\end{array}\quad \begin{array}{c}\Pi_{e+1}\\ R^e\, a\, a_1 \ldots a_e\end{array}}{a{:}\mathcal{M}^e A_1 \ldots A_e}\ \mathcal{M}^e\mathrm{I}\ ,$$

where Π_i is the derivation $\Gamma_i, \Delta_i \vdash_{\mathrm{N}(\mathcal{L})} a_i{:}A_i$, for $1 \leq i \leq e$; Π_{e+1} is the derivation $\Delta_{e+1} \vdash_{\mathrm{N}(\mathcal{L})} R^e\, a\, a_1 \ldots a_e$; $\Gamma = \bigcup_{1 \leq i \leq e} \Gamma_i$ and $\Delta = \bigcup_{1 \leq i \leq e+1} \Delta_i$. Assume $\models^{\mathfrak{M}} (\Gamma, \Delta)$. Then, from the induction hypotheses we obtain $\models^{\mathfrak{M}} a_1{:}A_1, \ldots, \models^{\mathfrak{M}} a_e{:}A_e$ and $\models^{\mathfrak{M}} R^e\, a\, a_1 \ldots a_e$, and thus $\models^{\mathfrak{M}} a{:}\mathcal{M}^e A_1 \ldots A_e$ from the definition of $\models^{\mathfrak{M}}$.

For $\mathcal{M}^e$E, let Π be the derivation

$$\dfrac{\begin{matrix}\Pi_1\\ a{:}\mathcal{M}^e A_1 \ldots A_e\end{matrix} \qquad \begin{matrix}[a_1{:}A_1]\cdots[a_e{:}A_e]\ [R^e\, a\, a_1 \ldots a_e]\\ \Pi_2\\ b{:}B\end{matrix}}{b{:}B}\ \mathcal{M}^e\mathrm{E} .$$

That is, Π is $\Gamma, \Delta \vdash_{\mathrm{N}(\mathcal{L})} b{:}B$, where, by the restriction on $\mathcal{M}^e$E, the labels $a_1, \ldots, a_e$ do not occur in (Γ, Δ) and are different from a and b. Moreover, Π_1 is the derivation $\Gamma, \Delta \vdash_{\mathrm{N}(\mathcal{L})} a{:}\mathcal{M}^e A_1 \ldots A_e$, and Π_2 is the derivation $\Gamma \cup \{a_1{:}A_1, \ldots, a_e{:}A_e\}, \Delta \cup \{R^e\, a\, a_1 \ldots a_e\} \vdash_{\mathrm{N}(\mathcal{L})} b{:}B$. By the induction hypothesis for Π_1, we have that $\Gamma, \Delta \models^{\mathfrak{M}} a{:}\mathcal{M}^e A_1 \ldots A_e$, and thus, from the definition of $\models^{\mathfrak{M}}$, there exist $b_1, \ldots, b_e$ such that $\models^{\mathfrak{M}} b_1{:}A_1, \ldots, \models^{\mathfrak{M}} b_e{:}A_e$ and $\models^{\mathfrak{M}} R^e\, a\, b_1 \ldots b_e$. We can then extend Γ and Δ to $\Gamma' = \Gamma \cup \{a'_1{:}A_1, \ldots, a'_e{:}A_e\}$ and $\Delta' = \Delta \cup \{R^e\, a\, a'_1 \ldots a'_e\}$ for arbitrary $a'_1, \ldots, a'_e \notin (\Gamma, \Delta)$, and from the induction hypothesis for Π_2 we conclude $\Gamma, \Delta \models^{\mathfrak{M}} b{:}B$.

Consider an application of the rule $\neg$I,

$$\dfrac{\begin{matrix}[a^*{:}A]\\ \Pi_1\\ b{:}\bot\!\!\!\bot\end{matrix}}{a{:}\neg A}\ \neg\mathrm{I} ,$$

where Π_1 is the derivation $\Gamma_1, \Delta \vdash_{\mathrm{N}(\mathcal{L})} b{:}\bot\!\!\!\bot$, with $\Gamma_1 = \Gamma \cup \{a^*{:}A\}$. The induction hypothesis is $\Gamma_1, \Delta \vdash_{\mathrm{N}(\mathcal{L})} b{:}\bot\!\!\!\bot$ implies $\Gamma_1, \Delta \models b{:}\bot\!\!\!\bot$. We assume $\models^{\mathfrak{M}} (\Gamma, \Delta)$ and prove $\models^{\mathfrak{M}} a{:}\neg A$. Since $\not\models^{\mathfrak{M}} b{:}\bot\!\!\!\bot$, from the induction hypothesis we obtain $\not\models^{\mathfrak{M}} \Gamma_1$, and therefore $\not\models^{\mathfrak{M}} \{a^*{:}A\}$. We conclude $\models^{\mathfrak{M}} a{:}\neg A$ from the definition of $\models^{\mathfrak{M}}$.

Consider an application of the rule $\neg$E,

$$\dfrac{\begin{matrix}\Pi_1\\ a{:}\neg A\end{matrix} \quad \begin{matrix}\Pi_2\\ a^*{:}A\end{matrix}}{b{:}\bot\!\!\!\bot}\ \neg\mathrm{E} ,$$

where Π_1 and Π_2 are the derivations $\Gamma_1, \Delta_1 \vdash_{\mathrm{N}(\mathcal{L})} a{:}\neg A$ and $\Gamma_2, \Delta_2 \vdash_{\mathrm{N}(\mathcal{L})} a^*{:}A$, with $\Gamma = \Gamma_1 \cup \Gamma_2$ and $\Delta = \Delta_1 \cup \Delta_2$. The induction hypotheses are $\Gamma_1, \Delta_1 \vdash_{\mathrm{N}(\mathcal{L})} a{:}\neg A$ implies $\Gamma_1, \Delta_1 \models a{:}\neg A$, and $\Gamma_2, \Delta_2 \vdash_{\mathrm{N}(\mathcal{L})} a^*{:}A$ implies $\Gamma_2, \Delta_2 \models a^*{:}A$. Assume $\models^{\mathfrak{M}} (\Gamma, \Delta)$. Then, from the induction hypotheses we obtain $\models^{\mathfrak{M}} a{:}\neg A$ and $\models^{\mathfrak{M}} a^*{:}A$, and thus $\models^{\mathfrak{M}} b{:}\bot\!\!\!\bot$ from the definition of $\models^{\mathfrak{M}}$.

Consider an application of the rule $\bot\!\!\!\bot$Ei,

$$\dfrac{\begin{matrix}\Pi_1\\ b{:}\bot\!\!\!\bot\end{matrix}}{a{:}A}\ \bot\!\!\!\bot\mathrm{Ei} ,$$

where Π_1 is the derivation $\Gamma, \Delta \vdash_{\mathrm{N}(\mathcal{L})} b{:}\bot\!\!\!\bot$. The induction hypothesis is $\Gamma, \Delta \vdash_{\mathrm{N}(\mathcal{L})} b{:}\bot\!\!\!\bot$ implies $\Gamma, \Delta \models b{:}\bot\!\!\!\bot$. Since $\not\models^{\mathfrak{M}} b{:}\bot\!\!\!\bot$, from the induction hypothesis we obtain $\not\models^{\mathfrak{M}} (\Gamma, \Delta)$, and thus we conclude that $\Gamma, \Delta \models a{:}A$.

For $\bot\!\!\!\bot$Ec, consider

$$\dfrac{\begin{matrix}[a{:}\neg A]\\ \Pi_1\\ b{:}\bot\!\!\!\bot\end{matrix}}{a^*{:}A}\ \bot\!\!\!\bot\mathrm{Ec}$$

where Π_1 is the derivation $\Gamma_1, \Delta \vdash_{\mathrm{N}(\mathcal{L})} b{:}\perp\!\!\!\perp$, with $\Gamma_1 = \Gamma \cup \{a{:}\neg A\}$. The induction hypothesis is $\Gamma_1, \Delta \vdash_{\mathrm{N}(\mathcal{L})} b{:}\perp\!\!\!\perp$ implies $\Gamma_1, \Delta \models^{\mathfrak{M}} b{:}\perp\!\!\!\perp$. We assume $\models^{\mathfrak{M}} (\Gamma, \Delta)$ and prove $\models^{\mathfrak{M}} a^*{:}A$. Since $\not\models^{\mathfrak{M}} b{:}\perp\!\!\!\perp$, from the induction hypothesis we obtain $\not\models^{\mathfrak{M}} \Gamma_1$, and therefore $\not\models^{\mathfrak{M}} \{a{:}\neg A\}$. We conclude $\models^{\mathfrak{M}} a^*{:}A$ from the definition of $\models^{\mathfrak{M}}$. ■

3.2.2 Completeness

Completeness follows by a Henkin-style proof, where a canonical model

$$\mathfrak{M}^C = (\mathfrak{W}^C, o^C, \mathfrak{R}^{uC}, \mathfrak{R}^{eC}, *^C, \mathfrak{V}^C)$$

is built to show the contrapositives of the conditions in Definition 3.2.1, i.e.

$$\Delta \nvdash_{\mathrm{N}(\mathcal{L})} R_i\, a\, a_1 \ldots a_n \text{ implies } \Delta \not\models^{\mathfrak{M}^C} R_i\, a\, a_1 \ldots a_n\,,$$

and

$$\Gamma, \Delta \nvdash_{\mathrm{N}(\mathcal{L})} a{:}A \text{ implies } \Gamma, \Delta \not\models^{\mathfrak{M}^C} a{:}A\,.$$

In standard completeness proofs for 'unlabelled' non-classical logics (with respect to a Kripke-style semantics), a counter-model for underivable formulas is built by defining a notion of maximality for sets of formulas, and then using an extension result (such as the Lindenbaum Lemma, the Zorn Lemma or the Belnap Extension Lemma, e.g. [77]) to show that every set of formulas is contained in some maximal set; the canonical model is then obtained by repeated applications of the extension lemma. There are several possible definitions of maximality that can be considered, depending on the logic. For instance, maximality can be defined in terms of consistency (as is usually done, and as we did in §2.2.2, for propositional modal logics), in terms of notions weaker than consistency for paraconsistent logics such as relevance logics, or we can simply build the canonical model by extending disjoint 'theory – counter-theory pairs' [2, 77, 80].

The latter approach is more general than the other ones as it does not rely on negation and thus applies also to positive fragments. We take here a similar approach, but instead of introducing counter-theories, we start by defining what it means for a proof context (Γ, Δ) to be maximal with respect to an underivable lwff $a{:}A$. Then, given the presence of labelled formulas and explicit assumptions on the relations between the labels, i.e. the rwffs in Δ, we modify the Lindenbaum Lemma (see Lemma 3.2.6 below) to extend (Γ, Δ) to a single proof context $(\Gamma^\bullet, \Delta^\bullet)$ maximal with respect to $a{:}A$, where maximality is 'globally' checked also against the additional assumptions in Δ. The elements of $\mathfrak{W}^C$ are then built by partitioning $\Gamma^\bullet$ and $\Delta^\bullet$ with respect to the labels, and the relations are defined by exploiting the information in $\Delta^\bullet$. Therefore only one application of the extension lemma is needed, in that we simultaneously build all elements of $\mathfrak{W}^C$. Moreover, and most importantly, our proof is independent of the details of the logic $\mathcal{L}$, since the same procedure applies to any fragment of any logic.

Definition 3.2.4 *For any system* $\mathrm{N}(\mathcal{L}) = \mathrm{N}(\mathcal{B}) + \mathrm{N}(\mathcal{T})$, *let* $\Delta_{\mathrm{N}(\mathcal{L})}$ *be the* deductive closure of Δ under $\mathrm{N}(\mathcal{L})$, *i.e.*

$$\Delta_{\mathrm{N}(\mathcal{L})} =_{def} \{R_i\, a\, a_1 \ldots a_n \mid \Delta \vdash_{\mathrm{N}(\mathcal{L})} R_i\, a\, a_1 \ldots a_n\}\,.$$

■

Like for propositional modal logics, note that, by Fact 3.1.10,

- $\Delta_{\mathrm{N}(\mathcal{L})} = \{R_i\, a\, a_1 \ldots a_n \mid \Delta \vdash_{\mathrm{N}(\mathcal{T})} R_i\, a\, a_1 \ldots a_n\}$,
- $\Gamma, \Delta \vdash_{\mathrm{N}(\mathcal{L})} \varphi$ iff $\Gamma, \Delta_{\mathrm{N}(\mathcal{L})} \vdash_{\mathrm{N}(\mathcal{L})} \varphi$, and
- $\Delta_{\mathrm{N}(\mathcal{L})}$ might be empty when Δ is empty.

Definition 3.2.5 *A proof context* (Γ, Δ) *is* maximal with respect to $a{:}A$ *iff*

(i) $\Delta = \Delta_{\mathrm{N}(\mathcal{L})}$, *and*
(ii) $b{:}B \notin (\Gamma, \Delta)$ *iff* $\Gamma \cup \{b{:}B\}, \Delta \vdash_{\mathrm{N}(\mathcal{L})} a{:}A$. ■

Note that, when (Γ, Δ) is maximal with respect to $a{:}A$, both $a{:}A \notin (\Gamma, \Delta)$ and $\Gamma, \Delta \nvdash_{\mathrm{N}(\mathcal{L})} a{:}A$. Moreover, $b{:}\perp\!\!\!\perp \notin (\Gamma, \Delta)$ and $\Gamma, \Delta \nvdash_{\mathrm{N}(\mathcal{L})} b{:}\perp\!\!\!\perp$ for any b, for otherwise $\Gamma, \Delta \vdash_{\mathrm{N}(\mathcal{L})} a{:}A$. Also note that $\Gamma, \Delta \vdash_{\mathrm{N}(\mathcal{L})} b{:}B$ iff $\Gamma, \Delta_{\mathrm{N}(\mathcal{L})} \vdash_{\mathrm{N}(\mathcal{L})} b{:}B$, and that $\Delta \vdash_{\mathrm{N}(\mathcal{L})} R_i\, a\, a_1 \ldots a_n$ iff $\Delta_{\mathrm{N}(\mathcal{L})} \vdash_{\mathrm{N}(\mathcal{L})} R_i\, a\, a_1 \ldots a_n$.

In the Lindenbaum lemma for first-order logic, a maximally consistent set of formulas is inductively built by adding for every formula $\exists x(P)$ a *witness* to its truth, namely a formula $P[c/x]$ for some new constant c. A similar procedure applies here in the case of existential non-local operators: if the addition of $w{:}\mathcal{M}^e A_1 \ldots A_e$ does not yield a derivation of $a{:}A$, then we also add $t_1{:}A_1, \ldots, t_e{:}A_e$ and $R^e\, w\, t_1 \ldots t_e$, for some new $t_1, \ldots, t_e$, which act as *witness worlds* to the truth of $w{:}\mathcal{M}^e A_1 \ldots A_e$. This ensures that the proof context $(\Gamma^\bullet, \Delta^\bullet)$ is maximal with respect to $a{:}A$, as shown in Lemma 3.2.6 below. As a comparison, recall from §2.2.2 that in the standard completeness proof for unlabelled propositional modal logics one shows instead that if $w \in \mathfrak{W}^C$ and $\models^{\mathfrak{M}^C} w{:}\Diamond A$, then, by the extension lemma, $\mathfrak{W}^C$ also contains a world w' accessible from w that serves as a witness to the truth of $w{:}\Diamond A$, i.e. $\models^{\mathfrak{M}^C} w'{:}A$.

Lemma 3.2.6 *If* $\Gamma, \Delta \nvdash_{\mathrm{N}(\mathcal{L})} a{:}A$, *then* (Γ, Δ) *can be extended to a proof context* $(\Gamma^\bullet, \Delta^\bullet)$ *that is maximal with respect to* $a{:}A$.

Proof We first extend the language of $\mathrm{N}(\mathcal{L})$ with infinitely many new constants for witness worlds. Systematically let t range over the new constants for witness worlds, and w range over labels (including a and 0) and over the new constants; t and w may be subscripted. Let $l_1, l_2, \ldots$ be an enumeration of all lwffs in the extended language. Starting from $(\Gamma_0, \Delta_0) = (\Gamma, \Delta)$, we inductively build a sequence of proof contexts by defining $(\Gamma_{i+1}, \Delta_{i+1})$ as follows:

- if $\Gamma_i \cup \{l_{i+1}\}, \Delta_i \vdash_{\mathrm{N}(\mathcal{L})} a{:}A$, then $(\Gamma_{i+1}, \Delta_{i+1}) = (\Gamma_i, \Delta_i)$
- if $\Gamma_i \cup \{l_{i+1}\}, \Delta_i \nvdash_{\mathrm{N}(\mathcal{L})} a{:}A$, then

- if l_{i+1} is $w{:}\,\mathcal{M}^e A_1 \ldots A_e$, then we add witnesses to its truth, i.e. for $t_1, \ldots, t_e \notin (\Gamma_i \cup \{w{:}\mathcal{M}^e A_1 \ldots A_e\}, \Delta_i)$,

$$\begin{array}{rcl} \Gamma_{i+1} & = & \Gamma_i \cup \{w{:}\mathcal{M}^e A_1 \ldots A_e, t_1{:}A_1, \ldots, t_e{:}A_e\} \\ \Delta_{i+1} & = & \Delta_i \cup \{R^e\, w\, t_1 \ldots t_e\} \end{array}$$

- if l_{i+1} is not $w{:}\mathcal{M}^e A_1 \ldots A_e$, then $(\Gamma_{i+1}, \Delta_{i+1}) = (\Gamma_i \cup \{l_{i+1}\}, \Delta_i)$

Every (Γ_i, Δ_i) is such that $\Gamma_i, \Delta_i \nvdash_{\mathrm{N}(\mathcal{L})} a{:}A$. To show this we show that

$$\text{if } \Gamma_i, \Delta_i \nvdash_{\mathrm{N}(\mathcal{L})} a{:}A \text{ then } \Gamma_{i+1}, \Delta_{i+1} \nvdash_{\mathrm{N}(\mathcal{L})} a{:}A\,.$$

The only non-trivial case is the addition of witnesses to the truth of $w{:}\,\mathcal{M}^e A_1 \ldots\ A_e$. Suppose that

$$\Gamma_i \cup \{w{:}\mathcal{M}^e A_1 \ldots A_e, t_1{:}A_1, \ldots, t_e{:}A_e\}, \Delta_i \cup \{R^e\, w\, t_1 \ldots t_e\} \vdash_{\mathrm{N}(\mathcal{L})} a{:}A$$

where $t_1, \ldots, t_e \notin (\Gamma_i \cup \{w{:}\mathcal{M}^e A_1 \ldots A_e\}, \Delta_i)$. Then we can apply $\mathcal{M}^e$E, and thus

$$\Gamma_i \cup \{w{:}\mathcal{M}^e A_1 \ldots A_e\}, \Delta_i \vdash_{\mathrm{N}(\mathcal{L})} a{:}A\,.$$

Contradiction.

Now define

$$\Gamma^\bullet = \bigcup_{i \geq 0} \Gamma_i \quad \text{and} \quad \Delta^\bullet = \bigcup_{i \geq 0} (\Delta_i)_{\mathrm{N}(\mathcal{L})}\,.$$

Then, $(\Gamma, \Delta) \in (\Gamma^\bullet, \Delta^\bullet)$ and $a{:}A \notin (\Gamma^\bullet, \Delta^\bullet)$. Moreover, $(\Gamma^\bullet, \Delta^\bullet)$ is maximal with respect to $a{:}A$. Condition (i) in Definition 3.2.5 is satisfied by definition of $\Delta^\bullet$, and we show that condition (ii) holds as well. $\Gamma^\bullet \cup \{b{:}B\}, \Delta^\bullet \nvdash_{\mathrm{N}(\mathcal{L})} a{:}A$ implies $b{:}B \in \Gamma^\bullet$ by construction. For the converse, assume that $b{:}B \in \Gamma^\bullet$. If $\Gamma^\bullet \cup \{b{:}B\}, \Delta^\bullet \vdash_{\mathrm{N}(\mathcal{L})} a{:}A$, then, since $\Gamma^\bullet, \Delta^\bullet \vdash_{\mathrm{N}(\mathcal{L})} b{:}B$, by transitivity of derivations we have that $\Gamma^\bullet, \Delta^\bullet \vdash_{\mathrm{N}(\mathcal{L})} a{:}A$. Contradiction. ■

When $\Delta \nvdash_{\mathrm{N}(\mathcal{L})} R_i\, a\, a_1 \ldots a_n$, then we simply extend Δ to $\Delta^\bullet = \Delta_{\mathrm{N}(\mathcal{L})}$, so that $R_i\, a\, a_1 \ldots a_n \notin \Delta^\bullet$, since by definition of the deductive closure $\Delta^\bullet \vdash_{\mathrm{N}(\mathcal{L})} R_i\, w\, w_1 \ldots w_n$ iff $R_i\, w\, w_1 \ldots w_n \in \Delta^\bullet$.

The following lemma states some properties of a proof context $(\Gamma^\bullet, \Delta^\bullet)$ maximal with respect to $a{:}A$.

Lemma 3.2.7 *Let $(\Gamma^\bullet, \Delta^\bullet)$ be maximal with respect to $a{:}A$. Then we have:*

(i) $\Delta^\bullet \vdash_{\mathrm{N}(\mathcal{L})} R_i\, a\, a_1 \ldots a_n$ *iff* $R_i\, a\, a_1 \ldots a_n \in \Delta^\bullet$.

(ii) $\Gamma^\bullet, \Delta^\bullet \vdash_{\mathrm{N}(\mathcal{L})} w{:}B$ *iff* $w{:}B \in \Gamma^\bullet$.

(iii) $w{:}\mathcal{M}^u A_1 \ldots A_u \in \Gamma^\bullet$ *iff* $R^u\, w\, w_1 \ldots w_u \in \Delta^\bullet$ *and* $w_1{:}A_1 \in \Gamma^\bullet$ *and ... and* $w_{u-1}{:}A_{u-1} \in \Gamma^\bullet$ *imply* $w_u{:}A_u \in \Gamma^\bullet$*, for all* $w_1, \ldots, w_u$.

(iv) $w{:}\mathcal{M}^e A_1 \ldots A_e \in \Gamma^\bullet$ *iff* $R^e\, w\, w_1 \ldots w_e \in \Delta^\bullet$ *and* $w_1{:}A_1 \in \Gamma^\bullet$ *and ... and* $w_e{:}A_e \in \Gamma^\bullet$*, for some* $w_1, \ldots, w_e$.

(v) $w{:}\neg A \in \Gamma^\bullet$ *iff* $w^*{:}A \notin \Gamma^\bullet$.

(vi) $w{:}A_1 \wedge A_2 \in \Gamma^\bullet$ *iff* $w{:}A_1 \in \Gamma^\bullet$ *and* $w{:}A_2 \in \Gamma^\bullet$.

(vii) $w{:}A_1 \vee A_2 \in \Gamma^\bullet$ *iff* $w{:}A_1 \in \Gamma^\bullet$ *or* $w{:}A_2 \in \Gamma^\bullet$.

(viii) $w{:}A_1 \supset A_2 \in \Gamma^\bullet$ *iff* $w{:}A_1 \in \Gamma^\bullet$ *implies* $w{:}A_2 \in \Gamma^\bullet$.

Proof The proof of (i) is straightforward.

(ii) Suppose that $\Gamma^\bullet, \Delta^\bullet \vdash_{\mathrm{N}(\mathcal{L})} w{:}B$. If $w{:}B \notin \Gamma^\bullet$, then, since $(\Gamma^\bullet, \Delta^\bullet)$ is maximal with respect to $a{:}A$, $\Gamma^\bullet \cup \{w{:}B\}, \Delta^\bullet \vdash_{\mathrm{N}(\mathcal{L})} a{:}A$, and thus, by transitivity of derivations, $\Gamma^\bullet, \Delta^\bullet \vdash_{\mathrm{N}(\mathcal{L})} a{:}A$. Contradiction. The converse holds by definition.

(iii) Suppose that $w{:}\mathcal{M}^u A_1 \ldots A_u \in \Gamma^\bullet$. Then $\Gamma^\bullet, \Delta^\bullet \vdash_{\mathrm{N}(\mathcal{L})} w{:}\mathcal{M}^u A_1 \ldots A_u$ by (ii). Now if $R^u\, w\, w_1 \ldots w_u \in \Delta^\bullet$ and $w_1{:}A_1 \in \Gamma^\bullet$ and ... and $w_{u-1}{:}A_{u-1} \in \Gamma^\bullet$, we conclude $w_u{:}A_u \in \Gamma^\bullet$ by (i), (ii) and $\mathcal{M}^u$E. For the converse, assume that $w{:}\mathcal{M}^u A_1 \ldots A_u \notin \Gamma^\bullet$, and prove that there exist $w_1, \ldots, w_u$ such that $R^u\, w\, w_1 \ldots w_u \in \Delta^\bullet$ and $w_1{:}A_1 \in \Gamma^\bullet$ and ... and $w_{u-1}{:}A_{u-1} \in \Gamma^\bullet$ and $w_u{:}A_u \notin \Gamma^\bullet$. By (i) and (ii), the assumption yields

$$\Gamma^\bullet \cup \{w{:}\mathcal{M}^u A_1 \ldots A_u\}, \Delta^\bullet \vdash_{\mathrm{N}(\mathcal{L})} a{:}A\,.$$

Now if for all $w_1, \ldots, w_u$,

$$\Gamma^\bullet \cup \{w_1{:}A_1, \ldots, w_{u-1}{:}A_{u-1}\}, \Delta^\bullet \cup \{R^u\, w\, w_1 \ldots w_u\} \vdash_{\mathrm{N}(\mathcal{L})} w_u{:}A_u\,,$$

then, by $\mathcal{M}^u$I, we have $\Gamma^\bullet, \Delta^\bullet \vdash_{\mathrm{N}(\mathcal{L})} w{:}\mathcal{M}^u A_1 \ldots A_u$, and thus $\Gamma^\bullet, \Delta^\bullet \vdash_{\mathrm{N}(\mathcal{L})} a{:}A$ by transitivity of derivations. Contradiction.

(iv) Suppose that $R^e\, w\, w_1 \ldots w_e \in \Delta^\bullet$ and $w_1{:}A_1 \in \Gamma^\bullet$ and ... and $w_{e-1}{:}A_{e-1} \in \Gamma^\bullet$ imply $w_e{:}A_e \notin \Gamma^\bullet$, for all $w_1, \ldots, w_e$. Then, by (i) and (ii), we have

$$\Gamma^\bullet \cup \{w_1{:}A_1, \ldots, w_{e-1}{:}A_{e-1}\} \cup \{w_e{:}A_e\}, \Delta^\bullet \cup \{R^e\, w\, w_1 \ldots w_e\} \vdash_{\mathrm{N}(\mathcal{L})} a{:}A$$

for all $w_1, \ldots, w_e$. Now, if $w{:}\mathcal{M}^e A_1 \ldots A_e \in \Gamma^\bullet$, then, by (ii), $\Gamma^\bullet, \Delta^\bullet \vdash_{\mathrm{N}(\mathcal{L})} w{:}\mathcal{M}^e A_1 \ldots A_e$, and thus $\Gamma^\bullet, \Delta^\bullet \vdash_{\mathrm{N}(\mathcal{L})} a{:}A$ by $\mathcal{M}^e$E. Contradiction. For the converse suppose that $w{:}\mathcal{M}^e A_1 \ldots A_e \notin \Gamma^\bullet$. Then

$$\Gamma^\bullet \cup \{w{:}\mathcal{M}^e A_1 \ldots A_e\}, \Delta^\bullet \vdash_{\mathrm{N}(\mathcal{L})} a{:}A\,.$$

If for some $w_1, \ldots, w_e$, $R^e\, w\, w_1 \ldots w_e \in \Delta^\bullet$ and $w_1{:}A_1 \in \Gamma^\bullet$ and ... and $w_e{:}A_e \in \Gamma^\bullet$, then, by (i), (ii) and $\mathcal{M}^e$I, we have $\Gamma^\bullet, \Delta^\bullet \vdash_{\mathrm{N}(\mathcal{L})} w{:}\mathcal{M}^e A_1 \ldots A_e$, and thus $\Gamma^\bullet, \Delta^\bullet \vdash_{\mathrm{N}(\mathcal{L})} a{:}A$, by transitivity of derivations. Contradiction.

(v) Suppose that $w{:}\neg A_1 \in \Gamma^\bullet$. If also $w^*{:}A_1 \in \Gamma^\bullet$, then, by (ii) and $\neg$E, we have $\Gamma^\bullet, \Delta^\bullet \vdash_{\mathrm{N}(\mathcal{L})} b{:}\perp\!\!\!\perp$, and thus $\Gamma^\bullet, \Delta^\bullet \vdash_{\mathrm{N}(\mathcal{L})} a{:}A$. Contradiction. For the converse suppose that $w{:}\neg A_1 \notin \Gamma^\bullet$. Then

$$\Gamma^\bullet \cup \{w{:}\neg A_1\}, \Delta^\bullet \vdash_{\mathrm{N}(\mathcal{L})} a{:}A\,.$$

If $w^*{:}A_1 \notin \Gamma^\bullet$, then $\Gamma^\bullet \cup \{w^*{:}A_1\}, \Delta^\bullet \vdash_{\mathrm{N}(\mathcal{L})} a{:}A$, and thus $\Gamma^\bullet, \Delta^\bullet \vdash_{\mathrm{N}(\mathcal{L})} a{:}A$. Contradiction.

The proofs of (vi) and (vii) are straightforward, and we treat only (vii) as an example. Suppose that $w{:}A_1 \vee A_2 \in \Gamma^\bullet$. If $w{:}A_1 \notin \Gamma^\bullet$ and $w{:}A_2 \notin \Gamma^\bullet$, then

$$\Gamma^\bullet \cup \{w{:}A_1\}, \Delta^\bullet \vdash_{\mathrm{N}(\mathcal{L})} a{:}A \qquad \text{and} \qquad \Gamma^\bullet \cup \{w{:}A_2\}, \Delta^\bullet \vdash_{\mathrm{N}(\mathcal{L})} a{:}A\,,$$

and thus $\Gamma^\bullet, \Delta^\bullet \vdash_{\mathrm{N}(\mathcal{L})} a{:}A$ by (ii) and $\vee$E. Contradiction. For the converse suppose that $w{:}A_i \in \Gamma^\bullet$ for $i = 1$ or $i = 2$. If $w{:}A_1 \vee A_2 \notin \Gamma^\bullet$, then

$$\Gamma^\bullet \cup \{w{:}A_1 \vee A_2\}, \Delta^\bullet \vdash_{\mathrm{N}(\mathcal{L})} a{:}A\,.$$

By (ii) and $\vee$Ii for $i = 1$ or $i = 2$, the assumption yields $\Gamma^\bullet, \Delta^\bullet \vdash_{\mathrm{N}(\mathcal{L})} w{:}A_1 \vee A_2$, and thus $\Gamma^\bullet, \Delta^\bullet \vdash_{\mathrm{N}(\mathcal{L})} a{:}A$, by transitivity of derivations. Contradiction.

(viii) Suppose that $w{:}A_1 \supset A_2 \in \Gamma^\bullet$ and $w{:}A_1 \in \Gamma^\bullet$. If $w{:}A_2 \notin \Gamma^\bullet$, then

$$\Gamma^\bullet \cup \{w{:}A_2\}, \Delta^\bullet \vdash_{\mathrm{N}(\mathcal{L})} a{:}A\,.$$

By (ii) and $\supset$E, the assumptions yield $\Gamma^\bullet, \Delta^\bullet \vdash_{\mathrm{N}(\mathcal{L})} w{:}A_2$, and thus $\Gamma^\bullet, \Delta^\bullet \vdash_{\mathrm{N}(\mathcal{L})} a{:}A$ by transitivity of derivations. Contradiction. For the converse suppose that $w{:}A_1 \in \Gamma^\bullet$ implies $w{:}A_2 \in \Gamma^\bullet$. If $w{:}A_1 \supset A_2 \notin \Gamma^\bullet$, then

$$\Gamma^\bullet \cup \{w{:}A_1 \supset A_2\}, \Delta^\bullet \vdash_{\mathrm{N}(\mathcal{L})} a{:}A\,.$$

By (ii) and $\supset$I, the assumptions yield $\Gamma^\bullet, \Delta^\bullet \vdash_{\mathrm{N}(\mathcal{L})} w{:}A_1 \supset A_2$, and thus $\Gamma^\bullet, \Delta^\bullet \vdash_{\mathrm{N}(\mathcal{L})} a{:}A$ by transitivity of derivations. Contradiction. ■

We can now define the canonical model $\mathfrak{M}^C = (\mathfrak{W}^C, \mathrm{o}^C, \mathfrak{R}^{u^C}, \mathfrak{R}^{e^C}, *^C, \mathfrak{V}^C)$.

Definition 3.2.8 *Given a proof context* $(\Gamma^\bullet, \Delta^\bullet)$ *maximal with respect to* $a{:}A$*, we define the* canonical model $\mathfrak{M}^C$ *for the system* $\mathrm{N}(\mathcal{L})$ *as follows:*

- $\mathfrak{W}^C = \{w \mid w \Subset (\Gamma^\bullet, \Delta^\bullet)\}$*, where* $\mathrm{o}^C = 0$ *and* $w^{*^C} = w^*$*;*
- $(w, w_1, \ldots, w_u) \in \mathfrak{R}^{u^C}$ *iff* $R^u\, w\, w_1 \ldots w_u \in \Delta^\bullet$*, and* $(w, w_1, \ldots, w_e) \in \mathfrak{R}^{e^C}$ *iff* $R^e\, w\, w_1 \ldots w_e \in \Delta^\bullet$*;*
- $\mathfrak{V}^C(w, p) = 1$ *iff* $w{:}p \in \Gamma^\bullet$*.* ■

The standard definition of $\mathfrak{R}^{u^C}$, i.e.

$$\begin{aligned}&(w, w_1, \ldots, w_u) \in \mathfrak{R}^{u^C} \text{ iff}\\ &\qquad \{A_u \mid \mathcal{M}^u A_1 \ldots A_u \in w,\ A_1 \in w_1, \ldots, A_{u-1} \in w_{u-1}\} \subseteq w_u\,,\end{aligned} \tag{3.25}$$

is not applicable in our setting, since (3.25) does *not* imply $\vdash_{\mathrm{N}(\mathcal{L})} R^u\, w\, w_1 \ldots w_u$. We would therefore be unable to prove completeness for rwffs, since there would be cases where $\nvdash_{\mathrm{N}(\mathcal{L})} R^u\, w\, w_1 \ldots w_u$ but $(w, w_1, \ldots, w_u) \in \mathfrak{R}^{u^C}$ and thus $\models^{\mathfrak{M}^C} R^u\, w\, w_1 \ldots w_u$. Hence, we instead define $(w, w_1, \ldots, w_u) \in \mathfrak{R}^{u^C}$ iff $R^u\, w\, w_1 \ldots w_u \in \Delta^\bullet$; note that therefore $R^u\, w\, w_1 \ldots w_u \in \Delta^\bullet$ implies (3.25). An analogous observation holds for $\mathfrak{R}^{e^C}$. As a further comparison with the standard definition of the canonical model, note that the label w can be identified with the set of formulas $\{B \mid w{:}B \in \Gamma^\bullet\}$. Moreover, we immediately have:

Fact 3.2.9 $R_i\, w\, w_1 \ldots w_n \in \Delta^\bullet$ *iff* $\Delta^\bullet \models^{\mathfrak{M}^C} R_i\, w\, w_1 \ldots w_n$. ■

The deductive closure of $\Delta^\bullet$ ensures not only completeness for rwffs, as shown in Lemma 3.2.11 below, but also that the conditions on $\mathfrak{R}^{u^C}$ and $\mathfrak{R}^{e^C}$ are satisfied, so that $\mathfrak{M}^C$ is really a model for $\mathrm{N}(\mathcal{L})$. As an example, we show that if $\mathrm{N}(\mathcal{L})$ contains *assoc1* and *assoc2* for a ternary relation R^u, then $\mathfrak{R}^{u^C}$ is associative. Consider an arbitrary proof context (Γ, Δ), from which we build $\mathfrak{M}^C$. Assume $(a, b, x) \in \mathfrak{R}^{u^C}$ and $(x, c, d) \in \mathfrak{R}^{u^C}$. Then $R^u\, a\, b\, x \in \Delta^\bullet$ and $R^u\, x\, c\, d \in \Delta^\bullet$. But $\Delta^\bullet$ is deductively closed, and thus $R^u\, b\, c\, f(a,b,c,d,x) \in \Delta^\bullet$ and $R^u\, a\, f(a,b,c,d,x)\, d \in \Delta^\bullet$, by *assoc1* and *assoc2*. Hence, there exists a world x such that $(b, c, x) \in \mathfrak{R}^{u^C}$ and $(a, x, d) \in \mathfrak{R}^{u^C}$, and $\mathfrak{R}^{u^C}$ is indeed associative.

By Lemma 3.2.7 and Fact 3.2.9, it follows that:

Lemma 3.2.10 $w{:}B \in (\Gamma^\bullet, \Delta^\bullet)$ *iff* $\Gamma^\bullet, \Delta^\bullet \models^{\mathfrak{M}^C} w{:}B$.

Proof We proceed by induction on the grade of $w{:}B$, i.e. on the number of local and non-local operators that occur in B, and we treat only the step case where $w{:}B$ is $w{:}\mathcal{M}^u A_1 \ldots A_u$; the other cases follow analogously.

For the left-to-right direction, assume that $w{:}\mathcal{M}^u A_1 \ldots A_u \in \Gamma^\bullet$. Then, by Lemma 3.2.7, $R^u\, w\, w_1 \ldots w_u \in \Delta^\bullet$ and $w_1{:}A_1 \in \Gamma^\bullet$ and ... and $w_{u-1}{:}A_{u-1} \in \Gamma^\bullet$ imply $w_u{:}A_u \in \Gamma^\bullet$, for all $w_1, \ldots, w_u$. Fact 3.2.9 and the induction hypotheses yield $\Gamma^\bullet, \Delta^\bullet \models^{\mathfrak{M}^C} w_u{:}A_u$ for all $w_1, \ldots, w_u$ such that $\Delta^\bullet \models^{\mathfrak{M}^C} R^u\, w\, w_1 \ldots w_u$ and $\Gamma^\bullet, \Delta^\bullet \models^{\mathfrak{M}^C} w_1{:}A_1$ and ... and $\Gamma^\bullet, \Delta^\bullet \models^{\mathfrak{M}^C} w_{u-1}{:}A_{u-1}$, i.e. $\Gamma^\bullet, \Delta^\bullet \models^{\mathfrak{M}^C} w{:}\mathcal{M}^u A_1 \ldots A_u$ from the definition of truth.

For the right-to-left direction, assume that $w{:}\mathcal{M}^u A_1 \ldots A_u \notin \Gamma^\bullet$. Then, by Lemma 3.2.7, we have $R^u\, w\, w_1 \ldots w_u \in \Delta^\bullet$ and $w_1{:}A_1 \in \Gamma^\bullet$ and ... and $w_{u-1}{:}A_{u-1} \in \Gamma^\bullet$ and $w_u{:}A_u \notin \Gamma^\bullet$, for some $w_1, \ldots, w_u$. Fact 3.2.9 and the induction hypotheses yield $\Delta^\bullet \models^{\mathfrak{M}^C} R^u\, w\, w_1 \ldots w_u$ and $\Gamma^\bullet, \Delta^\bullet \models^{\mathfrak{M}^C} w_1{:}A_1$ and ... and $\Gamma^\bullet, \Delta^\bullet \models^{\mathfrak{M}^C} w_{u-1}{:}A_{u-1}$ and $\Gamma^\bullet, \Delta^\bullet \not\models^{\mathfrak{M}^C} w_u{:}A_u$, i.e. $\Gamma^\bullet, \Delta^\bullet \not\models^{\mathfrak{M}^C} w{:}\mathcal{M}^u A_1 \ldots A_u$ from the definition of truth. ■

We can now finally show that:

Lemma 3.2.11 $\mathrm{N}(\mathcal{L}) = \mathrm{N}(\mathcal{B}) + \mathrm{N}(\mathcal{T})$ *is complete.*

Proof (i) If $\Delta \nvdash_{\mathrm{N}(\mathcal{L})} R_i\, a\, a_1 \ldots a_n$, then $R_i\, a\, a_1 \ldots a_n \notin \Delta^\bullet$, and thus $\Delta^\bullet \not\models^{\mathfrak{M}^C} R_i\, a\, a_1 \ldots a_n$, by Fact 3.2.9. Hence, $\Delta \not\models^{\mathfrak{M}^C} R_i\, a\, a_1 \ldots a_n$.

(ii) If $\Gamma, \Delta \nvdash_{\mathrm{N}(\mathcal{L})} a{:}A$, then we extend (Γ, Δ) to a proof context $(\Gamma^\bullet, \Delta^\bullet)$ maximal with respect to $a{:}A$. Then, by Lemma 3.2.10, $\Gamma^\bullet, \Delta^\bullet \not\models^{\mathfrak{M}^C} a{:}A$, and thus $\Gamma, \Delta \not\models^{\mathfrak{M}^C} a{:}A$. ■

3.2.3 Positive fragments and interrelated relations

In §3.1 we argued that an unrestricted *monI* rule produces an unsound system in which intuitionistic and classical implication are equivalent, and that soundness is restored when applications of *monI* are restricted to persistent formulas. We show now that the soundness and completeness of our systems (Theorem 3.2.2) depends on another restriction we imposed in §3.1, namely that there are no a priori assumptions on the interrelationships of the different relations associated with universal and existential operators. If this restriction is withdrawn and the relations are interrelated, e.g. $\mathfrak{R}^u \subseteq \mathfrak{R}^e$, then incompleteness may arise.

To illustrate this, we consider positive fragments of (classical) modal logics. Without negation we cannot define $\diamond$ in terms of $\Box$ and derive the rules for $\diamond$. Indeed, there is no a priori reason why $\Box$ and $\diamond$ must be related at all. Therefore, we characterize the positive fragments containing both $\Box$ and $\diamond$ by the interrelationships between $R^\Box$ and $R^\diamond$, which are specified by a (possibly empty) collection of the Horn relational rules

$$\frac{x\,R^\diamond\,y}{x\,R^\Box\,y}\;(\diamond\Box) \qquad \text{and} \qquad \frac{x\,R^\Box\,y}{x\,R^\diamond\,y}\;(\Box\diamond) \quad .$$

Using these rules, we can prove theorems that relate $\Box$ and $\diamond$. For instance, using $(\diamond\Box)$ we can prove

$$x{:}(\diamond A \wedge \Box B) \supset \diamond(A \wedge B)\,, \tag{3.26}$$

and using $(\Box\diamond)$ we can prove

$$x{:}(\diamond A \supset \Box B) \supset \Box(A \supset B)\,. \tag{3.27}$$

That these theorems are provable is not surprising: correspondence theory provides a means of showing that (3.26) corresponds to the semantic condition $\mathfrak{R}^\diamond \subseteq \mathfrak{R}^\Box$ and that (3.27) corresponds to $\mathfrak{R}^\Box \subseteq \mathfrak{R}^\diamond$.

Now consider

$$x{:}\Box(A \vee B) \supset (\diamond A \vee \Box B)\,, \tag{3.28}$$

which corresponds to $\mathfrak{R}^\Box \subseteq \mathfrak{R}^\diamond$, and therefore is true in the models satisfying this property. By analysis of normal form proofs, see §3.3, we can show that (3.28) is not provable using $(\Box\diamond)$.[8] Hence, positive modal systems where $R^\Box$ and $R^\diamond$ are not independent but are related by $(\Box\diamond)$ are incomplete with respect to Kripke models $(\mathfrak{W}, \mathfrak{R}^\Box, \mathfrak{R}^\diamond, \mathfrak{V})$ where $\mathfrak{R}^\Box \subseteq \mathfrak{R}^\diamond$. This illustrates that:

Theorem 3.2.12 *If the relations R_i's associated with the operators are not independent, then there are positive fragments of (our ND systems for) non-classical logics that are incomplete with respect to the corresponding Kripke semantics.* ■

A similar problem holds for Hilbert-style presentations, as pointed out by Dunn in [80]; he ensures the completeness of the 'absolutely' positive fragment of modal logic (i.e. without negation and implication) by extending his Hilbert-style system with postulates equivalent to (3.26) and (3.28). Similarly, we could restore completeness

[8] This is because the proof of (3.28) requires properties of classical negation. Thus, instead of 'strengthening' the deduction system, we could try to restore completeness by adopting a semantics with a 'weaker' negation.

in our setting by giving up our claim to a fixed base system extended with relational theories, and adding a rule directly encoding (3.28), e.g.

$$\frac{}{x{:}\Box(A \vee B) \supset (\Diamond A \vee \Box B)} \; .$$

However, such a rule is not in the (philosophical) spirit of natural deduction since it does not contribute to the theory of meaning of the operators. Moreover, it complicates proof normalization arguments.

3.3 NORMALIZATION AND ITS CONSEQUENCES

In this section we generalize the results of §2.3 to show that each derivation of an lwff in $\mathrm{N}(\mathcal{B}) + \mathrm{N}(\mathcal{T})$ can be reduced to a normal form that does not contain unnecessary detours and satisfies a subformula property.

There are two possible forms of detours in a derivation and we eliminate them by the reduction operations defined below. For brevity, we consider again the restricted language of Definition 3.1.7 (with the operators $\wedge$, $\vee$, $\supset$, $\mathcal{M}^u$, $\mathcal{M}^e$, $\neg$ and $\perp\!\!\!\perp$), and we only show the part of the derivation where the reduction actually takes place; the missing parts remain unchanged.

The first, and simplest, form of detour is the application of an elimination rule immediately below the application of the corresponding introduction rule. That is, as we observed for propositional modal logics, if an lwff is introduced and then immediately eliminated, then we can avoid introducing it in the first place. Formally, Definition 2.3.3 generalizes straightforwardly to $\mathrm{N}(\mathcal{B}) + \mathrm{N}(\mathcal{T})$ as follows:

Definition 3.3.1 *Any lwff $a{:}A$ in a derivation is the root of a tree of rule applications leading back to assumptions. The lwffs in this tree other than $a{:}A$ we call* side lwffs *of $a{:}A$. A* maximal lwff *in a derivation is an lwff that is both the conclusion of an introduction rule and the major premise of an elimination rule.* ∎

Maximal lwffs are removed from a derivation by (finitely many applications of) *proper reductions*. There is one proper reduction for each operator. The proper reductions for universal and existential non-local operators are as follows, where the substitutions are allowed by the side conditions on $\mathcal{M}^u$I and $\mathcal{M}^e$E.

Proper reduction for $\mathcal{M}^u$:

$$\dfrac{\dfrac{\begin{matrix}[a_1{:}A_1]\cdots[a_{u-1}{:}A_{u-1}]\;[R^u\,a\,a_1\ldots a_u]\\ \Pi \\ a_u{:}A_u\end{matrix}}{a{:}\mathcal{M}^u A_1\ldots A_u}\,\mathcal{M}^u\mathrm{I} \qquad \begin{matrix}\Pi_1\\ b_1{:}A_1\end{matrix}\cdots\begin{matrix}\Pi_{u-1}\\ b_{u-1}{:}A_{u-1}\end{matrix}\quad \begin{matrix}\Pi_R\\ R^u\,a\,b_1\ldots b_u\end{matrix}}{b_u{:}A_u}\,\mathcal{M}^u\mathrm{E}$$

$$\rightsquigarrow \quad \begin{matrix}\begin{matrix}\Pi_1\\ b_1{:}A_1\end{matrix}\;\cdots\;\begin{matrix}\Pi_{u-1}\\ b_{u-1}{:}A_{u-1}\end{matrix}\quad\begin{matrix}\Pi_R\\ R^u\,a\,b_1\ldots b_u\end{matrix}\\ \Pi[b_1/a_1,\ldots,b_u/a_u]\\ b_u{:}A_u\end{matrix} \; .$$

Proper reduction for $\mathcal{M}^e$:

$$\dfrac{\dfrac{\begin{matrix}\Pi_1 \\ a_1{:}A_1\end{matrix} \quad \cdots \quad \begin{matrix}\Pi_e \\ a_e{:}A_e\end{matrix} \quad \begin{matrix}\Pi_R \\ R^e\, a\, a_1 \ldots a_e\end{matrix}}{a{:}\mathcal{M}^e A_1 \ldots A_e}\,\mathcal{M}^e\mathrm{I} \qquad \begin{matrix}[b_1{:}A_1] \cdots [b_e{:}A_e]\ [R^e\, a\, b_1 \ldots b_e] \\ \Pi \\ c{:}C\end{matrix}}{c{:}C}\,\mathcal{M}^e\mathrm{E}$$

$$\rightsquigarrow \quad \begin{matrix}\Pi_1 & & \Pi_e & \Pi_R \\ a_1{:}A_1 & \cdots & a_e{:}A_e & R^e\, a\, a_1 \ldots a_e \\ & \Pi[a_1/b_1, \ldots, a_e/b_e] & & \\ & c{:}C & & \end{matrix} \quad .$$

The proper reduction for $\supset$ is the same as the reduction (2.5), while the proper reductions for negation and for local operators can be easily adapted from the standard 'unlabelled' reductions, e.g. for negation:

$$\dfrac{\dfrac{\begin{matrix}[a^*{:}A] \\ \Pi_1 \\ b{:}\perp\!\!\!\perp\end{matrix}}{a{:}\neg A}\,\neg\mathrm{I} \qquad \begin{matrix}\Pi_2 \\ a^*{:}A\end{matrix}}{b{:}\perp\!\!\!\perp}\,\neg\mathrm{E} \quad \rightsquigarrow \quad \begin{matrix}\Pi_2 \\ a^*{:}A \\ \Pi_1 \\ b{:}\perp\!\!\!\perp\end{matrix} \quad .$$

Let us call *indirect rules* the rules $\mathcal{M}^e$E, $\vee$E, $\perp\!\!\!\perp$Ei and *monl*. The second form of detour arises when the conclusion of an indirect rule is the major premise of an elimination rule. Consider the different cases. At applications of $\mathcal{M}^e$E, occurrences of the same lwff appear immediately below each other, and this can constitute a detour in which lwffs that potentially interact in a proper reduction are too far apart. The same problem holds for applications of $\vee$E, and a similar one for applications of *monl*. Furthermore, when the conclusion of $\perp\!\!\!\perp$Ei is the major premise of an elimination, then we can easily show that the elimination is an unnecessary inference.

To remove this second form of detour we permute the order of application of indirect and elimination rules. Formally we define:

Definition 3.3.2 *A* permutable lwff *in a derivation is an lwff that is both the conclusion of an indirect rule and the major premise of an elimination rule.* ■

Permutable lwffs are removed from a derivation by (finitely many applications of) *permutative reductions.* The difference with respect to Prawitz [186, 187] is twofold. First, we explicitly define $\perp\!\!\!\perp$Ei to be an indirect rule, since, unlike Prawitz's $\perp$ elimination rule for intuitionistic logic (and unlike our rule $\perp$E for modal logics), we cannot restrict $\perp\!\!\!\perp$Ei to applications where the conclusion is an atomic lwff. For instance, to replace

$$\dfrac{\begin{matrix}\Pi \\ b{:}\perp\!\!\!\perp\end{matrix}}{a{:}\mathcal{M}^e A_1 \ldots A_e}\,\perp\!\!\!\perp\mathrm{Ei} \qquad \text{with} \qquad \dfrac{\dfrac{\begin{matrix}\Pi \\ b{:}\perp\!\!\!\perp\end{matrix}}{a_1{:}A_1}\,\perp\!\!\!\perp\mathrm{Ei} \ \cdots \ \dfrac{\begin{matrix}\Pi \\ b{:}\perp\!\!\!\perp\end{matrix}}{a_e{:}A_e}\,\perp\!\!\!\perp\mathrm{Ei} \quad \dfrac{\begin{matrix}\Pi \\ b{:}\perp\!\!\!\perp\end{matrix}}{R^e\, a\, a_1 \ldots a_e}\,(\dagger)}{a{:}\mathcal{M}^e A_1 \ldots A_e}\,\mathcal{M}^e\mathrm{I}$$

we would need a rule ($\dagger$) that would violate the separation between base system and relational theory. (See also the discussion on $\perp\!\!\!\perp$Ec before Lemma 3.3.5 below.)

Second, although it is not an elimination rule, we define *monI* to be an indirect rule since, like $\mathcal{M}^e$E, $\vee$E and $\perp\!\!\!\perp$Ei, it can interrupt a potential reduction.

As notation, we write

$$\dfrac{a{:}A \quad \Psi}{b{:}B}\,(r)$$

for an application of an elimination (or indirect) rule (r) with major premise $a{:}A$ and conclusion $b{:}B$, where Ψ represents the finite sequence of derivations of the minor premises of the rule. The (schematic) permutative reductions for $\mathcal{M}^e$E, $\vee$E and $\perp\!\!\!\perp$Ei are as follows.

Permutative reductions for $\mathcal{M}^e$E:

$$\dfrac{\dfrac{\begin{array}{c}\Pi\\ a{:}\mathcal{M}^e A_1\ldots A_e\end{array} \quad \begin{array}{c}[a{:}A_1]\ \cdots\ [a{:}A_e]\ [R^e\, a\, a_1 \ldots a_e]\\ \Pi_1\\ b{:}B\end{array}}{b{:}B}\,\mathcal{M}^e\mathrm{E} \qquad \Psi}{c{:}C}\,(r)$$

$$\rightsquigarrow \quad \dfrac{\begin{array}{c}\Pi\\ a{:}\mathcal{M}^e A_1\ldots A_e\end{array} \quad \dfrac{\begin{array}{c}[a{:}A_1]\ \cdots\ [a{:}A_e]\ [R^e\, a\, a_1 \ldots a_e]\\ \Pi_1\\ b{:}B\end{array} \qquad \Psi}{c{:}C}\,(r)}{c{:}C}\,\mathcal{M}^e\mathrm{E} \quad .$$

Permutative reductions for $\vee$E:

$$\dfrac{\dfrac{\begin{array}{c}\Pi\\ a{:}A \vee B\end{array} \quad \begin{array}{c}[a{:}A]\\ \Pi_1\\ c{:}C\end{array} \quad \begin{array}{c}[a{:}B]\\ \Pi_2\\ c{:}C\end{array}}{c{:}C}\,\vee\mathrm{E} \qquad \Psi}{d{:}D}\,(r)$$

$$\rightsquigarrow \quad \dfrac{\begin{array}{c}\Pi\\ a{:}A \vee B\end{array} \quad \dfrac{\begin{array}{c}[a{:}A]\\ \Pi_1\\ c{:}C\end{array} \quad \Psi}{d{:}D}\,(r) \quad \dfrac{\begin{array}{c}[a{:}B]\\ \Pi_2\\ c{:}C\end{array} \quad \Psi}{d{:}D}\,(r)}{d{:}D}\,\vee\mathrm{E} \quad .$$

Permutative reductions for $\perp\!\!\!\perp$Ei:

$$\dfrac{\dfrac{\begin{array}{c}\Pi\\ b{:}\perp\!\!\!\perp\end{array}}{a{:}A}\,\perp\!\!\!\perp\mathrm{Ei} \qquad \Psi}{c{:}C}\,(r) \quad \rightsquigarrow \quad \dfrac{\begin{array}{c}\Pi\\ b{:}\perp\!\!\!\perp\end{array}}{c{:}C}\,\perp\!\!\!\perp\mathrm{Ei} \quad .$$

The permutative reductions for *monI* are more complex and we consider them in detail. First, note that since *monI* can be only applied to persistent formulas, we need not consider permuting it with $\supset$ E (cf. the discussions in §3.1.1 and §3.1.3). Now let $A \wedge B$ and $A \vee B$ be persistent formulas. In the permutative reductions of *monl* with applications of $\wedge$E or $\vee$E, the application of *monl* is 'pushed' to lwffs of smaller grade, e.g.

$$\dfrac{\dfrac{\begin{matrix}\Pi_1\\ a{:}A\vee B\end{matrix}\quad \begin{matrix}\Pi_2\\ a\sqsubseteq b\end{matrix}}{b{:}A\vee B}\,monl \qquad \begin{matrix}[b{:}A]\\ \Pi_3\\ c{:}C\end{matrix}\quad \begin{matrix}[b{:}B]\\ \Pi_4\\ c{:}C\end{matrix}}{c{:}C}\,\vee\mathrm{E}$$

$$\leadsto \quad \dfrac{\begin{matrix}\Pi_1\\ a{:}A\vee B\end{matrix}\qquad \begin{matrix}\dfrac{[a{:}A]\quad \begin{matrix}\Pi_2\\ a\sqsubseteq b\end{matrix}}{b{:}A}\,monl\\ \Pi_3\\ c{:}C\end{matrix}\qquad \begin{matrix}\dfrac{[a{:}B]\quad \begin{matrix}\Pi_2\\ a\sqsubseteq b\end{matrix}}{b{:}B}\,monl\\ \Pi_4\\ c{:}C\end{matrix}}{c{:}C}\,\vee\mathrm{E} \quad .$$

The permutative reductions of *monl* with $\perp\!\!\!\perp$Ei or $\perp\!\!\!\perp$Ec simply result in the deletion of the application of *monl*, e.g.

$$\dfrac{\dfrac{\begin{matrix}[a{:}\neg A]\\ \Pi_1\\ c{:}\perp\!\!\!\perp\end{matrix}\quad \begin{matrix}\Pi_2\\ c\sqsubseteq b\end{matrix}}{b{:}\perp\!\!\!\perp}\,monl}{a^*{:}A}\,\perp\!\!\!\perp\mathrm{Ec} \quad \leadsto \quad \dfrac{\begin{matrix}[a{:}\neg A]\\ \Pi_1\\ c{:}\perp\!\!\!\perp\end{matrix}}{a^*{:}A}\,\perp\!\!\!\perp\mathrm{Ec} \quad .$$

In the permutative reduction of *monl* with $\mathcal{M}^u$E the application of *monl* is 'pushed' to rwffs, i.e. it is replaced with an application of *mon*$R(1)$:

$$\dfrac{\dfrac{\begin{matrix}\Pi\\ a{:}\mathcal{M}^u A_1\ldots A_u\end{matrix}\quad \begin{matrix}\Pi_0\\ a\sqsubseteq b\end{matrix}}{b{:}\mathcal{M}^u A_1\ldots A_u}\,monl \qquad \begin{matrix}\Pi_1\\ b_1{:}A_1\end{matrix}\cdots\begin{matrix}\Pi_{u-1}\\ b_{u-1}{:}A_{u-1}\end{matrix}\qquad \begin{matrix}\Pi_R\\ R^u\, b\, b_1\ldots b_u\end{matrix}}{b_u{:}A_u}\,\mathcal{M}^u\mathrm{E}$$

$$\leadsto \quad \dfrac{\begin{matrix}\Pi\\ a{:}\mathcal{M}^u A_1\ldots A_u\end{matrix}\quad \begin{matrix}\Pi_1\\ b_1{:}A_1\end{matrix}\cdots\begin{matrix}\Pi_{u-1}\\ b_{u-1}{:}A_{u-1}\end{matrix}\qquad \dfrac{\begin{matrix}\Pi_R\\ R^u\, b\, b_1\ldots b_u\end{matrix}\quad \begin{matrix}\Pi_0\\ a\sqsubseteq b\end{matrix}}{R^u\, a\, b_1\ldots b_u}\,monR(1)}{b_u{:}A_u}\,\mathcal{M}^u\mathrm{E} \quad .$$

In the permutative reduction of *monl* with $\mathcal{M}^e$E the application of *monl* is 'transformed' into applications of the monotony rules in the subderivation Π_1 of the minor premise (where we 'substitute' a for b using $a \sqsubseteq b$ and the monotony rules):

$$\dfrac{\dfrac{\begin{matrix}\Pi\\ a{:}\mathcal{M}^e A_1 \dots A_e\end{matrix} \quad \begin{matrix}\Pi_0\\ a \sqsubseteq b\end{matrix}}{b{:}\mathcal{M}^e A_1 \dots A_e}\,monl \qquad \begin{matrix}[b_1{:}A_1]\cdots[b_e{:}A_e]\;[R^e\, b\, b_1 \dots b_e]\\ \Pi_1\\ c{:}C\end{matrix}}{c{:}C}\,\mathcal{M}^e\mathrm{E}$$

$$\rightsquigarrow \quad \dfrac{\begin{matrix}\Pi\\ a{:}\mathcal{M}^e A_1 \dots A_e\end{matrix} \qquad \begin{matrix}[b_1{:}A_1]\cdots[b_e{:}A_e] \quad [R^e\, a\, b_1 \dots b_e]\\ \Pi_1[a/b]\\ c{:}C\end{matrix}}{c{:}C}\,\mathcal{M}^e\mathrm{E} \quad .$$

When we permute *monl* with itself we exploit the transitivity of the partial order, which is an instance of $monR(n)$ with $n = 3$ (or, equivalently, $n = 2$ when $a \sqsubseteq b$ is defined as $R\,0\,a\,b$):

$$\dfrac{\dfrac{\begin{matrix}\Pi_1\\ a{:}A\end{matrix} \quad \begin{matrix}\Pi_2\\ a \sqsubseteq b\end{matrix}}{b{:}A}\,monl \qquad \begin{matrix}\Pi_3\\ b \sqsubseteq c\end{matrix}}{c{:}A}\,monl \quad \rightsquigarrow \quad \dfrac{\begin{matrix}\Pi_1\\ a{:}A\end{matrix} \qquad \dfrac{\begin{matrix}\Pi_2\\ a \sqsubseteq b\end{matrix} \quad \begin{matrix}\Pi_3\\ b \sqsubseteq c\end{matrix}}{a \sqsubseteq c}\,monR(n)}{c{:}A}\,monl \quad .$$

We are now in a position to state our desired normalization results. We first generalize Definition 2.3.4 to:

Definition 3.3.3 *A derivation is in* normal form (*is a* normal derivation) *iff it contains no maximal lwffs and no permutable lwffs.* ■

Then we consider the three systems in Table 3.1. For $\mathrm{N}(\mathcal{ML})$ and $\mathrm{N}(\mathcal{JL})$ we have:

Lemma 3.3.4 *Every derivation of $a{:}A$ from Γ, Δ in $\mathrm{N}(\mathcal{ML})$ or $\mathrm{N}(\mathcal{JL})$ reduces to a derivation in normal form.* ■

Note first that derivations in Horn relational theories $\mathrm{N}(\mathcal{T})$ cannot introduce maximal or permutable lwffs. The lemma then follows by a straightforward modification of the well-known proof for minimal and intuitionistic logic given originally by Prawitz in [186], and also found in many textbooks, e.g. [221, 230]. The proof relies on the identification of particular sequences of formulas (the threads and tracks of Definition 2.3.8 in §2.3.1.1 and a straightforward adaptation of Prawitz's *segments*) to show that each application of proper and permutative reductions reduces a suitable well-ordered measure on normal derivations. Thus, the reduction process must eventually terminate with a derivation free of maximal and permutable lwffs.

Before proving analogous results for $\mathrm{N}(\mathcal{CL})$, let us perform a simplification that will allow us to consider (as is standardly done) a simplified language. The rules for classical non-local negation and $\perp\!\!\!\perp$ allow us to define for each existential modal operator $\mathcal{M}^e$, with associated relation R^e, a dual universal modal operator $\mathcal{M}^u_e$, with associated relation R^u_e, while retaining completeness (see the discussion on the possible incompleteness of positive modal logics in §3.2.3). In particular we define:[9]

$$a{:}\neg\mathcal{M}^u_e A_1 \ldots A_{e-1}\neg A_e \quad \text{iff} \quad a{:}\mathcal{M}^e A_1 \ldots A_e$$

and

$$R^u_e\, a^*\, a_1 \ldots a_{e-1}\, a^*_e \quad \text{iff} \quad R^e\, a\, a_1 \ldots a_{e-1}\, a_e\,.$$

To illustrate that this is correct, i.e. that $\mathcal{M}^e$ and $\mathcal{M}^u_e$ are really interdefinable, we take $\mathcal{M}^u_e$ as primitive and derive the rules for $\mathcal{M}^e$, e.g. for $\mathcal{M}^e$E:

$$\dfrac{\dfrac{a{:}\neg\mathcal{M}^u_e A_1 \ldots A_{e-1}\neg A_e \qquad \dfrac{\dfrac{\dfrac{\begin{array}{c}[a_1{:}A_1]^2 \cdots [a_{e-1}{:}A_{e-1}]^2\ [a_e{:}A_e]^1\ [R^u_e\, a^*\, a_1 \ldots a_{e-1}\, a^*_e]^2 \\ \Pi \\ b{:}B \end{array} \qquad [b^*{:}\neg B]^3}{c{:}\bot\!\!\bot}\;\neg\text{E}}{a^*_e{:}\neg A_e}\;\neg\text{I}^1}{a^*{:}\mathcal{M}^u_e A_1 \ldots A_{e-1}\neg A_e}\;\mathcal{M}^u_e\text{I}^2}{d{:}\bot\!\!\bot}\;\neg\text{E}}{b{:}B}\;\bot\!\!\bot\text{Ec}^3$$

where, for brevity, we have identified a^{**} with a instead of explicitly using the rules $**$i and $**$c. Hence we can safely replace $\mathcal{M}^e$ and R^e with $\mathcal{M}^u_e$ and R^u_e. Analogously, we can define disjunction in terms of conjunction, and with these replacements we obtain the system N($\mathcal{CL}'$), which is adequate for representing a non-classical logic with a classical treatment of negation.

Considering this simplified language (with the operators $\wedge$, $\supset$, $\mathcal{M}^u$, $\mathcal{M}^u_e$, $\neg$ and $\bot\!\!\bot$) allows us to reduce applications of $\bot\!\!\bot$Ec to instances where the conclusion is atomic, by showing that any application of $\bot\!\!\bot$Ec with a non-atomic consequence can be replaced with a derivation in which $\bot\!\!\bot$Ec is applied only to lwffs of smaller grade. For instance, again identifying a^{**} with a,

$$\dfrac{\begin{array}{c}[a{:}\neg\mathcal{M}^u A_1 \ldots A_u] \\ \Pi \\ b{:}\bot\!\!\bot\end{array}}{a^*{:}\mathcal{M}^u A_1 \ldots A_u}\;\bot\!\!\bot\text{Ec}$$

[9]Note that this is equivalent to defining $R^e\, a\, a_1 \ldots a_{e-1}\, a_e$ iff $R^u_e\; a_1 \ldots a_{e-1}\, a_e\, a$ and adding *switching* rules for both R^e and R^u_e, e.g.

$$\dfrac{R^u_e\, a_1 \ldots a_{e-1}\, a_e\, a}{R^u_e\, a^*\, a_1 \ldots a_{e-1}\, a^*_e}\;.$$

This is, for instance, the case in relevance logics [79], where fusion ($\circ$) and relevant implication ($\rightarrow$) are associated with the one and the same R and $a{:}A \circ B$ is shown equivalent to $a{:}\neg(A \rightarrow \neg B)$ by means of switching.

is replaced with

$$
\cfrac{
  \cfrac{
    \cfrac{
      \cfrac{
        [a_u^*{:}\neg A_u]^2 \qquad
        \cfrac{[a^*{:}\mathcal{M}^u A_1 \ldots A_u]^1 \; [a_1{:}A_1]^3 \cdots [a_{u-1}{:}A_{u-1}]^3 \; [R^u\, a^*\, a_1 \ldots a_u]^3}{a_u{:}A_u}\,\mathcal{M}^u\mathrm{E}
      }{c{:}\bot\!\!\bot}\,\neg\mathrm{E}
    }{a{:}\neg\mathcal{M}^u A_1 \ldots A_u}\,\neg\mathrm{I}^1
    \\ \Pi \\
    b{:}\bot\!\!\bot
  }{a_u{:}A_u}\,\bot\!\!\bot\mathrm{Ec}^2
}{a^*{:}\mathcal{M}^u A_1 \ldots A_u}\,\mathcal{M}^u\mathrm{I}^3
$$
.

Therefore, in the case of $\mathrm{N}(\mathcal{CL}')$ the only permutative reductions that need to be considered are those for *monl*, and, by analogy with Lemma 3.3.4, we have:

Lemma 3.3.5 *Every derivation of* $a{:}A$ *from* Γ, Δ *in* $\mathrm{N}(\mathcal{CL}')$ *reduces to a derivation in normal form.* ■

As we showed in §2.3.1, one of the main advantages of normal derivations is that they possess a well-defined structure that has several desirable properties. In particular, in any of the three families of systems we considered, the two parts of each system are strictly separated: derivations of lwffs may depend on derivations of rwffs, but not vice versa. As a consequence, any derivation of an lwff is structured as a central derivation in the base system 'decorated' with subderivations in the relational theory, which attach onto the central derivation through instances of $\mathcal{M}^u$E, $\mathcal{M}^e$I, or *monl*. Moreover, when in normal form, the structure of the central derivation in the base system can be further characterized by identifying particular sequences of lwffs (our adaptations of Prawitz's threads, tracks and segments), and showing that in these sequences there is an ordering on inferences.[10] By exploiting this ordering, we can then show a subformula property for all three families of systems.

We first generalize Definitions 2.3.10 and 2.3.12, for the restricted language of Definition 3.1.7, as follows.

Definition 3.3.6 *B is a* subformula *of A iff (i) A is B; or (ii) A is $A_1 \wedge A_2$, $A_1 \vee A_2$, $A_1 \supset A_2$, $\neg A_1$, $\mathcal{M}^u A_1 \ldots A_u$, or $\mathcal{M}^e A_1 \ldots A_e$, and B is a subformula of one of the A_i's. We say that $b{:}B$ is a (labelled) subformula of $a{:}A$ iff B is a subformula of A.* ■

Definition 3.3.7 *Given a derivation $\Gamma, \Delta \vdash a{:}A$, let $\mathcal{S}$ be the set of subformulas of the formulas in $\{C \mid c{:}C \in \Gamma \cup \{a{:}A\}$ for some $c\}$, i.e. $\mathcal{S}$ is the set consisting of the subformulas of the assumptions Γ and the goal $a{:}A$.*

We say that a derivation $\Gamma, \Delta \vdash a{:}A$ in $\mathrm{N}(\mathcal{ML})$ or $\mathrm{N}(\mathcal{JL})$ satisfies the subformula property *iff for all lwffs $b{:}B$ used in the derivation, $B \in \mathcal{S}$.*

We say that a derivation $\Gamma, \Delta \vdash a{:}A$ in $\mathrm{N}(\mathcal{CL}')$ satisfies the subformula property *iff for all lwffs $b{:}B$ used in the derivation, (i) $B \in \mathcal{S}$; or (ii) B is an assumption*

[10]To restrict further the structure of normal derivations it is then interesting to study the eliminability of *monl* from the systems.

$\neg D$ or $D \to \perp\!\!\!\perp$ discharged by $\perp\!\!\!\perp$Ec *and $D \in \mathcal{S}$; or (iii) B is an occurrence of $\perp\!\!\!\perp$ immediately below an assumption $\neg D$ or $D \to \perp\!\!\!\perp$ discharged by* $\perp\!\!\!\perp$Ec *and $D \in \mathcal{S}$; or (iv) B is an occurrence of $\perp\!\!\!\perp$ obtained by an application of* $\perp\!\!\!\perp$Ec *that does not discharge any assumption.* ■

Then it follows that:

Lemma 3.3.8 *Every normal derivation of $a{:}A$ from Γ, Δ in* $\mathrm{N}(\mathcal{ML})$, $\mathrm{N}(\mathcal{JL})$ *or* $\mathrm{N}(\mathcal{CL}')$ *satisfies the subformula property.* ■

To summarize, we can generalize Theorem 2.3.14 and its commentary as follows.

Theorem 3.3.9 *Our labelled ND systems have the following properties.*

(i) *The deduction machinery is minimal: the systems formalize a minimum fragment of first-order logic required by the semantics of propositional non-classical logics with Horn axiomatizable properties of the relations and of the $*$ function.*

(ii) *Derivations are strictly separated: derivation of lwffs may depend, via rules for non-local operators, on derivations of rwffs, but not vice versa.*

(iii) *Derivations normalize: derivations of lwffs have a well-structured normal form that satisfies the subformula property.* ■

For comparison, consider again the semantic embedding approach: a propositional non-classical logic is encoded as a 'suitable' (e.g. intuitionistic or classical) first-order theory by axiomatizing an appropriate definition of truth, but all structure is lost as propositions and relations are flattened into first-order formulas, and derivations of propositions are mingled with derivations of relations.

However, in exchange for the extra structure in our systems there are limits to the generality of the formulation: the properties in Theorem 3.3.9 depend on design decisions we have made, in particular, the use of Horn relational theories. This, of course, places stronger limitations on what we can formalize than a semantic embedding in first-order logic. Consider, for instance, the relevance logic RM, obtained by extending the logic R with the postulate

$$\forall a \forall b \forall c (R\, a\, b\, c \supset (R\, 0\, a\, c \vee R\, 0\, b\, c))\,, \tag{3.29}$$

which corresponds to the 'mingle' axiom schema $A \to (A \to A)$, A16 in Table 3.2. We cannot directly present RM because (3.29) is not formalizable as a set of Horn rules. This is a design decision. Consider the alternatives. Analogously to §2.3.1, we can extend our deduction machinery by providing rules for a full first-order relational theory and explicitly add (3.29) as an axiom schema. However, if we then maintain (ii) of Theorem 3.3.9 we lose completeness with respect to the semantics, since by analysis of normal form proofs we can show that $0{:}A \to (A \to A)$ is not provable. Alternatively, we can regain completeness by giving up (ii), by identifying *falsum* in the first-order relational theory with $\perp\!\!\!\perp$, i.e. adopting a *universal* $\perp\!\!\!\perp$. However, the resulting system is then essentially equivalent to semantic embedding and we lose (i); this follows by a straightforward generalization of the results for modal ND systems with universal falsum in §2.3.3.

But there is another reason why the latter solution is not satisfactory: since it is based on the $\perp\!\!\!\perp$ rules, it does not apply to positive fragments. For these (and also for full logics), we can regain completeness by again giving up (ii) to introduce rules similar to Simpson's 'geometric rules' [216], e.g. we can formalize (3.29) with the rule

$$\dfrac{R\,a\,b\,c \qquad \begin{matrix}[R\,0\,a\,c]\\ \vdots \\ d{:}A\end{matrix} \qquad \begin{matrix}[R\,0\,b\,c]\\ \vdots \\ d{:}A\end{matrix}}{d{:}A} \; .$$

4 LABELLED NATURAL DEDUCTION SYSTEMS FOR QUANTIFIED MODAL LOGICS

In the previous chapters we have given a framework based on labelled deduction that provides a systematic solution to the problem of finding uniform and modular presentations of propositional non-classical logics. Here we consider quantified modal logics [89, 104, 141] as a significant case study of the additional complexity introduced by quantifiers with respect to the range of possible logics and semantics for them. (Other quantified non-classical logics, e.g. quantified relevance logics, can be presented similarly.) In this case we must choose not only properties of the accessibility relation in the Kripke frame, as in the propositional case, but also how the domains of individuals change between worlds; for example, do the domains vary arbitrarily (varying domains), or do the same objects exist in every world (constant domains), or are objects possibly created (increasing domains) or destroyed (decreasing domains) when moving to accessible worlds?

These two choices can be made independently, which results in a two-dimensional space of possible logics. (Other dimensions are possible, e.g. non-rigid designators [89, 104]; we consider here only the rigid case.) This space has often been explored in a piecemeal fashion and there has been a lack of uniformity in the formalization of deduction systems and in the way their associated metatheoretical results, in particular completeness, are proved. Consider the following aspects.

First, different deduction systems are employed. Quantified modal logics are typically presented by using Hilbert systems extending those for the propositional case, but the standard quantifier rules automatically require the domains of a semantics to be increasing [89, p. 426], and this restricts the class of logics that are formalizable in a modular way. This problem can be solved by modifying either the deduction system (e.g. by adopting the rules of 'free logic'), or the semantics (e.g. by introducing 'truth value gaps'), see [104, 141]. These techniques, however, are imperfect in that none provides a general and uniform solution. For example, the rules of free logic don't provide modular completeness proofs: different strategies must be adopted for different conditions on the domains.

Second, incompleteness with respect to Kripke semantics is common. Simply adding quantifier rules to a Hilbert system complete for a propositional modal logic may not result in a system complete with respect to the corresponding extension of the semantics. Moreover, minor changes to a complete quantified modal logic, e.g. changing the conditions on the domains, can produce incompleteness. For instance, there are Hilbert systems for logics with the *Barcan Formula* (BF, $\forall x(\Box A) \supset \Box\forall x(A)$) that are incomplete, while those without it are complete, and vice versa; e.g. $\mathrm{H}(\mathrm{QS4.2} + \mathrm{BF})$ is incomplete although $\mathrm{H}(\mathrm{QS4.2})$ is complete [141].

Third, metatheoretical results are not proved in a uniform way. Often, even for related logics, completeness proofs or counter-examples must be devised ad hoc, using different techniques. For example, the standard canonical model technique fails for $\mathrm{H}(\mathrm{QS4.2})$, but we can prove completeness with respect to Kripke semantics using the 'subordination method' [61, p. 175].

Quantified modal logics also raise special challenges when we begin actually to prove theorems with them. Many propositional modal logics are decidable, so proof search can be automated; see §12 and [87, 89, 232]. In the quantified case, however, even when we restrict ourselves to terms built from constants and variables, as is often done [141], and as we do here, the resulting modal logics are undecidable. Thus if we want to use them, it is desirable to have deduction systems that 'naturally' support the interactive construction of proofs and that possess properties, such as normalization of derivations and the subformula property, which restrict the search space for proofs.

We extend the development of §2 to the quantified case and thereby provide solutions to the above problems: we give a natural deduction presentation of quantified modal logics that is modular in two dimensions, reflecting the two degrees of freedom discussed above. As before, it is based on a fixed base system (now $\mathrm{N}(\mathrm{QK})$, for quantified K) where extensions are made by independently instantiating two separate theories: a relational theory (as before), and a *domain theory*, which formalizes the behavior of the domains of quantification. That is, in the domain theory we reason about *labelled terms*, $w{:}t$, expressing the existence of term t at world w. Thus $\vdash w{:}\forall x(A)$ iff $\vdash w{:}A[t/x]$ for all t such that $\vdash w{:}t$. This formulation naturally suggests that we adopt quantifier rules similar to those of *free logic* [28], and we show below that the previously mentioned problems for Hilbert-style quantified modal logics based on free logic do not apply in our approach.[1] By appropriate instantiation of these two theories,

[1]There is also another important respect in which our approach differs from the standard ones based on free logic. In the latter, the existence of a term at a particular world is not an independent 'judgement' like $w{:}t$, but it is expressed by the atomic modal formula $E(t)$, which has to be explicitly considered in the completeness proof [104, p. 279].

we formalize the predicate extensions (with varying, increasing, decreasing or constant domains) of the propositional modal logics we presented in §2.

The metatheoretical properties of our ND systems extend as well: we give modular proofs of soundness and completeness by extending the canonical model construction of §2.2 to account for the explicit formalization of the properties of the domains of quantification. This means that our quantified modal ND systems are sound and complete with respect to the appropriate Kripke semantics, and thus equivalent to the corresponding Hilbert systems only when these are themselves complete with respect to the same semantics. We also show that the proof-theoretical results for propositional modal systems (in particular, normalization and the subformula property) carry over to the quantified case. Hence, proof search can be restricted and the effectiveness of theorem proving improved. Finally, we discuss tradeoffs in formalizations of the base system and the theories extending it, and show not only that the results for propositional modal systems carry over to the quantified case, but also that new tradeoffs must be considered.

The remainder of this chapter is organized as follows. In §4.1 we extend our framework to present quantified modal logics by formalizing the base ND system and the theories extending it. In §4.2 we prove that our systems are sound and complete with respect to the corresponding Kripke semantics. In §4.3 we prove that derivations in our systems normalize and investigate the consequences of this result. In §5.3 we will then present the Isabelle encodings of our systems, give applications, and demonstrate their correctness.

4.1 A MODULAR PRESENTATION OF QUANTIFIED MODAL LOGICS

We extend Definitions 2.1.1, 2.1.3 and 2.1.4 and Notations 2.1.2 and 2.1.5 as follows.

Definition 4.1.1 *Let W be a set of* labels *and R a binary relation over W. If w and w' are labels, then wRw' is a* relational formula (rwff)*. If t is a constant c or a variable x, then $w{:}t$ is* labelled term (lterm)*. If A is a modal formula built from atomic propositions (i.e. predicates applied to terms, e.g. $P(t)$) and the connectives, modal operator and quantifier $\bot$, $\supset$, $\Box$ and $\forall$, then $w{:}A$ is a* labelled formula (lwff)*. Other connectives, modal operators and quantifiers can be defined in the usual manner, e.g. $\sim A =_{def} A \supset \bot$, $\Diamond A =_{def} \sim\Box\sim A$ and $\exists x(A) =_{def} \sim\forall x(\sim A)$.* ■

Definition 4.1.2 *The* grade *of an lwff $w{:}A$, in symbols $grade(w{:}A)$, is the number of times $\supset$, $\Box$ and $\forall$ occur in A.* ■

Notation 4.1.3 In order to simplify our notation, we will omit brackets whenever no confusion can arise, and we adopt the convention that $\sim$, $\Box$, $\Diamond$, $\forall$ and $\exists$ are of equal binding strength and bind tighter than $\wedge$, which binds tighter than $\vee$, which binds tighter than $\supset$.

For the rest of this chapter, we assume that the variable w ranges over labels, t ranges over terms, and $A, B, \ldots$ range over quantified modal formulas. Further, let Γ, Δ and Θ be, respectively, arbitrary sets of lwffs, $\{w_1{:}A_1, \ldots, w_n{:}A_n\}$, rwffs, $\{w_1Rw_2, \ldots, w_lRw_m\}$, and lterms, $\{w_1{:}t_1, \ldots, w_j{:}t_j\}$. All variables may be annotated with subscripts or superscripts. Finally, we adopt the standard notation of

$$\dfrac{\begin{matrix}[w_i{:}A \supset \bot] \\ \vdots \\ w_j{:}\bot\end{matrix}}{w_i{:}A}\ \bot\text{E} \qquad \dfrac{\begin{matrix}[w{:}A] \\ \vdots \\ w{:}B\end{matrix}}{w{:}A \supset B}\ \supset\text{I} \qquad \dfrac{w{:}A \supset B \quad w{:}A}{w{:}B}\ \supset\text{E}$$

$$\dfrac{\begin{matrix}[w_i R w_j] \\ \vdots \\ w_j{:}A\end{matrix}}{w_i{:}\Box A}\ \Box\text{I} \qquad \dfrac{w_i{:}\Box A \quad w_i R w_j}{w_j{:}A}\ \Box\text{E} \qquad \dfrac{\begin{matrix}[w{:}t] \\ \vdots \\ w{:}A[t/x]\end{matrix}}{w{:}\forall x(A)}\ \forall\text{I} \qquad \dfrac{w{:}\forall x(A) \quad w{:}t}{w{:}A[t/x]}\ \forall\text{E}$$

In $\Box$I, w_j is different from w_i and does not occur in any assumption on which $w_j{:}A$ depends other than $w_i R w_j$. In $\forall$I, t does not occur in any assumption on which $w{:}A[t/x]$ depends other than $w{:}t$.

Figure 4.1. The rules of $\mathrm{N(QK)}$

predicate logic, e.g. $A[t/x]$ denotes the substitution of the term t for the variable x in the formula A. ∎

4.1.1 *The base system* $\mathrm{N(QK)}$

The rules given in Figure 4.1 determine $\mathrm{N(QK)}$, the base ND system presenting the quantified modal logic QK. (That $\mathrm{N(QK)}$ presents QK is proved in Theorem 4.2.5 below.) The rules for $\forall$ are a labelled version of the rules of free logic [28], and, as in free logic, $w{:}\forall x(A) \supset \exists x(A)$ is provable only under the assumption $w{:}t$, stating that the domain of quantification of w is non-empty; see §4.3. Note the symmetry between the rules for $\Box$ and those for $\forall$; this reinforces the role of $\Box$, and of modal logics in general, "as a replacement for the more powerful machinery of quantified classical logic, at least in some cases" [89, p. 377]. The same symmetry holds between the derived rules for $\Diamond$ and $\exists$ given in Figure 4.2; the rules for $\Diamond$ are derived like in Example 2.1.14, and the derivations of the rules for $\exists$ are given in Example 4.1.8 below.

4.1.2 *Relational theories*

Different quantified modal logics are obtained from the base logic QK by placing conditions on the accessibility relation in the Kripke frame; e.g., like in the propositional case, we get the logic QT from QK by adding that the relation is reflexive, and then QS4 from QT by further adding transitivity. We present particular logics by extending our base system $\mathrm{N(QK)}$ with relational theories axiomatizing properties of R. However, as we argued before, some properties of R can only be expressed using higher-order logic, although for other properties first-order logic, or even fragments of it, is enough. We showed in §2.3 that there are tradeoffs in formalization: different

$$\frac{w_j{:}A \quad w_i R\, w_j}{w_i{:}\Diamond A}\ \Diamond\mathrm{I} \qquad\qquad \frac{w{:}A[t/x] \quad w{:}t}{w{:}\exists x(A)}\ \exists\mathrm{I}$$

$$\frac{w_i{:}\Diamond A \qquad \begin{matrix}[w_j{:}A]\ [w_i R\, w_j] \\ \vdots \\ w_k{:}B\end{matrix}}{w_k{:}B}\ \Diamond\mathrm{E} \qquad\qquad \frac{w_i{:}\exists x(A) \qquad \begin{matrix}[w_i{:}A[t/x]]\ [w_i{:}t] \\ \vdots \\ w_j{:}B\end{matrix}}{w_j{:}B}\ \exists\mathrm{E}$$

In $\Diamond$E, w_j is different from w_i and w_k, and does not occur in any assumption on which the upper occurrence of $w_k{:}B$ depends other than $w_j{:}A$ and $w_i R\, w_j$. In $\exists$E, t does not occur in any assumption on which the upper occurrence of $w_j{:}B$ depends other than $w_i{:}A[t/x]$ and $w_i{:}t$.

Figure 4.2. The derived rules for $\Diamond$ and $\exists$

choices require different formalizations of the base modal system and have different metatheoretical properties. In the previous chapters we settled on relations axiomatizable in terms of Horn clauses, a choice we repeat here; we discuss the implications of this decision in §4.3.

We choose to admit precisely those properties of R that can be formalized as a collection of *Horn relational rules*, i.e. rules of the form

$$\frac{p_1 R\, q_1 \quad \cdots \quad p_m R\, q_m}{p_0 R\, q_0}$$

where $m \geq 0$, and the p_i and q_i are terms built from labels $w_1, \ldots, w_n$ and function symbols (recall that some properties of R, e.g. seriality and convergency, can be expressed as relational rules only after a conservative extension of our theories with Skolem function constants).[2] A *Horn relational theory* $\mathrm{N}(\mathcal{T})$ is a theory generated by a set of such rules.

Relational rules suffice to present the predicate extensions of a large family of common propositional modal logics; in Table 4.1 we recall some of the properties that are instances of restricted (i, j, m, n) convergency and the corresponding Horn relational rules, written with the new notation.

The ND system $\mathrm{N}(\mathrm{Q}\mathcal{L}) = \mathrm{N}(\mathrm{QK}) + \mathrm{N}(\mathcal{T})$for the quantified modal logic $\mathrm{Q}\mathcal{L}$ is obtained by extending $\mathrm{N}(\mathrm{QK})$ with a Horn relational theory $\mathrm{N}(\mathcal{T})$; this extension is represented by the horizontal arrows in Figure 4.3.

Extending Notation 2.1.10, we refer to the system $\mathrm{N}(\mathrm{QK})+\mathrm{N}(\mathcal{T})$ also as $\mathrm{N}(\mathrm{QK}Ax)$, where Ax is a string consisting of the standard names of the characteristic axiom

[2]Also recall that in the presence of Skolem function constants we must appropriately extend our language and rules to distinguish between atomic and composite labels.

Table 4.1. Some properties of R, corresponding characteristic axiom schemas and Horn relational rules

Property	*Axiom schema*	*Horn relational rule*
Seriality	D: $\Box A \supset \Diamond A$	$\dfrac{}{w_i R f(w_i)}\ ser$
Reflexivity	T: $\Box A \supset A$	$\dfrac{}{w_i R w_i}\ refl$
Transitivity	4: $\Box A \supset \Box\Box A$	$\dfrac{w_i R w_j \quad w_j R w_k}{w_i R w_k}\ trans$
Euclideaness	5: $\Diamond A \supset \Box\Diamond A$	$\dfrac{w_i R w_j \quad w_i R w_k}{w_j R w_k}\ eucl$
Convergency	2: $\Diamond\Box A \supset \Box\Diamond A$	$\dfrac{w_i R w_j \quad w_i R w_k}{w_j R\, g(w_i, w_j, w_k)}\ conv1$ $\dfrac{w_i R w_j \quad w_i R w_k}{w_k R\, g(w_i, w_j, w_k)}\ conv2$

Where f and g are (Skolem) function constants.

schemas corresponding to the relational rules generating $\mathrm{N}(\mathcal{T})$. Then, for example, N(QKT4) is a synonym of N(QS4). Various combinations of relational rules define therefore predicate extensions of propositional modal logics, including QD, QT, QB, QS4, QS4.2, QKD45 and QS5, which are respectively presented by the systems N(QKD), N(QKT), N(QKTB), N(QKT4), N(QKT4.2), N(QKD45) and N(QKT5).

4.1.3 Domain theories

So far, we have made no commitments to the relationship between the domains of quantification in the different worlds; in this case we say that the domains of $\mathrm{N(QK)} + \mathrm{N}(\mathcal{T})$ are *varying*. We can then place constraints on them; e.g. requiring that, when we move from a world to another world accessible from it, objects persist (the domains are *increasing*), are not created (and possibly deleted, i.e. the domains are *decreasing*), or stay the same (the domains are both increasing and decreasing, i.e. *constant*). The conditions for increasing and decreasing domains can be respectively formalized by the (Horn) rules

$$\frac{w_i R w_j \quad w_i{:}t}{w_j{:}t}\ id \qquad \text{and} \qquad \frac{w_i R w_j \quad w_j{:}t}{w_i{:}t}\ dd \quad .$$

Different combinations of these rules define different labelled ND systems for common quantified modal logics.

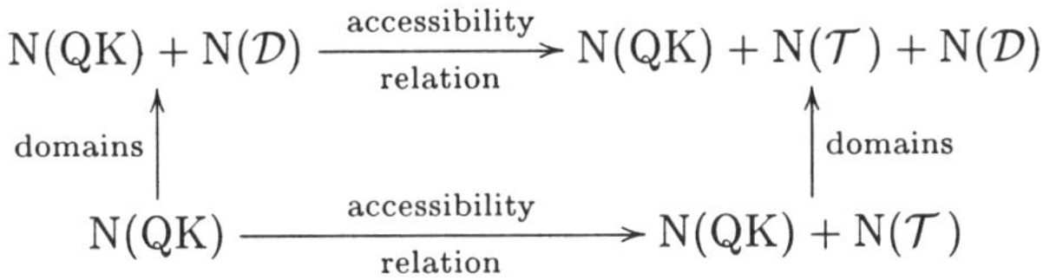

Figure 4.3. Extensions of N(QK)

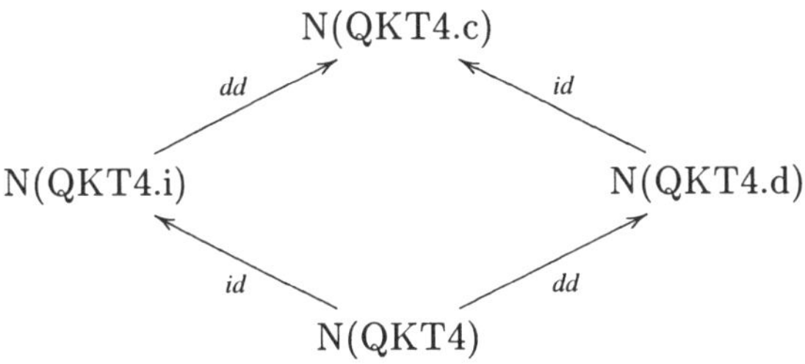

Figure 4.4. The systems N(QKT4.l)

Definition 4.1.4 *The labelled ND system* $\mathrm{N}(\mathrm{Q}\mathcal{L}) = \mathrm{N}(\mathrm{QK}) + \mathrm{N}(\mathcal{T}) + \mathrm{N}(\mathcal{D})$ *is obtained by extending* $\mathrm{N}(\mathrm{QK}) + \mathrm{N}(\mathcal{T})$ *with a given theory* $\mathrm{N}(\mathcal{D})$ *of the domains of quantification (or* domain theory, *for short), generated by a subset of* $\{id, dd\}$*; this extension is represented by the vertical arrows in Figure 4.3.* ■

This yields the two-dimensional uniformity of the deduction system motivated above. (Uniform proofs of soundness and completeness are given in §4.2.)

Notation 4.1.5 We extend the above notational conventions and refer to the system $\mathrm{N}(\mathrm{QK})+\mathrm{N}(\mathcal{T})+\mathrm{N}(\mathcal{D})$ also as $\mathrm{N}(\mathrm{QK}Ax.l)$, where l represents the conditions imposed on the domains. We write $\mathrm{N}(\mathrm{QK}Ax)$ when $\mathrm{N}(\mathcal{D})$ is empty, as done above; $\mathrm{N}(\mathrm{QK}Ax.\mathrm{i})$ or $\mathrm{N}(\mathrm{QK}Ax.\mathrm{d})$ when $\mathrm{N}(\mathcal{D})$ is generated by *id* or *dd*, respectively; $\mathrm{N}(\mathrm{QK}Ax.\mathrm{c})$ when $\mathrm{N}(\mathcal{D})$ is generated by *id* and *dd*.[3] ■

We can therefore formalize one of four related logics simply by instantiating $\mathrm{N}(\mathcal{D})$; e.g., as shown in Figure 4.4, we can specify N(QKT4), i.e. N(QS4), with domains

[3]We here consider, as is standard, constant domains only for worlds connected by the accessibility relation. The case where all worlds, even unconnected ones, share the same domain, can be formalized by the rule

$$\frac{w_i : t}{w_j : t}$$

from which both *id* and *dd* can be derived.

that are varying, N(QKT4), increasing, N(QKT4.i), decreasing, N(QKT4.d), or constant, N(QKT4.c).

This is *not* the case in the standard Hilbert presentations of quantified modal logics, where the domains are committed to being increasing, since the standard rules for $\forall$ automatically enforce the *Converse Barcan Formula* CBF,

$$\Box\forall x(A) \supset \forall x(\Box A),$$

which corresponds to the increasing domains condition [89, p. 426]. Constant domains are then obtained by further adding as an axiom schema the *Barcan Formula* BF,

$$\forall x(\Box A) \supset \Box\forall x(A),$$

which corresponds to the decreasing domains condition. Hilbert-style systems for logics with varying domains can be given by substituting the classical quantifier rules with the rules of free logic, as done by Garson in [104]. However, Garson also shows that his completeness proof fails for some logics, e.g. for QB; we return to this at the end of §4.2.

Some particular quantified modal logics with varying domains can also be formalized by systems that keep the classical quantifier rules, e.g. by using free variables as disguised universal quantifiers and restricting the necessitation rule to closed sentences [151], or by adopting a semantics with truth value gaps [141]. However, none of these techniques provides uniform deduction systems (or semantics) since it is not clear how to generalize them to other logics. For a detailed discussion of the limits of these systems see [104].

We now extend Definition 2.1.11 and Fact 2.1.12; Notation 2.1.13 for derivations in propositional modal systems generalizes straightforwardly to derivations in $\mathrm{N(QK)} + \mathrm{N}(\mathcal{T}) + \mathrm{N}(\mathcal{D})$.

Definition 4.1.6 *A* derivation of an lwff, rwff or lterm φ from a set of rwffs Γ, a set of rwffs Δ and a set of lterms Θ in a ND system $\mathrm{N}(\mathrm{Q}\mathcal{L}) = \mathrm{N(QK)} + \mathrm{N}(\mathcal{T}) + \mathrm{N}(\mathcal{D})$ *is a tree formed using the rules in* $\mathrm{N}(\mathrm{Q}\mathcal{L})$*, ending with* φ *and depending only on* $\Gamma \cup \Delta \cup \Theta$*. We write* $\Gamma, \Delta, \Theta \vdash_{\mathrm{N}(\mathrm{Q}\mathcal{L})} \varphi$ *when* φ *can be so derived. A derivation of* φ *in* $\mathrm{N}(\mathrm{Q}\mathcal{L})$ *depending on the empty set,* $\vdash_{\mathrm{N}(\mathrm{Q}\mathcal{L})} \varphi$*, is a* proof of φ in $\mathrm{N}(\mathrm{Q}\mathcal{L})$*, and we then say that* φ *is a* $\mathrm{N}(\mathrm{Q}\mathcal{L})$-theorem. ■

Fact 4.1.7 *Due to the separations enforced between the base system, the relational theory, and the domain theory, in* $\mathrm{N}(\mathrm{Q}\mathcal{L}) = \mathrm{N(QK)} + \mathrm{N}(\mathcal{T}) + \mathrm{N}(\mathcal{D})$ *we have that:*

(i) $\Gamma, \Delta, \Theta \vdash_{\mathrm{N}(\mathrm{Q}\mathcal{L})} w_i R w_j$ *iff* $\Delta \vdash_{\mathrm{N}(\mathcal{T})} w_i R w_j$.

(ii) $\Gamma, \Delta, \Theta \vdash_{\mathrm{N}(\mathrm{Q}\mathcal{L})} w{:}t$ *iff* $\Delta, \Theta \vdash_{\mathrm{N}(\mathcal{T})+\mathrm{N}(\mathcal{D})} w{:}t$. ■

That is, while lwffs are derived from lwffs, rwffs *and* lterms, i.e. $\Gamma, \Delta, \Theta \vdash_{\mathrm{N}(\mathrm{Q}\mathcal{L})} w{:}A$, (i) rwffs are derived from rwffs *alone*, and (ii) lterms are derived from rwffs and lterms *but not* from lwffs.

We conclude this section with some examples of derivations; corresponding Isabelle proofs are given in §5.3.1.

Example 4.1.8 We begin by deriving the rules for $\exists$ using the rules of N(QK), where the side condition on the application of $\exists$ E follows from the condition on the application of $\forall$ I.

$$\dfrac{w{:}A[t/x] \quad w{:}t}{w{:}\exists x(A)}\,\exists \mathrm{I} \quad \rightsquigarrow \quad \dfrac{\dfrac{\dfrac{[w{:}\forall x(\sim A)]^1 \quad w{:}t}{w{:}\sim A[t/x]}\,\forall \mathrm{E} \qquad w{:}A[t/x]}{w{:}\bot}\,{\sim}\mathrm{E}}{w{:}\sim\forall x(\sim A)}\,{\sim}\mathrm{I}^1 \quad . \tag{4.1}$$

$$\dfrac{w_i{:}\exists x(A) \qquad \begin{matrix}[w_i{:}A[t/x]]\ [w_i{:}t]\\ \Pi \\ w_j{:}B\end{matrix}}{w_j{:}B}\,\exists \mathrm{E}$$

$$\rightsquigarrow \quad \dfrac{\dfrac{w_i{:}\sim\forall x(\sim A) \qquad \dfrac{\dfrac{\dfrac{\dfrac{[w_j{:}B\supset\bot]^3 \qquad \begin{matrix}[w_i{:}A[t/x]]^1\ [w_i{:}t]^2\\ \Pi \\ w_j{:}B\end{matrix}}{w_j{:}\bot}\,\supset\mathrm{E}}{w_i{:}\bot}\,\bot\mathrm{E}}{w_i{:}\sim A[t/x]}\,{\sim}\mathrm{I}^1}{w_i{:}\forall x(\sim A)}\,\forall\mathrm{I}^2}{w_i{:}\bot}\,{\sim}\mathrm{E}}{w_j{:}B}\,\bot\mathrm{E}^3 \quad . \tag{4.2}$$

Note that the symmetry between $\Box$ and $\forall$ and between $\Diamond$ and $\exists$ is reflected also in these derivations, which are 'symmetrical' to the derivations (2.1) and (2.2) of the rules for $\Diamond$ given in Example 2.1.14 (except for the necessary application of $\bot$E, or *gf*, in (2.1)).

As a further example, we show that CBF is a theorem of (any extension of) N(QK.i):[4]

$$\dfrac{\dfrac{\dfrac{\dfrac{\dfrac{[w{:}\Box\forall x(A)]^3 \quad [wRw_1]^1}{w_1{:}\forall x(A)}\,\Box\mathrm{E} \qquad \dfrac{[wRw_1]^1 \quad [w{:}t]^2}{w_1{:}t}\,id}{w_1{:}A[t/x]}\,\forall\mathrm{E}}{w{:}\Box A[t/x]}\,\Box\mathrm{I}^1}{w{:}\forall x(\Box A)}\,\forall\mathrm{I}^2}{w{:}\Box\forall x(A)\supset\forall x(\Box A)}\,\supset\mathrm{I}^3 \quad .$$

■

[4]Note that the assumption $w{:}t$ is discharged by the application of $\forall$ I. Informally, CBF is not a theorem of N(QK) because *id* is missing and the application of $\forall$ I at world w cannot discharge $w_1{:}t$; a formal proof of this can be given by exploiting the normalization results we establish in §4.3 to show that there is no normal proof (and, a fortiori, no proof at all) of CBF in N(QK).

In a similar manner, we can prove BF in N(QK.d),

$$\vdash_{\mathrm{N(QK.d)}} w{:}\forall x(\Box A) \supset \Box\forall x(A)\,,$$

and other examples usually considered in standard texts:

$$\vdash_{\mathrm{N(QKB.i)}} w{:}\forall x(\Box A) \supset \Box\forall x(A)\,, \tag{4.3}$$

$$\vdash_{\mathrm{N(QK.d)}} w{:}\Diamond\exists x(A) \supset \exists x(\Diamond A)\,, \tag{4.4}$$

$$\vdash_{\mathrm{N(QK.i)}} w{:}\exists x(\Diamond A) \supset \Diamond\exists x(A)\,, \tag{4.5}$$

$$\vdash_{\mathrm{N(QK.i)}} w{:}\Diamond\forall x(A) \supset \forall x(\Diamond A)\,, \tag{4.6}$$

$$\vdash_{\mathrm{N(QK.i)}} w{:}\exists x(\Box A) \supset \Box\exists x(A)\,. \tag{4.7}$$

Some remarks. The rules *id* and *dd* are interderivable when the rule

$$\frac{w_i R\, w_j}{w_j R\, w_i}\ \mathit{symm}$$

is present, i.e. when the accessibility relation is symmetric (recall that symmetry corresponds to the modal axiom schema B: $A \supset \Box\Diamond A$). (4.3) shows that a quantified modal logic with a symmetric accessibility relation and with increasing domains, e.g. N(QKB.i), validates BF and has therefore constant domains; similarly we can show that CBF is a theorem of N(QKB.d). By (4.4) and (4.5), $\Diamond\exists x(A)$ and $\exists x(\Diamond A)$ are equivalent in N(QK.c); by analysis of normal form proofs, see §4.3, we can show that they are equivalent only in systems with constant domains. Similarly, we can show that, as is the case in Hilbert systems, the converses of (4.6) and (4.7) are not provable even when *dd* is added as a rule.

4.2 SOUNDNESS AND COMPLETENESS

We extend the definitions and results we gave for propositional modal logics in §2.2 to introduce a Kripke semantics for our systems and prove that any system $\mathrm{N}(\mathrm{Q}\mathcal{L})$ obtained by extending N(QK) with a Horn relational theory $\mathrm{N}(\mathcal{T})$ and a domain theory $\mathrm{N}(\mathcal{D})$ is sound and complete with respect to the corresponding semantics.

Definition 4.2.1 *A* (Kripke) model *for* $\mathrm{N}(\mathrm{Q}\mathcal{L})$ *is a tuple* $\mathfrak{M} = (\mathfrak{W}, \mathfrak{R}, \mathfrak{D}, \mathfrak{q}, \mathfrak{a})$, *where* $\mathfrak{W}$ *is a non-empty set of worlds;* $\mathfrak{R} \subseteq \mathfrak{W} \times \mathfrak{W}$; $\mathfrak{D}$ *is a set of objects;* $\mathfrak{q}$ *is a mapping assigning to each member* w *of* $\mathfrak{W}$ *some subset of* $\mathfrak{D}$, *the* domain of quantification *of* w; $\mathfrak{a}$ *is a function interpreting the terms and predicate letters by assigning to them the corresponding kind of intensions with respect to* $\mathfrak{W}$ *and* $\mathfrak{D}$. $\mathfrak{a}(w,t)$ *is an element of* $\mathfrak{D}$, *and for a predicate letter* P *of arity* n, $\mathfrak{a}(w,P)$ *is a set of ordered* n*-tuples,* $\langle a_1, \ldots, a_n\rangle$, *where each* $a_i \in \mathfrak{D}$. *We say that* $\mathfrak{M}$ *has some property of binary relations iff* $\mathfrak{R}$ *has that property. Moreover, for every* $w_i, w_j \in \mathfrak{W}$ *such that* $(w_i, w_j) \in \mathfrak{R}$, *the domains of* $\mathfrak{M}$ *are:* increasing *iff* $\mathfrak{q}(w_i) \subseteq \mathfrak{q}(w_j)$; decreasing *iff* $\mathfrak{q}(w_i) \supseteq \mathfrak{q}(w_j)$; *and* constant *iff* $\mathfrak{q}(w_i) = \mathfrak{q}(w_j)$. *Otherwise, the domains are* varying. ■

Note that we only consider rigid designators [89, 104], where $\mathfrak{a}$ is such that $\mathfrak{a}(w_i,t) = \mathfrak{a}(w_j,t)$ for all $w_i, w_j \in \mathfrak{W}$. Moreover, like in the propositional case, our models do not contain functions corresponding to possible Skolem functions in the signature; when such constants are present, the appropriate Skolem expansion of the model is required [230, p. 137]. As before, we identify labels with identically named worlds.

Definition 4.2.2 *Given a set of lwffs* Γ*, a set of rwffs* Δ *and a set of lterms* Θ*, we call the ordered triple* (Γ, Δ, Θ) *a* proof context*. When* $\Gamma_1 \subseteq \Gamma_2$*,* $\Delta_1 \subseteq \Delta_2$ *and* $\Theta_1 \subseteq \Theta_2$*, we write* $(\Gamma_1, \Delta_1, \Theta_1) \subseteq (\Gamma_2, \Delta_2, \Theta_2)$ *and say that* $(\Gamma_1, \Delta_1, \Theta_1)$ is included in $(\Gamma_2, \Delta_2, \Theta_2)$*. We write* $w{:}A \in (\Gamma, \Delta, \Theta)$ *when* $w{:}A \in \Gamma$*;* $wRw' \in (\Gamma, \Delta, \Theta)$ *when* $wRw' \in \Delta$*; and* $w{:}t \in (\Gamma, \Delta, \Theta)$ *when* $w{:}t \in \Theta$*. Finally, we say that a label* w occurs in *the proof context* (Γ, Δ, Θ)*, in symbols* $w \Subset (\Gamma, \Delta, \Theta)$*, if there exists an* A *such that* $w{:}A \in \Gamma$*, or a* w' *such that* $wRw' \in \Delta$ *or* $wRw' \in \Delta$*, or a* t *such that* $w{:}t \in \Theta$*.* $t \Subset (\Gamma, \Delta, \Theta)$ *is defined analogously.* ■

We now define truth for ground lterms, rwffs and lwffs, where truth for lterms indicates definedness, truth for rwffs indicates accessibility, and quantifiers are treated in each world as ranging over the domain of that world only.

Definition 4.2.3 *We define a ground lterm, rwff or lwff* φ *to be* true *in a model* $\mathfrak{M}$*, in symbols* $\models^{\mathfrak{M}} \varphi$*, as follows. First we ensure, as is standard [167, 213], that we have a name for each object in the domain* $\mathfrak{D}$ *of* $\mathfrak{M}$ *by extending, if necessary, the class of terms with a new constant* c_o *for each* $o \in \mathfrak{D}$*, and then extending* $\mathfrak{a}$ *so that* $\mathfrak{a}(w, \mathrm{c}_o) = o$*. Then we define* $\models^{\mathfrak{M}}$ *to be the smallest relation satisfying:*

$\models^{\mathfrak{M}} w{:}t$	*iff*	$\mathfrak{a}(w,t) \in \mathfrak{q}(w)$;
$\models^{\mathfrak{M}} w_i R w_j$	*iff*	$(w_i, w_j) \in \mathfrak{R}$;
$\models^{\mathfrak{M}} w{:}P(t_1, \ldots, t_n)$	*iff*	$\langle \mathfrak{a}(w,t_1), \ldots, \mathfrak{a}(w,t_n) \rangle \in \mathfrak{a}(w, P)$;
$\models^{\mathfrak{M}} w{:}A \supset B$	*iff*	$\models^{\mathfrak{M}} w{:}A$ *implies* $\models^{\mathfrak{M}} w{:}B$;
$\models^{\mathfrak{M}} w{:}\Box A$	*iff*	*for all* w_i, $\models^{\mathfrak{M}} w R w_i$ *implies* $\models^{\mathfrak{M}} w_i{:}A$;
$\models^{\mathfrak{M}} w{:}\forall x(A)$	*iff*	*for all* t, $\models^{\mathfrak{M}} w{:}t$ *implies* $\models^{\mathfrak{M}} w{:}A[t/x]$.

When $\models^{\mathfrak{M}} \varphi$*, we say that* φ *is* true *in* $\mathfrak{M}$*. By extension:*

$\models^{\mathfrak{M}} \Gamma$	*means that*	$\models^{\mathfrak{M}} w{:}A$ *for all* $w{:}A \in \Gamma$;
$\models^{\mathfrak{M}} \Delta$	*means that*	$\models^{\mathfrak{M}} w_i R w_j$ *for all* $w_i R w_j \in \Delta$;
$\models^{\mathfrak{M}} \Theta$	*means that*	$\models^{\mathfrak{M}} w{:}t$ *for all* $w{:}t \in \Theta$;
$\models^{\mathfrak{M}} (\Gamma, \Delta, \Theta)$	*means that*	$\models^{\mathfrak{M}} \Gamma$, $\models^{\mathfrak{M}} \Delta$ *and* $\models^{\mathfrak{M}} \Theta$;
$\Delta \models^{\mathfrak{M}} w_i R w_j$	*means that*	$\models^{\mathfrak{M}} \Delta$ *implies* $\models^{\mathfrak{M}} w_i R w_j$;
$\Delta \models w_i R w_j$	*means that*	$\Delta \models^{\mathfrak{M}} w_i R w_j$ *for all* $\mathfrak{M}$;
$\Delta, \Theta \models^{\mathfrak{M}} w{:}t$	*means that*	$\models^{\mathfrak{M}} \Delta$ *and* $\models^{\mathfrak{M}} \Theta$ *imply* $\models^{\mathfrak{M}} w{:}t$;
$\Delta, \Theta \models w{:}t$	*means that*	$\Delta, \Theta \models^{\mathfrak{M}} w{:}t$ *for all* $\mathfrak{M}$;
$\Gamma, \Delta, \Theta \models^{\mathfrak{M}} x{:}A$	*means that*	$\models^{\mathfrak{M}} (\Gamma, \Delta, \Theta)$ *implies* $\models^{\mathfrak{M}} x{:}A$;
$\Gamma, \Delta, \Theta \models x{:}A$	*means that*	$\Gamma, \Delta, \Theta \models^{\mathfrak{M}} x{:}A$ *for all* $\mathfrak{M}$.

■

Truth for lwffs built using other operators can be defined in the usual manner, e.g.

$$\begin{aligned} \models^{\mathfrak{M}} w{:}\exists x(A) \quad &\text{iff} \quad \not\models^{\mathfrak{M}} w{:}\forall x(\sim A) \\ &\text{iff} \quad \text{for some } t\,, \models^{\mathfrak{M}} w{:}t \text{ and } \models^{\mathfrak{M}} w{:}A[t/x] \end{aligned}$$

since $\not\models^{\mathfrak{M}} w{:}\bot$ for every w, by Definition 4.2.3.

The explicit embedding of properties of the models, and the capability of explicitly reasoning about them, via lterms and rwffs, require us to consider soundness and completeness also for lterms and rwffs, where we show that $\Delta \vdash_{\mathrm{N(Q\mathcal{L})}} w_i R w_j$ iff $\Delta \models w_i R w_j$, and that $\Delta, \Theta \vdash_{\mathrm{N(Q\mathcal{L})}} w{:}t$ iff $\Delta, \Theta \models w{:}t$.

Definition 4.2.4 *The system* $\mathrm{N(Q\mathcal{L})} = \mathrm{N(QK)} + \mathrm{N}(\mathcal{T}) + \mathrm{N}(\mathcal{D})$ *is* sound *iff*

(i) $\Delta \vdash_{\mathrm{N(Q\mathcal{L})}} w_i R w_j$ *implies* $\Delta \models w_i R w_j$,
(ii) $\Delta, \Theta \vdash_{\mathrm{N(Q\mathcal{L})}} w{:}t$ *implies* $\Delta, \Theta \models w{:}t$, *and*
(iii) $\Gamma, \Delta, \Theta \vdash_{\mathrm{N(Q\mathcal{L})}} w{:}A$ *implies* $\Gamma, \Delta, \Theta \models w{:}A$.

$\mathrm{N(Q\mathcal{L})}$ *is* complete *iff the converses hold, i.e. iff*

(i) $\Delta \models w_i R w_j$ *implies* $\Delta \vdash_{\mathrm{N(Q\mathcal{L})}} w_i R w_j$,
(ii) $\Delta, \Theta \models w{:}t$ *implies* $\Delta, \Theta \vdash_{\mathrm{N(Q\mathcal{L})}} w{:}t$, *and*
(iii) $\Gamma, \Delta, \Theta \models w{:}A$ *implies* $\Gamma, \Delta, \Theta \vdash_{\mathrm{N(Q\mathcal{L})}} w{:}A$. ■

By Lemma 4.2.6 and Lemma 4.2.14 below, we have:

Theorem 4.2.5 $\mathrm{N(Q\mathcal{L})} = \mathrm{N(QK)} + \mathrm{N}(\mathcal{T}) + \mathrm{N}(\mathcal{D})$ *is sound and complete.* ■

4.2.1 Soundness

We extend Lemma 2.2.6.

Lemma 4.2.6 $\mathrm{N(Q\mathcal{L})} = \mathrm{N(QK)} + \mathrm{N}(\mathcal{T}) + \mathrm{N}(\mathcal{D})$ *is sound.*

Proof Throughout the proof let $\mathfrak{M} = (\mathfrak{W}, \mathfrak{R}, \mathfrak{D}, \mathfrak{q}, \mathfrak{a})$ be an arbitrary model for $\mathrm{N(Q\mathcal{L})}$. We proceed by induction on the structure of the $\mathrm{N(Q\mathcal{L})}$-derivations. The base cases, e.g. $w{:}A \in (\Gamma, \Delta, \Theta)$, are trivial. There is a step case for each inference rule of $\mathrm{N(Q\mathcal{L})}$, and we treat only *id*, $\forall$I and $\forall$E as representative cases; the cases for the other rules follow analogously (in particular, the cases for Horn relational rules, for $\bot$E and for the $\Box$ rules are a straightforward adaptation of the corresponding propositional cases in Lemma 2.2.6).

Assume that the domains of $\mathfrak{M}$ are increasing and consider an application of the rule *id*,

$$\dfrac{\begin{matrix}\Pi_1 & \Pi_2 \\ w_i R w_j & w_i{:}t\end{matrix}}{w_j{:}t}\ id \quad ,$$

where Π_1 and Π_2 are the derivations $\Delta_1 \vdash_{\mathrm{N(Q\mathcal{L})}} w_i R w_j$ and $\Delta_2, \Theta \vdash_{\mathrm{N(Q\mathcal{L})}} w_i{:}t$, with $\Delta = \Delta_1 \cup \Delta_2$. Assume $\models^{\mathfrak{M}} \Delta$ and $\models^{\mathfrak{M}} \Theta$. Then, from the induction hypotheses we obtain $\models^{\mathfrak{M}} w_i R w_j$ and $\models^{\mathfrak{M}} w_i{:}t$. Since the domains of $\mathfrak{M}$ are increasing, we conclude $\models^{\mathfrak{M}} w_j{:}t$ by Definition 4.2.3.

The cases for $\forall$ can be obtained from the ones for $\Box$ by exploiting the symmetry between the rules for $\Box$ and $\forall$. Consider an application of the rule $\forall$I,

$$\dfrac{\begin{array}{c}[w{:}t]\\ \Pi\\ w{:}A[t/x]\end{array}}{w{:}\forall x(A)}\ \forall\mathrm{I}\ ,$$

where Π is the derivation $\Gamma, \Delta, \Theta_1 \vdash_{\mathrm{N}(\mathcal{QL})} w{:}A[t/x]$, with $\Theta_1 = \Theta \cup \{w{:}t\}$. By the induction hypothesis, $\Gamma, \Delta, \Theta_1 \vdash_{\mathrm{N}(\mathcal{QL})} w{:}A[t/x]$ implies $\Gamma, \Delta, \Theta_1 \models w{:}A[t/x]$. Assume $\models^{\mathfrak{M}} (\Gamma, \Delta, \Theta)$. Considering the restriction on the application of $\forall$I, we can extend Θ to $\Theta' = \Theta \cup \{w{:}t'\}$ for an arbitrary $t' \notin (\Gamma, \Delta, \Theta)$, and assume $\models^{\mathfrak{M}} \Theta'$. Since $\models^{\mathfrak{M}} \Theta'$ implies $\models^{\mathfrak{M}} \Theta_1$, from the induction hypothesis we obtain $\models^{\mathfrak{M}} w{:}A[t/x]$, that is $\models^{\mathfrak{M}} w{:}A[t'/x]$ for an arbitrary $t' \notin (\Gamma, \Delta, \Theta)$ such that $\models^{\mathfrak{M}} w{:}t'$. We conclude $\models^{\mathfrak{M}} w{:}\forall x(A)$ by Definition 4.2.3.

Consider an application of the rule $\forall$E,

$$\dfrac{\begin{array}{c}\Pi_1\\ w{:}\forall x(A)\end{array} \quad \begin{array}{c}\Pi_2\\ w{:}t\end{array}}{w{:}A[t/x]}\ \forall\mathrm{E}\ ,$$

where Π_1 and Π_2 are the derivations $\Gamma, \Delta_1, \Theta_1 \vdash_{\mathrm{N}(\mathcal{QL})} w{:}\forall x(A)$ and $\Delta_2, \Theta_2 \vdash_{\mathrm{N}(\mathcal{QL})} w{:}t$, with $\Delta = \Delta_1 \cup \Delta_2$ and $\Theta = \Theta_1 \cup \Theta_2$. Assume $\models^{\mathfrak{M}} (\Gamma, \Delta, \Theta)$. Then, from the induction hypotheses we obtain $\models^{\mathfrak{M}} w{:}\forall x(A)$ and $\models^{\mathfrak{M}} w{:}t$, and thus $\models^{\mathfrak{M}} w{:}A[t/x]$ by Definition 4.2.3. ■

4.2.2 Completeness

For simplicity, we extend the completeness proof that we gave for propositional modal logics in §2.2.2, instead of the more general proof for arbitrary propositional non-classical logics given in §3.2.2.

Completeness follows by a Henkin-style proof, where a canonical model

$$\mathfrak{M}^C = (\mathfrak{W}^C, \mathfrak{R}^C, \mathfrak{D}^C, \mathfrak{q}^C, \mathfrak{a}^C)$$

is built to show the contrapositives of the conditions in Definition 4.2.4, i.e.

$$\Delta \nvdash_{\mathrm{N}(\mathcal{QL})} w_i R w_j \text{ implies } \Delta \not\models^{\mathfrak{M}^C} w_i R w_j\,, \tag{4.8}$$

$$\Delta, \Theta \nvdash_{\mathrm{N}(\mathcal{QL})} w{:}t \text{ implies } \Delta, \Theta \not\models^{\mathfrak{M}^C} w{:}t\,, \tag{4.9}$$

$$\Gamma, \Delta, \Theta \nvdash_{\mathrm{N}(\mathcal{QL})} w{:}A \text{ implies } \Gamma, \Delta, \Theta \not\models^{\mathfrak{M}^C} w{:}A\,. \tag{4.10}$$

In particular, given the presence of labelled formulas and explicit assumptions on the relations between the labels and their domains of quantification (i.e. Δ and Θ), in our 'quantified' version of the Lindenbaum lemma (Lemma 4.2.9 below) we consider a 'global' saturated set of labelled formulas, where consistency is also checked against the additional assumptions in Δ and Θ, instead of the usual saturated sets of unlabelled formulas. Moreover, we extend Definition 2.2.9 as follows.

Definition 4.2.7 *For any system* $\mathrm{N}(\mathrm{Q}\mathcal{L}) = \mathrm{N}(\mathrm{QK}) + \mathrm{N}(\mathcal{T}) + \mathrm{N}(\mathcal{D})$ *and proof context* (Γ, Δ, Θ), *let* $\Delta_{\mathrm{N}(\mathrm{Q}\mathcal{L})}$ *be the* deductive closure of Δ under $\mathrm{N}(\mathrm{Q}\mathcal{L})$, *i.e.*

$$\Delta_{\mathrm{N}(\mathrm{Q}\mathcal{L})} =_{def} \{w_i R w_j \mid \Delta \vdash_{\mathrm{N}(\mathrm{Q}\mathcal{L})} w_i R w_j\},$$

and let $\Theta_{\mathrm{N}(\mathrm{Q}\mathcal{L}),\Delta}$ *be the deductive closure of* Θ *under* $\mathrm{N}(\mathrm{Q}\mathcal{L})$ *with respect to* Δ, *i.e.*

$$\Theta_{\mathrm{N}(\mathrm{Q}\mathcal{L}),\Delta} =_{def} \{w{:}t \mid \Delta, \Theta \vdash_{\mathrm{N}(\mathrm{Q}\mathcal{L})} w{:}t\}.$$

■

This allows us to generalize to the quantified case the notion of maximal consistency of a propositional proof context, Definition 2.2.10, as follows.

Definition 4.2.8 *A proof context* (Γ, Δ, Θ) *is* saturated *iff*

(i) (Γ, Δ, Θ) *is consistent, i.e.* $\Gamma, \Delta, \Theta \nvdash_{\mathrm{N}(\mathrm{Q}\mathcal{L})} w{:}\bot$ *for every* w;
(ii) $\Delta = \Delta_{\mathrm{N}(\mathrm{Q}\mathcal{L})}$ *and* $\Theta = \Theta_{\mathrm{N}(\mathcal{L}),\Delta}$;
(iii) *for every* w *and every* A, *either* $w{:}A \in \Gamma$ *or* $w{:} \sim A \in \Gamma$;
(iv) *for every* w, *if* $\Gamma, \Delta, \Theta \vdash_{\mathrm{N}(\mathrm{Q}\mathcal{L})} w{:}t$ *implies* $\Gamma, \Delta, \Theta \vdash_{\mathrm{N}(\mathrm{Q}\mathcal{L})} w{:}A[t/x]$ *for every* t, *then* $\Gamma, \Delta, \Theta \vdash_{\mathrm{N}(\mathrm{Q}\mathcal{L})} w{:}\forall x(A)$; *and*
(v) *for every* w, *if* $\Gamma, \Delta, \Theta \vdash_{\mathrm{N}(\mathrm{Q}\mathcal{L})} wRw_i$ *implies* $\Gamma, \Delta, \Theta \vdash_{\mathrm{N}(\mathrm{Q}\mathcal{L})} w_i{:}B$ *for every* w_i, *then* $\Gamma, \Delta, \Theta \vdash_{\mathrm{N}(\mathrm{Q}\mathcal{L})} w{:}\Box B$. ■

In the Lindenbaum lemma for first-order logic, a saturated set of formulas is inductively built by adding for every formula $\sim \forall x(A)$ a *witness* to its truth, namely a formula $\sim A[c/x]$ for some new individual constant c. This ensures that the set is ω-complete, a property equivalent to condition (iv) in Definition 4.2.8. A similar procedure applies here not only for every lwff $w{:} \sim \forall x(A)$, but also for every lwff $w{:} \sim \Box A$, cf. condition (v) in Definition 4.2.8. That is, together with $w{:} \sim \Box A$, we consistently add $v{:} \sim A$ and wRv for some new v, which acts as a *witness world* to the truth of $w{:} \sim \Box A$. This ensures that the saturated proof context (Γ, Δ, Θ) is such that $w{:}\Box B \in (\Gamma, \Delta, \Theta)$ iff $wRw_i \in (\Gamma, \Delta, \Theta)$ implies $w_i{:}B \in (\Gamma, \Delta, \Theta)$ for every w_i, as shown in Lemma 4.2.10 below.[5]

Lemma 4.2.9 *Every consistent proof context* (Γ, Δ, Θ) *can be extended to a saturated proof context* $(\Gamma^*, \Delta^*, \Theta^*)$.

Proof We first extend the language of $\mathrm{N}(\mathrm{Q}\mathcal{L})$ with infinitely many new constants for witness terms and witness worlds. Systematically let t range over the original terms, s

[5]Recall from §2.2.2 that in the standard completeness proof for unlabelled modal logics, $\mathfrak{W}^C$ is defined to be the set of all saturated sets, and it is possible to show that if $w \in \mathfrak{W}^C$ and $\models^{\mathfrak{M}^C} w{:} \sim \Box A$, then $\mathfrak{W}^C$ also contains a world w' accessible from w that serves as a witness world to the truth of $w{:} \sim \Box A$, i.e. $\models^{\mathfrak{M}^C} w'{:} \sim A$.

range over the new constants for witness terms, and r range over both. Analogously, let w range over labels, v range over the new constants for witness worlds, and u range over both. All these may be subscripted. Let $l_1, l_2, \ldots$ be an enumeration of all lwffs in the extended language. Starting from $(\Gamma_0, \Delta_0, \Theta_0) = (\Gamma, \Delta, \Theta)$, we inductively build a sequence of consistent proof contexts by defining $(\Gamma_{i+1}, \Delta_{i+1}, \Theta_{i+1})$ to be:

- $(\Gamma_i, \Delta_i, \Theta_i)$, if $(\Gamma_i \cup \{l_{i+1}\}, \Delta_i, \Theta_i)$ is inconsistent; else
- $(\Gamma_i \cup \{l_{i+1}\}, \Delta_i, \Theta_i)$, if l_{i+1} is neither $u{:} \sim\Box A$ nor $u{:} \sim\forall x(A)$; else
- $(\Gamma_i \cup \{u{:} \sim\forall x(A), u{:} \sim A[s/x]\}, \Delta_i, \Theta_i \cup \{u{:}s\})$, for an $s \notin (\Gamma_i \cup \{u{:} \sim\forall x(A)\}, \Delta_i, \Theta_i)$, if l_{i+1} is $u{:} \sim\forall x(A)$; else
- $(\Gamma_i \cup \{u{:} \sim\Box A, v{:} \sim A\}, \Delta_i \cup \{uRv\}, \Theta_i)$, for a $v \notin (\Gamma_i \cup \{u{:} \sim\Box A\}, \Delta_i, \Theta_i)$, if l_{i+1} is $u{:} \sim\Box A$.

Now define

$$\Gamma^* = \bigcup_{i\geq 0} \Gamma_i, \quad \Delta^* = \bigcup_{i\geq 0} (\Delta_i)_{\mathrm{N}(\mathcal{QL})}, \quad \text{and} \quad \Theta^* = \bigcup_{i\geq 0} (\Theta_i)_{\mathrm{N}(\mathcal{QL}),\Delta}.$$

It immediately follows that the proof context $(\Gamma^*, \Delta^*, \Theta^*)$ is saturated, since it satisfies all the conditions in Definition 4.2.8. ■

The following lemma states some properties of saturated proof contexts.

Lemma 4.2.10 *Let* $(\Gamma^*, \Delta^*, \Theta^*)$ *be a saturated proof context. Then*

(i) $\Delta^* \vdash_{\mathrm{N}(\mathcal{QL})} u_i R u_j$ *iff* $u_i R u_j \in \Delta^*$.
(ii) $\Delta^*, \Theta^* \vdash_{\mathrm{N}(\mathcal{QL})} u{:}r$ *iff* $u{:}r \in \Theta^*$.
(iii) $\Gamma^*, \Delta^*, \Theta^* \vdash_{\mathrm{N}(\mathcal{QL})} u{:}A$ *iff* $u{:}A \in \Gamma^*$.
(iv) $u{:}A \supset B \in \Gamma^*$ *iff* $u{:}A \in \Gamma^*$ *implies* $u{:}B \in \Gamma^*$.
(v) $u_i{:}\Box B \in \Gamma^*$ *iff* $u_i R u_j \in \Delta^*$ *implies* $u_j{:}B \in \Gamma^*$ *for all* u_j.
(vi) $u{:}\forall x(A) \in \Gamma^*$ *iff* $u{:}r \in \Theta^*$ *implies* $u{:}A[r/x] \in \Gamma^*$ *for all* r.

Proof (i), (ii) and (iii) follow immediately by definition and Fact 4.1.7. We only treat (vi); (iv) follows like case (iii) in Lemma 2.2.12, and the proof of (v) can be easily obtained from the proof of (vi) by exploiting the symmetry between $\Box$ and $\forall$ (or by generalizing case (iv) in Lemma 2.2.12). For the left-to-right direction of (vi) suppose that $u{:}\forall x(A) \in \Gamma^*$. Then, by (iii), we have $\Gamma^*, \Delta^*, \Theta^* \vdash_{\mathrm{N}(\mathcal{QL})} u{:}\forall x(A)$, and, by $\forall$E, we have $\Delta^*, \Theta^* \vdash_{\mathrm{N}(\mathcal{QL})} u{:}r$ implies $\Gamma^*, \Delta^*, \Theta^* \vdash_{\mathrm{N}(\mathcal{QL})} u{:}A[r/x]$ for all r. By (i), (ii) and (iii), conclude $u{:}r \in \Theta^*$ implies $u{:}A[r/x] \in \Gamma^*$ for all r. For the converse, suppose that $w{:}\forall x(A) \notin \Gamma^*$. Then $u{:} \sim\forall x(A) \in \Gamma^*$ and, by the construction of $(\Gamma^*, \Delta^*, \Theta^*)$, there exists an r such that $u{:}r \in \Theta^*$ and $u{:}A[r/x] \notin \Gamma^*$. ■

We can now define the canonical model $\mathfrak{M}^C = (\mathfrak{W}^C, \mathfrak{R}^C, \mathfrak{D}^C, \mathfrak{q}^C, \mathfrak{a}^C)$.

Definition 4.2.11 *Given a saturated proof context* $(\Gamma^*, \Delta^*, \Theta^*)$, *we define the* canonical model $\mathfrak{M}^C$ *for the system* $\mathrm{N}(\mathcal{QL})$ *as follows:*

- $\mathfrak{W}^C = \{u \mid u \Subset (\Gamma^*, \Delta^*, \Theta^*)\}$;
- $(u_i, u_j) \in \mathfrak{R}^C$ *iff* $u_i R u_j \in \Delta^*$;
- $\mathfrak{a}^C(u, r) = r$, *and* $\langle r_1, \ldots, r_n \rangle \in \mathfrak{a}^C(u, P)$ *iff* $u{:}P(r_1, \ldots, r_n) \in \Gamma^*$, *for* P *an* n*–ary predicate;*
- $\mathfrak{q}^C(u) = \{\mathfrak{a}^C(u, r) \mid u{:}r \in \Theta^*\}$;
- $\mathfrak{D}^C = \bigcup_{u \Subset (\Gamma^*, \Delta^*, \Theta^*)} \mathfrak{q}^C(u)$. ■

Like for the propositional case, the standard definition of $\mathfrak{R}^C$, i.e.

$$(u_i, u_j) \in \mathfrak{R}^C \text{ iff } \{A \mid \Box A \in u_i\} \subseteq u_j\,,$$

is not applicable in our setting, since $\{A \mid \Box A \in u_i\} \subseteq u_j$ does *not* imply $\vdash_{\mathrm{N(Q}\mathcal{L})}$ $u_i R u_j$. We would therefore lose completeness for rwffs, since there would be cases, e.g. if $\mathrm{N(Q}\mathcal{L}) = \mathrm{N(QK)}$ and $\Delta = \{\}$, where $\nvdash_{\mathrm{N(Q}\mathcal{L})} u_i R u_j$ but $(u_i, u_j) \in \mathfrak{R}^C$ and thus $\models^{\mathfrak{M}^C} u_i R u_j$. Hence, we instead define $(u_i, u_j) \in \mathfrak{R}^C$ iff $u_i R u_j \in \Delta^*$; note that therefore $u_i R u_j \in \Delta^*$ implies $\{A \mid \Box A \in u_i\} \subseteq u_j$. As a further comparison with the standard definition, note that in the canonical model the label u can be identified with the set of formulas $\{A \mid u{:}A \in \Gamma^*\}$.

The deductive closures of Δ^* and Θ^* ensure not only completeness for rwffs and lterms, but also that the conditions on $\mathfrak{R}^C$ and $\mathfrak{D}^C$ are satisfied, so that $\mathfrak{M}^C$ is really a model for $\mathrm{N(Q}\mathcal{L})$. For example, it is straightforward to show, like we did in the propositional case, that if $\mathrm{N}(\mathcal{T})$ includes *conv1* and *conv2*, then $\mathfrak{R}^C$ is convergent. Analogously, if $\mathrm{N}(\mathcal{D})$ includes, e.g., *id*, it follows that the domains of $\mathfrak{M}^C$ are increasing. Moreover, we immediately have that:

Fact 4.2.12

(i) $u_i R u_j \in (\Gamma^*, \Delta^*, \Theta^*)$ *iff* $\Delta^* \models^{\mathfrak{M}^C} u_i R u_j$.

(ii) $u{:}r \in (\Gamma^*, \Delta^*, \Theta^*)$ *iff* $\Delta^*, \Theta^* \models^{\mathfrak{M}^C} u{:}r$. ■

By Lemma 4.2.10 and Fact 4.2.12, it follows that:

Lemma 4.2.13 $u{:}A \in (\Gamma^*, \Delta^*, \Theta^*)$ *iff* $\Gamma^*, \Delta^*, \Theta^* \models^{\mathfrak{M}^C} u{:}A$.

Proof We proceed by induction on the grade of $u{:}A$, and we treat only the step case where $u{:}A$ is $u{:}\forall x(B)$; the other cases follow analogously.[6] For the left-to-right direction, assume $u{:}\forall x(B) \in \Gamma^*$. Then, by Lemma 4.2.10, $u{:}r \in \Theta^*$ implies $u{:}A[r/x] \in \Gamma^*$, for all r. Hence, by the induction hypothesis and Fact 4.2.12, we obtain $\Gamma^*, \Delta^*, \Theta^* \models^{\mathfrak{M}^C} u{:}B[r/x]$ for all r such that $\Delta^*, \Theta^* \models^{\mathfrak{M}^C} u{:}r$, and thus $\Gamma^*, \Delta^*, \Theta^* \models^{\mathfrak{M}^C} u{:}\forall x(B)$ by Definition 4.2.3. For the converse, assume $u{:}\sim$

[6]As above, the case for $u{:}\forall x(B)$ can be obtained from the case for $u{:}\Box B$, cf. Lemma 2.2.15 for the propositional case, by exploiting the symmetry between $\Box$ and $\forall$.

$\forall x(B) \in \Gamma^*$. Then, by Lemma 4.2.10, $u{:}r \in \Theta^*$ and $u{:} \sim B[r/x] \in \Gamma^*$, for some r. Fact 4.2.12 and the induction hypothesis yield $\Delta^*, \Theta^* \models^{\mathfrak{M}^C} u{:}r$ and $\Gamma^*, \Delta^*, \Theta^* \models^{\mathfrak{M}^C} u{:} \sim B[r/x]$, i.e. $\Gamma^*, \Delta^*, \Theta^* \models^{\mathfrak{M}^C} u{:} \sim\forall x(B)$ by Definition 4.2.3. ∎

It is now a simple matter to show (4.8), (4.9) and (4.10), analogously to Lemma 2.2.16, and thus prove that:

Lemma 4.2.14 $\mathrm{N}(\mathrm{Q}\mathcal{L}) = \mathrm{N}(\mathrm{QK}) + \mathrm{N}(\mathcal{T}) + \mathrm{N}(\mathcal{D})$ *is complete.* ∎

Some remarks and comparisons are in order. Our proof is modular: the same method applies uniformly to every system $\mathrm{N}(\mathrm{Q}\mathcal{L})$. As remarked above, this is *not* the case for the completeness proof of Hilbert-style systems for quantified modal logics based on free logic [104, 141]. Garson himself points out that his proof "lacks generality" [104, pp. 280-281], since (i) it does not work for logics with constant domains, and (ii) it is not general with respect to the underlying propositional modal logic (although there are tricks one can use to overcome the difficulties for particular systems). As we have shown, none of these problems applies in our approach.

Most importantly, being complete, our systems are adequate presentations of the Kripke semantics, and are thus equivalent to the corresponding Hilbert systems only when these are themselves complete with respect to it. For example, by the results in [141], N(QKT42.i) is equivalent to the Hilbert system H(QS4.2) since they are both complete with respect to reflexive, transitive and convergent Kripke models with increasing domains, but N(QKT42.c) is not equivalent to the Hilbert system H(QS4.2 + BF), since the latter is incomplete with respect to reflexive, transitive and convergent Kripke models with constant domains.

4.3 NORMALIZATION AND ITS CONSEQUENCES

We extend our results of §2.3 to show that derivations in $\mathrm{N}(\mathrm{QK}) + \mathrm{N}(\mathcal{T}) + \mathrm{N}(\mathcal{D})$ have additional properties: derivations of lwffs can be reduced to a normal form that does not contain unnecessary detours and satisfies a subformula property. This provides us with positive results, such as alternative proofs of the consistency of our systems and restricted search space for proofs. It also allows us to establish negative results, such as how incompleteness can arise; we show how analysis of normal forms provides a basis for investigating tradeoffs in formalizations also in the quantified case.

Recall that any lwff $w{:}A$ in a derivation is the root of a tree of rule applications leading back to assumptions. The lwffs in this tree other than $w{:}A$ we call *side lwffs* of $w{:}A$. A *maximal lwff* in a derivation is an lwff that is both the conclusion of an introduction rule and the major premise of an elimination rule. A maximal lwff constitutes a detour in a derivation, and we remove it by the application of the corresponding *proper reduction*. Three possible configurations (for $\supset$, $\Box$ and $\forall$) result in a maximal lwff in a derivation. We have already given the proper reductions for $\supset$

and $\Box$, albeit in the propositional case, in (2.5) and (2.6). The proper reduction for $\forall$ is

$$\dfrac{\dfrac{\begin{matrix}[w{:}t_i]^1\\ \Pi_1\\ w{:}A[t_i/x]\end{matrix}}{w{:}\forall x(A)}\,\forall\mathrm{I}^1 \quad \begin{matrix}\Pi_2\\ w{:}t_j\end{matrix}}{w{:}A[t_j/x]}\,\forall\mathrm{E} \quad \rightsquigarrow \quad \begin{matrix}\Pi_2\\ w{:}t_j\\ \Pi_1[t_j/t_i]\\ w{:}A[t_j/x]\end{matrix}$$

where $\Pi[t_j/t_i]$ is obtained from Π by systematically substituting t_j for t_i, possibly with a renaming of the variables to avoid clashes. Note that Π_2 is empty when the domain theory is empty. Like for (classical) propositional modal systems, we define:

Definition 4.3.1 *A derivation is in* normal form (*is a* normal derivation) *iff it contains no maximal lwffs.* ■

Following §2.3, we can straightforwardly show that each proper reduction reduces a suitable well-ordered measure on derivations. Hence, the reduction process must eventually terminate with a derivation free of maximal lwffs. We have:

Lemma 4.3.2 *Every derivation of $w{:}A$ from Γ, Δ, Θ in* $\mathrm{N(QK)} + \mathrm{N}(\mathcal{T}) + \mathrm{N}(\mathcal{D})$ *reduces to a derivation in normal form.*

Proof First, note that derivations in the Horn theories $\mathrm{N}(\mathcal{T})$ and $\mathrm{N}(\mathcal{D})$ cannot introduce maximal lwffs. Then, consider a derivation Π of $w{:}A$ from Γ, Δ, Θ in $\mathrm{N(QK)} + \mathrm{N}(\mathcal{T}) + \mathrm{N}(\mathcal{D})$, and from the set of maximal lwffs of Π pick some $w_i{:}B$ that has the highest grade and has maximal lwffs only of lower grade as side lwffs. Let Π' be the reduction of Π at $w_i{:}B$. Π' is also a derivation of $w{:}A$ from Γ, Δ, Θ in $\mathrm{N(QK)} + \mathrm{N}(\mathcal{T}) + \mathrm{N}(\mathcal{D})$ and no new maximal lwff as large, or larger than $w_i{:}B$ has been introduced. Hence, by a finite number of similar reductions we obtain a derivation of $w{:}A$ from Γ, Δ, Θ in $\mathrm{N(QK)} + \mathrm{N}(\mathcal{T}) + \mathrm{N}(\mathcal{D})$ containing no maximal lwffs. ■

We can now exploit Lemma 4.3.2 to show that derivations in $\mathrm{N}(\mathrm{Q}\mathcal{L}) = \mathrm{N(QK)} + \mathrm{N}(\mathcal{T}) + \mathrm{N}(\mathcal{D})$ have a well-defined structure. We start by observing that, given Fact 4.1.7, we can strictly separate $\mathrm{N}(\mathrm{Q}\mathcal{L})$-derivations involving lwffs, rwffs and lterms as follows.

Fact 4.3.3 *Consider derivations in* $\mathrm{N}(\mathrm{Q}\mathcal{L}) = \mathrm{N(QK)} + \mathrm{N}(\mathcal{T}) + \mathrm{N}(\mathcal{D})$.

(i) *A derivation of an lwff can depend on a derivation of an rwff (via an application of* $\Box$E*), but not vice versa.*

(ii) *A derivation of an lwff can depend on a derivation of an lterm (via an application of* $\forall$E*), but not vice versa.*

(iii) *A derivation of an lterm can depend on a derivation of an rwff (via an application of id or dd), but not vice versa.* ■

As a consequence, any derivation of an lwff is structured as a central derivation Π in the base system N(QK) 'decorated' with (i) subderivations in the relational theory, which attach onto Π through instances of □E, and (ii) subderivations in the domain theory, which attach onto Π through instances of ∀E. Moreover, the structure of the central N(QK)-derivation Π, when in normal form, can be further characterized by identifying particular sequences of lwffs (i.e. threads and tracks, as in §2.3), and showing that in these sequences there is an ordering on inferences. By exploiting this ordering, we can show a subformula property for all extensions of N(QK).

Definition 4.3.4 *B is a* subformula *of A iff (i) A is B; or (ii) A is $A_1 \supset A_2$ and B is a subformula of A_1 or A_2; or (iii) A is $\Box A_1$ and B is a subformula of A_1; or (iv) A is $\forall x(A_1)$ and B is a subformula of $A_1[t/x]$ for some t. We say that $w'{:}B$ is a (labelled) subformula of $w{:}A$ iff B is a subformula of A.* ■

Definition 4.3.5 *Given a derivation $\Gamma, \Delta, \Theta \vdash w_i{:}A$, let $\mathcal{S}$ be the set of subformulas of the formulas in $\{C \mid w_k{:}C \in \Gamma \cup \{w_i{:}A\}$ for some $w_k\}$, i.e. $\mathcal{S}$ is the set consisting of the subformulas of the assumptions Γ and of the conclusion $w_i{:}A$. We say that $\Gamma, \Delta, \Theta \vdash w_i{:}A$ satisfies the* subformula property *iff for all lwffs $w_j{:}B$ in the derivation (i) $B \in \mathcal{S}$; or (ii) B is an assumption $D \supset \bot$ discharged by an application of $\bot$E, where $D \in \mathcal{S}$; or (iii) B is an occurrence of $\bot$ obtained by $\supset$E from an assumption $D \supset \bot$ discharged by an application of $\bot$E, where $D \in \mathcal{S}$; or (iv) B is an occurrence of $\bot$ obtained by an application of $\bot$E that does not discharge any assumption (i.e. an occurrence of $\bot$ obtained by an application of gf).* ■

Lemma 4.3.6 *Every normal derivation of $w{:}A$ from Γ, Δ, Θ in* N(QK) + N($\mathcal{T}$) + N($\mathcal{D}$) *satisfies the subformula property.* ■

To summarize, we can generalize Theorem 2.3.14 and its commentary as follows.

Theorem 4.3.7 *Our labelled ND systems have the following properties.*

(i) *The deduction machinery is minimal: labelled ND systems formalize a minimum fragment of first-order logic required by the semantics of quantified modal logics with Horn axiomatizable properties of the relations.*
(ii) *Derivations are strictly separated as in Fact 4.3.3.*
(iii) *Derivations normalize: the derivations of lwffs have a well-structured normal form that satisfies the subformula property.* ■

For comparison, consider again the semantic embedding approach: a quantified modal logic is encoded as a first-order theory by axiomatizing an appropriate definition of truth, but all structure is lost as relations, predicates and terms are flattened into first-order formulas, and derivations of predicates are mingled with derivations of relations and terms.

From Theorem 4.3.7, standard corollaries follow; for example, our ND systems are consistent since there is no introduction rule for $\bot$. We can also exploit the existence of normal forms to design equivalent cut-free sequent systems and automate proof search.[7] However, in exchange for this extra structure there are limits to the generality of the formulation: the properties in Theorem 4.3.7 depend on design decisions we have made, in particular, the use of Horn theories. This, of course, limits what we can formalize in comparison to a semantic embedding in first-order logic. There are tradeoffs in the possible formalizations: if we remove these limitations by introducing first-order (or even higher-order) theories of the accessibility relation and of the domains of quantification, then, in general, to achieve complete presentations we must give up the properties in Theorem 4.3.7. In particular, we must give up the ability to separate derivations so that reasoning can be factored into interacting theories, and instead retreat to systems where derivations arbitrarily mix lwffs, rwffs and lterms. Such liberalized systems essentially amount to a direct formalization (embedding) of the semantics in first-order logic.

To illustrate this, we consider problems that appear only in the quantified case, namely the tradeoffs in ND presentations of logics with first-order domain theories; the tradeoffs for first-order relational theories discussed in §2.3 generalize straightforwardly to the quantified case. Before doing this, however, let us briefly discuss when and why first-order domain theories might be of interest.

As we have shown above, all the domain properties commonly considered, i.e. that the domains are varying, increasing, decreasing or constant, can be easily axiomatized by Horn clauses. However, in particular applications of quantified modal logics, we might want to consider more complicated properties. For example, we might want to state explicitly that some object does not exist in a world w. Or we might want to refine the increasing domains property by specifying the size of the increment, e.g. that there are at least n 'new' objects. Such properties require a full first-order (or even higher-order) domain theory. Analogously to the case of relational theories, it is not conceptually difficult to introduce first-order domain theories.[8] We just need to introduce a standard first-order natural deduction system for reasoning about labelled terms built using the operators $\emptyset$ (falsum), $\sqsupset$ (implies), $\bigsqcap$ (for all); lterms over other connectives and quantifiers, e.g. $-$ (not), $\sqcap$ (and), $\sqcup$ (or), $\bigsqcup$ (exists), and corresponding rules are defined as usual. Note that we use the same operators that we used for relational theories in §2.3.2; since here we do not consider such relational theories, no confusion should arise.

The particular properties of the domains are then added as axioms or rules directly in their full form, i.e. the *first-order domain theory* $\mathrm{N}(\mathcal{D}_{\mathrm{F}}) = \mathrm{N}_{\mathcal{D}} + \mathcal{C}_{D}$ is obtained by adding a collection $\mathcal{C}_D$ of such axioms or rules to the first-order ND system for labelled terms $\mathrm{N}_{\mathcal{D}}$. The rules of $\mathrm{N}_{\mathcal{D}}$ are given in Figure 4.5, where the τ_i range over

[7] In §6 we give cut-free labelled sequent systems for propositional and quantified non-classical logics, and then in Part II we show that in the sequent systems for certain propositional modal logics we can bound applications of the contraction rules and thus establish decidability. This will not be the case for quantified modal logics (since we cannot bound the use of universally quantified subformulas), but still the existence of partitioned normal forms allows us to restrict the search space during theorem proving.

[8] Note that the possibility of expressing complex properties of the domains of quantification in our systems provides another advantage of our approach with respect to Hilbert-style axiomatizations, since it is often difficult, if not impossible, to give Hilbert-style axioms corresponding to such properties.

$$\dfrac{\begin{matrix}[w_i{:}\tau \sqsupset \emptyset] \\ \vdots \\ w_j{:}\emptyset\end{matrix}}{w_i{:}\tau}\ \emptyset\text{E} \qquad \dfrac{\begin{matrix}[w{:}\tau_1] \\ \vdots \\ w{:}\tau_2\end{matrix}}{w{:}\tau_1 \sqsupset \tau_2}\ \sqsupset\text{I} \qquad \dfrac{w{:}\tau_1 \sqsupset \tau_2 \quad w{:}\tau_1}{w{:}\tau_2}\ \sqsupset\text{E}$$

$$\dfrac{w{:}\tau}{w{:}\bigsqcap x(\tau)}\ \bigsqcap\text{I} \qquad \dfrac{w{:}\bigsqcap x(\tau)}{w{:}\tau[t/x]}\ \bigsqcap\text{E}$$

In $\bigsqcap$I, x must not occur free in any open assumption on which $w{:}\tau$ depends.

Figure 4.5. The rules of $\mathrm{N}_{\mathcal{D}}$

terms (note that $\emptyset$E is a 'global falsum', and we can therefore formulate also local and universal variants).

For example, to state that the domain of each world contains at least one term we add the rule

$$\dfrac{}{w{:}\bigsqcup x(x)}\ \textit{non-empty} \quad .$$

The non-emptiness of the domains is a property expressible as a Horn rule, since we can express it as

$$\dfrac{}{w{:}c(w)}$$

where c is a Skolem function constant. However, it is interesting to consider it in its full (unskolemized) form, since even this very simple property gives rise to a tradeoff between expressivity, completeness and metatheoretical properties of our systems.

As remarked in §4.1, from free logic [28] we know that *non-empty* corresponds to the axiom schema

$$w{:}\forall x(A) \supset \exists x(A)\,. \tag{4.11}$$

Therefore there should be a proof of it in the extension of N(QK) with a first-order domain theory $\mathrm{N}(\mathcal{D}_\mathrm{F}) = \mathrm{N}_{\mathcal{D}} + \mathcal{C}_D$, where $\mathcal{C}_D$ consists of *non-empty* as the only property. Moreover, since normalization in $\mathrm{N(QK)} + \mathrm{N}_{\mathcal{D}} + \mathcal{C}_D$ can be shown by extending Lemmas 4.3.2 and 4.3.6, if there is a proof of (4.11), then there is a normal one satisfying the subformula property. But reasoning backwards from (4.11), we see that we need a proof of $w{:}t$ from *non-empty*:

$$\dfrac{\dfrac{\dfrac{[\forall x(A)]^1 \quad w{:}t}{w{:}A[t/x]}\ \forall\text{E} \qquad w{:}t}{w{:}\exists x(A)}\ \exists\text{I}}{w{:}\forall x(A) \supset \exists x(A)}\ \supset\text{I}^1 \quad .$$

However, such a proof cannot exist: we can only use *non-empty* as the major premise in an application of the derived rule $\bigsqcup$E,

$$\dfrac{w_i{:}\bigsqcup x(x) \qquad \begin{array}{c}[w_i{:}x]\\ \vdots \\ w_j{:}t\end{array}}{w_j{:}t}\,\bigsqcup\mathrm{E} \qquad \rightsquigarrow \qquad \dfrac{\dfrac{w_i{:}-\bigsqcap x(-x) \qquad \dfrac{\dfrac{\dfrac{[w_j{:}-t]^2 \qquad \begin{array}{c}[w_i{:}x]^1\\ \vdots \\ w_j{:}t\end{array}}{w_j{:}\emptyset}\,-\mathrm{E}}{w_i{:}-x}\,-\mathrm{I}^1}{w_i{:}\bigsqcap x(-x)}\,\bigsqcap\mathrm{I}}{w_i{:}\emptyset}\,-\mathrm{E}}{w_j{:}t}\,\emptyset\mathrm{E}^2 ,$$

where the side condition of $\bigsqcup$E requires that x must not occur free in $w_j{:}t$ or in any assumption on which the upper occurrence of $w_j{:}t$ depends other than $w_i{:}x$; in particular, $w_j{:}t$ cannot be $w_i{:}x$.[9]

Thus, we cannot derive $w{:}t$ by *non-empty* and $\bigsqcup$E, and (4.11) is not provable in $\mathrm{N(QK)} + \mathrm{N}_{\mathcal{D}} + \mathcal{C}_D$. As a consequence, $\mathrm{N(QK)} + \mathrm{N}_{\mathcal{D}} + \mathcal{C}_D$ is not complete with respect to its corresponding semantics (in which (4.11) is a valid formula), and we have:

Theorem 4.3.8 *There are systems* $\mathrm{N(QK)}+\mathrm{N}(\mathcal{T})+\mathrm{N}(\mathcal{D}_{\mathrm{F}})$*, with* $\mathrm{N}(\mathcal{D}_{\mathrm{F}}) = \mathrm{N}_{\mathcal{D}}+\mathcal{C}_D$*, that are incomplete with respect to the corresponding Kripke models with domains of quantification defined by a collection* $\mathcal{C}_D$ *of first-order axioms.* ■

As for first-order relational theories, we can restore completeness by giving up the separations in our systems. Specifically, we need rules that allow us to propagate, in either direction, inconsistency (falsum) between the base system and the domain theory extending it. The addition of the rules

$$\dfrac{w_i{:}\bot}{w_j{:}\emptyset}\,\mathit{uft}_1 \qquad \text{and} \qquad \dfrac{w_j{:}\emptyset}{w_i{:}\bot}\,\mathit{uft}_2$$

(*universal falsum for terms*) allows us to mingle derivations of lwffs with derivations of lterms, and we can then prove (4.11) as follows:

$$\dfrac{\dfrac{}{w{:}\bigsqcup x(x)}\,\mathit{non\text{-}empty} \qquad \dfrac{\dfrac{\dfrac{[w{:}\forall x(A)]^1 \quad [w{:}x]^2}{w{:}A[x/x]}\,\forall\mathrm{E} \qquad [w{:}x]^2}{w{:}\exists x(A)}\,\exists\mathrm{I}}{w{:}\forall x(A) \supset \exists x(A)}\,\supset\mathrm{I}^1}{w{:}\forall x(A) \supset \exists x(A)}\,\bigsqcup\mathrm{E}_{lwff}{}^2 .$$

[9]The derivations of the rules

$$\dfrac{\begin{array}{c}[w{:}\tau]\\ \vdots \\ w{:}\emptyset\end{array}}{w{:}-\tau}\,-\mathrm{I} \qquad \text{and} \qquad \dfrac{w{:}-\tau \quad w{:}\tau}{w{:}\emptyset}\,-\mathrm{E}$$

are straightforward (analogously to the derivations of $\sim$I and $\sim$E).

Note that we have used a derived rule of the domain theory, $\bigsqcup \mathrm{E}_{lwff}$, to infer an lwff; the derivation of $\bigsqcup \mathrm{E}_{lwff}$ requires using uft_1 and uft_2, namely:

$$\dfrac{w{:}\bigsqcup x(\tau) \qquad \begin{array}{c}[w{:}\tau]\\ \vdots \\ w{:}A\end{array}}{w{:}A}\bigsqcup \mathrm{E}_{lwff} \quad \rightsquigarrow \quad \dfrac{\dfrac{\dfrac{w{:}-\bigsqcap x(-\tau) \qquad \dfrac{\dfrac{\dfrac{\dfrac{[w{:}\sim A]^2 \qquad \begin{array}{c}[w{:}\tau]^1\\ \vdots \\ w{:}A\end{array}}{w{:}\bot}\sim\mathrm{E}}{w{:}\emptyset}uft_1}{w{:}-\tau}-\mathrm{I}^1}{w{:}\bigsqcap x(-\tau)}\bigsqcap \mathrm{I}}{w{:}\emptyset}-\mathrm{E}}{w{:}\bot}uft_2}{w{:}A}\bot\mathrm{E}^2$$

where $\bigsqcup \mathrm{E}_{lwff}$ has the side condition that x does not occur free in $w{:}A$ or in any assumption on which the upper occurrence of $w{:}A$ depends other than $w{:}\tau$.

Hence, to restore completeness not only have we lost the separation of derivations, but also the other good metatheoretical properties in Theorem 4.3.7, in exchange for a system in which, like in semantic embedding, derivations of lwffs are mingled with derivations of rwffs and lterms. In fact, by defining a suitable mapping between derivations, like we did in Definition 2.3.19 for propositional modal logics, we can show that the quantified system with uft_1 and uft_2 is essentially equivalent to the usual semantic embedding of quantified modal logics in first-order logic.

5 ENCODING LABELLED NON-CLASSICAL LOGICS IN ISABELLE

We have used the generic theorem prover Isabelle [181] to implement the non-classical logis we presented. The logical basis of Isabelle is a natural deduction presentation of minimal implicational predicate logic with universal quantification over all higher-types [179].[1] We call this metalogic *Meta*, and to prevent object/meta confusion we use $\bigwedge$ to represent *Meta*'s universal quantifier and $\Rightarrow$ for implication.

An object logic is encoded in Isabelle by declaring a theory, composed of a signature and axioms, which are formulas in the language of *Meta*. The axioms are used to establish the validity of judgements, which are assertions about syntactic objects declared in the signature. Derivations are constructed by deduction in the metalogic.

5.1 ENCODING PROPOSITIONAL MODAL LOGICS

5.1.1 Implementation

We begin by declaring a theory $Meta_{\mathrm{N(K)}}$, which encodes the base propositional modal ND system N(K).

[1] Isabelle's logic also contains equality (that of the λ-calculus under α, β, and η-conversion), but we do not need to consider this, since, in the analysis of derivations in the metalogic, we shall identify terms with their $\beta\eta$ normal forms. This is possible as terms in our metatheories are terms in the simply-typed λ-calculus (with additional function constants) and every term can be reduced to a normal form that is unique up to α-conversion.

```
K = Pure +          (*  K extends Pure (Isabelle's metalogic)  *)
                    (* with the following signature and axioms *)

types (* Definition of type constructors *)
  label,o 0

arities (* Addition of the arity 'logic' to the existing types *)
  label, o :: logic

consts
  (* Logical operators *)
  falsum   :: "o"
  imp      :: "[o, o] => o"               ("_ --> _" [25,26] 26)
  not      :: "o => o"                    ("~ _" [40] 40)
  box      :: "o => o"                    ("[]_" [50] 50)
  dia      :: "o => o"                    ("<>_" [50] 50)

  (* Judgements *)
  LF       :: "[label, o] => prop"        ("(_ : _)" [0,0] 100)
  RF       :: "[label, label] => prop"    ("(_ R _)" [0,0] 100)

rules
  (* Axioms representing the object-level rules *)
  falsumE  "(x: A --> falsum ==> y: falsum) ==> x:A"
  impI     "(x:A ==> x:B) ==> x: A --> B"
  impE     "x: A --> B ==> x:A ==> x:B"
  boxI     "(!!y. (x R y ==> y:A)) ==> x:[]A"
  boxE     "x:[]A ==> x R y ==> y:A"

  (* Definitions *)
  not_def  "x: ~A == x: A --> falsum"
  dia_def  "x: <>A == x: ~([](~A))"

end
```

Figure 5.1. Isabelle encoding of $\mathrm{N(K)}$

The signature of $Meta_{\mathrm{N(K)}}$ declares two types *label* and *o*, which denote labels and unlabelled modal formulas, respectively. Logical operators are declared as typed constants over this signature, e.g. *box* of type $o \Rightarrow o$. There are two judgements, which correspond to predicate symbols in the metalogic: $\mathcal{LF}$ and $\mathcal{RF}$, which stand for 'Labelled Formula' and 'Relational Formula'. $\mathcal{LF}(x, A)$ and $\mathcal{RF}(x, y)$ respectively express the judgements that $x{:}A$ is a provable lwff and that $x R y$ is a provable rwff.[2] The axioms for $\mathcal{LF}$ are a direct axiomatization of the rules in Figure 2.1; for example,

[2]Indexed judgements similar to ours have been adopted in other encodings of some non-classical logics in Logical Frameworks, e.g. modal and dynamic logics in [10, 18, 133, 216].

for $\Box$I we give the axiom

$$\bigwedge x \bigwedge A((\bigwedge y(\mathcal{RF}(x,y) \Rightarrow \mathcal{LF}(y,A))) \Rightarrow \mathcal{LF}(x,\Box A)).$$

Figure 5.1 contains our entire Isabelle declaration for the theory $Meta_{\mathrm{N(K)}}$. Some brief explanations are in order.[3] First, we use `typewriter font` for displaying concrete Isabelle syntax which has come from actual Isabelle sessions: `Pure` stands for Isabelle's metalogic *Meta*, and `K` for $Meta_{\mathrm{N(K)}}$. The operators `!!` and `==>` are concrete syntax in Isabelle for universal quantification ($\bigwedge$) and implication ($\Rightarrow$) in *Meta*. `not_def` and `dia_def` are the definitions (using meta-equality `==`) of the logical operators $\sim$ and $\Diamond$; other operators can be defined similarly. The use of mixfix annotations, declared with information for Isabelle's parser, allows us to abbreviate `imp` with `-->`, `not` with `~`, `box` with `[]`, `dia` with `<>`, `LF(x,A)` with `x:A`, and `RF(x,y)` with `x R y`. Finally, note that, in axioms, free variables are implicitly outermost universally quantified, that comments are added between '`(*`' and '`*)`', and that there is additional information present to fix notation and help Isabelle's parser.

A system $\mathrm{N}(\mathcal{L}) = \mathrm{N(K)} + \mathrm{N}(\mathcal{T})$ is encoded by extending $Meta_{\mathrm{N(K)}}$ with the theory $Meta_{\mathrm{N}(\mathcal{T})}$, which encodes $\mathrm{N}(\mathcal{T})$.[4] The axioms for $\mathcal{RF}$ are given by directly translating Horn relational rules to axioms in *Meta*: each rule corresponds to an iterated (Curried) implication where the assumptions of the rule together imply the conclusion.

Theories in Isabelle correspond to instances of an abstract datatype in the ML programming language [180] and Isabelle provides means for creating elements of these types, extending them, and combining them. We use these facilities to combine and extend our modal theories. This is best illustrated by an example. `KT` (for $Meta_{\mathrm{N(T)}}$) is obtained by extending `K` with the axiom `refl`; this is specified as follows.

```
KT = K +
rules
  refl        "x R x"
end
```

Again, recall that outermost quantifiers are left implicit, so the above is shorthand for adding `!!x. x R x` as an axiom to `K`. Similarly, `K4` is formed by extending `K` with `trans`.

```
K4 = K +
rules
  trans       "x R y ==> y R z ==> x R z"
end
```

[3]Further details on Isabelle syntax and theory declarations can be found in [181]. Note that we here employ the release 'Isabelle98'; other releases, available from the world-wide-web pages of Isabelle, may require small changes to our encodings.

[4]Note that we could also extend $Meta_{\mathrm{N(K)}}$ by adding constants and rules that encode new logical operators, as we do, e.g., below where we introduce $\leftrightarrow$.

We may now obtain `KT4`, i.e. `S4`, by similarly extending `KT` (or `K4` or `K`); alternatively, we may apply the ML-function `merge_theories` to `KT` and `K4`. As remarked above, `KT4` inherits theorems and derived rules from its ancestor logics. As an example, consider the `KT4`-theorem `x: []A <-> [][]A`. The formulas `x: []A --> [][]A` and `x: [][]A --> []A` are theorems of `K4` and `KT`, respectively:

$$\dfrac{\dfrac{\dfrac{\dfrac{[x{:}\Box A]^3 \quad \dfrac{[x R\, y]^2 \quad [y R\, z]^1}{x R\, z}\,\textit{trans}}{z{:}A}\,\Box\mathrm{E}}{y{:}\Box A}\,\Box\mathrm{I}^1}{x{:}\Box\Box A}\,\Box\mathrm{I}^2}{x{:}\Box A \supset \Box\Box A}\,\supset\mathrm{I}^3 \qquad\qquad \dfrac{\dfrac{[x{:}\Box\Box A]^1 \quad \dfrac{}{x R\, x}\,\textit{refl}}{x{:}\Box A}\,\Box\mathrm{E}}{x{:}\Box\Box A \supset \Box A}\,\supset\mathrm{I}^1 \quad . \tag{5.1}$$

In §5.1.3 below, we show how these theorems are interactively proved in Isabelle in the corresponding theories and then applied to conclude:

$$\dfrac{\begin{matrix}\vdots \\ x{:}\Box A \supset \Box\Box A\end{matrix} \quad \begin{matrix}\vdots \\ x{:}\Box\Box A \supset \Box A\end{matrix}}{x{:}\Box A \leftrightarrow \Box\Box A} \leftrightarrow \mathrm{I} \quad .$$

Note that this requires adding a definition of `<->` to our theory, which can be done in the standard way, e.g.

```
iff_def  "x: A <-> B == x: ~((A --> B) --> ~ (B --> A))"
```

We also need to add, or derive, an axiom

```
iffI    "x: A --> B ==> (x: B --> A ==> x: A <-> B)"
```

for the rule $\leftrightarrow$ I.

As a further example of theory definition, `K2` is obtained by extending `K` with the constant function symbol `g` and with the axioms `conv1` and `conv2`, i.e.

```
K2 = K +
consts
  g          :: "[label,label,label] => label"
rules
  conv1      "x R y ==> x R z ==> y R g(x,y,z)"
  conv2      "x R y ==> x R z ==> z R g(x,y,z)"
end
```

In §5.1.3, we use this theory to prove $x{:}\Diamond\Box A \supset \Box\Diamond A$, the characteristic axiom of `K2`. The examples we work through in Isabelle should help convince the reader that the approach we have taken to interactive theorem proving for modal and other non-classical logics is both simple and flexible. In particular, it supports hierarchical structuring of theories and inheritance of theorems between them.

5.1.2 Faithfulness and adequacy

When one logic encodes another, correctness of the encoding must be shown. A technique established with the Edinburgh Logical Framework [125] is to demonstrate a correspondence between derivations in the object logic and derivations in the metalogic by considering certain normal forms for derivations in the metalogic. Given $\Gamma = \{x_1{:}A_1, \ldots, x_n{:}A_n\}$ and $\Delta = \{x_1 R y_1, \ldots, x_m R y_m\}$, in what follows we abuse notation and write $\mathcal{LF}(\Gamma)$ and $\mathcal{RF}(\Delta)$ for the sets $\{\mathcal{LF}(x_1, A_1), \ldots, \mathcal{LF}(x_n, A_n)\}$ and $\{\mathcal{RF}(x_1, y_1), \ldots, \mathcal{RF}(x_m, y_m)\}$.

Definition 5.1.1 $Meta_{\mathrm{N}(\mathcal{L})}$ *is* faithful (with respect to $\mathrm{N}(\mathcal{L})$) *iff*

(i) $\mathcal{RF}(\Delta) \vdash \mathcal{RF}(x, y)$ *in* $Meta_{\mathrm{N}(\mathcal{L})}$ *implies* $\Delta \vdash xRy$ *in* $\mathrm{N}(\mathcal{L})$, *and*
(ii) $\mathcal{LF}(\Gamma), \mathcal{RF}(\Delta) \vdash \mathcal{LF}(x, A)$ *in* $Meta_{\mathrm{N}(\mathcal{L})}$ *implies* $\Gamma, \Delta \vdash x{:}A$ *in* $\mathrm{N}(\mathcal{L})$.

$Meta_{\mathrm{N}(\mathcal{L})}$ *is* adequate (with respect to $\mathrm{N}(\mathcal{L})$) *iff the converses hold, i.e. iff*

(i) $\Delta \vdash xRy$ *in* $\mathrm{N}(\mathcal{L})$ *implies* $\mathcal{RF}(\Delta) \vdash \mathcal{RF}(x, y)$ *in* $Meta_{\mathrm{N}(\mathcal{L})}$, *and*
(ii) $\Gamma, \Delta \vdash x{:}A$ *in* $\mathrm{N}(\mathcal{L})$ *implies* $\mathcal{LF}(\Gamma), \mathcal{RF}(\Delta) \vdash \mathcal{LF}(x, A)$ *in* $Meta_{\mathrm{N}(\mathcal{L})}$. ■

By Lemma 5.1.3 and Lemma 5.1.4 below, we have:

Theorem 5.1.2 $Meta_{\mathrm{N}(\mathcal{L})}$ *is faithful and adequate.* ■

Lemma 5.1.3 $Meta_{\mathrm{N}(\mathcal{L})}$ *is faithful.*

Proof Following Prawitz and our definitions in §2.3.1, we call a *thread* a sequence of formulas in a derivation in the metalogic leading from some assumption to the conclusion. A *track* is the initial segment of a thread ending at either the first minor premise of a $\Rightarrow$E rule encountered, or the conclusion of the derivation if no such minor premise occurs. Paulson [179] shows that derivations in $Meta$ have an *expanded normal form* (which in the case of $Meta$ amounts to derivations in normal form where the *minimal formulas* of the tracks are atomic). Since this result immediately extends to $Meta_{\mathrm{N}(\mathcal{L})}$, in the following we exploit the fact that derivations in $Meta_{\mathrm{N}(\mathcal{L})}$ have an expanded normal form in which there are no maximal formulas and each track has a minimal formula of the form $\mathcal{LF}(x, A)$ or $\mathcal{RF}(x, y)$. The proof proceeds by induction on the size of the expanded normal form of $Meta_{\mathrm{N}(\mathcal{L})}$-derivation of $\mathcal{RF}(x, y)$ from $\mathcal{RF}(\Delta)$, or of $\mathcal{LF}(x, A)$ from $\mathcal{LF}(\Gamma)$ and $\mathcal{RF}(\Delta)$.

In the base case for (ii), if $\mathcal{LF}(x, A)$ follows from an assumption in $\mathcal{LF}(\Gamma)$, then $x{:}A$ is an assumption in Γ, so we trivially have $\Gamma, \Delta \vdash_{\mathrm{N}(\mathcal{L})} x{:}A$. We conclude similarly in the base case for (i), i.e. when $\mathcal{RF}(x, y)$ follows from an assumption in $\mathcal{RF}(\Delta)$.

In the step case, a track begins with an axiom followed by a sequence of elimination rules. We proceed by showing that the application of each axiom in $Meta_{\mathrm{N}(\mathcal{L})}$ corresponds to an object level inference in $\mathrm{N}(\mathcal{L})$. All of the cases are simple and we give two representative ones below: a Horn axiom from $Meta_{\mathrm{N}(\mathcal{T})}$ and the axiom `boxI` from $Meta_{\mathrm{N(K)}}$.

Consider a Horn axiom of the relational theory corresponding to $Meta_{\mathrm{N}(\mathcal{T})}$. The $Meta_{\mathrm{N}(\mathcal{L})}$-derivation must comprise a sequence of $\bigwedge$E steps, one for each quantifier, followed by a sequence of $\Rightarrow$E steps, one for each premise. For concreteness, consider the axiom `conv1`, where xRy is $vRg(u, v, w)$ for some u, v and w. The $Meta_{\mathrm{N}(\mathcal{L})}$-derivation must have the structure shown at the top of Figure 5.2. $\mathrm{N}(\mathcal{L})$-derivations of

$$
\dfrac{\dfrac{\dfrac{\dfrac{\bigwedge x \bigwedge y \bigwedge z(\mathcal{RF}(x,y) \Rightarrow (\mathcal{RF}(x,z) \Rightarrow \mathcal{RF}(y, g(x,y,z))))}{\bigwedge y \bigwedge z(\mathcal{RF}(u,y) \Rightarrow (\mathcal{RF}(u,z) \Rightarrow \mathcal{RF}(y, g(u,y,z))))}\bigwedge\mathrm{E}}{\bigwedge z(\mathcal{RF}(u,v) \Rightarrow (\mathcal{RF}(u,z) \Rightarrow \mathcal{RF}(v, g(u,v,z))))}\bigwedge\mathrm{E}}{\mathcal{RF}(u,v) \Rightarrow (\mathcal{RF}(u,w) \Rightarrow \mathcal{RF}(v, g(u,v,w)))}\bigwedge\mathrm{E} \quad \begin{array}{c}\vdots\\ \mathcal{RF}(u,v)\end{array}}{\mathcal{RF}(u,w) \Rightarrow \mathcal{RF}(v, g(u,v,w))}\Rightarrow\mathrm{E} \quad \begin{array}{c}\vdots\\ \mathcal{RF}(u,w)\end{array}}{\mathcal{RF}(v, g(u,v,w))}\Rightarrow\mathrm{E}
$$

$$
\dfrac{\dfrac{\dfrac{\bigwedge x \bigwedge A((\bigwedge y(\mathcal{RF}(x,y) \Rightarrow \mathcal{LF}(y,A))) \Rightarrow \mathcal{LF}(x,\Box A))}{\bigwedge A((\bigwedge y(\mathcal{RF}(z,y) \Rightarrow \mathcal{LF}(y,A))) \Rightarrow \mathcal{LF}(z,\Box A))}\bigwedge\mathrm{E}}{(\bigwedge y(\mathcal{RF}(z,y) \Rightarrow \mathcal{LF}(y,B))) \Rightarrow \mathcal{LF}(z,\Box B)}\bigwedge\mathrm{E} \quad \dfrac{\dfrac{\begin{array}{c}[\mathcal{RF}(z,y)]^1\\ \vdots\\ \mathcal{LF}(y,B)\end{array}}{\mathcal{RF}(z,y) \Rightarrow \mathcal{LF}(y,B)}\Rightarrow\mathrm{I}^1}{\bigwedge y(\mathcal{RF}(z,y) \Rightarrow \mathcal{LF}(y,B))}\bigwedge\mathrm{I}}{\mathcal{LF}(z,\Box B)}\Rightarrow\mathrm{E}
$$

Figure 5.2. The metalevel derivations formalizing *conv1* and $\Box$I

$uR\,v$ and $uR\,w$ from Γ and Δ are given by the induction hypotheses, so that applying *conv1* gives a N($\mathcal{L}$)-derivation of $vR\,g(u,v,w)$ from Γ and Δ.

In the case of `boxI`, let $x{:}A$ be $z{:}\Box B$ for some z and B. The $Meta_{\mathrm{N}(\mathcal{L})}$-derivation must have the structure shown at the bottom of Figure 5.2. It contains a $Meta_{\mathrm{N}(\mathcal{L})}$-derivation of $\bigwedge y(\mathcal{RF}(z,y) \Rightarrow \mathcal{LF}(y,B))$ from $\mathcal{LF}(\Gamma)$ and $\mathcal{RF}(\Delta)$, which, by expanded normal form, consists of a $Meta_{\mathrm{N}(\mathcal{L})}$-derivation of $\mathcal{LF}(y,B)$ from $\mathcal{LF}(\Gamma)$ and $\mathcal{RF}(\Delta) \cup \mathcal{RF}(z,y)$, where y is not free in the assumptions, followed first by a $\Rightarrow$I, discharging the assumption $\mathcal{RF}(x,y)$, and then by a $\bigwedge$I. A N($\mathcal{L}$)-derivation of $y{:}B$ from Γ and $\Delta \cup \{zRy\}$, where y is not free in the assumptions, is given by the induction hypothesis, so that applying $\Box$I gives a N($\mathcal{L}$)-derivation of $z{:}\Box B$ from Γ and Δ. ■

Lemma 5.1.4 *$Meta_{\mathrm{N}(\mathcal{L})}$ is adequate.*

Proof The proof proceeds by induction on the structure of the N($\mathcal{L}$)-derivation of $xR\,y$ from Δ, or of $x{:}A$ from Γ and Δ. The base cases are trivial, and we treat only two step cases as examples.

In the first case, a relational rule has been applied. Consider the case of *conv1*. $xR\,y$ is $vR\,g(u,v,w)$, and *conv1* is applied to N($\mathcal{L}$)-derivations of $uR\,v$ and $uR\,w$ from Γ and Δ. $Meta_{\mathrm{N}(\mathcal{L})}$-derivations of $\mathcal{RF}(u,v)$ and $\mathcal{RF}(u,w)$ from $\mathcal{LF}(\Gamma)$ and $\mathcal{RF}(\Delta)$ are given by the induction hypotheses, and we conclude by building a $Meta_{\mathrm{N}(\mathcal{L})}$-derivation like that at the top of Figure 5.2.

In the second case, we consider the propositional and the modal rules, i.e. the rules of N(K), individually. For example, for $\Box$I, $x{:}A$ is $z{:}\Box B$ and $\Box$I is applied to a N($\mathcal{L}$)-derivation of $y{:}B$ from Γ and $\Delta \cup \{zRy\}$, where y is not free in the assumptions. A $Meta_{\mathrm{N}(\mathcal{L})}$-derivation of $\mathcal{LF}(y, B)$ from $\mathcal{LF}(\Gamma)$ and $\mathcal{RF}(\Delta) \cup \mathcal{RF}(z, y)$, where y is not free in the assumptions, i.e. a $Meta_{\mathrm{N}(\mathcal{L})}$-derivation of $\bigwedge y(\mathcal{RF}(z,y) \Rightarrow \mathcal{LF}(y, B))$ from $\mathcal{LF}(\Gamma)$ and $\mathcal{RF}(\Delta)$, is given by the induction hypothesis. We conclude by building a $Meta_{\mathrm{N}(\mathcal{L})}$-derivation like that at the bottom of Figure 5.2. ■

5.1.3 *Isabelle proof session*

We now illustrate Isabelle proofs for the examples given in §5.1.1. Some brief background is required; see [181] for a full account.

5.1.3.1 Background. Isabelle manipulates *rules* (metatheorems). A rule is a formula

```
!! v1 ... vm. A1 ==> ... ==> (An ==> A)
```

which is also displayed as

```
!! v1 ... vm. [| A1; ...; An|] ==> A
```

Rules represent proof states where `A` is the goal to be established and the `Ai`'s are the subgoals to be proved. Under this view, an initial proof state has the form `A ==> A`, i.e. it has one subgoal, namely `A`. The final proof state *is itself* the desired theorem (no subgoals are left). Isabelle supports proof construction through *higher-order resolution*, which is roughly analogous to resolution in Prolog. That is, given a proof state with subgoal `B` and a rule as above, then, treating the `vi`'s of the rule as variables for unification, we higher-order unify `A` with `B`. If this succeeds, then the unification yields a substitution σ, and the proof state is updated replacing `B` with the subgoals `A1`, ..., `An` and applying σ to the whole proof state. This resolution step can be justified by a sequence of proof steps in the metalogic. Although rules are formalized in a natural deduction style, they may be read as intuitionistic sequents where the `Ai`'s are the hypotheses. Isabelle has procedures which apply rules in a way that maintains this 'illusion' of working with sequents.

5.1.3.2 Derivations. To prove the equivalence of $\Box A$ and $\Box\Box A$ in the theory `S4` encoding N(S4), we begin by proving the left-to-right direction in the subtheory `K4`. The following proof, which is taken verbatim from an Isabelle session with the exception of minor pretty-printing and omission of diagnostic output, corresponds to the first derivation given in (5.1). We begin with the desired goal.

```
> goal K4.thy "x:[]A --> [][]A";
x : []A --> [][]A
 1. x : []A --> [][]A
```

On the first line, at the Isabelle prompt '>', we use the command `goal` to state the theory we are using and the theorem to be proved. Isabelle responds with the next

2 lines, which give the goal to be proved, and what subgoals (in this case the goal itself) must be established to prove it. We proceed by applying our rule for implication introduction, `impI`, declared in Figure 5.1. The command `by (rtac impI 1)` directs Isabelle to apply `impI` using resolution (`rtac`) to the first subgoal. Isabelle responds with the new subgoal.

```
> by (rtac impI 1);
x : []A --> [][]A
 1. x : []A ==> x : [][]A
```

If we read the subgoal as a sequent, we must now show `x : [][]A` under the assumption `x : []A`. We proceed with two applications of `boxI`, each of which gives us new relational assumptions, followed by `boxE` (using `etac`, which first applies `rtac` and then unifies the first assumption of the rule `boxE` against an assumption in the subgoal).

```
> by (rtac boxI 1);
x : []A --> [][]A
 1. !!y. [| x : []A; x R y |] ==> y : []A

> by (rtac boxI 1);
x : []A --> [][]A
 1. !!y ya. [| x : []A; x R y; y R ya |] ==> ya : A

> by (etac boxE 1);
x : []A imp [][]A
 1. !!y ya. [| x R y; y R ya |] ==> x R ya
```

The theory `K4` extends `K` with the transitivity of R. Applying transitivity using `etac`, we obtain

```
> by (etac trans 1);
x : []A --> [][]A
 1. !!y ya. y R ya ==> y R ya
```

This leaves only one subgoal, which we prove by assumption using `atac`; Isabelle then reports that we have finished the proof.

```
> by (atac 1);
x : []A --> [][]A
No subgoals!
```

We can now name this theorem (we name it `BoxImpliesBoxBox` using the command `qed`) and use it in subsequent proofs; Isabelle provides *unknowns*, written with a ? prefix, that may be instantiated later during unification.

```
> qed "BoxImpliesBoxBox";
val BoxImpliesBoxBox = "?x : []?A --> [][]?A"
```

The proof of the converse direction in the theory `KT` directly mirrors the second derivation in (5.1); we give it here without further comment.

```
> goal KT.thy "x:[][]A --> []A";
```

```
x : [][]A --> []A
 1. x : [][]A --> []A

> by (rtac impI 1);
x : [][]A --> []A
 1. x : [][]A ==> x : []A

> by (etac boxE 1);
x : [][]A --> []A
 1. x R x

> by (rtac refl 1);
x : [][]A --> []A
No subgoals!

> qed "BoxBoxImpliesBox";
val BoxBoxImpliesBox = "?x : []?A --> ?A"
```

Having proved both directions, we may now combine them to prove the equivalence in `KT4`, i.e. `S4`, which has inherited both theorems from its ancestors, and which we assume to contain an axiom `iffI` encoding the (possibly derived) rule $\leftrightarrow$ I.

```
> goal KT4.thy "x:[]A <-> [][]A";
x : []A <-> [][]A
 1. x : []A <-> [][]A

> by (rtac iffI 1);
x : []A <-> [][]A
 1. x : []A --> [][]A
 2. x : [][]A --> []A

> by (rtac BoxImpliesBoxBox 1);
x : []A <-> [][]A
 1. x : [][]A --> []A

> by (rtac BoxBoxImpliesBox 1);
x : []A <-> [][]A
No subgoals!
```

In our Isabelle implementation we can also derive new rules. To illustrate this, we derive the rules for $\sim$ and $\Diamond$ following the N(K)-derivations we gave in Example 2.1.14. To derive $\sim$ I, we call `goalw` with a list of definitions (in this case only `not_def`) and the appropriate metalevel formula. `goalw` has two effects: (i) it returns a list consisting of the rule's premises (in this case a one-element list, which we bind to the ML identifier `prem` using `val`), and (ii) it applies the definitions specified in the list as meta-rewrite rules to the subgoal and the premises. Specifically, we type

```
> val [prem] =
   goalw K.thy [not_def] "(x:A ==> x: falsum) ==> x:~A";
```

and Isabelle responds with the lines

```
x : ~ A
 1. x : A --> falsum
val prem = "x : A ==> x : falsum [x : A ==> x : falsum]" : thm
```

which give respectively the goal to be proved, what subgoals must be established to prove it (in this case the goal itself, rewritten using `not_def`), and the binding of the identifier `prem` to the premise. We then apply `impI` and use `rtac` to resolve the subgoal using the premise.

```
> by (rtac impI 1);
x : ~ A
 1. x : A ==> x : falsum

> by (rtac prem 1);
x : ~ A
 1. x : A ==> x : A
```

This leaves us with one subgoal, which we prove by assumption; we name this theorem (derived rule) `notI` in order to use it in subsequent proofs.

```
> by (atac 1);
x : ~ A
No subgoals!

> qed "notI";
val notI = "(?x : ?A ==> ?x : falsum) ==> ?x : ~ ?A" : thm
```

We derive $\sim$E analogously, but this time we have two premises, enclosed in `[| ... |]` to represent nested meta-implications, which we respectively bind to the ML identifiers `major` and `minor`.

```
> val [major,minor] =
   goalw K.thy [not_def] "[| x:~A; x:A |]  ==> x:falsum";
x : falsum
 1. x : falsum
val major = "x : A --> falsum  [x : ~ A]" : thm
val minor = "x : A  [x : A]" : thm
```

The rest of the proof is straightforward:

```
> by (rtac impE 1);
x : falsum
 1. x : ?A --> falsum
 2. x : ?A

> by (rtac major 1);
x : falsum
 1. x : A

> by (rtac minor 1);
x : falsum
```

```
No subgoals!

> qed "notE";
val notE = "[| ?x : ~ ?A; ?x : ?A |] ==> ?x : falsum" : thm
```

Using notI and notE and the definition dia_def, we can derive the rules for ◇, directly mirroring the derivations (2.1) and (2.2) in Example 2.1.14. To shorten the proofs, we make use of some of Isabelle's built-in *tacticals*: EVERY executes commands in sequence, THEN composes two commands, and REPEAT applies commands as many times as possible.

```
> val [major,minor] =
   goalw K.thy [dia_def] "[| y:A; x R y |] ==> x: <>A";
x : <>A
 1. x : ~ [](~ A)
val major = "y : A  [y : A]" : thm
val minor = "x R y  [x R y]" : thm

> by (EVERY [rtac notI 1, rtac falsumE 1, rtac notE 1]);
x : <>A
 1. [| x : [](~ A); x : falsum --> falsum |] ==> ?y1 : ~ ?A2
 2. [| x : [](~ A); x : falsum --> falsum |] ==> ?y1 : ?A2

> by (rtac major 2);
x : <>A
 1. [| x : [](~ A); x : falsum --> falsum |] ==> y : ~ A

> by (etac boxE 1);
x : <>A
 1. x : falsum --> falsum ==> x R y

> by (rtac minor 1);
x : <>A
No subgoals!

> qed "diaI";
val diaI = "[| ?y : ?A; ?x R ?y |] ==> ?x : <>?A" : thm

> val [major,minor] = goalw K.thy [dia_def]
   "[| x:<>A ; (!!y. y:A ==> x R y ==> z:B) |] ==> z:B";
z : B
 1. z : B
val major = "x : ~ [](~ A)  [x : <>A]" : thm
val minor = "[| ?y : A; x R ?y |] ==> z : B
   [!!y. [| y : A; x R y |] ==> z : B]" :  thm

> by (rtac falsumE 1 THEN rtac notE 1);
z : B
 1. z : B --> falsum ==> ?y : ~ ?A1
 2. z : B --> falsum ==> ?y : ?A1
```

```
> by (EVERY [rtac major 1, rtac boxI 1, rtac notI 1,
             rtac falsumE 1, etac impE 1, rtac minor 1,
             REPEAT (atac 1)]);
z : B
No subgoals!

> qed "diaE";
val diaE = "[| ?x : <>?A; !!y. [| y : ?A; ?x R y |]
   ==> ?z : ?B |] ==> ?z : ?B" : thm
```

As a final example, we exploit these derived rules to prove the characteristic axiom schema for N(K2) based on the extension of `K` given in §5.1.1. The proof directly mirrors the N(K2)-proof (2.4) given in Example 2.1.14.

```
> goal K2.thy "x: <>[]A --> []<>A";
x : <>[]A --> []<>A
 1. x : <>[]A --> []<>A

> by (EVERY [rtac impI 1, rtac boxI 1, etac diaE 1, rtac diaI 1,
             etac boxE 1, etac conv2 1, atac 1, etac conv1 1,
             atac 1];
x : <>[]A --> []<>A
No subgoals!
```

5.2 ENCODING PROPOSITIONAL NON-CLASSICAL LOGICS

5.2.1 *Implementation and Isabelle proof session*

Since the implementation issues are not significantly different from the simpler case for propositional modal logics described above, we give only a brief overview.

Recall that a logic is encoded in Isabelle using a theory composed of a signature and axioms, which are formulas in the language of *Meta*, and that proving theorems in the encoded logic means simply proving theorems with these axioms in the metalogic. As an example, Figure 5.3 contains $Meta_{\mathrm{N(R+)}}$, `Rplus` in concrete syntax, which encodes the system N(R$^+$) given in Table 3.4. Note that we could also obtain $Meta_{\mathrm{N(R+)}}$ by extending a theory for a suitable base system, e.g. a theory $Meta_{\mathrm{N(B+)}}$ for the base system N(B$^+$).

The signature of `Rplus` declares two types `label` and `o`, for labels and unlabelled formulas of relevance logic. Logical operators are declared as typed constants over this signature; e.g. `act` (for the actual world) of type `label`, `inc` (for incoherence, i.e. $\perp\!\!\!\perp$) of type `o`, and `imp` (for relevant implication) of type `[o,o] => o`, i.e. `o => (o => o)`. There are two judgements, encoded as predicates: first, `LF(a,A)`, for provable lwffs, which we abbreviate to `a:A`; second, `RF(a,b,c)`, for provable rwffs, which we abbreviate to `R a b c`. The axioms for `LF` and `RF` correspond directly to the rules in §3 (recall that in the axioms free variables are implicitly outermost universally quantified).

We may now extend `Rplus` by adding axioms, which reflect the discussion in §3.1.4. The encoding `JR` of N($\mathcal{J}$R) is obtained by extending `Rplus` with axioms for an intuitionistic treatment of negation, i.e.

```
Rplus = Pure +    (* Rplus extends Pure (Isabelle's metalogic) *)
                  (* with the following signature and axioms   *)

types   (* Definition of type constructors *)
  label, o 0

arities (* Addition of the arity 'logic' to the existing types *)
  label, o :: logic

consts  (* Labels, Logical operators, Judgements *)
  act  :: "label"
  f2   :: "[label,label,label,label,label] => label"
  f3   :: "[label,label,label,label,label] => label"
  f4   :: "[label,label,label] => label"

  and  :: "[o, o] => o"        (infixr 35)
  or   :: "[o, o] => o"        (infixr 30)
  imp  :: "[o, o] => o"        (infixr 25)

  LF   :: "[label, o] => prop"           ("(_ : _)" [0,0] 100)
  RF   :: "[label, label, label] => prop" ("(R _ _ _)"
                                           [0,0,0] 100)

rules
  (* Base system *)
  conjI   "[| a:A; a:B |] ==> a: A and B"
  conjE1  "a: A and B ==> a:A"
  conjE2  "a: A and B ==> a:B"
  disjI1  "a:A ==> a: A or B"
  disjI2  "a:B ==> a: A or B"
  disjE   "[| a: A or B;  a:A ==> c:C;  a:B ==> c:C |] ==> c:C"
  impI    "[| !!b c. [| b:A; R a b c |] ==> c:B |]
           ==> a: A imp B"
  impE    "[| a: A imp B; b:A; R a b c |] ==> c:B"
  monl    "[| a:A; R act a b |] ==> b:A"

  (* Properties of the compossibility relation R *)
  monR1      "[| R a b c; R act x a |] ==> R x b c"
  monR2      "[| R a b c; R act x b |] ==> R a x c"
  monR3      "[| R a b c; R act c x |] ==> R a b x"
  iden       "R act a a"
  suff1      "[| R a b x; R x c d |] ==> R a c f2(a,b,c,d,x)"
  suff2      "[| R a b x; R x c d |] ==> R b f2(a,b,c,d,x) d"
  assoc1     "[| R a b x; R x c d |] ==> R b c f3(a,b,c,d,x)"
  assoc2     "[| R a b x; R x c d |] ==> R a f3(a,b,c,d,x) d"
  cont1      "R a b c ==> R a b f4(a,b,c)"
  cont2      "R a b c ==> R f4(a,b,c) b c"
  specassert "R a act a"
  comm       "R a b c ==> R b a c"
end
```

Figure 5.3. Isabelle encoding of $\mathrm{N}(\mathrm{R}^{+})$

```
JR = Rplus +
consts
  inc    :: "o"
  neg    :: "o => o"               ("~_" [40] 40)
  star   :: "label => label"       ("_*" [40] 40)
rules
  negI     "(a*: A ==> b: inc) ==> a: ~A"
  negE     "[| a: ~A; a*: A |] ==> b: inc"
  incEi    "b: inc ==> a: A"
  inv      "R a b c ==> R a c* b*"
  stari    "R act a a**"
end
```

Then we can further add an axiom encoding the rule *int* to obtain an encoding of the ND system N(J) for intuitionistic logic as in Proposition 3.1.11, or we can add 'classical' negation rules to obtain R, the encoding of N(R), i.e.

```
R = JR +
rules
  incEc    "(a: ~A ==> b: inc) ==> a*: A"
  starc    "R act a** a"
end
```

(Alternatively, we can encode N(R) by directly extending `Rplus`.) Using this encoding we can, for example, prove `act : ~~A imp A` in N(R) as follows; cf. the proof (3.24) of $0{:}\neg\neg A \to A$ given in Example 3.1.12.

```
> goal R.thy "act : ~~A imp A";
act : ~~A imp A
 1. act : ~~A imp A
```

We begin by instructing Isabelle to apply implication introduction using resolution to the first (and only) subgoal.

```
> by (rtac impI 1);
act : ~~A imp A
 1. !!b c. [| b : ~~A; R act b c |] ==> c : A
```

We now apply `monl` and dispose of the second subgoal using the rule `starc`.

```
> by ((rtac monl 1) THEN (rtac starc 2));
act : ~~A imp A
 1. !!b c. [| b : ~~A; R act b c |] ==> c** : A
```

Next we apply rules as in the proof in Example 3.1.12 (recall that `EVERY` executes commands in sequence and that `atac` proves a subgoal by assumption after solving for unknowns).

```
> by (EVERY [rtac incEc 1, rtac negE 1, atac 2]);
act : ~~A imp A
 1. !!b c. [| b : ~~A; R act b c; c* : ~A |] ==> c : ~~A
```

```
> by (rtac monl 1);
act : ~~A imp A
 1. !!b c. [| b : ~~A; R act b c; c* : ~A |] ==> ?a5(b, c) : ~~A
 2. !!b c. [| b : ~~A; R act b c; c* : ~A |]
    ==> R act ?a5(b, c) c
```

This leaves us with two subgoals, which are both proved by assumption, simplifying `?a5(b, c)` to `b`.

```
> by (REPEAT (atac 1));
act : ~~A imp A
No subgoals!
```

As a second example, we give a R-derivation of one of the contraposition rules, reflecting the derivation (3.23) in Example 3.1.12. In this case, at the Isabelle prompt, we state the theory we are using, the lwff to be proved (`act : ~B imp ~A`) and the premise from which it follows (`act : A imp B`); the premise is bound to the ML identifier `prem`.

```
> val [prem] = goal R.thy "act : A imp B ==> act : ~B imp ~A";
act : ~B imp ~A
 1. act : ~B imp ~A
val prem = "act : A imp B  [act : A imp B]" : thm

> by (EVERY [rtac impI 1, rtac negI 1, rtac negE 1, atac 1,
             rtac impE 1]);
act : ~B imp ~A
 1. !!b c. [| b : ~B; R act b c; c* : A |]
     ==> ?a3(b, c) : ?A3(b, c) imp B
 2. !!b c. [| b : ~B; R act b c; c* : A |]
     ==> ?b3(b, c) : ?A3(b, c)
 3. !!b c. [| b : ~B; R act b c; c* : A |]
     ==> R ?a3(b, c) ?b3(b, c) b*
```

We resolve the first subgoal using the premise:

```
> by (rtac prem 1);
act : ~B imp ~A
 1. !!b c. [| b : ~B; R act b c; c* : A |] ==> ?b3(b, c) : A
 2. !!b c. [| b : ~B; R act b c; c* : A |] ==> R act ?b3(b, c) b*
```

This leaves us with two subgoals. The first is proved by assumption, simplifying `?b3(b, c)` to `c*`.

```
> by (atac 1);
act : ~B imp ~A
 1. !!b c. [| b : ~B; R act b c; c* : A |] ==> R act c* b*
```

We conclude the proof by first exploiting the inversion property of R, which is stronger than the antitonicity rule that we exploited in (3.23), and then proving the remaining subgoal by assumption.

```
> by ((rtac inv 1) THEN (atac 1));
act : ~B imp ~A
No subgoals!
```

As a final example, we encode the derivation of the relational rule *idem* given in Table 3.4, using which we could then, for example, prove the axiom schema $0{:}(A \wedge (A \to B)) \to B$.

```
> goal Rplus.thy "R a a a";
R a a a
 1. R a a a

> by (rtac monR1 1);
R a a a
 1. R ?a a a
 2. R act a ?a

> by (rtac cont2 1);
R a a a
 1. R ?a1 a a
 2. R act a f4(?a1, a, a)

> by (rtac cont1 2);
R a a a
 1. R act a a
 2. R act a a

> by (rtac iden 1);
R a a a
 1. R act a a

> by (rtac iden 1);
R a a a
No subgoals!
```

5.2.2 Faithfulness and adequacy

By reasoning about our encoding and the metalogic *Meta* we can prove that $Meta_{\mathrm{N(R)}}$ corresponds to the original N(R). Like for propositional modal logics in §5.1.2, we do this in two parts, by showing *adequacy*, that any proof in N(R) can be reconstructed in $Meta_{\mathrm{N(R)}}$, and *faithfulness*, that we can recover from a derivation in $Meta_{\mathrm{N(R)}}$ a proof in N(R) itself. Formally, we generalize Definition 5.1.1 as follows; note that, given $\Gamma = \{a_1{:}A_1, \ldots, a_n{:}A_n\}$ and $\Delta = \{R\, a_1\, b_1\, c_1, \ldots, R\, a_m\, b_m\, c_m\}$, we abuse notation and write $\mathcal{LF}(\Gamma)$ and $\mathcal{RF}(\Delta)$ for the sets $\{\mathcal{LF}(a_1, A_1), \ldots, \mathcal{LF}(a_n, A_n)\}$ and $\{\mathcal{RF}(a_1, b_1, c_1), \ldots, \mathcal{RF}(a_m, b_m, c_m)\}$.

Definition 5.2.1 *$Meta_{\mathrm{N(R)}}$ is* faithful *(with respect to* N(R)*) iff*

(i) $\mathcal{RF}(\Delta) \vdash \mathcal{RF}(a,b,c)$ *in* $Meta_{\mathrm{N(R)}}$ *implies* $\Delta \vdash R\, a\, b\, c$ *in* N(R), *and*
(ii) $\mathcal{LF}(\Gamma), \mathcal{RF}(\Delta) \vdash \mathcal{LF}(a, A)$ *in* $Meta_{\mathrm{N(R)}}$ *implies* $\Gamma, \Delta \vdash a{:}A$ *in* N(R).

$Meta_{\mathrm{N(R)}}$ *is* adequate (with respect to N(R)) *iff the converses hold, i.e. iff*

(i) $\Delta \vdash R\,a\,b\,c$ *in* N(R) *implies* $\mathcal{RF}(\Delta) \vdash \mathcal{RF}(a,b,c)$ *in* $Meta_{\mathrm{N(R)}}$, *and*
(ii) $\Gamma, \Delta \vdash a{:}A$ *in* N(R) *implies* $\mathcal{LF}(\Gamma), \mathcal{RF}(\Delta) \vdash \mathcal{LF}(a, A)$ *in* $Meta_{\mathrm{N(R)}}$. ■

Adequacy is easy to show, because the rules of N(R) map directly onto the axioms of $Meta_{\mathrm{N(R)}}$. A simple inductive argument on the structure of proofs in N(R) establishes this (as in Lemma 5.1.4). Faithfulness is more complex, since there is no such simple mapping in this direction: arbitrary derivations in $Meta_{\mathrm{N(R)}}$ do not map directly onto proofs in N(R). Instead we use proof-theoretical properties of *Meta*: any derivation in *Meta* is equivalent to another in expanded normal form, see [179] and Lemma 5.1.3. Thus, given a derivation in $Meta_{\mathrm{N(R)}}$ we can, by induction over the size of its expanded normal form, find a derivation in N(R). This establishes faithfulness; moreover, this proof is constructive: it not only tells us that there is a proof in N(R), it also provides an effective method for finding one.

Faithfulness and adequacy of theories $Meta_{\mathrm{N}(\mathcal{L})}$ with respect to systems $\mathrm{N}(\mathcal{L})$ for other propositional non-classical logics $\mathcal{L}$ follow analogously. Thus, we can generalize Theorem 5.1.2 to:

Theorem 5.2.2 $Meta_{\mathrm{N}(\mathcal{L})}$ *is faithful and adequate.* ■

5.3 ENCODING QUANTIFIED MODAL LOGICS

5.3.1 Implementation and Isabelle proof session

Again, we give only a brief overview. We extend $Meta_{\mathrm{N(QK)}}$ with theories $Meta_{\mathrm{N}(\mathcal{T})}$ and $Meta_{\mathrm{N}(\mathcal{D})}$ to encode the corresponding systems $\mathrm{N(QK)} + \mathrm{N}(\mathcal{T}) + \mathrm{N}(\mathcal{D})$.

Figure 5.4 contains `QK`, our entire Isabelle declaration for N(QK). The signature of `QK` declares two types `label` and `o`, for labels and unlabelled modal formulas. The type variable `'a` has the default sort `term`, which we declare to be a subclass of `logic`. Logical operators are declared as typed constants over this signature, e.g. the universal quantifier `All` of type `('a => o) => o`. The variable-binding construct `binder` lets us write universal quantification in Isabelle concrete syntax as `ALL x. A(x)` (of type `o`), where `x` is a bound variable of type `'a` of sort `term`, and the body `A(x)` has type `o`. Similarly, the concrete syntax for the existential quantifier `Ex` of type `('a => o) => o` is `EX x. A(x)`.

There are three judgements: `LF(w,A)` for provable lwffs, `RF(w,v)` for provable rwffs, and `LT(w,t)` for provable labelled terms, which we abbreviate with `w E t` using mixfix annotations. These judgements respectively correspond to the predicate symbols $\mathcal{LF}$, $\mathcal{RF}$ and $\mathcal{LT}$ in the metalogic. `not_def` and `dia_def` are the definitions (using `==`) of the logical operators $\sim$ and $\Diamond$; `ex_def` is the definition of the existential quantifier. Other operators can be defined similarly.

We now illustrate Isabelle proofs for the examples given in §4.1. We begin by deriving the rules for $\exists$ following the N(QK)-derivations of Example 4.1.8.

To derive `exI` we type

```
> val [major,minor] = goalw QK.thy [ex_def]
    "[| w:A(t); w E t |] ==> (w:EX x. A(x))";
```

```
QK = Pure +          (* QK extends Pure (Isabelle's metalogic)  *)
                     (* with the following signature and axioms *)

classes
  term < logic

default
  term

types  (* Definition of type constructors *)
  label, o 0

arities (* Addition of the arity 'logic' to the existing types *)
  label, o :: logic

consts
  (* Logical operators *)
  falsum   :: "o"
  imp      :: "[o, o] => o"                ("_ --> _" [25] 25)
  not      :: "o => o"                     ("~ _" [40] 40)
  box      :: "o=> o"                      ("[]_" [50] 50)
  dia      :: "o=> o"                      ("<>_" [50] 50)
  All      :: "('a => o) => o"             (binder "ALL " 10)
  Ex       :: "('a => o) => o"             (binder "EX " 10)

  (* Judgements *)
  LF       :: "[label, o] => prop"         ("(_ : _)" [0,0] 100)
  RF       :: "[label, label] => prop"     ("(_ R _)" [0,0] 100)
  LT       :: "[label, 'a] => prop"        ("(_ E _)" [0,0] 100)

rules
  (* Axioms representing the object-level rules *)
  falsumE  "(w:A --> falsum ==> v: falsum) ==> w:A"
  impI     "(w:A ==> w:B) ==> w:(A --> B)"
  impE     "w: A --> B ==> w:A ==> w:B"
  boxI     "(!!v. (w R v ==> v:A)) ==> w:([]A)"
  boxE     "w:[]A ==> w R v ==> v:A"
  allI     "(!!t. (w E t ==> w: A(t))) ==> (w: ALL x.A(x))"
  allE     "w: ALL x. A(x) ==> w E t ==> w:A(t)"

  (* Definitions *)
  not_def  "w: ~A == w: A --> falsum"
  dia_def  "w: <>A == w: ~([](~A))"
  ex_def   "w: EX x. A(x) == w: ~(ALL x. ~A(x))"

end
```

Figure 5.4. Isabelle encoding of $\mathrm{N}(\mathrm{QK})$

and Isabelle responds with four lines, which give the goal to be proved, what subgoals must be established to prove it (in this case the goal itself, rewritten using `ex_def`), and the binding of the ML identifiers `major` and `minor` to the two premises of the rule

```
w : EX x. A(x)
  1. w : ~ (ALL x. ~ A(x))
val major = "w : A(t)  [w : A(t)]" : thm
val minor = "w E t  [w E t]" : thm
```

The rest of the derivation is straightforward, assuming that we have derived the rules for negation introduction and elimination, $\sim$I and $\sim$E, which we can do exactly like in the propositional case in §5.1.3.

```
> by (rtac notI 1);
w : EX x. A(x)
 1. w : ALL x. ~ A(x) ==> w : falsum

> by (rtac notE 1);
w : EX x. A(x)
 1. w : ALL x. ~ A(x) ==> w : ~ ?A1
 2. w : ALL x. ~ A(x) ==> w : ?A1

> by (rtac major 2);
w : EX x. A(x)
 1. w : ALL x. ~ A(x) ==> w : ~ A(t)

> by (etac allE 1);
w : EX x. A(x)
 1. w E t

> by (rtac minor 1);
w : EX x. A(x)
No subgoals!

> qed "exI";
val exI = "[| ?w : ?A(?t); ?w E ?t |] ==> ?w : EX x. ?A(x)" : thm
```

The rule `exE` is derived analogously and we give it without further comments.

```
> val [major,minor] = goalw QK.thy [ex_def]
   "[| w:EX x. A(x); (!!t. w:A(t) ==> w E t ==> v:B) |] ==> v:B";
v : B
 1. v : B
val major = "w : ~ (ALL x. ~ A(x))  [w : EX x. A(x)]" : thm
val minor = "[| w : A(?t); w E ?t |] ==> v : B
 [!!t. [| w : A(t); w E t |] ==> v : B]" : thm

> by (EVERY [rtac falsumE 1, rtac notE 1, rtac major 1,
             rtac allI 1, rtac notI 1, rtac falsumE 1,
             etac impE 1, rtac minor 1, REPEAT (atac 1)]);
v : B
```

```
No subgoals!

> qed "exE";
val exE = "[| ?w : EX x. ?A(x);
 !!t. [| ?w : ?A(t); ?w E t |] ==> ?v : ?B |] ==> ?v : ?B" : thm
```

We can now extend the encoding of N(QK) with encodings of relational and domain theories. For example, for N(QK.i) we extend `QK` with an encoding of *id*, i.e.

```
QKi = QK +
rules
  id        "[| w E t; w R v |] ==> v E t"
end
```

Then we can, e.g., use `exI` and `exE`, which `QKi` inherits from `QK`, to prove the N(QK.i)-theorem

$$w{:}\exists x(\Box A) \supset \Box\exists x(A)$$

as follows.

```
> goal QKi.thy "w: (EX x. ([]A(x))) --> [](EX x. A(x))";
w : (EX x. []A(x)) --> [](EX x. A(x))
 1. w : (EX x. []A(x)) --> [](EX x. A(x))

> by (EVERY [rtac impI 1, etac exE 1, rtac boxI 1, rtac exI 1,
             etac boxE 1, atac 1, rtac id 1, REPEAT (atac 1)]);
w : (EX x. []A(x)) --> [](EX x. A(x))
No subgoals!
```

As a final example, we show that CBF is a theorem of N(QKB.i). First we extend `QKi` with an encoding of *symm*, i.e.

```
QKBi = QKi +
rules
  symm      "w R v ==> v R w"
end
```

and then we exploit `id` (inherited from `QKi`) and `symm` to prove $w{:}\forall x(\Box A) \supset \Box\forall x(A)$ as follows.

```
> goal QKBi.thy "w: (ALL x. ([]A(x))) --> [](ALL x. A(x))";
w : (ALL x. []A(x)) --> [](ALL x. A(x))
 1. w : (ALL x. []A(x)) --> [](ALL x. A(x))

> by (EVERY [rtac impI 1, rtac boxI 1, rtac allI 1, rtac boxE 1,
             rtac allE 1, atac 3, atac 1, rtac id 1, atac 1,
             rtac symm 1, atac 1]);
w : (ALL x. []A(x)) --> [](ALL x. A(x))
No subgoals!
```

5.3.2 Faithfulness and adequacy

$Meta_{\mathrm{N(Q\mathcal{L})}} = Meta_{\mathrm{N(QK)}} + Meta_{\mathrm{N}(\mathcal{T})} + Meta_{\mathrm{N}(\mathcal{D})}$ corresponds to $\mathrm{N(Q\mathcal{L})} = \mathrm{N(QK)} + \mathrm{N}(\mathcal{T}) + \mathrm{N}(\mathcal{D})$. Like for propositional modal logics in §5.1.2, we prove this in two parts, by showing *adequacy*, that any proof in $\mathrm{N(Q\mathcal{L})}$ can be reconstructed in $Meta_{\mathrm{N(Q\mathcal{L})}}$, and *faithfulness*, that we can recover from a derivation in $Meta_{\mathrm{N(Q\mathcal{L})}}$ a proof in $\mathrm{N(Q\mathcal{L})}$ itself. Formally, we extend Definition 5.1.1 to quantified modal logics as follows; note that given $\Gamma = \{w_1{:}A_1, \ldots, w_n{:}A_n\}$, $\Delta = \{w_1 R\, w_2, \ldots, w_l R\, w_m\}$ and $\Theta = \{w_1{:}t_1, \ldots, w_j{:}t_j\}$, we abuse notation and write $\mathcal{LF}(\Gamma)$ for

$$\{\mathcal{LF}(w_1, A_1), \ldots, \mathcal{LF}(w_n, A_n)\},$$

$\mathcal{RF}(\Delta)$ for

$$\{\mathcal{RF}(w_1, w_2), \ldots, \mathcal{RF}(w_l, w_m)\},$$

and $\mathcal{LT}(\Theta)$ for

$$\{\mathcal{LT}(w_1, t_1), \ldots, \mathcal{LT}(w_n, t_n)\}.$$

Definition 5.3.1 $Meta_{\mathrm{N(Q\mathcal{L})}}$ *is* faithful (with respect to $\mathrm{N(Q\mathcal{L})}$) *iff*

(i) $\mathcal{RF}(\Delta) \vdash \mathcal{RF}(w, v)$ *in* $Meta_{\mathrm{N(Q\mathcal{L})}}$ *implies* $\Delta \vdash w R\, v$ *in* $\mathrm{N(Q\mathcal{L})}$,
(ii) $\mathcal{RF}(\Delta), \mathcal{LT}(\Theta) \vdash \mathcal{LT}(w, t)$ *in* $Meta_{\mathrm{N(Q\mathcal{L})}}$ *implies* $\Delta, \Theta \vdash w{:}t$ *in* $\mathrm{N(Q\mathcal{L})}$, *and*
(iii) $\mathcal{LF}(\Gamma), \mathcal{RF}(\Delta), \mathcal{LT}(\Theta) \vdash \mathcal{LF}(w, A)$ *in* $Meta_{\mathrm{N(Q\mathcal{L})}}$ *implies* $\Gamma, \Delta, \Theta \vdash w{:}A$ *in* $\mathrm{N(Q\mathcal{L})}$.

$Meta_{\mathrm{N(Q\mathcal{L})}}$ *is* adequate (with respect to $\mathrm{N(Q\mathcal{L})}$) *iff the converses hold, i.e. iff*

(i) $\Delta \vdash w R\, v$ *in* $\mathrm{N(Q\mathcal{L})}$ *implies* $\mathcal{RF}(\Delta) \vdash \mathcal{RF}(w, v)$ *in* $Meta_{\mathrm{N(Q\mathcal{L})}}$,
(ii) $\Delta, \Theta \vdash w{:}t$ *in* $\mathrm{N(Q\mathcal{L})}$ *implies* $\mathcal{RF}(\Delta), \mathcal{LT}(\Theta) \vdash \mathcal{LT}(w, t)$ *in* $Meta_{\mathrm{N(Q\mathcal{L})}}$, *and*
(iii) $\Gamma, \Delta, \Theta \vdash w{:}A$ *in* $\mathrm{N(Q\mathcal{L})}$ *implies* $\mathcal{LF}(\Gamma), \mathcal{RF}(\Delta), \mathcal{LT}(\Theta) \vdash \mathcal{LF}(w, A)$ *in* $Meta_{\mathrm{N(Q\mathcal{L})}}$. ■

By a straightforward generalization of Theorem 5.1.2, it follows that:

Theorem 5.3.2 $Meta_{\mathrm{N(Q\mathcal{L})}}$ *is faithful and adequate.* ■

6 LABELLED SEQUENT SYSTEMS FOR NON-CLASSICAL LOGICS

We show that our normalizing labelled natural deduction systems yield equivalent *cut-free labelled sequent systems* that

(i) allow us to present non-classical logics in a uniform and modular way;
(ii) are decomposed into two separated parts: a base system fixed for related logics, and a labelling algebra, which we extend to generate particular logics;
(iii) contain left and right rules for each logical operator (except for falsum $\bot$ and incoherence $\perp\!\!\!\perp$), independent of the relation(s) R_i and of the other operators;
(iv) satisfy a subformula property; and
(v) provide the basis of a general proof-theoretical method for bounding the complexity of the decision problem for propositional non-classical logics.

Following the development of the previous chapters, in §6.1 we introduce cut-free labelled sequent systems for (classical) propositional modal logics, and then, in §6.2, we formalize generalizations for other non-classical logics, restrictions to minimal and intuitionistic subsystems, and extensions to the quantified case. In §6.3 we show that normalizing natural deduction systems and cut-free sequent systems are intertranslatable; this guarantees that our sequent systems are sound and complete with respect to the corresponding Kripke semantics.

In Part II we then show how to use our sequent systems to establish decidability and complexity results, and consider some propositional modal logics as examples. As we discuss in the introductory chapter, §8, the fact that derivations in our ND systems can be reduced to a normal form that has a well-defined structure, and satisfies a subformula property, provides only a first step towards establishing decidability of the logics presented that way. Additional steps are required, such as bounding the number of times a particular formula may be assumed or discharged. This kind of proof-theoretical analysis is more easily performed when logics are presented using sequent systems, which allow a finer grained control of structural information via their structural rules. More specifically, we show that for certain logics we can restrict applications of the structural rules of our sequent systems, so that we can bound the number of times a given formula appears in a given proof (the number of different formulas that can appear in a proof is already bounded by the subformula property). This, combined with an analysis of the accessibility relation of the corresponding Kripke frames, yields decision procedures with bounded space requirements for a number of modal (and other non-classical) logics.

6.1 LABELLED SEQUENT SYSTEMS FOR PROPOSITIONAL MODAL LOGICS

We introduce partitioned, cut-free, labelled sequent systems for propositional modal logics: the logics are presented by modularly extending the labelling algebra of the fixed base sequent system S(K).

We begin with some terminology and notation. Let formulas be defined as for propositional modal ND systems, Definitions 2.1.1 and 2.1.3, possibly with the addition of Skolem function constants (and in that case with our usual convention about atomic and composite labels). That is, if R is a binary relation over a set of labels W, x and y are labels, and A is a propositional modal formula built from propositional variables and the operators $\bot$, $\supset$ and $\Box$, then xRy is a *relational formula* (*rwff*) and $x{:}A$ is a *labelled formula* (*lwff*). As before, other logical operators can be defined in the usual manner, e.g. ${\sim}A =_{def} A \supset \bot$ and $\Diamond A =_{def} {\sim}\Box{\sim}A$.

Further, let Γ and Δ, possibly annotated, vary over finite *multiset* of lwffs and rwffs respectively. We write Γ, Δ for the union of Γ and Δ, and $\Gamma, x{:}A$ $[\Delta, xRy]$ for the union of Γ $[\Delta]$ with the singleton multiset $\{x{:}A\}$ $[\{xRy\}]$. A *sequent* is an ordered pair of the form

$$\langle \Delta\,; xRy \rangle\,, \quad \text{written } \Delta \vdash xRy,$$

or

$$\langle \Gamma, \Delta\,; \Gamma' \rangle\,, \quad \text{written } \Gamma, \Delta \vdash \Gamma',$$

where $\vdash$ is a new meta-logical symbol, the *sequent symbol*, not to be confused with the derivability symbol for ND systems.

Let S, possibly annotated, vary over sequents. We call the left hand side of S, i.e. the multiset(s) of formulas Δ or Γ, Δ, the *antecedent* of S, and the right hand side of S, i.e. the rwff xRy or the multiset of lwffs Γ', the *succedent* of S. The intuitive semantic reading of S is: 'if all the formulas in the antecedent are true, then some (or the) formula in the succedent is true'. Formally, where $\mathfrak{M} = (\mathfrak{W}, \mathfrak{R}, \mathfrak{V})$ as in Definition 2.2.1, *truth for an rwff or lwff* φ *in a model* $\mathfrak{M}$, $\models^{\mathfrak{M}} \varphi$, is the smallest relation $\models^{\mathfrak{M}}$ satisfying:

$$\begin{array}{lll}
\models^{\mathfrak{M}} xRy & \text{iff} & (x,y) \in \mathfrak{R}\,; \\
\models^{\mathfrak{M}} x{:}p & \text{iff} & \mathfrak{V}(x,p) = 1\,; \\
\models^{\mathfrak{M}} x{:}A \supset B & \text{iff} & \models^{\mathfrak{M}} x{:}A \text{ implies } \models^{\mathfrak{M}} x{:}B\,; \\
\models^{\mathfrak{M}} x{:}\Box A & \text{iff} & \text{for all } y\,, \models^{\mathfrak{M}} xRy \text{ implies } \models^{\mathfrak{M}} y{:}A\,.
\end{array}$$

When $\models^{\mathfrak{M}} \varphi$, we say that φ is *true* in $\mathfrak{M}$. By extension:

- the sequent $\Delta \vdash xRy$ is true in a model $\mathfrak{M}$ iff, whenever $\models^{\mathfrak{M}} uRv$ for every $uRv \in \Delta$, then $\models^{\mathfrak{M}} xRy$;
- the sequent $\Gamma, \Delta \vdash \Gamma'$ is true in a model $\mathfrak{M}$ iff, whenever $\models^{\mathfrak{M}} x{:}A$ for every $x{:}A \in \Gamma$ and $\models^{\mathfrak{M}} uRv$ for every $uRv \in \Delta$, then $\models^{\mathfrak{M}} y{:}B$ for some $y{:}B \in \Gamma'$.

Sequent systems can be understood as meta-calculi for the corresponding natural deduction systems (see [106, 186, 221] and Theorem 6.3.1 below). Indeed, the two possible forms of a sequent directly correspond to the separations enforced in our labelled ND systems: $\Delta \vdash xRy$ expresses that rwffs follow only from other rwffs, and $\Gamma, \Delta \vdash \Gamma'$ expresses that lwffs follow from lwffs and rwffs.[1]

Formally, a *sequent system* (or *sequent calculus*) is a collection of *axioms* (also called *initial sequents*) and *rules*. For example, the axioms and rules given in Figure 6.1 determine S(K), the sequent system presenting the modal logic K. (That S(K) presents K is a consequence of the equivalence of S(K) and N(K), which is proved in Theorem 6.3.1.) In the following we will use also the derived rules given in Figure 6.2, especially the rules for $\sim$, which follow immediately from the axioms, weakening and the rules for $\supset$; the derivations of the $\Diamond$ rules are given in Example 6.1.3.

Some remarks about terminology.

Definition 6.1.1 *We call the sequent below the line in a rule the* conclusion *of the rule, and the sequents above the line the* premises *of the rule. We call the formulas, lwffs and rwffs, which pass through the application of a rule unchanged (they appear in the premises and in the conclusion) the* parametric formulas *of the rule. The formula contracted or introduced in the conclusion of the rule is called the* principal formula *of the rule, and the formulas from which the principal formula derives are the* active formulas*; we also say for short that a rule introduces or contracts (if the rule is a contraction) its principal formula.* ∎

The rules of a sequent system are divided in two categories: *logical rules* and *structural rules*. Logical rules define the behavior of logical operators (they introduce instances of the operators of the logic), while structural rules alter on the structure of the sequents, adding new formulas (weakening, in symbols W ...) or deleting duplicated

[1] Note also that since $\Delta \vdash xRy$ has the form of minimal or intuitionistic sequents, in the sense that only one rwff appears in the succedent, our sequent systems do not contain rules for weakening or contraction of rwffs on the right (see also §6.2). It follows that we could consider two kinds of sequents, $\vdash^r$ for $\Delta \vdash^r xRy$ and $\vdash^l$ for $\Gamma, \Delta \vdash^l \Gamma'$, where the rules for $\vdash^l$ 'employ' those for $\vdash^r$ but not vice versa.

Axioms:

$$\frac{}{x{:}A \vdash x{:}A}\ \mathrm{AXl} \qquad \frac{}{xRy \vdash xRy}\ \mathrm{AXr} \qquad \frac{}{y{:}\bot \vdash x{:}A}\ \bot\mathrm{L}$$

Logical rules:

$$\frac{\Gamma, \Delta \vdash \Gamma', x{:}A \quad x{:}B, \Gamma, \Delta \vdash \Gamma'}{x{:}A \supset B, \Gamma, \Delta \vdash \Gamma'}\ \supset\mathrm{L} \qquad \frac{x{:}A, \Gamma, \Delta \vdash \Gamma', x{:}B}{\Gamma, \Delta \vdash \Gamma', x{:}A \supset B}\ \supset\mathrm{R}$$

$$\frac{\Delta \vdash xRy \quad y{:}A, \Gamma, \Delta \vdash \Gamma'}{x{:}\Box A, \Gamma, \Delta \vdash \Gamma'}\ \Box\mathrm{L} \qquad \frac{\Gamma, \Delta, xRy \vdash \Gamma', y{:}A}{\Gamma, \Delta \vdash \Gamma', x{:}\Box A}\ \Box\mathrm{R}$$

In $\Box$R, the (atomic) label y does not occur in $\Gamma, \Delta \vdash \Gamma', x{:}\Box A$.

Structural rules:

$$\frac{\Gamma, \Delta \vdash \Gamma'}{\Gamma, \Delta, xRy \vdash \Gamma'}\ \mathrm{WrL} \qquad \frac{\Delta, xRy, xRy \vdash uRv}{\Delta, xRy \vdash uRv}\ \mathrm{CrL}$$

$$\frac{\Gamma, \Delta \vdash \Gamma'}{x{:}A, \Gamma, \Delta \vdash \Gamma'}\ \mathrm{WlL} \qquad \frac{\Gamma, \Delta \vdash \Gamma'}{\Gamma, \Delta \vdash \Gamma', x{:}A}\ \mathrm{WlR}$$

$$\frac{x{:}A, x{:}A, \Gamma, \Delta \vdash \Gamma'}{x{:}A, \Gamma, \Delta \vdash \Gamma'}\ \mathrm{ClL} \qquad \frac{\Gamma, \Delta \vdash \Gamma', x{:}A, x{:}A}{\Gamma, \Delta \vdash \Gamma', x{:}A}\ \mathrm{ClR}$$

Figure 6.1. The axioms and rules of $\mathrm{S(K)}$

$$\frac{\Gamma, \Delta \vdash \Gamma', x{:}A}{x{:}\sim A, \Gamma, \Delta \vdash \Gamma'}\ \sim\mathrm{L} \qquad \frac{x{:}A, \Gamma, \Delta \vdash \Gamma'}{\Gamma, \Delta \vdash \Gamma', x{:}\sim A}\ \sim\mathrm{R}$$

$$\frac{x{:}A, x{:}B, \Gamma, \Delta \vdash \Gamma'}{x{:}A \wedge B, \Gamma, \Delta \vdash \Gamma'}\ \wedge\mathrm{L} \qquad \frac{\Gamma, \Delta \vdash \Gamma', x{:}A \quad \Gamma, \Delta \vdash \Gamma', x{:}B}{\Gamma, \Delta \vdash \Gamma', x{:}A \wedge B}\ \wedge\mathrm{R}$$

$$\frac{x{:}A, \Gamma, \Delta \vdash \Gamma' \quad x{:}B, \Gamma, \Delta \vdash \Gamma'}{x{:}A \vee B, \Gamma, \Delta \vdash \Gamma'}\ \vee\mathrm{L} \qquad \frac{\Gamma, \Delta \vdash \Gamma', x{:}A, x{:}B}{\Gamma, \Delta \vdash \Gamma', x{:}A \vee B}\ \vee\mathrm{R}$$

$$\frac{y{:}A, \Gamma, \Delta, xRy \vdash \Gamma'}{x{:}\Diamond A, \Gamma, \Delta \vdash \Gamma'}\ \Diamond\mathrm{L} \qquad \frac{\Delta \vdash xRy \quad \Gamma, \Delta \vdash \Gamma', y{:}A}{\Gamma, \Delta \vdash \Gamma', x{:}\Diamond A}\ \Diamond\mathrm{R}$$

In $\Diamond$L, the (atomic) label y does not occur in $x{:}\Diamond A, \Gamma, \Delta \vdash \Gamma'$.

Figure 6.2. Some derived rules of $\mathrm{S(K)}$

formulas (contraction, in symbols C...). We write ...r... and ...l... to distinguish weakenings and contractions of rwffs and lwffs. Logical and structural rules are further divided into *left rules*, ...L, and *right rules*, ...R, depending on which side of $\vdash$ the principal formula appears.

One of the advantages of our labelled sequent systems over standard ones for modal (and other non-classical) logics is that each logical operator is characterized by a left and a right rule. In particular, this holds for $\Box$ (and $\Diamond$). Using the terminology of Wansing [233, 235], our rules for $\Box$ are

- *separated*: $\Box$L and $\Box$R are independent of the other operators, as well as independent of the properties of R (this allows us to view the rules as specifying the meaning of $\Box$, as is philosophically required of sequent and ND systems);

- *symmetric*: $\Box$L introduces $x{:}\Box A$ in the antecedent of the conclusion, $\Box$R introduces $x{:}\Box A$ in the succedent of the conclusion;

- *explicit*: only one boxed formula $\Box A$ is introduced in the conclusion.

Furthermore, in our systems, $\Box$ is properly interrelated with $\Diamond$: if we take both $\Box$ and $\Diamond$ as primitive, then we can easily prove $\vdash x{:}\Box A \leftrightarrow \sim\Diamond\sim A$ and $\vdash x{:}\Diamond A \leftrightarrow \sim\Box\sim A$. Note also that, as for ND systems, there is a close correspondence between the rules for $\Box$ and $\supset$; this holds since we express $x{:}\Box A$ as the metalevel implication $xRy \Rightarrow y{:}A$ for an arbitrary world y accessible from x.

Standard modal sequent systems, e.g. [87, 119, 233, 238], do not possess all of these properties. Consider, for example, the standard system SS(K) given in Figure 6.3, where Σ and Σ' are multisets of formulas and $\Box\Gamma$ denotes the multiset $\{\Box B \mid B \in \Gamma\}$. Note that SS(K) contains no structural rules: weakening is built into (AX) and (K), and contraction can be shown to be *admissible* (Zeman [238] first explicitly gives contraction rules, and then shows how to eliminate them).[2] Also note that the traditional negation rules, ($\sim$L) and ($\sim$R), can be replaced by the axiom

$$\frac{}{\bot, \Sigma \vdash \Sigma', A}\ (\bot\mathrm{L})$$

where $\sim$ is then defined in terms of $\supset$ and $\bot$. Most importantly, while (K) is a separated rule, it is neither symmetric nor explicit (it is only *weakly symmetric* and *weakly explicit* [233]). In the terminology of consequence relations, (K) is an *impure* rule, since it carries a non-local side condition on the complete set of assumptions (while $\Box$R is a pure rule as it has a local side condition).[3] Moreover, in our labelled systems we extend the fixed base system S(K) with a labelling algebra consisting only of relational rules (see below). SS(K), on the other hand, is extended with rules for $\Box$ that depend on the properties of the accessibility relation in the corresponding frames, i.e. the behavior of modal operators is not independent of the details of the

---

[2] That is, while the contraction rules are not derived rules, no new theorems become provable by their addition. For a technical discussion of admissible and derived rules see, e.g., [131, 220].

[3] See §1 and Avron's discussion of degrees of impurity of rules in [6, §5.5].

$$\frac{}{A, \Sigma \vdash \Sigma', A}\ (\text{AX}) \qquad \frac{\Sigma \vdash \Sigma', A}{\sim A, \Sigma \vdash \Sigma'}\ (\sim\text{L}) \qquad \frac{A, \Sigma \vdash \Sigma'}{\Sigma \vdash \Sigma', \sim A}\ (\sim\text{R})$$

$$\frac{\Sigma \vdash \Sigma', A \quad B, \Sigma \vdash \Sigma'}{A \supset B, \Sigma \vdash \Sigma'}\ (\supset\text{L}) \qquad \frac{A, \Sigma \vdash \Sigma', B}{\Sigma \vdash \Sigma', A \supset B}\ (\supset\text{R}) \qquad \frac{\Gamma \vdash A}{\Sigma, \Box\Gamma \vdash \Box A, \Sigma'}\ (\text{K})$$

Figure 6.3. The axioms and rules of SS(K)

Kripke frame providing their semantics. For example, SS(K) can be extended with non-explicit rules like

$$\frac{A, \Box A, \Sigma \vdash \Sigma'}{\Box A, \Sigma \vdash \Sigma'}\ (\text{T})\,, \qquad \frac{\Gamma, \Box\Gamma \vdash A}{\Sigma, \Box\Gamma \vdash \Box A, \Sigma'}\ (\text{K4})\,, \quad \text{or} \quad \frac{\Box\Gamma \vdash A}{\Sigma, \Box\Gamma \vdash \Box A, \Sigma'}\ (\text{S4})\,.$$

The standard sequent system SS(T) is then SS(K) plus (T), while SS(K4) is obtained from SS(K) by replacing (K) with (K4), and SS(S4) is obtained from SS(K) by replacing (K) with (T) and (S4); alternatively, we can obtain SS(S4) by extending SS(T) or SS(K4). (Note that actually the rule (K) becomes redundant when we add the rule (K4) or (T) and (S4).)

Rules for other systems, e.g. SS(K5), SS(K45) and SS(S5), require considerably more ingenuity, and have been given various formulations, e.g. [87, 115, 119, 163, 214, 233, 235]. For example, Shvarts [214] formalizes SS(K45) by extending the propositional rules with the non-explicit, 'left-and-right' rule

$$\frac{\Box\Sigma_1, \Sigma_2 \vdash \Box\Sigma_3, \Sigma_4}{\Box\Sigma_1, \Box\Sigma_2 \vdash \Box\Sigma_3, \Box\Sigma_4}$$

where Σ_4 contains at most one formula.

Their being difficult to invent is not the only problem of standard sequent rules; as observed by Kripke [150], the equivalences between $\Box A$ and $\sim \Diamond \sim A$, and $\Diamond A$ and $\sim \Box \sim A$ do not follow from many standard rules.

We will consider again standard sequent systems in §7 and in Part II, in which we also show that our rules provide a proof-theoretical justification (and a refinement of some) of the standard rules. We now adapt common terminology to define derivations and proofs of sequents in an arbitrary labelled sequent system $\mathrm{S}(\mathcal{L})$.

Definition 6.1.2 *Let a* branch *be a sequence of sequents, written vertically and separated by horizontal lines. A* derivation *of a sequent S in a system* $\mathrm{S}(\mathcal{L})$ *is a finite tree of branches, growing upwards, where each sequent except the* root *of the tree is obtained from the one below it by a (backwards) application of one of the rules of* $\mathrm{S}(\mathcal{L})$*. We call the root sequent the* end-sequent, *and we say that a sequent is a* leaf *of the tree iff it is the top-most sequent in a branch. A branch is* closed *iff all of its leaves are axioms* *(initial sequents) and is* open *otherwise. A derivation of S in* $\mathrm{S}(\mathcal{L})$ *is a* proof *of S in* $\mathrm{S}(\mathcal{L})$ *iff all of its branches are closed. An lwff or rwff φ of the logic $\mathcal{L}$ is a* theorem *of* $\mathrm{S}(\mathcal{L})$ *(or, simply, a* $\mathrm{S}(\mathcal{L})$-theorem*) iff the sequent $\vdash \varphi$ is provable in* $\mathrm{S}(\mathcal{L})$. ■

Analogously with ND systems, we also call a derivation [proof] in $\mathrm{S}(\mathcal{L})$ a $\mathrm{S}(\mathcal{L})$-*derivation* [$\mathrm{S}(\mathcal{L})$-*proof*], and we omit the '$\mathrm{S}(\mathcal{L})$' when the details of the particular logic are not relevant or are clear from context. We systematically use Π, possibly annotated, to range over derivations of sequents, and write

$$\begin{array}{c} S_1 \ldots S_n \\ \Pi \\ S \end{array}$$

to specify that the sequent S follows from sequents $S_1, \ldots, S_n$ by the derivation Π. Further, we sometimes combine derivations graphically, e.g. we can combine

$$\begin{array}{c} S_1 \ldots S_m \\ \Pi_1 \\ S_n \end{array} \quad \text{and} \quad \begin{array}{c} S_n \;\; S_o \ldots S_z \\ \Pi_2 \\ S \end{array} \quad \text{to} \quad \begin{array}{c} S_1 \ldots S_m \\ \Pi_1 \\ S_n \;\; S_o \ldots S_z \\ \Pi_2 \\ S \end{array} \; .$$

For brevity, we denote a sequence of applications of the rule (r) in a derivation with vertical dots labelled with (r), e.g.

$$\begin{array}{c} \Delta_2, xRy \vdash xRy \\ \vdots \;\mathrm{WrL} \\ \Delta_1, \Delta_2, xRy \vdash xRy \end{array} \quad \text{or, more generally,} \quad \begin{array}{c} \Gamma_1, \Delta_1 \vdash \Gamma_1' \\ \vdots \;\mathrm{W} \\ \Gamma_1, \Gamma_2, \Delta_1, \Delta_2 \vdash \Gamma_1', \Gamma_2' \end{array}$$

where W stands for one or more applications of the weakening rules. Moreover, given a derivation

$$\begin{array}{c} S_1 \ldots S_n \\ \Pi \\ S \end{array} \quad \text{we write} \quad \begin{array}{c} S_1' \ldots S_n' \\ \Pi^\dagger \\ S' \end{array}$$

to denote that in $\Pi^\dagger$ we apply to $S_1' \ldots S_n'$ the same sequence of rules (with the same principal and active formulas) applied to $S_1 \ldots S_n$ in Π. That is, Π and $\Pi^\dagger$ differ only in their parametric formulas. For example, let

$$\begin{array}{c} x{:}A, \Gamma, \Delta \vdash \Gamma', x{:}B \quad x{:}C, \Gamma, \Delta \vdash \Gamma' \\ \Pi \\ x{:}(A \supset B) \supset C, \Gamma, \Delta \vdash \Gamma' \end{array}$$

be

$$\dfrac{\dfrac{x{:}A, \Gamma, \Delta \vdash \Gamma', x{:}B}{\Gamma, \Delta \vdash \Gamma', x{:}A \supset B} \supset\mathrm{R} \qquad x{:}C, \Gamma, \Delta \vdash \Gamma'}{x{:}(A \supset B) \supset C, \Gamma, \Delta \vdash \Gamma'} \supset\mathrm{L} \; .$$

Then we write

$$\begin{array}{c} x{:}A, \Gamma_1, \Delta_1 \vdash \Gamma_1', x{:}B \quad x{:}C, \Gamma_1, \Delta_1 \vdash \Gamma_1' \\ \Pi^\dagger \\ x{:}(A \supset B) \supset C, \Gamma_1, \Delta_1 \vdash \Gamma_1' \end{array}$$

to denote

$$\dfrac{\dfrac{x{:}A,\Gamma_1,\Delta_1 \vdash \Gamma_1', x{:}B}{\Gamma_1,\Delta_1\vdash\Gamma_1',x{:}A\supset B}\supset\text{R} \qquad x{:}C,\Gamma_1,\Delta_1\vdash\Gamma_1'}{x{:}(A\supset B)\supset C,\Gamma_1,\Delta_1\vdash\Gamma_1'}\supset\text{L} \quad .$$

Example 6.1.3 As examples of derivations and proofs, we use the rules of S(K) and the derived $\sim$ rules to derive the rules for $\Diamond$, cf. (2.1) and (2.2), and prove the S(K)-theorem $x{:}\Box(A \supset B) \supset (\Box A \supset \Box B)$. Note that the side condition on the application of $\Diamond$L, that the (atomic) label y does not occur in $x{:}\Diamond A, \Gamma, \Delta \vdash \Gamma'$, follows from the condition on the application of $\Box$R.

$$\dfrac{\begin{matrix}\Pi\\ y{:}A,\Gamma,\Delta,xRy\vdash\Gamma'\end{matrix}}{x{:}\Diamond A,\Gamma,\Delta\vdash\Gamma'}\Diamond\text{L} \quad\rightsquigarrow\quad \dfrac{\dfrac{\dfrac{\begin{matrix}\Pi\\ y{:}A,\Gamma,\Delta,xRy\vdash\Gamma'\end{matrix}}{\Gamma,\Delta,xRy\vdash\Gamma',y{:}\sim A}\sim\text{R}}{\Gamma,\Delta\vdash\Gamma',x{:}\Box\sim A}\Box\text{R}}{x{:}\sim\Box\sim A,\Gamma,\Delta\vdash\Gamma'}\sim\text{L}$$

$$\dfrac{\begin{matrix}\Pi_1\\ \Delta\vdash xRy\end{matrix} \quad \begin{matrix}\Pi_2\\ \Gamma,\Delta\vdash\Gamma',y{:}A\end{matrix}}{\Gamma,\Delta\vdash\Gamma',x{:}\Diamond A}\Diamond\text{R} \quad\rightsquigarrow\quad \dfrac{\dfrac{\begin{matrix}\Pi_1\\ \Delta\vdash xRy\end{matrix} \quad \dfrac{\begin{matrix}\Pi_2\\ \Gamma,\Delta\vdash\Gamma',y{:}A\end{matrix}}{y{:}\sim A,\Gamma,\Delta\vdash\Gamma'}\sim\text{L}}{x{:}\Box\sim A,\Gamma,\Delta\vdash\Gamma'}\Box\text{L}}{\Gamma,\Delta\vdash\Gamma',x{:}\sim\Box\sim A}\sim\text{R}$$

$$\dfrac{\dfrac{\dfrac{\dfrac{\dfrac{\overline{xRy\vdash xRy}\,\text{AXr} \quad \dfrac{\overline{xRy\vdash xRy}\,\text{AXr} \quad \dfrac{\dfrac{\dfrac{\overline{y{:}A\vdash y{:}A}\,\text{AXl}}{y{:}A\vdash y{:}B,y{:}A}\text{WlL} \quad \dfrac{\overline{y{:}B\vdash y{:}B}\,\text{AXl}}{y{:}B,y{:}A\vdash y{:}B}\text{WlL}}{y{:}A\supset B,y{:}A\vdash y{:}B}\supset\text{L}}{y{:}A\supset B,y{:}A,xRy\vdash y{:}B}\text{WrL}}{y{:}A\supset B,x{:}\Box A,xRy\vdash y{:}B}\Box\text{L}}{x{:}\Box(A\supset B),x{:}\Box A,xRy\vdash y{:}B}\Box\text{L}}{x{:}\Box(A\supset B),x{:}\Box A\vdash x{:}\Box B}\Box\text{R}}{x{:}\Box(A\supset B)\vdash x{:}\Box A\supset\Box B}\supset\text{R}}{\vdash x{:}\Box(A\supset B)\supset(\Box A\supset\Box B)}\supset\text{R} \quad .$$

■

Note that in S(K) the provability of a sequent $\Delta \vdash xRy$ reduces to a question of membership: $\Delta \vdash xRy$ is provable iff $xRy \in \Delta$. This is because the labelling algebra of S(K) is essentially empty; the axiom AXr and the rules WrL and CrL are needed in S(K) simply to manipulate assumptions of rwffs. In fact, as is sometimes the case for other sequent systems, including standard systems for modal logics [119, 221, 238], all of the structural rules can be absorbed into the axioms and the logical rules of S(K). For example, we can dispense with weakenings by building them into 'extended' axioms of the form

$$\overline{x{:}A,\Gamma,\Delta\vdash\Gamma',x{:}A}\;, \qquad \overline{\Delta,xRy\vdash xRy}\;, \quad\text{and}\quad \overline{y{:}\bot,\Gamma,\Delta\vdash\Gamma',x{:}A}\;,$$

and we can dispense with contractions by building them into 'extended' logical rules of the form

$$\frac{x{:}A, \Gamma, \Delta \vdash \Gamma', x{:}A \supset B, x{:}B}{\Gamma, \Delta \vdash \Gamma', x{:}A \supset B} \quad \text{and} \quad \frac{\Delta \vdash xRy \quad y{:}A, x{:}\Box A, \Gamma, \Delta \vdash \Gamma'}{x{:}\Box A, \Gamma, \Delta \vdash \Gamma'} \,.$$

It is then a trivial matter to show that these extensions yield equivalent sequent systems.

The provability of $\Delta \vdash xRy$ requires more than testing membership when we extend S(K) with relational rules to obtain sequent systems for other logics. To formalize these systems, we can directly import definitions and results given for natural deduction systems (including the 'space' of systems with local, global or universal falsum analyzed in §2.3). In particular, in the following we will focus on *Horn relational sequent theories*, often dropping the adjective 'Horn', with the implicit understanding that arbitrary first-order (or even higher-order) relational theories can be introduced like for ND systems. For example, for transitivity, we can extend the labelling algebra of a system with

$$\frac{\Delta \vdash xRy \quad \Delta \vdash yRz}{\Delta \vdash xRz} \,, \qquad \text{or add} \qquad \overline{\vdash \prod x \prod y \prod z (xRy \sqcap yRz \sqsupset xRz)}$$

together with a full first-order sequent system with rules for $\prod$, $\sqcap$, $\sqcup$ and other operators (by analogy with the system N_R in the propositional case).

Definition 6.1.4 *A sequent system* $\mathrm{S}(\mathcal{L}) = \mathrm{S}(\mathrm{K}) + \mathrm{S}(\mathcal{T})$ *is obtained by extending* S(K) *with a* Horn relational sequent theory $\mathrm{S}(\mathcal{T})$, *a collection of* Horn relational sequent rules. ■

Table 6.1 contains some examples of Horn relational rules for propositional modal sequent systems, corresponding to the rules given in Table 2.2.[4] Then, for example, the systems S(KD), S(KT), S(KTB), S(KT4) and S(KT5), and their synonyms S(D), S(T), S(B), S(S4) and S(S5), present the modal logics D, T, B, S4, and S5 (we prove this in §6.3 where we show the equivalence of normalizing natural deduction systems and cut-free sequent systems).

Note that we use the same names for the relational rules of ND and sequent systems; it will always be clear from context which system is meant. Most importantly, all of our Horn relational sequent rules have a common form: they operate only 'on the right', i.e.

Fact 6.1.5 *The principal rwff of each Horn relational sequent rule is introduced in the succedent of the conclusion.* ■

Hence, these rules only affect applications of $\Box$L, for which they provide new possible active rwffs. Recall that, symmetrically, in ND systems a relational rule introduces an rwff that can serve as minor premise in an application of $\Box$E. This fact is of crucial importance for the substructural analysis that we perform in the following chapters (see, for example, Lemma 8.2.1, Proposition 8.2.9, Theorem 9.1.1 and Corollary 9.1.2).

[4] The relational rules with empty premises, e.g. *ser* and *refl*, can also be seen as 'relational axioms'.

Table 6.1. Some properties of R and corresponding Horn relational sequent rules

Property	*Horn relational rule*
Seriality	$\dfrac{}{\vdash x R f(x)}\ ser$
Reflexivity	$\dfrac{}{\vdash x R x}\ refl$
Symmetry	$\dfrac{\Delta \vdash x R y}{\Delta \vdash y R x}\ symm$
Transitivity	$\dfrac{\Delta \vdash x R y \quad \Delta \vdash y R z}{\Delta \vdash x R z}\ trans$
Euclideaness	$\dfrac{\Delta \vdash x R y \quad \Delta \vdash x R z}{\Delta \vdash z R y}\ eucl$
Convergency	$\dfrac{\Delta \vdash x R y \quad \Delta \vdash x R z}{\Delta \vdash y R g(x,y,z)}\ conv1 \qquad \dfrac{\Delta \vdash x R y \quad \Delta \vdash x R z}{\Delta \vdash z R g(x,y,z)}\ conv2$

Where f and g are (Skolem) function constants.

Example 6.1.6 As further examples of derivations, which exhibit the use of relational rules and of contraction, we prove the S(T)-theorem $x_1{:} \sim\Box\sim(B \supset \Box B)$ and the S(K4)-theorem $x_1{:}\Box\sim\Box B \supset \Box\sim\Box\Box B$ as follows:

$$
\dfrac{\dfrac{\dfrac{\dfrac{}{\vdash x_1 R x_1}\ refl \qquad \dfrac{\dfrac{\dfrac{\dfrac{\dfrac{}{x_1 R x_2 \vdash x_1 R x_2}\ \mathrm{AXr} \qquad \dfrac{\dfrac{\dfrac{\dfrac{}{x_2{:}B \vdash x_2{:}B}\ \mathrm{AXl}}{x_1{:}B, x_2{:}B, x_1 R x_2 \vdash x_2{:}B, x_2{:}\Box B}\ \vdots\,\mathrm{W}}{x_1{:}B, x_1 R x_2 \vdash x_2{:}B, x_2{:}B \supset \Box B}\ \supset\mathrm{R}}{x_2{:}\sim(B \supset \Box B), x_1{:}B, x_1 R x_2 \vdash x_2{:}B}\ \sim\mathrm{L}}{x_1{:}\Box\sim(B \supset \Box B), x_1{:}B, x_1 R x_2 \vdash x_2{:}B}\ \Box\mathrm{L}}{x_1{:}\Box\sim(B \supset \Box B), x_1{:}B \vdash x_1{:}\Box B}\ \Box\mathrm{R}}{x_1{:}\Box\sim(B \supset \Box B) \vdash x_1{:}B \supset \Box B}\ \supset\mathrm{R}}{x_1{:}\sim(B \supset \Box B), x_1{:}\Box\sim(B \supset \Box B) \vdash}\ \sim\mathrm{L}}{x_1{:}\Box\sim(B \supset \Box B), x_1{:}\Box\sim(B \supset \Box B) \vdash}\ \Box\mathrm{L}}{x_1{:}\Box\sim(B \supset \Box B) \vdash}\ \mathrm{ClL}}{\vdash x_1{:}\sim\Box\sim(B \supset \Box B)}\ \sim\mathrm{R}
\tag{6.1}
$$

$$
\dfrac{\dfrac{}{x_1 R x_2 \vdash x_1 R x_2}\,\text{AXr} \qquad \dfrac{\dfrac{\dfrac{\Pi}{\Delta \vdash x_1 R x_3} \qquad \dfrac{\dfrac{\dfrac{\dfrac{}{x_2 R x_3 \vdash x_2 R x_3}\,\text{AXr}}{\Delta \vdash x_2 R x_3}\,\text{WrL} \qquad \dfrac{\dfrac{}{x_3{:}\Box B \vdash x_3{:}\Box B}\,\text{AXl}}{x_3{:}\Box B, \Delta \vdash x_3{:}B, x_3{:}\Box B}\,\vdots\,\text{W}}{x_2{:}\Box\Box B, \Delta \vdash x_3{:}B, x_3{:}\Box B}\,\Box\text{L}}{x_3{:}\,{\sim}\Box B, x_2{:}\Box\Box B, \Delta \vdash x_3{:}B}\,{\sim}\text{L}}{x_1{:}\Box\,{\sim}\Box B, x_2{:}\Box\Box B, \Delta \vdash x_3{:}B}\,\Box\text{L}}{\dfrac{x_1{:}\Box\,{\sim}\Box B, x_2{:}\Box\Box B, x_1 R x_2 \vdash x_2{:}\Box B}{x_2{:}\,{\sim}\Box B, x_1{:}\Box\,{\sim}\Box B, x_2{:}\Box\Box B, x_1 R x_2 \vdash}\,{\sim}\text{L}}\,\Box\text{R}}{\dfrac{\dfrac{\dfrac{\dfrac{x_1{:}\Box\,{\sim}\Box B, x_1{:}\Box\,{\sim}\Box B, x_2{:}\Box\Box B, x_1 R x_2 \vdash}{x_1{:}\Box\,{\sim}\Box B, x_2{:}\Box\Box B, x_1 R x_2 \vdash}\,\text{ClL}}{x_1{:}\Box\,{\sim}\Box B, x_1 R x_2 \vdash x_2{:}\,{\sim}\Box\Box B}\,{\sim}\text{R}}{x_1{:}\Box\,{\sim}\Box B \vdash x_1{:}\Box\,{\sim}\Box\Box B}\,\Box\text{R}}{\vdash x_1{:}\Box\,{\sim}\Box B \supset \Box\,{\sim}\Box\Box B}\,{\supset}\text{R}}\,\Box\text{L}
\tag{6.2}
$$

where $\Delta = \{x_1 R x_2, x_2 R x_3\}$ and Π is

$$
\dfrac{\dfrac{\dfrac{}{x_1 R x_2 \vdash x_1 R x_2}\,\text{AXr}}{x_1 R x_2, x_2 R x_3 \vdash x_1 R x_2}\,\text{WrL} \qquad \dfrac{\dfrac{}{x_2 R x_3 \vdash x_2 R z}\,\text{AXr}}{x_1 R x_2, x_2 R x_3 \vdash x_2 R x_3}\,\text{WrL}}{x_1 R x_2, x_2 R x_3 \vdash x_1 R x_3}\,\textit{trans} .
$$

■

Note that the contractions in (6.1) and (6.2) are indispensable; in fact, in §10 and §11 we show that these two end-sequents cannot be proved without (at least) one application of ClL.

6.2 LABELLED SEQUENT SYSTEMS FOR NON-CLASSICAL LOGICS

Parallel to Chapter 3, we generalize our cut-free labelled sequent systems to present other non-classical logics. For example, the universal non-local operator $\mathcal{M}^u$ of arity u associated with a u+1–ary relation R^u is characterized by the following left and right logical rules:

$$
\dfrac{\Delta \vdash R^u\, a\, a_1 \ldots a_u \quad \Gamma, \Delta \vdash \Gamma', a_1{:}A_1 \quad \cdots \quad \Gamma, \Delta \vdash \Gamma', a_{u-1}{:}A_{u-1} \quad a_u{:}A_u, \Gamma, \Delta \vdash \Gamma'}{a{:}\mathcal{M}^u A_1 \ldots A_u, \Gamma, \Delta \vdash \Gamma'}\,\mathcal{M}^u\text{L}
$$

$$
\dfrac{a_1{:}A_1, \ldots, a_{u-1}{:}A_{u-1}, \Gamma, \Delta, R^u\, a\, a_1 \ldots a_u \vdash \Gamma', a_u{:}A_u}{\Gamma, \Delta \vdash \Gamma', a{:}\mathcal{M}^u A_1 \ldots A_u}\,\mathcal{M}^u\text{R}
$$

where, in $\mathcal{M}^u$R, the labels $a_1, \ldots, a_u$ are all different from a and each other, and do not occur in $\Gamma, \Delta \vdash \Gamma', a{:}\mathcal{M}^u A_1 \ldots A_u$. Then, e.g., the sequent system S(R) for the relevance logic R contains, among others, the following rules for relevant implication

($\to$) and non-local negation ($\neg$),

$$\dfrac{\Delta \vdash R\,a\,b\,c \quad \Gamma,\Delta \vdash \Gamma', b{:}A \quad c{:}B,\Gamma,\Delta \vdash \Gamma'}{a{:}A \to B,\Gamma,\Delta \vdash \Gamma'}\to\mathrm{L}$$

$$\dfrac{b{:}A,\Gamma,\Delta, R\,a\,b\,c \vdash \Gamma', c{:}B}{\Gamma,\Delta \vdash \Gamma', a{:}A \to B}\to\mathrm{R} \qquad \dfrac{\Gamma,\Delta \vdash \Gamma', a^*{:}A}{a{:}\neg A,\Gamma,\Delta \vdash \Gamma'}\neg\mathrm{L} \qquad \dfrac{a^*{:}A,\Gamma,\Delta \vdash \Gamma'}{\Gamma,\Delta \vdash \Gamma', a{:}\neg A}\neg\mathrm{R}$$

$$\dfrac{\Delta \vdash R\,a\,b\,c}{\Delta \vdash R\,a\,c^*\,b^*}\,\mathit{inv} \qquad \dfrac{}{\vdash R\,0\,a\,a^{**}}\,{**}\mathrm{i} \qquad \dfrac{}{\vdash R\,0\,a^{**}\,a}\,{**}\mathrm{c}$$

where, in $\to$R, the labels b and c are different from a and each other, and do not occur in $\Gamma,\Delta \vdash \Gamma', a{:}A \to B$.

These sequent rules correspond to the ND rules given in §3, and they allow us, for example, to derive a contraposition rule (cf. the derivation (3.23) in Example 3.1.12, where we used antitonicity instead of the stronger inversion property):

$$\dfrac{\dfrac{\dfrac{\dfrac{}{R\,0\,a\,b \vdash R\,0\,a\,b}\mathrm{AXr}}{R\,0\,a\,b \vdash R\,0\,b^*\,a^*}\mathit{inv} \quad \dfrac{\begin{array}{c}\dfrac{}{b^*{:}A \vdash b^*{:}A}\mathrm{AXl}\\ \vdots\ \mathrm{W}\\ b^*{:}A, a{:}\neg B, R\,0\,a\,b \vdash b^*{:}A\end{array}}{a{:}\neg B, R\,0\,a\,b \vdash b{:}\neg A, b^*{:}A}\neg\mathrm{R} \quad \dfrac{\begin{array}{c}\dfrac{}{a^*{:}B \vdash a^*{:}B}\mathrm{AXl}\\ \vdots\ \mathrm{W}\\ a^*{:}B, R\,0\,a\,b \vdash b{:}\neg A, a^*{:}B\end{array}}{a^*{:}B, a{:}\neg B, R\,0\,a\,b \vdash b{:}\neg A}\neg\mathrm{L}}{a{:}\neg B, 0{:}A \to B, R\,0\,a\,b \vdash b{:}\neg A}\to\mathrm{L}}{0{:}A \to B \vdash 0{:}\neg B \to \neg A}\to\mathrm{R} \; .$$

As for labelled ND systems, labelled sequent systems for various families of non-classical logics are obtained by extending a suitable fixed base system with collections of (Horn) rules axiomatizing semantic properties.

We can also give systems with local and universal falsum, or consider systems with different treatments of negation, e.g. minimal, intuitionistic, classical or ortho negation. We can even consider 'full' intuitionistic versions of non-classical logics. For example, mirroring Simpson's work on intuitionistic modal logics [216], we can present an intuitionistic version of the modal logic K as the subsystem of S(K) obtained by

(i) explicitly adding the rules for $\wedge$, $\vee$ and $\Diamond$ given in Figure 6.2,

(ii) restricting the sequents to contain at most one formula in the succedent, and

(iii) replacing $\supset$L with a labelled version of the standard intuitionistic left implication rule:

$$\dfrac{\Gamma,\Delta \vdash x{:}A \quad x{:}B,\Gamma,\Delta \vdash y{:}C}{x{:}A \to B,\Gamma,\Delta \vdash y{:}C}\to\mathrm{L} \quad .$$

Two remarks. First, the restriction (ii) rules out contraction on the right. Second, the addition (i) is needed since $\wedge$, $\vee$ and $\Diamond$ must now be taken as primitive logical operators (e.g. $\Diamond$ is not definable in terms of $\Box$) and their rules are not derivable anymore. However, in the resulting system we can still derive the rules for $\sim$, as well as the rules given by Simpson.

Finally, we can consider also extensions to the quantified case. For example, for quantified modal logics with varying domains we simply need to appropriately modify the definitions of formulas and sequents to introduce a base sequent system S(QK) containing the rules

$$\frac{\Delta, \Theta \vdash w{:}t \quad w{:}A[t/x], \Gamma, \Delta, \Theta \vdash \Gamma'}{w{:}\forall x(A), \Gamma, \Delta, \Theta \vdash \Gamma'}\ \forall\mathrm{L} \quad \text{and} \quad \frac{\Gamma, \Delta, \Theta, w{:}t \vdash \Gamma' w{:}A[t/x]}{\Gamma, \Delta, \Theta \vdash \Gamma', w{:}\forall x(A)}\ \forall\mathrm{R}\ ,$$

where Θ is a multiset of lterms and $\forall$ R has the side condition that t does not occur in $\Gamma, \Delta, \Theta \vdash \Gamma', w{:}\forall x(A)$. Then we can formalize increasing, decreasing and constant domains by appropriately adding the rules

$$\frac{\Delta \vdash w_i R w_j \quad \Delta, \Theta \vdash w_i{:}t}{\Delta, \Theta \vdash w_j{:}t}\ id \quad \text{and} \quad \frac{\Delta \vdash w_i R w_j \quad \Delta, \Theta \vdash w_j{:}t}{\Delta, \Theta \vdash w_i{:}t}\ dd\ .$$

6.3 EQUIVALENCE OF LABELLED NATURAL DEDUCTION AND SEQUENT SYSTEMS

Normalizing natural deduction systems and cut-free sequent systems are closely related [185, 186, 221, 239]. In this section, we show that $\mathrm{S}(\mathcal{L})$ and $\mathrm{N}(\mathcal{L})$ are intertranslatable, from which it follows that our cut-free labelled sequent systems are sound and complete with respect to the corresponding Kripke semantics.

We prove this equivalence for classical propositional modal logics, i.e. for $\mathrm{S}(\mathcal{L}) = \mathrm{S}(\mathrm{K}) + \mathrm{S}(\mathcal{T})$ and $\mathrm{N}(\mathcal{L}) = \mathrm{N}(\mathrm{K}) + \mathrm{N}(\mathcal{T})$, and then discuss extensions and restrictions to other systems.

Theorem 6.3.1 *Let* $\Lambda = \{x_i{:} \sim B_i \mid x_i{:} \sim B_i \in \Gamma\}$, *say* $\Lambda = \{x_1{:} \sim B_1, \ldots, x_n{:} \sim B_n\}$, *and let* $\Psi = \Gamma \setminus \Lambda$. *Then we have:*

(i) $\Delta \vdash_{\mathrm{N}(\mathcal{L})} xRy$ *iff the sequent* $\Delta \vdash xRy$ *is provable in* $\mathrm{S}(\mathcal{L})$.

(ii) $\Gamma, \Delta \vdash_{\mathrm{N}(\mathcal{L})} x{:}A$ *iff the sequent* $\Psi, \Delta \vdash x{:}A, x_1{:}B_1, \ldots, x_n{:}B_n$ *is provable in* $\mathrm{S}(\mathcal{L})$..

Proof We adapt and extend the proof given by Prawitz in [186, App. A]. Claim (i) follows trivially by induction on the length of the derivations, applying the same rules in both systems.

((ii), *left-to-right*) We proceed by induction on the length of the normal $\mathrm{N}(\mathcal{L})$-derivation of $x{:}A$ from Γ, Δ. The base case, $x{:}A \in \Gamma$, is trivial. There is one step case for each rule of $\mathrm{N}(\mathcal{L})$ and we distinguish three main cases, depending on the form of the rules.

(*Case 1*) If the last rule in Π is an introduction, then we apply the corresponding right rule in $\mathrm{S}(\mathcal{L})$. Consider, for example, an application of $\Box$I,

$$\begin{array}{c} \Gamma\ \Delta\ [xRy] \\ \Pi_1 \\ \dfrac{y{:}A}{x{:}\Box A}\ \Box\mathrm{I} \end{array} ,$$

where y is different from x and does not occur in any assumption on which $y{:}A$ depends other than xRy. Π_1 is a $\mathrm{N}(\mathcal{L})$-derivation of $y{:}A$ from Γ, Δ and xRy. By the induction hypothesis, there is then a $S(\mathcal{L})$-proof Π_1' of $\Psi, \Delta, xRy \vdash y{:}A, x_1{:}B_1, \ldots, x_n{:}B_n$, where y is different from x and does not occur in $\Psi, \Delta, x_1{:}B_1, \ldots, x_n{:}B_n$. We conclude by an application of $\Box$R, i.e.

$$\dfrac{\begin{array}{c}\Pi_1' \\ \Psi, \Delta, xRy \vdash y{:}A, x_1{:}B_1, \ldots, x_n{:}B_n\end{array}}{\Psi, \Delta \vdash x{:}\Box A, x_1{:}B_1, \ldots, x_n{:}B_n}\ \Box\text{R} \quad .$$

We conclude analogously when the last rule in Π is an application of $\supset$I, i.e.

$$\dfrac{\begin{array}{c}\Gamma\ \Delta\ [x{:}A] \\ \Pi_1 \\ x{:}C\end{array}}{x{:}A \supset C}\ \supset\text{I} \quad \rightsquigarrow \quad \dfrac{\begin{array}{c}\Pi_1' \\ \Psi, x{:}A, \Delta \vdash x{:}C, x_1{:}B_1, \ldots, x_n{:}B_n\end{array}}{\Psi, \Delta \vdash x{:}A \supset C, x_1{:}B_1, \ldots, x_n{:}B_n}\ \supset\text{R} \quad .$$

(*Case 2*) Suppose that the last rule in Π is an application of $\bot$E, i.e.

$$\dfrac{\begin{array}{c}[x{:}\sim A]\ \Gamma\ \Delta \\ \Pi_1 \\ y{:}\bot\end{array}}{x{:}A}\ \bot\text{E} \quad ,$$

where Π_1 is a $N(\mathcal{L})$-derivation of $y{:}\bot$ from $x{:}\sim A, \Gamma, \Delta$. By the induction hypothesis, there is a $S(\mathcal{L})$-proof Π_1' of the sequent $\Psi, \Delta \vdash y{:}\bot, x{:}A, x_1{:}B_1, \ldots, x_n{:}B_n$. Now reason on $y{:}\bot$ in Π_1'. It can only be the result either of an application of WlR or of an axiom. In the first case, we delete the weakening from Π_1 to obtain the desired $S(\mathcal{L})$-proof Π_1'' of $\Psi, \Delta \vdash x{:}A, x_1{:}B_1, \ldots, x_n{:}B_n$. In the second case, Π_1' has the form

$$\begin{array}{c}\dfrac{}{z{:}\bot \vdash y{:}\bot}\ \bot\text{L (or AXl if } z = y) \\ \Pi_2 \\ \Psi, \Delta \vdash y{:}\bot, x{:}A, x_1{:}B_1, \ldots, x_n{:}B_n\end{array} \quad . \tag{6.3}$$

Then the occurrence of $y{:}\bot$ in the succedent is parametric in Π_2, and Π_2 must contain at least one weakening introducing some labelled subformula of some formula in the end-sequent. Thus, we conclude by simply transforming the initial $\bot$L (or AXl) to introduce this other lwff. For example, if (6.3) is

$$\begin{array}{c}\dfrac{}{y{:}\bot \vdash y{:}\bot}\ \text{AXl} \\ \Pi_3 \\ \dfrac{\Psi_1, \Delta_1 \vdash y{:}\bot}{\Psi_1, \Delta_1 \vdash y{:}\bot, x_1{:}B_1}\ \text{WlR} \\ \Pi_4 \\ \Psi, \Delta \vdash y{:}\bot, x{:}A, x_1{:}B_1, \ldots, x_n{:}B_n\end{array}$$

we transform this to obtain the desired $S(\mathcal{L})$-proof

$$\begin{array}{c}\dfrac{}{y{:}\bot \vdash x_1{:}B_1}\ \bot\text{L} \\ \Pi_3^\dagger \\ \Psi_1, \Delta_1 \vdash x_1{:}B_1 \\ \Pi_4^\dagger \\ \Psi, \Delta \vdash x{:}A, x_1{:}B_1, \ldots, x_n{:}B_n\end{array} \quad .$$

Note that if $y{:}\bot = x_i{:}B_i$ for some $1 \leq i \leq n$, we can alternatively conclude by an application of ClR. In fact, we could have dispensed with the second subcase by showing that the restrictions that $A \neq \bot$ in $\bot$L and AXl yield an equivalent sequent system.

(*Case 3*) The last rule in Π is an elimination. Then we apply the corresponding left rule in $\mathrm{S}(\mathcal{L})$. Let τ be an lwff-thread in Π that contains no minor premise, so that, by the structure of normal $\mathrm{N}(\mathcal{L})$-derivations (Lemma 2.3.11), the introduction part of τ is empty.[5] Hence, no assumption can be discharged in τ, and the first lwff occurring in τ, say $y{:}C$, belongs to Γ and is the major premise of an elimination rule. We consider the different cases for this rule.

If it is $\supset$E, then $y{:}C = y{:}C_1 \supset C_2$ and Π has the form

$$\cfrac{y{:}C_1 \supset C_2 \quad \begin{array}{c}\Pi_1\\ y{:}C_1\end{array}}{\begin{array}{c}y{:}C_2\\ \Pi_2\\ x{:}A\end{array}}\supset\mathrm{E} \quad .$$

Since no assumption is discharged at any formula occurrence in τ, $y{:}C_1$ cannot depend on assumptions other than those on which the end-formula $x{:}A$ of Π depends. Hence, Π_1 is a $\mathrm{N}(\mathcal{L})$-derivation of $y{:}C_1$ from Γ, Δ. Furthermore, Π_2 is a $\mathrm{N}(\mathcal{L})$-derivation of $x{:}A$ from $y{:}C_2, \Gamma, \Delta$. By the induction hypotheses, there are $\mathrm{S}(\mathcal{L})$-proofs of the sequents $\Psi, \Delta \vdash x_1{:}B_1, \ldots, x_n{:}B_n, y{:}C_1$ and $y{:}C_2, \Psi, \Delta \vdash x{:}A, x_1{:}B_1, \ldots, x_n{:}B_n$, where $y{:}C_1 \supset C_2 \in \Psi$. Thus, by an application of $\supset$L and a contraction of $y{:}C_1 \supset C_2$, we obtain the desired $\mathrm{S}(\mathcal{L})$-proof of $\Psi, \Delta \vdash x{:}A, x_1{:}B_1, \ldots, x_n{:}B_n$.

If it is $\Box$E, then $y{:}C = y{:}\Box C_1$ and Π has the form

$$\cfrac{y{:}\Box C_1 \quad \begin{array}{c}\Pi_1\\ yR\,z\end{array}}{\begin{array}{c}z{:}C_1\\ \Pi_2\\ x{:}A\end{array}}\Box\mathrm{E} \quad .$$

Since no assumption is discharged at any formula occurrence in τ, $yR\,z$ cannot depend on assumptions other than those on which the end-formula $x{:}A$ of Π depends. Hence, Π_1 is a $\mathrm{N}(\mathcal{L})$-derivation of $yR\,z$ from Δ. Furthermore, Π_2 is a $\mathrm{N}(\mathcal{L})$-derivation of $x{:}A$ from $z{:}C_1, \Gamma, \Delta$. By the induction hypotheses, there are $\mathrm{S}(\mathcal{L})$-proofs of the sequents $\Delta \vdash yR\,z$ and $z{:}C_1, \Psi, \Delta \vdash x{:}A, x_1{:}B_1, \ldots, x_n{:}B_n$, where $y{:}\Box C_1 \in \Psi$. Thus, by an application of $\Box$L and a contraction of $y{:}\Box C_1$, we obtain the desired $\mathrm{S}(\mathcal{L})$-proof of $\Psi, \Delta \vdash x{:}A, x_1{:}B_1, \ldots, x_n{:}B_n$.

((ii), *right-to-left*) We proceed by induction on the length of the $\mathrm{S}(\mathcal{L})$-proof Π of the sequent. The base case is when the sequent is an axiom. AXl corresponds to a $\mathrm{N}(\mathcal{L})$-derivation consisting of the single assumption A, and $\bot$L corresponds to

$$\frac{y{:}\bot}{x{:}A}\bot\mathrm{E} \quad .$$

[5]Recall from Definition 2.3.8 that an lwff-thread is a sequence of formulas $\varphi_1, \ldots, \varphi_n$ in a derivation where each φ_i is an lwff. An lwff-thread containing no minor premise is also a track.

There is one step for each rule of $\mathrm{S}(\mathcal{L})$, and we distinguish three main cases.

(*Case 1*) If the last rule in Π is a structural rule, then we conclude by exploiting the fact that ND systems admit equivalents of weakening and contraction. For example,

$$\frac{\Psi, \Delta \vdash x{:}A, x_1{:}B_1, \ldots, x_n{:}B_n}{y{:}C, \Psi, \Delta \vdash x{:}A, x_1{:}B_1, \ldots, x_n{:}B_n}\ \mathrm{WlL}$$

and

$$\frac{\Psi, \Delta \vdash x{:}A, x_1{:}B_1, \ldots, x_n{:}B_n}{\Psi, \Delta \vdash x{:}A, x_1{:}B_1, \ldots, x_n{:}B_n, y{:}C}\ \mathrm{WlR}$$

correspond to transforming

$$\begin{array}{c}\Gamma\ \Delta\\ \Pi\\ x{:}A\end{array} \quad \text{to} \quad \begin{array}{c}y{:}C\ \Gamma\ \Delta\\ \Pi\\ x{:}A\end{array} \quad \text{and} \quad \begin{array}{c}y{:}\sim C\ \Gamma\ \Delta\\ \Pi\\ x{:}A\end{array}$$

respectively, while contractions correspond to identifications of *assumption classes* in ND systems; see, e.g., [221, p. 59].

(*Case 2*) If the last rule in Π is a left logical rule, then we apply the corresponding elimination rule in $\mathrm{N}(\mathcal{L})$. Consider, for example, an application of $\Box$L,

$$\frac{\Delta \vdash xRy \quad y{:}A, \Psi, \Delta \vdash z{:}C, x_1{:}B_1, \ldots, x_n{:}B_n}{x{:}\Box A, \Psi, \Delta \vdash z{:}C, x_1{:}B_1, \ldots, x_n{:}B_n}\ \Box\mathrm{L} \quad .$$

By the induction hypotheses, there are $\mathrm{N}(\mathcal{L})$-derivations

$$\begin{array}{c}\Delta\\ \Pi_1\\ xRy\end{array} \quad \text{and} \quad \begin{array}{c}y{:}A\ \Gamma\ \Delta\\ \Pi_2\\ z{:}C\end{array} \quad .$$

Then we apply $\Box$E to obtain the desired $\mathrm{N}(\mathcal{L})$-derivation of $z{:}C$ from $x{:}\Box A, \Gamma, \Delta$, i.e. we combine the derivations as follows:

$$\begin{array}{c}\dfrac{x{:}\Box A \quad \begin{array}{c}\Delta\\ \Pi_1\\ xRy\end{array}}{y{:}A}\ \Box\mathrm{E} \quad \Gamma\ \Delta\\ \Pi_2\\ z{:}C\end{array} \quad .$$

If the last rule in Π is an application of $\supset$L,

$$\frac{\Psi, \Delta \vdash z{:}D, x_1{:}B_1, \ldots, x_n{:}B_n, x{:}A \quad x{:}C, \Psi, \Delta \vdash z{:}D, x_1{:}B_1, \ldots, x_n{:}B_n}{x{:}A \supset C, \Psi, \Delta \vdash z{:}D, x_1{:}B_1, \ldots, x_n{:}B_n}\ \supset\mathrm{L} \quad ,$$

we conclude analogously, i.e.

$$\begin{array}{c}\dfrac{x{:}A \supset C \quad \begin{array}{c}\Gamma\ \Delta\\ \Pi_1\\ x{:}A\end{array}}{x{:}C}\ \supset\mathrm{E} \quad \Gamma\ \Delta\\ \Pi_2\\ z{:}D\end{array} \quad .$$

(*Case 3*) If the last rule in Π is a right logical rule, then we apply the corresponding introduction rule in $\mathrm{N}(\mathcal{L})$. Consider, for example, an application of $\Box$R,

$$\frac{\Psi, \Delta, xRy \vdash x_1{:}B_1, \ldots, x_n{:}B_n, y{:}A}{\Psi, \Delta \vdash x_1{:}B_1, \ldots, x_n{:}B_n, x{:}\Box A}\ \Box\mathrm{R}\quad,$$

where y is different from x and does not occur in $\Psi, \Delta, x_1{:}B_1, \ldots, x_n{:}B_n$. By the induction hypothesis, there is a $\mathrm{N}(\mathcal{L})$-derivation Π_1 of $y{:}A$ from Γ, Δ, xRy, where y is different from x and does not occur in any assumption on which $y{:}A$ depends other than xRy. Then we apply $\Box$I to obtain the desired $\mathrm{N}(\mathcal{L})$-derivation of $x{:}\Box A$ from Γ, Δ, i.e.

$$\begin{array}{c} \Gamma\ \Delta\ [xRy] \\ \Pi_1 \\ \dfrac{y{:}A}{x{:}\Box A}\ \Box\mathrm{I} \end{array} \quad .$$

We conclude analogously when the last rule in Π is an application of $\supset$R, i.e.

$$\frac{x{:}A, \Psi, \Delta \vdash x_1{:}B_1, \ldots, x_n{:}B_n, x{:}C}{\Psi, \Delta \vdash x_1{:}B_1, \ldots, x_n{:}B_n, x{:}A \supset C}\ \supset\mathrm{R} \quad \rightsquigarrow \quad \begin{array}{c} [x{:}A]\ \Gamma\ \Delta \\ \Pi_1 \\ \dfrac{x{:}C}{x{:}A \supset C}\ \supset\mathrm{I} \end{array} \quad .$$

This concludes the proof of the theorem. ∎

The proof proceeds along the same lines for sequent systems for other non-classical logics with a 'classical' non-local negation, provided that we restrict our attention to 'primitive' rules like we did for S(K), where we omitted the derived rules for $\vee$ and $\Diamond$; this is analogous to considering normalization for the system $\mathrm{N}(\mathcal{CL}')$ with a simplified language, as we did in §3.3.

For intuitionistic and minimal subsystems we prove a slightly different theorem, since for these systems there is no need to push negated formulas on the other side of $\vdash$. In fact, by a standard [186] modification of the proof of Theorem 6.3.1 we can show that for propositional modal systems with an intuitionistic or minimal treatment of negation we have the following theorem.

Theorem 6.3.2 *Let* $\mathrm{N}(\mathcal{L})$ *be* $\mathrm{N}(\mathcal{J}\mathrm{K}) + \mathrm{N}(\mathcal{T})$ *or* $\mathrm{N}(\mathcal{M}\mathrm{K}) + \mathrm{N}(\mathcal{T})$, *and* $\mathrm{S}(\mathcal{L})$ *be* $\mathrm{S}(\mathcal{J}\mathrm{K}) + \mathrm{S}(\mathcal{T})$ *or* $\mathrm{S}(\mathcal{M}\mathrm{K}) + \mathrm{S}(\mathcal{T})$, *respectively. Then we have:*

(i) $\Delta \vdash_{\mathrm{N}(\mathcal{L})} xRy$ *iff the sequent* $\Delta \vdash xRy$ *is provable in* $\mathrm{S}(\mathcal{L})$.

(ii) $\Gamma, \Delta \vdash_{\mathrm{N}(\mathcal{L})} x{:}A$ *iff the sequent* $\Gamma, \Delta \vdash x{:}A$ *is provable in* $\mathrm{S}(\mathcal{L})$. ∎

Consider now again labelled sequent systems for propositional modal logics (with a classical negation); analogous results hold for our sequent systems for other non-classical logics. From Theorem 6.3.1 and the soundness and completeness of $\mathrm{N}(\mathcal{L}) = \mathrm{N}(\mathrm{K}) + \mathrm{N}(\mathcal{T})$ with respect to the corresponding Kripke semantics (Theorem 2.2.5), it immediately follows that our cut-free sequent systems are sound and complete with respect to same semantics.

Corollary 6.3.3 $\mathrm{S}(\mathcal{L}) = \mathrm{S}(\mathrm{K}) + \mathrm{S}(\mathcal{T})$ *is sound and complete.* ■

Another important consequence of Theorem 6.3.1 is the admissibility of the rule

$$\frac{\Gamma_1, \Delta_1 \vdash \Gamma_1', x{:}A \quad x{:}A, \Gamma_2, \Delta_2 \vdash \Gamma_2'}{\Gamma_1, \Gamma_2, \Delta_1, \Delta_2 \vdash \Gamma_1', \Gamma_2'}\ cut$$

in any sequent system $\mathrm{S}(\mathcal{L})$. That is, although *cut* is not a derived rule of $\mathrm{S}(\mathcal{L})$, no new theorems become provable by its addition. This follows from Theorem 6.3.1 because it holds trivially that if $\Gamma_1, \Delta_1 \vdash_{\mathrm{N}(\mathcal{L})} x{:}A$ and $x{:}A, \Gamma_2, \Delta_2 \vdash_{\mathrm{N}(\mathcal{L})} y{:}B$, then also $\Gamma_1, \Gamma_2, \Delta_1, \Delta_2 \vdash_{\mathrm{N}(\mathcal{L})} y{:}B$. Graphically, we combine

$$\begin{array}{c}\Gamma_1\ \Delta_1\\ \Pi_1\\ x{:}A\end{array} \quad \text{and} \quad \begin{array}{c}x{:}A\ \Gamma_2\ \Delta_2\\ \Pi_2\\ y{:}B\end{array} \quad \text{to} \quad \begin{array}{c}\Gamma_1\ \Delta_1\\ \Pi_1\\ x{:}A\quad \Gamma_2\ \Delta_2\\ \Pi_2\\ y{:}B\end{array}$$

possibly renaming some labels to avoid variable clashes.

Note that instead of showing that *cut* is admissible as a consequence of Theorem 6.3.1, we could have followed Gentzen [106], who first proved the completeness of his sequent systems *with cut*, and then that applications of *cut* can be eliminated from a given proof.

Although *cut* provides a powerful tool for shortening and reusing proofs, its (unrestricted) addition to a system $\mathrm{S}(\mathcal{L})$ would require us to control its application if we want to establish the decidability of $\mathcal{L}$. To illustrate this briefly, suppose that, as is usually done, we apply the rules *backwards* (i.e. upwards) to build a proof *top-down*, starting with the end-sequent and working towards the axioms. Since *cut* is always applicable, no matter what its end-sequent looks like, it follows that potentially infinite branches exist and top-down proof search may not terminate.

Moreover, each backwards application of *cut* may introduce an arbitrary formula A, so that its addition would spoil an important feature of our systems: our labelled sequent systems satisfy the *subformula property*, in the sense that in any proof of a sequent $\Gamma, \Delta \vdash \Gamma'$, only labelled subformulas of Γ and Γ' occur; cf. Definition 2.3.10 and Lemma 2.3.13 for the propositional case.[6] Therefore, if in our systems we build proofs top-down, at each rule application other than a contraction we obtain lwffs of smaller grade (or we just delete lwffs using weakening). That this is the case follows immediately by the form of our rules, observing that each active lwff is a subformula of the principal labelled formula of the rule. Contraction, on the other hand, duplicates

[6]Note that eliminating the rule

$$\frac{\Sigma_1 \vdash \Sigma_2, A \quad A, \Sigma_1 \vdash \Sigma_2}{\Sigma_1 \vdash \Sigma_2}\ (cut)$$

where the Σ_i's are multisets of formulas, is not the only way to control its application in standard sequent systems for modal and other logics. Common alternatives include reducing it to *analytic cut* where we 'cut' only subformulas of the goal, e.g. [35, 62, 68], or to applications in which we cut complex but 'controlled' formulas built from subformulas according to a specific *superformula principle*, e.g. [87, 120].

formulas instead of simplifying them, and is thus always applicable, with, again, the consequence that proof search may not terminate.

Let us consider in more detail the problems with contraction, as this also helps motivate the substructural analysis that we perform in Part II.[7] We can, as is generally done, view our sequent systems as *refutation systems*, and associate the progressive (backwards) construction of a derivation to the progressive construction of a model. For concreteness, consider the case of propositional modal logics: in our framework we associate the derivation

$$\begin{array}{c}\Pi \\ S_0 = \Gamma_0, \Delta_0 \vdash \Gamma_0'\end{array}$$

to the progressive construction of a (partial) model $\mathfrak{M} = (\mathfrak{W}, \mathfrak{R}, \mathfrak{V})$ such that for each sequent $S_i = \Gamma_i, \Delta_i \vdash \Gamma_i'$ in Π, with $i \geq 0$,

- the worlds of $\mathfrak{M}$ are connected according to Δ_i, i.e. $(x, y) \in \mathfrak{R}$ iff $\Delta_i \vdash xR\,y$,
- $\mathfrak{M}$ satisfies all lwffs $x{:}A \in \Gamma_i$, i.e. $\models^{\mathfrak{M}} x{:}A$, and
- $\mathfrak{M}$ falsifies all lwffs $x{:}B \in \Gamma_i'$, i.e. $\not\models^{\mathfrak{M}} x{:}B$.

Then we have:

- if S_0 is provable, then $\mathfrak{M}$ is inconsistent (i.e. it contains an inconsistent world),
- if S_0 is not provable, then $\mathfrak{M}$ is a counter-model for it.

Note that $\mathfrak{M}$ is partial in the sense that the truth values of some propositional variables might be missing from the model, but we can univocally determine these values from the values of the composite formulas of S_i they appear in (e.g. $\models^{\mathfrak{M}} x{:} \sim p$, for p a propositional variable, implies $\not\models^{\mathfrak{M}} x{:}p$, i.e. $\mathfrak{V}(x, p) = 0$).

Consider, for example, the proof (6.1) of the S(T)-theorem $x{:} \sim\Box\sim(A \supset \Box A)$. We can represent the inconsistent model $\mathfrak{M}$ spawned by (6.1) (inconsistent since $\models^{\mathfrak{M}} y{:} \sim A$ and $\models^{\mathfrak{M}} y{:}A$) with the following diagram

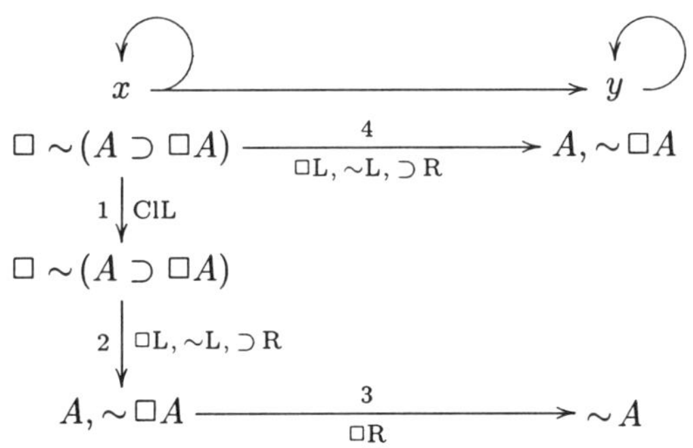

---

[7] Contraction is not the only problematic aspect of proof search, as we must also consider permutability of rules, and, if present, we must also control applications of relational rules and of *monl*; see §8 and §12, and the discussion on $S(B^+)$ in §13.1.4.

As notation, we connect worlds, built by applications of $\Box$R, according to the Δ_i's in the proof (6.1), and we write beneath each world the formulas that are true in it, i.e. the formulas in the antecedents of the sequents in the proof and the negation of the formulas in the succedents. We use numbered and indexed arrows to represent applications of rules with principal formula $x{:}\Box A$ and the local (i.e. propositional) reasoning following them. In other words, reading (6.1) backwards, we write $\Box \sim (A \supset \Box A)$ below x as the result of the initial $\sim$ R; then: 1 represents the application of ClL, 2 the lowest application of $\Box$L and the applications of $\sim$L and $\supset$R following it, 3 the application of $\Box$R, and 4 the uppermost application of $\Box$L and the applications of $\sim$L and $\supset$R following it.

Let now A be a formula that is not trivially provable (i.e. assume that A is not a propositional tautology) and consider an attempted proof of the non-theorem $x{:}\Box(A \supset \Box A)$ in S(K)

$$\dfrac{\dfrac{\dfrac{y{:}A, xRy, yRz \vdash z{:}A}{y{:}A, xRy \vdash y{:}\Box A}\,\Box\mathrm{R}}{xRy \vdash y{:}A \supset \Box A}\,\supset\mathrm{R}}{\vdash x{:}\Box(A \supset \Box A)}\,\Box\mathrm{R}$$

and its associated 'putative' counter-model

$$\begin{array}{ccccc} x & \longrightarrow & y & \longrightarrow & z \\ \sim\Box(A \supset \Box A) & \xrightarrow[\Box\mathrm{R},\, \supset\mathrm{R}]{1} & A, \sim\Box A & \xrightarrow[\Box\mathrm{R}]{2} & \sim A \end{array} .$$

The subformula property tells us that a S(K)-proof of $S = \vdash x{:}\Box(A \supset \Box A)$ must have this form. Thus, to disprove S, we must check whether no rule can be applied to $y{:}A, xRy, yRz \vdash z{:}A$ to yield a closed branch. When this is the case, then the diagram really is a counter-model of S in S(K). However, by looking at the rules, we immediately see that $y{:}A, xRy, yRz \vdash z{:}A$ could be the result of a contraction or of a weakening. Since there are only a finite number of possible weakenings, which we can try out in turn, to establish the claim that S is not provable, we must show that contraction is eliminable in S(K); this amounts to showing that if a theorem is provable in S(K), then it has a proof in which there are no applications of contraction. This investigation is the main topic of Part II, in which we first perform a substructural analysis of some of our modal sequent systems, bounding applications of contraction to bound the complexity of the decision problem of the corresponding modal logics, and then discuss the generalizations required for other non-classical logics.

Before moving on to this, however, we conclude this part of the book by summarizing our results up to now and discussing some related work.

7 DISCUSSION

We have given a framework that provides uniform and modular presentations and implementations of families of non-classical logics in terms of labelled deduction systems: logics in a family are presented as extensions of a fixed base system, consisting of rules for local and non-local operators, with theories comprised of (Horn) rules formalizing the properties of the relations connecting worlds in the underlying Kripke-style semantics and, in the case of quantified logics, the way domains of individuals change between worlds. The previous chapters demonstrate, we think, that our systems fit well into the Logical Framework setting (our Isabelle encodings provide a simple and natural environment for interactive proof development that supports hierarchical structuring), and have modular metatheoretical properties, in particular soundness and completeness, and normalization of derivations and a subformula property, which we can exploit to delineate advantages and limitations of our approach.[1]

7.1 RELATED WORK

Throughout the chapters we described various problems that arise in traditional approaches to non-classical logics based on Hilbert-style presentations, and showed that such problems are not encountered in our labelled presentations. We now compare our work with some related approaches based on standard deduction systems, based

[1] Although we have not explicitly given an Isabelle encoding of our sequent systems, this can be done fairly straightforwardly, e.g. by adapting Pfenning's [183] representation of sequents in a Logical Framework.

on implicit or explicit labelling, or based on embeddings of non-classical logics in predicate logic.

7.1.1 Standard deduction systems

Prawitz [186] gives a natural deduction rule for $\Box$ introduction in S4 and S5 with the non-local side condition that all the assumptions on which it depends are 'modal' (i.e. their main operator is $\Box$), in the case of S4, or 'modal' formulas or their negation, in the case of S5.[2] (Prawitz also gives similar rules for relevant implication.)

The main problem with Prawitz's systems, besides this 'impurity' of the $\Box$ introduction rule (see [6, 9]), is that it is unclear how to generalize or restrict them to present other logics. As mentioned in §1, a solution to the impurity problem is given in [9, § 4.4], where a deduction system for S4 is factored into two ordinary, pure, single-conclusioned consequence relations. Unfortunately, the result is fairly far removed from the standard presentations based on accessibility relations or characteristic axioms, and there is no attempt to modularize structure or correctness: only a particular modal logic is analyzed and it is not apparent how to generalize the results in a uniform way.

Another approach to the formalization of non-local conditions in a Logical Framework is to manage assumptions explicitly with sequents (as done, e.g., in the encodings of the modal logics T, S4 and S4.3 that are part of the Isabelle system distribution). In fact, several standard sequent (or tableaux) systems have been proposed and studied, especially for modal logics, e.g. [46, 87, 115, 119, 120, 174, 190, 238]. Although these systems can be employed for automated theorem proving [129], they are often unsatisfactory from a proof-theoretical point of view as they are based on rules that may require ingenuity in their invention and application.[3] (In the following chapters, we show how a substructural analysis of our labelled sequent systems naturally yields justifications and, in some cases, refinements of some of these rules.)

Moreover, standard sequent systems lack, in general, modularity (sometimes 'extensions' require deleting some rules while others are added), and in some cases desirable proof-theoretical properties such as the subformula property and the eliminability of the *cut* rule (although for, e.g, several modal logics one can fortunately show that only certain 'superformulas' are needed or that only 'analytic' or 'semi-analytic' *cut* is required). In fact, Wansing [233, p. 128] summarizes the situation for propositional modal logics when he notes that:

---

[2]That is, making the correspondence with the rules of standard modal sequent systems (see §6) more evident, the $\Box$ introduction rule has the form

$$\begin{array}{c}\Gamma\\ \vdots\\ \dfrac{A}{\Box A}\ \Box\mathrm{I}\end{array}$$

where for S4 all the formulas in the set of open assumptions Γ have a $\Box$ as their main operator, and for S5 the formulas in Γ either have a $\Box$ as their main operator, or are the negation of formulas of this form.

[3]Furthermore, these rules fail to meet the philosophical requirements at the basis of natural deduction, e.g. the independence of the logical operators, as discussed in §6 and, in more detail, in [233, 235].

> In contrast to the axiomatic approach, the standard sequent-style proof theory for normal modal logics fails to be 'modular', and the very mechanism behind the small range of known possible variations is not very clear. One might be inclined to agree with Segerberg's [46, p. 30] remark (in connection with natural deduction systems for modal logics) that 'only exceptional systems ... seem to be characterizable in terms of reasonably simple rules'. [...] Apart from the absence of symmetric and explicit introduction rules for $\Box$ and $\Diamond$, the problem is that it is not quite clear which parameters could be *systematically* modified so as to obtain characteristic sequent rules. [his emphasis]

Therefore, alternative, 'natural', deduction systems have been proposed, which systematically attempt to restore modularity by extending the standard presentations with additional information. We can roughly classify these approaches in terms of the nature of this information, syntactic or semantic, and we now briefly discuss some of the proposed systems.

7.1.2 Non-standard deduction systems

We begin by considering non-standard deduction systems in which additional (syntactic) metatheoretical information is employed. Avron [7, 8], Benevides and Maibaum [29], Cerrato [53], Došen [73], Martini and Masini [157, 158], among others, have devised different non-standard deduction systems for modal and other non-classical logics that have in common the use of a 'higher-level' deduction system. For example, in [53] Cerrato proposes modal sequent systems based on the introduction of 'metamodalities' that communicate with the usual modal operators by means of ingenious rules, and in [157] Martini and Masini give a two-dimensional generalization of the notion of sequent that asserts provability between two-dimensional sequences of formulas, instead of the usual consequence relation between two sequences of formulas. These approaches provide elegant formalizations of several modal logics but their generalization to other logics is not immediate.

Another metatheoretical approach is based on Belnap's *display logic* [26], which provides a framework for the 'Gentzenization' of non-classical logics [27, 149, 195, 233, 235]. In this approach, different families of related logics are presented by extending a fixed set of logical rules with collections of particular structural rules formalizing the behavior of 'structural modalities' (i.e. structural modal operators). Thus, the display logic approach bares some similarity with our systems, in which the rules for logical operators are never changed, and all changes are made in the collections of relational or domain rules; we return to this below.

7.1.3 Implicit and explicit labelling

Based on Kripke's semantic tableaux for modal logics [150], Fitch [86], among others, proposed a 'semantic' style of natural deduction, where the additional information can be used either implicitly by employing nested derivations, or explicitly by extending the (object) language.

In the implicit case, derivations are structured as trees consisting of a main derivation that communicates with subordinate derivations according to metatheoretical conditions, so that different communication rules yield systems for different logics. Examples of several such deduction systems for non-classical logics are given in [1, 2, 38, 39, 77, 87, 215], but, while effective, also these systems suffer in some cases from a lack of modularity, and ingenuity may be required in inventing the appropriate communication rules.

These problems can be solved, at least partially, by explicitly encoding additional semantic information in the syntax of the deduction systems. This can be done in various ways, ranging from the adoption of labels (or *prefixes*) representing possible worlds (as suggested by Kripke and Fitch), e.g. [17, 90, 66], to 'full' semantic embeddings of non-classical logics into predicate logic, e.g. [171]. Several such possibilities have been studied and new ones are frequently proposed. In fact, we could say that our work, especially when compared with full translations, is an analysis of the minimal semantic information, in other words the minimal *partial* translation, needed to formulate deduction systems for non-classical logics in a uniform and modular way.

It is however worth pointing out that given the commitment to a particular semantics, be it the standard Kripke semantics that we employ or the alternative (algebraic or neighborhood) semantics that have been proposed for non-classical logics, e.g. [60, 141, 154, 194, 223], it is 'philosophically' questionable whether the use of labels, prefixes or other forms of semantic translation indeed yields 'natural' deduction systems for these logics. We briefly return to this in §14, and observe here that detailed discussions championing the semantic view of modal and other non-classical logics can be found both in Gabbay's *Labelled Deductive Systems* book [90] and in Blackburn's papers on *Hybrid Languages* [30, 31].

Our work is inspired by the Labelled Deductive Systems (LDS) approach proposed and developed by Gabbay [90] and several others as a general and unifying methodology for presenting almost any logic, e.g. [4, 41, 42, 43, 44, 45, 63, 67, 70, 97, 121, 202, 203]. For example, *Compiled Labelled Deductive Systems* (CLDS), developed by Broda and Russo [43, 44, 202, 203], build upon LDS to formalize uniform (and abductive) systems for families of modal and other non-classical logics.

To support the desired generality, the LDS (and CLDS) metatheory and presentations are based on a notion of diagrams and logic data-bases, which are manipulated by rules (for propositional and first-order logical operators) with multiple premises and conclusions. For example, for $\Diamond$ elimination [90, p. 49] gives a rule

$$\frac{t{:}\Diamond A}{\text{Create a new point } s \text{ with } t < s \text{ and deduce } s{:}A} \tag{7.1}$$

the application of which updates a modal data-base with the new conclusion; a rule to the same effect is given in [45, 202, 203]. The formal details are different from our 'pure' ND presentation, which comprises the rule for $\Diamond$ elimination given in Figure 2.2, i.e.

$$\frac{x{:}\Diamond A \qquad \begin{array}{c}[y{:}A]\ [xRy]\\ \vdots\\ z{:}B\end{array}}{z{:}B}\ \Diamond\mathrm{E}$$

We can state this difference even more sharply by quoting Blackburn [31], who, while discussing the relationships between LDS and Hybrid Languages [3, 30, 32, 33, 34, 211, 222], and focussing in particular on Gabbay's rule (7.1), points out that:

> [...] Gabbay proceeds by manipulating labels *metalinguistically* (in effect, he makes use of a programming language containing expressions such as '*create*', '*and*', 'R', ':', and a supply of labels, to manipulate object language formulas) whereas [...] we work with an object language rich enough to state the required deduction step. [his emphasis]

The object language of our $\mathrm{N}(\mathcal{L})$ and $\mathrm{S}(\mathcal{L})$ is rich enough as well: labels, ':', 'R' and rwffs *are* all part of the language of our labelled deduction systems. Leaving a more detailed analysis of the relationships between our systems and Hybrid Languages for future work, we here simply remark that while there are some important differences (e.g. Hybrid Languages are in fact far more expressive than labelled systems like ours), there are also several significant similarities besides the one we just mentioned. For example, the intuitions and techniques underlying the 'hybrid completeness proof' in [33] are very close to those underlying our completeness proof for systems $\mathrm{N}(\mathcal{L})$, although the proofs were developed in complete independence. We expect there to be cross-fertilization of ideas and results in the future.

There is another difference between our and the LDS approach that is worth emphasizing. In our work, we have identified an important property of the structured presentation of logics, their combination, and extension. Namely, there is tension between modularity and extensibility: a narrow interface between the base system and the separate labelling algebra (i.e. the relational and the domain theories) provides well-defined structural properties but limits extensions. This separation is critical: it is only when we attempt to modularize and separate theories formally, and define a precise interface between them, that we see that only limited extensibility is actually possible. Of course, in formalizing particular LDS or CLDS, one could similarly separate theories. The precise nature of this would be reflected in the rules chosen for propagating results between data-bases. It should be the case that if these rules enforce a similar separation, then one will encounter similar limitations to those reported here. That is, the problems we identify have some generality and should appear in other labelled deduction frameworks where theories are separated and results are communicated in a limited way between them. In fact, the purely semantic view taken in [45, 202, 203], i.e. the explicit adoption of what we call universal falsum, results in systems close to the semantic embedding approach, to which we have compared our work throughout the chapters and which we discuss again below.[4]

LDS have been further developed in various other directions. For example, in [63], D'Agostino and Gabbay give labelled tableaux for substructural logics based on algebraic semantics. Their rules support automated proof search, but are not easy to

[4]An example of an approach in which, like with our local falsum, local inconsistency does not imply global inconsistency, is the work of Giunchiglia and Serafini [114], who show that particular 'multicontext systems', where (indexed) formulas are translated between contexts using 'bridge rules', define the same classes of provable formulas as certain common propositional modal logics. Based on [114], Ghidini and Serafini [107] have then introduced first-order systems that allow for the formalization of distributed knowledge representation with local inconsistency. These systems are, in general, radically different from ours, and not comparable in more detail.

recast as ordinary pure ND rules (e.g. the general closure rule they give depends on arbitrarily many formulas). The systems of [63] have then been extended with the modal operators $\Box$ and $\Diamond$ in [65, 67] to investigate 'modal substructural implication logics', while in [42] they are extended to include 'resource abduction' in labelled natural deduction systems for substructural logics.

Another promising research direction based on LDS aims at formalizing 'goal-oriented deduction systems' for non-classical logics. As a preliminary report of a larger project [98], in [97] Gabbay and Olivetti introduce uniform goal-oriented deduction systems for the implicational fragment of several modal logics. These systems behave well from the point of view of proof theory: they are analytical and satisfy cut-admissibility. Moreover, we believe that the analysis of the conditions under which an original goal must be re-asked in modal goal-oriented systems is closely related to the analysis of applications in our systems of contractions and other structural rules, which we perform in Part II.

The kind of labelled natural deduction presentation we employ is close also to the work of Simpson [216] even though his focus, proof techniques and applications are based on using labelling as a means of investigating intuitionistic versions of propositional modal logics, and his metatheoretical considerations are quite different. Moreover, his relations have no independent theory with which one can work: Simpson treats relational formulas only as assumptions in inferences of labelled formulas via his 'geometric' rules, which are derivable in our systems. Investigating presentations of intuitionistic versions of modal logics using our systems (see §6.2) will reveal further similarities and differences between our approach and his.

Our work is also closely related to, and influenced by, the algebraic approach proposed by Dunn (see [79] and the references there), who introduces *gaggle theory* as an abstraction of Boolean algebras with operators [145], where n–ary operators are interpreted by means of n+1–ary relations. Gaggle theory yields a space of algebras where the standard Kripke semantics for a particular logic is obtained by manipulating the gaggle presentation at the level of the canonical model, as opposed to instantiating the appropriate relational theory as in our approach. For instance, an analysis of the canonical model shows how to reduce the ternary relation associated with the binary intuitionistic implication to the more customary partial order on possible worlds. This algebraic approach is extremely powerful, but does not lend itself well to direct implementation; however, with appropriate simplifications or by combination with display logic, as in [193, 195, 235], this may be possible.

We have already mentioned above that Fitting [87] has studied standard sequent systems for modal logics. Extending Kripke [150], Fitting [87, 88, 89], among others, e.g. [13, 52, 141, 167], investigated also prefixed sequent and tableaux systems. For example, for quantified modal logics he first gives 'standard' systems for non-symmetric logics with increasing domains, and then, in order to capture the other conditions, he extends his systems by introducing prefixes that represent possible worlds. These allow him to formulate systems for several modal logics (including symmetric logics like S5) with varying, increasing or constant domains. In Fitting's prefixed systems, the different properties of the domains are expressed by imposing different (metatheoretical) side conditions on the applicability of the quantifier rules; analogously, the properties of the accessibility relation require different (metatheoretical) side conditions on the rules for the modal operators. (In contrast, in our systems we add relational or domain rules to a fixed base system.)

The main disadvantage of Fitting's prefixed systems, apart from the fact that they don't capture decreasing domains, is that their formalizations may require ingenuity, and that the rules for the modal operators and quantifiers can be quite awkward, since they carry non-local side conditions on the complete set of assumptions. These prefixed systems could however be modified to cover decreasing domains, and we believe that one of the best ways for doing so is to replace the standard quantifier rules with rules similar to ours.

Fitting's prefixed systems have been refined and extended by several researchers. For example, Massacci [159, 160] (but see also Goré's survey [120]) gives modular prefixed tableaux systems for a wide range of modal logics, including the ones we considered here. The main characteristic of Massacci's systems is their being *single-step*: modal formulas are prefixed with a non-empty sequence of integers (naming possible worlds) and rules are such that the prefixes of the premises and of the conclusion are at most 'one step away'. In other words, the rules do not require an explicit accessibility relation and a relational theory for reasoning about it, but code them implicitly in the prefixes by concatenating integer sequences. We return to this for further comparison in §13.[5]

7.1.4 Translations and semantic embeddings

Clearly, our work is related also to approaches based on semantic embeddings, in which a formula of non-classical logic is translated into a formula of predicate logic, and shown to be valid (or not) in a theory formalizing the semantics of the modal operators and domains of quantification. Several translation methods have been proposed, e.g. the standard *relational* translation (see [170] for references), but also *functional* [5, 83, 126, 169] or *semi-functional* [168] translations. For example, according to the relational method, the modal formula $\Box(A \land B)$, where A and B are propositional variables and w_i is the actual world, would be translated into a first-order formula like

$$\forall w_j.\, R(w_i, w_j) \supset (A(w_j) \land B(w_j))$$

where there would be additional axioms characterizing the accessibility relation R and the domains of quantification (cf. Definition 2.3.19). Ohlbach [170], for example, provides a general framework for carrying out such translations and reasoning about their soundness and completeness; translations are defined by morphisms on formulas and these are shown sound and complete by providing morphisms on interpretations.

[5]Other novel related work is the one by Castilho, Fariñas del Cerro, Gasquet and Herzig [49], who formalize 'modal tableaux with propagation and structural rules'. These tableaux systems are based on acyclic graphs, instead of trees, and have the single-step systems of Massacci as a special case. Moreover, they allow for presentations of logics based on the axiom 5 that respect the subformula property, as opposed to the superformula principle common in standard systems for non-analytic logics such as K5 (cf. the discussion in §8).

As we remarked when discussing first-order labelling algebras in §2, §3 and §4, our work differs from approaches based on semantic embedding with respect to the nature of the translations, the metatheoretical properties that hold, and how they are proved. First, we separate, rather than combine, reasoning about formulas, relations and terms. With semantic embeddings there is, by design, no formal distinction between formulas, relations and terms, or separation between relational and first-order reasoning. (On the other hand, this is what allows semantic embedding to present a wider range of logics.) Second, rather than using interpretation morphisms and building on top of the semantics of first-order logic, we directly define deduction systems for non-classical logics and show, using a parameterized canonical model construction, that these systems are sound and complete. Finally, our proofs have normal forms with separated reasoning and a subformula property, while the translation approach has no separation and the normal forms are those of derivations in first-order logic.

This is not to say that semantic embedding is not interesting or useful. On the contrary, it can be efficiently used for automated theorem proving. Moreover, the structure that is missing from our point of view, and the advantages that we gain from it, can be made up for by alternative analyses, at least in some cases. So, for example, Schmidt and others [101, 205, 206, 207] have shown how to exploit an *optimized functional translation* to employ resolution as a decision procedure for many propositional modal logics; this is achieved by first translating modal formulas into a fragment of monadic first-order logic with function symbols, and then showing that the depth of terms can be bounded. We return to this for further comparison in §13.

We conclude by mentioning the work of Orlowska [176, 177, 178], who introduced tableaux-like relational systems for relevance, modal and intuitionistic logics, by translating formulas into relations, which are then proved by a process of decomposition. Although the metalogic is different, relational logic instead of predicate logic, this method is close to semantic embedding, since formulas of the logic and relations from the Kripke semantics are treated in a uniform way as relations.

II Substructural and complexity analysis of modal sequent systems

8 INTRODUCTION AND PRELIMINARIES

8.1 INTRODUCTION

In Part I we introduced a labelled deduction framework for presenting modal and other non-classical logics in a uniform and modular way. In particular, in §2 and §6, we showed that for a large family of propositional modal logics, essentially those with accessibility relations axiomatizable using Horn-clauses, e.g. K, T, K4, S4, etc., we can decompose our labelled deduction (ND or sequent) systems into two separated parts: a base system, fixed for all logics in the family, and a labelling algebra, which we extend to generate systems for particular logics. Now we use our framework to develop a proof-theoretical method for bounding the computational complexity of the decision problem for a number of these logics.

Our method is inspired by the observation that the decidability of a logic follows from a sound and complete, cut-free and contraction-free, sequent system presentation satisfying a subformula property (or some sort of superformula property), since in such a presentation both the depth of possible proofs and the possible sequents appearing in them are bounded. The decidability of propositional classical logic, for instance, can easily be proved this way.

In general, however, it is not possible to eliminate cut and/or the contraction rules from a sequent system if we want to retain its completeness with respect to the logic it presents.[1] But, as we argued in §6.3, since we build proofs *backwards* (i.e. 'top-down', starting with the end-sequent and working towards the axioms), these structural rules are always applicable so that their 'uncontrolled' application may give rise to infinite branches and proof search may not terminate. Hence, in order to establish decidability, we must find a way of controlling, i.e. *bounding*, their application.

Suppose therefore that we approach the problem of establishing the decidability of a logic $\mathcal{L}$ by using our labelled sequent systems $S(\mathcal{L})$. For concreteness, let us consider modal logics. We know that each modal $S(\mathcal{L})$ is a sound and complete system for the corresponding logic $\mathcal{L}$. And $S(\mathcal{L})$ is cut-free. Hence, in contrast to Hilbert-style axiomatizations, proofs in $S(\mathcal{L})$ satisfy a *subformula property*, which restricts the formulas that may appear in proofs to subformulas of the formula we are trying to prove.

However, the subformula property alone is not sufficient to establish decidability: if subformulas may appear (be duplicated) infinitely often, we cannot bound the size of branches in an attempted proof. To fully bound the space of possible proofs we must thus also bound how often subformulas may appear in sequents, and thereby bound the length of branches. In other words, the problem we have to tackle when using $S(\mathcal{L})$ is that of controlling duplications of formulas caused by backwards application of the contraction rules

$$\frac{x{:}A, x{:}A, \Gamma, \Delta \vdash \Gamma'}{x{:}A, \Gamma, \Delta \vdash \Gamma'}\ \text{ClL} \qquad \text{and} \qquad \frac{\Gamma, \Delta \vdash \Gamma', x{:}A, x{:}A}{\Gamma, \Delta \vdash \Gamma', x{:}A}\ \text{ClR}\ ,$$

as well as CrL.

Contraction is, in general, similarly required in standard modal sequent systems [87, 238]; as we have seen, left contraction is embedded in standard rules such as

$$\frac{A, \Box A, \Sigma \vdash \Sigma'}{\Box A, \Sigma \vdash \Sigma'}\ (\text{T})\ , \qquad \frac{\Gamma, \Box\Gamma \vdash A}{\Sigma, \Box\Gamma \vdash \Box A, \Sigma'}\ (\text{K4})\ , \quad \text{and} \quad \frac{\Box\Gamma \vdash A}{\Sigma, \Box\Gamma \vdash \Box A, \Sigma'}\ (\text{S4})\ ,$$

and much effort has been devoted to the design of contraction-free modal systems, e.g. [55, 119, 120, 137, 138].[2]

An analogous problem exists also in standard sequent systems for first-order logic: we cannot say, in general, how often a universally quantified formula must be instantiated, which is equivalent to being unable to recursively bound the number of times that a formula must be contracted. In fact, many modal logics can be seen as a

---

[1]Typical examples of this are standard sequent systems for modal logics such as the system for S5 given by Ohnishi and Matsumoto [174], which requires both cut and contraction rules. In recent years, however, several cut-free sequent systems have been devised for S5 and related modal logics; for more detailed discussions of these systems see, for example, [87, 119, 120, 233, 235].

[2]Note that contraction is similarly embedded in the left implication rule

$$\frac{C \to D, \Gamma \vdash C \quad D, \Gamma \vdash E}{C \to D, \Gamma \vdash E}\ (\to\text{L})$$

of standard sequent systems for propositional intuitionistic logic. In fact, the techniques applied for modal logics by Hudelmaier in [137, 138] are closely related to those that he originally devised, in parallel with Dyckhoff [81], for propositional intuitionistic logic in [136]. We discuss related work in more detail in §13.

'halfway-house' between propositional and predicate logic, an observation reflected in the labelled deduction systems we have developed, which correspond to strict (proof-theoretical) subsystems of first-order logic. On the other hand, a number of modal logics are known, on independent (semantic) grounds, to be decidable. Hence, if we are to analyze decidability and complexity of modal logics proof-theoretically using sequent systems, as opposed to semantic methods such as the finite model property, we must bound the application of these structural rules.

In the following chapters, we take advantage of the reduced complexity of our sequent systems to show that, at least in some cases, it is indeed possible to bound, or even eliminate, applications of the contraction rules, and thus provide decision procedures with bounded space requirements.[3] More specifically, we show that our labelled sequent systems enable a combinatorial analysis of possible proofs, which we can use to obtain Polynomial Space (PSPACE, for short) upper-bounds for the decision problem for some of the modal (and other non-classical) logics they present.

Our method factors into a collection of general properties shared by families of logics (and systems), and a supplementary analysis of the distinguishing qualities of particular logics. This supplementary analysis itself factors into a collection of independent subproblems, centered on bounding the applications of contraction rules and the complexity of reasoning about the underlying accessibility relation.

Our analysis proceeds thus in several steps. Before considering, as examples, the modal systems S(K), S(T), S(K4) and S(S4) in detail in the next three chapters, in the rest of this chapter we introduce useful notation and terminology, and prove preliminary results, which we divide in logic-independent results (§8.2.1) and logic-dependent ones (§8.2.2). In the light of these results, we then consider the unique features of particular logics, and show that, for theoremhood, i.e. for proofs of end-sequents of the form $\vdash x_1{:}D$, contraction can be completely eliminated in S(K) (Theorem 9.1.1), while in S(T), S(K4) and S(S4) (Theorems 10.1.4 and 11.2.5) it can be restricted to a number of applications polynomially bounded above in the size of the end-sequent we are trying to prove.[4]

We then apply these results to analyze the computational complexity of these systems. In §12 we establish separate bounds on reasoning in the relational theory and combine our results to bound the depth of proofs and the size of sequents arising in them, and thus, by 'resource aware' programming, provide PSPACE upper-bounds for the decision problem of logics that can be analyzed in this way. In particular, we give a $O(n \log n)$-space procedure for K, a $O(n^2 \log n)$-space procedure for T, and $O(n^4 \log n)$-space (or, possibly, $O(n^3 \log n)$-space) procedures for K4 and S4.

[3] A legitimate question to ask at this point is why we employ labelled sequent systems instead of labelled ND systems. In fact, we have shown in §2 that derivations in our ND systems can be reduced to a normal form that has a well-defined structure and satisfies a subformula property. This provides a first step towards establishing decidability of the modal logics presented that way, but additional steps are required, such as bounding the number of times a particular formula may be assumed or discharged. This kind of proof-theoretical analysis is more easily performed when logics are presented using sequent systems, which allow a finer grained control of structural information via their structural rules.

[4] These contraction bounds and the resulting space bounds revise some of the bounds of [20, 23, 24].

While the space bounds we arrive at using our method are not necessarily new or 'optimal', they compare well with the best currently known [138]; we give a detailed comparison in §13. What is new is the use of labelled deduction systems to provide a framework for complexity bounds combined with the analysis of contraction for particular logics in this general setting. We view this as a first step towards a general method for both applying proof-theory to the analysis of the decision problem for families of modal and other non-classical logics, e.g. [231], and for implementing decision procedures for these logics.

Moreover, as a by-product of our substructural analysis, in §9.2, §10.2, §11.3 and §11.4 we are able to give proof-theoretical justifications, and in some cases partial refinements, of the rules of the corresponding standard sequent systems (in contrast to the usual semantic justifications). Specifically, we first show that in our systems we can obtain labelled equivalents (as derived or admissible rules) of the standard rules. We then use these labelled rules to compare our systems with standard ones (showing them equivalent via proof transformations), thereby illustrating the advantages of the labelled approach in the way it allows us to analyze substructural properties and 'control resources' (specifically, applications of the contraction rules).

8.2 PRELIMINARY RESULTS

8.2.1 *Logic-independent results*

We now give results that hold for all modal systems $\mathrm{S}(\mathcal{L}) = \mathrm{S}(\mathrm{K}) + \mathrm{S}(\mathcal{T})$, where $\mathrm{S}(\mathcal{T})$ is a Horn relational sequent theory. We begin by introducing notation and terminology.

We call *lwff-rules* the rules that have an lwff as principal formula, namely WlL, WlR, ClL, ClR, $\supset$ L, $\supset$ R, $\Box$L and $\Box$R. Suppose now that a sequent S is derived (reasoning forwards) from sequents $S_1, \ldots, S_n$ by first applying the lwff-rule (r_1) and then applying the lwff-rule (r_2), where

(i) each of the premises of (r_2) results from an application of (r_1),
(ii) each application of (r_1) introduces or contracts the same lwff, and
(iii) the lwff introduced or contracted by (r_1) is parametric in the application(s) of (r_2).[5]

We then say that (r_1) is *permutable over* (r_2), or that (r_1) *permutes over* (r_2), if the original inference may be replaced by one in which the sequent S is derived from $S_1, \ldots, S_n$ by applying first (r_2) and then (r_1). By extension, we say that a derivation or proof Π *permutes to* Π', if Π' is obtained from Π by one or more permutations. For example,

$$\dfrac{\dfrac{u{:}A, \Gamma, \Delta, xRy \vdash \Gamma', u{:}B, y{:}C}{\Gamma, \Delta, xRy \vdash \Gamma', u{:}A \supset B, y{:}C}\ \supset\mathrm{R}}{\Gamma, \Delta \vdash \Gamma', u{:}A \supset B, x{:}\Box C}\ \Box\mathrm{R} \tag{8.1}$$

[5] In other words, (ii) means that all applications of (r_1) have the same principal formula $x{:}A$, and (iii) means that this $x{:}A$ is parametric in the subsequent applications of (r_2).

permutes to

$$\dfrac{\dfrac{u{:}A, \Gamma, \Delta, xRy \vdash \Gamma', u{:}B, y{:}C}{u{:}A, \Gamma, \Delta \vdash \Gamma', u{:}B, x{:}\Box C}\ \Box\mathrm{R}}{\Gamma, \Delta \vdash \Gamma', u{:}A \supset B, x{:}\Box C}\ \supset\mathrm{R} \tag{8.2}$$

since y is different from u by the condition on the application of □R. We can also reverse this permutation, i.e. (8.2) permutes to (8.1).

Lemma 8.2.1 *Every lwff-rule is permutable over any other lwff-rule, with the exception of □L, which is permutable over every lwff-rule other than □R.* ■

This follows immediately by inspection of the rules, where □L does not always permute over □R since □L may have the same active rwff xRy as □R.[6] More concrete examples of this situation are given in the following chapters, e.g. in the derivation (9.2) in the proof of Theorem 9.1.1. There are, however, cases in which □L permutes over □R, most notably when the active rwff of □L is introduced by reflexivity or seriality. For example

$$\dfrac{\dfrac{\begin{array}{c}\dfrac{}{\vdash xRx}\ \mathit{refl}\\ \vdots\ \mathrm{WrL}\\ \Delta, uRv \vdash xRx\end{array} \quad x{:}A, \Gamma, \Delta, uRv \vdash \Gamma', v{:}B}{x{:}\Box A, \Gamma, \Delta, uRv \vdash \Gamma', v{:}B}\ \Box\mathrm{L}}{x{:}\Box A, \Gamma, \Delta \vdash \Gamma', u{:}\Box B}\ \Box\mathrm{R}$$

permutes to

$$\dfrac{\begin{array}{c}\dfrac{}{\vdash xRx}\ \mathit{refl}\\ \vdots\ \mathrm{WrL}\\ \Delta \vdash xRx\end{array} \quad \dfrac{x{:}A, \Gamma, \Delta, uRv \vdash \Gamma', v{:}B}{x{:}A, \Gamma, \Delta \vdash \Gamma', u{:}\Box B}\ \Box\mathrm{R}}{x{:}\Box A, \Gamma, \Delta \vdash \Gamma', u{:}\Box B}\ \Box\mathrm{L}$$

since v is different from x by the condition on the application of □R.

As a first step in the substructural analysis of labelled modal sequent systems, recall that in §6.1 we have shown that by analogy with standard results for unlabelled sequent systems, e.g. [138, 221, 238], all of the structural rules of $\mathrm{S(K)} + \mathrm{S}(\mathcal{T})$ can be built into the axioms and the logical rules. A substructural analysis similar to ours could be performed for these equivalent systems. However, we choose not to follow this approach as the fine grained investigation of contraction needed to bound its application is more easily performed when our systems explicitly contain structural rules. Indeed, to simplify the analysis, we restrict, instead of generalize, the axioms and rules of our systems.

[6] An analogous restriction holds for the quantifier rules ∀L and ∀R of standard (unlabelled) sequent systems for first-order logic, e.g. [221], where ∀L does not always permute over ∀R as they may share the same eigenvariable.

We begin by restricting instances of axioms.

Fact 8.2.2 *Every sequent provable in* S($\mathcal{L}$) *has a proof in which all the axioms employed contain no logical operators, i.e. the formula A in each application of* AXl *and* $\perp$L *is atomic.* ■

It is easy to see that this restriction yields equivalent systems. For example, we can replace

$$\dfrac{}{x{:}\Box B \vdash x{:}\Box B}\ \text{AXl} \qquad \text{with} \qquad \dfrac{\dfrac{\dfrac{}{xRy \vdash xRy}\ \text{AXr} \quad \dfrac{\dfrac{}{y{:}B \vdash y{:}B}\ \text{AXl}}{y{:}B, xRy \vdash y{:}B}\ \text{WrL}}{x{:}\Box B, xRy \vdash y{:}B}\ \Box\text{L}}{x{:}\Box B \vdash x{:}\Box B}\ \Box\text{R}$$

and, similar to Lemma 2.3.2, we can replace

$$\dfrac{}{y{:}\perp \vdash x{:}\Box B}\ \perp\text{L} \qquad \text{with} \qquad \dfrac{\dfrac{\dfrac{}{y{:}\perp \vdash z{:}B}\ \perp\text{L}}{y{:}\perp, xRz \vdash z{:}B}\ \text{WrL}}{y{:}\perp \vdash x{:}\Box B}\ \Box\text{R} \ .$$

We make another simplifying assumption. Instead of (and equivalent to) dispensing with weakening by building it into ‘extended’ axioms, we transform proofs so that all applications of WrL occur immediately below the axioms. Indeed, by the permutability of the rules (all rules permute over WrL), it follows that:

Fact 8.2.3 *Every sequent provable in* S($\mathcal{L}$) *has a backwards proof in which all applications of* WrL *in the proof immediately precede the axioms.* ■

We henceforth assume that all proofs have been so transformed (except when explicitly noted otherwise).[7]

Let us now consider contraction. While in the following we show that applications of the rule ClL can be eliminated in S(K) and bounded in S(T), S(K4) and S(S4), applications of the rules CrL and ClR in these modal systems can be eliminated once and for all. Indeed, CrL is eliminable in every labelled sequent system that extends S(K) with a Horn relational theory S($\mathcal{T}$) (and, more generally, in every non-classical system S($\mathcal{L}$) = S($\mathcal{B}$) + S($\mathcal{T}$) where S($\mathcal{T}$) is a Horn theory).

---

[7]Note that each branch of a proof so transformed contains a maximal multiset of rwffs, $\Delta_{\max}$, such that for each Δ_i occurring in the branch we have $\Delta_i \subseteq \Delta_{\max}$. That is, each branch has the form

$$\begin{array}{c} \text{axiom(s)} \\ \Pi_2 \\ \Gamma_1, \Delta_{\max} \vdash \Gamma_1' \\ \Pi_1 \\ \Gamma, \Delta \vdash \Gamma' \end{array}$$

where Π_2 contains all applications of WrL in the branch and no applications of $\Box$R, and Π_1 contains all applications of $\Box$R in the branch and no applications of WrL, so that $\Delta_i \subseteq \Delta_{\max}$ for each Δ_i occurring in Π_1 or Π_2, including Δ.

Lemma 8.2.4 *The rule* CrL *is eliminable in* $\mathrm{S}(\mathcal{L}) = \mathrm{S}(\mathrm{K}) + \mathrm{S}(\mathcal{T})$, *where* $\mathrm{S}(\mathcal{T})$ *is a Horn relational theory.* ■

This follows by the separation that we have enforced between base system and the relational theories extending it: contracted rwffs can only be introduced in the antecedent of a sequent by applications of the rule WrL, since (cf. Fact 6.1.5) relational rules introduce rwffs only in the succedent of the conclusion. Therefore, we just need to delete both the CrL and the corresponding WrL, e.g. we transform

$$\dfrac{\dfrac{\begin{matrix}\Pi_2\\ \Delta_2 \vdash uRv\end{matrix}}{\begin{matrix}\Delta_2, xRy \vdash uRv\\ \Pi_1\\ \Delta_1, xRy, xRy \vdash uRv\end{matrix}}\text{WrL}}{\Delta_1, xRy \vdash uRv}\text{CrL} \qquad \text{to} \qquad \begin{matrix}\Pi_2\\ \Delta_2 \vdash uRv\\ \Pi_1^{\dagger}\\ \Delta_1, xRy \vdash uRv\end{matrix}$$

where, as in §6.1, we have used † to denote that the derivations Π and $\Pi^{\dagger}$ differ only in their parametric formulas.

We conclude this section with two further definitions: first, we extend the definition of subformula, Definition 2.3.10, to distinguish positive and negative occurrences of subformulas, and then we define when an occurrence of an lwff is weak in a proof (or in a branch of it).

Definition 8.2.5 *We inductively define that a subformula B of A* occurs positive [negative] in A, *in symbols* $B \Subset_+ A$ [$B \Subset_- A$], *as follows:*

- *if* $B = A$, *then* $B \Subset_+ A$;
- *if* $B \supset C \Subset_- A$, *then* $B \Subset_+ A$ *and* $C \Subset_- A$;
- *if* $B \supset C \Subset_+ A$, *then* $B \Subset_- A$ *and* $C \Subset_+ A$;
- *if* $\Box B \Subset_+ A$, *then* $B \Subset_+ A$;
- *if* $\Box B \Subset_- A$, *then* $B \Subset_- A$.

We say that $y{:}B$ *occurs positive [negative] in* $x{:}A$ *iff* B *occurs positive [negative] in* A. *By extension, we inductively define that an lwff* $y{:}B$ occurs positive [negative] in a multiset of lwffs Γ, *in symbols* $y{:}B \Subset_+ \Gamma$ [$y{:}B \Subset_- \Gamma$], *as follows:*

- *if* $y{:}B \in \Gamma$, *then* $y{:}B \Subset_+ \Gamma$;
- *if* $y{:}B \Subset_+ x{:}A$ *and* $x{:}A \Subset_+ \Gamma$, *then* $y{:}B \Subset_+ \Gamma$;
- *if* $y{:}B \Subset_- x{:}A$ *and* $x{:}A \Subset_- \Gamma$, *then* $y{:}B \Subset_+ \Gamma$;
- *if* $y{:}B \Subset_+ x{:}A$ *and* $x{:}A \Subset_- \Gamma$, *then* $y{:}B \Subset_- \Gamma$;
- *if* $y{:}B \Subset_- x{:}A$ *and* $x{:}A \Subset_+ \Gamma$, *then* $y{:}B \Subset_- \Gamma$.

Finally, we inductively define that an lwff $y{:}B$ occurs positive [negative] in a sequent $S = \Gamma, \Delta \vdash \Gamma'$, *in symbols* $y{:}B \Subset_+ S$ [$y{:}B \Subset_- S$], *as follows:*

- *if* $y{:}B \Subset_+ \Gamma$, *then* $y{:}B \Subset_+ S$;
- *if* $y{:}B \Subset_- \Gamma'$, *then* $y{:}B \Subset_+ S$;
- *if* $y{:}B \Subset_- \Gamma$, *then* $y{:}B \Subset_- S$;
- *if* $y{:}B \Subset_+ \Gamma'$, *then* $y{:}B \Subset_- S$.

We will also write $A[\![B]\!]_+$ *to specify that* $B \Subset_+ A$, *and* $A[\![B]\!]_-$ *to specify that* $B \Subset_- A$. ■

Note that our definition classifies formulas on the left [right] of the sequent symbol $\vdash$ as positive [negative], and reflects the standard interpretation of sequent systems as refutation systems in which the progressive (backwards) construction of a proof is associated with the progressive construction of a partial (counter-)model. In other words, $y{:}B \Subset_+ S$ and $y{:}B \Subset_- S$ correspond to the possible interpretation of sequent systems as refutation systems discussed in §6.3. Namely, $y{:}B \Subset_+ S$ if $y{:}B$ occurs positive in the antecedent of S (the lwffs of which are satisfied by the model $\mathfrak{M}$ corresponding to the attempted proof of S), or if $y{:}B$ occurs negative in the succedent of S (the lwffs of which are falsified by $\mathfrak{M}$).

Definition 8.2.6 *Consider a proof* Π *of a sequent* S *in* $\mathrm{S(K)} + \mathrm{S}(\mathcal{T})$, *and let* $x{:}A$ *be a particular occurrence of an lwff in* S. *We say that* $x{:}A$ *is* weak *in* Π *when it is weak in every branch* $\mathcal{B}$ *of* Π, *where* $x{:}A$ *is weak in a branch* $\mathcal{B}$ *of* Π *when*

(i) $x{:}A$ *is introduced by weakening in* $\mathcal{B}$, *or*

(ii) $x{:}A$ *is* $x{:}B \supset C$ *and* $x{:}B$ *and* $x{:}C$ *are both weak in* $\mathcal{B}$, *or*

(iii) $x{:}A$ *is* $x{:}\Box B$ *and is introduced by an application of* $\Box\mathrm{L}$ *in* $\mathcal{B}$,

$$\dfrac{\begin{array}{cc}\Pi_1 & \Pi_2 \\ \Delta_1 \vdash xRy & y{:}B, \Gamma_1, \Delta_1 \vdash \Gamma_1'\end{array}}{x{:}\Box B, \Gamma_1, \Delta_1 \vdash \Gamma_1'}\,\Box\mathrm{L} \qquad \begin{array}{c}\vdots \\ x{:}\Box B, \Gamma, \Delta \vdash \Gamma' \\ \vdots \\ S\end{array} ,$$

and $y{:}B$ *is weak in* Π_2, *or*

(iv) $x{:}A$ *is* $x{:}\Box B$ *and is introduced by an application of* $\Box\mathrm{R}$ *in* $\mathcal{B}$,

$$\dfrac{\begin{array}{c}\Pi_1 \\ \Gamma_1, \Delta_1, xRy \vdash \Gamma_1', y{:}B\end{array}}{\Gamma_1, \Delta_1 \vdash \Gamma_1', x{:}\Box B}\,\Box\mathrm{R} \qquad \begin{array}{c}\vdots \\ \Gamma, \Delta \vdash \Gamma', x{:}\Box B \\ \vdots \\ S\end{array} ,$$

and $y{:}B$ *is weak in* Π_1. ■

In other words, $x{:}A$ is weak in $\mathcal{B}$ if it is introduced by weakening in $\mathcal{B}$, or if so are introduced all of its subformulas of smallest grade that appear in $\mathcal{B}$ (in the sense that the lwffs active in the rules introducing $x{:}A$ are themselves weak in $\mathcal{B}$). If $x{:}A$ is not weak, then we sometimes say that $x{:}A$ *leads to axioms*.

8.2.2 Logic-dependent results

We now focus our attention on particular modal sequent systems. To show that ClR is eliminable in some of these systems (in particular, in S(K), S(T), S(K4), and S(S4); cf. Corollary 8.2.13), we prove additional results that will also be useful later for the analysis of ClL. We begin by introducing a modal analogue of the disjunction property of propositional intuitionistic logic.

One of the consequences of cut-elimination in sequent systems for propositional intuitionistic logic is the *disjunction property* [221, 229], if $\vdash A \vee B$ is provable then so is $\vdash A$ or $\vdash B$. The property holds also under hypotheses,

$$\text{if } \Gamma \vdash A \vee B \text{ is provable, then so is } \Gamma \vdash A \text{ or } \Gamma \vdash B\,,$$

provided that the hypotheses in Γ are *Harrop formulas* [229], where the class $\mathcal{H}$ of Harrop formulas is inductively defined by:

(i) $p \in \mathcal{H}$ for p a propositional variable,
(ii) $A \wedge B \in \mathcal{H}$ if $A \in \mathcal{H}$ and $B \in \mathcal{H}$, and
(iii) $A \to B \in \mathcal{H}$ if $B \in \mathcal{H}$, where $\to$ is intuitionistic implication.

Since $C \vee D \notin \mathcal{H}$, the restriction to Harrop formulas ensures that no disjunctive formula $C \vee D$ occurs positive in Γ, i.e. $C \vee D \notin_+ \Gamma$. This guarantees that $\Gamma \vdash A \vee B$ is not the conclusion of an application of the (branching) intuitionistic left disjunction rule ($\vee$L), e.g.

$$\frac{C, \Gamma_1 \vdash A \vee B \quad D, \Gamma_1 \vdash A \vee B}{C \vee D, \Gamma_1 \vdash A \vee B}\ (\vee \mathrm{L})$$

where $\Gamma = \{C \vee D\} \cup \Gamma_1$. For example, the disjunction property fails for $\Gamma = \{B \vee A\}$, since we can prove $B \vee A \vdash A \vee B$ but neither $B \vee A \vdash A$ nor $B \vee A \vdash B$ are provable.

Note that ($\vee$L) is the only rule that endangers the disjunction property in systems for propositional intuitionistic logic since intuitionistic sequents are single-conclusioned and the intuitionistic left implication rule (with built-in contraction) has the form

$$\frac{C \to D, \Gamma \vdash C \quad D, \Gamma \vdash E}{C \to D, \Gamma \vdash E}\ (\to \mathrm{L})$$

and it therefore suffices to require that $D \in \mathcal{H}$.

The disjunction property does not hold in our modal sequent systems since they are extensions of propositional classical logic. However, an analogous property holds for boxed formulas in many of our modal systems, including S(K), S(T), S(K4) and S(S4). For these systems we can easily check semantically that:

$$\text{if } \vdash x{:}\Box A, x{:}\Box B \text{ is provable, then so is } \vdash x{:}\Box A \text{ or } \vdash x{:}\Box B\,.$$

To generalize this property to hold under hypotheses, in Definition 8.2.7 we define an analogue of the branchings caused by formulas that are not Harrop: since in our systems $\vee$ is defined in terms of $\supset$ and $\bot$, we need to consider only the branchings caused by applications of the classical (i.e. with more than one lwff in the succedent) left implication rule with principal formula labelled with x or with a predecessor of x.[8]

Definition 8.2.7 *The class of* x-branching lwffs with respect to a multiset of rwffs Δ in $\mathrm{S}(\mathcal{L}) = \mathrm{S}(\mathrm{K}) + \mathrm{S}(\mathcal{T})$, *in symbols* $\mathfrak{B}_{\mathrm{S}(\mathcal{L})}(x, \Delta)$, *is inductively defined by*

(i) $x{:}A \supset B \in \mathfrak{B}_{\mathrm{S}(\mathcal{L})}(x, \Delta)$,

(ii) $u{:}A \supset B \in \mathfrak{B}_{\mathrm{S}(\mathcal{L})}(x, \Delta)$ *if* $\Delta \vdash u R x$ *is provable in* $\mathrm{S}(\mathcal{L})$, *and*

(iii) $u{:}\Box A \in \mathfrak{B}_{\mathrm{S}(\mathcal{L})}(x, \Delta)$ *if there exists a* v *such that* $\Delta \vdash u R v$ *is provable in* $\mathrm{S}(\mathcal{L})$ *and* $v{:}A \in \mathfrak{B}_{\mathrm{S}(\mathcal{L})}(x, \Delta)$.

By extension, given a sequent $\Gamma, \Delta \vdash \Gamma'$, *we define that*

(i) Γ *is* x-branching with respect to Δ in $\mathrm{S}(\mathcal{L})$ *if* $w{:}A \in \mathfrak{B}_{\mathrm{S}(\mathcal{L})}(x, \Delta)$ *for some* $w{:}A \in_{+} \Gamma$, *and*

(ii) Γ' *is* x-branching with respect to Δ in $\mathrm{S}(\mathcal{L})$ *if* $w{:}A \in \mathfrak{B}_{\mathrm{S}(\mathcal{L})}(x, \Delta)$ *for some* $w{:}A \in_{-} \Gamma'$. ■

We also need the following definition.

Definition 8.2.8 *Consider a branch of a proof in* $\mathrm{S}(\mathcal{L}) = \mathrm{S}(\mathrm{K}) + \mathrm{S}(\mathcal{T})$. *We call a* chain *a sequence of worlds (labels)* $x_1, x_2, x_3, \ldots$ *where* x_{i+1} *is a successor of* x_i *for each* i *(in the sense that each* x_{i+1} *has been generated in the branch by an application of* $\Box$R, *so that there is a* Δ_i *in the branch such that* $x_i R x_{i+1} \in \Delta_i$*).*

Let now y *and* z *be two distinct successors of* x *generated by two applications of* $\Box$R *in the branch, i.e.*

$$\begin{array}{c} \vdots \\ \dfrac{\Gamma_2, \Delta_2, x R z \vdash \Gamma_2', z{:}A_2}{\Gamma_2, \Delta_2 \vdash \Gamma_2', x{:}\Box A_2}\ \Box\mathrm{R} \\ \vdots \\ \dfrac{\Gamma_1, \Delta_1, x R y \vdash \Gamma_1', y{:}A_1}{\Gamma_1, \Delta_1 \vdash \Gamma_1', x{:}\Box A_1}\ \Box\mathrm{R} \\ \vdots \\ S \end{array}$$

We say that y *and* z *are* independent in $\mathrm{S}(\mathcal{L})$ *if we cannot prove in* $\mathrm{S}(\mathcal{T})$ *that* y *accesses* z *or that* z *accesses* y. *That is, if there is no* Δ_j *(in some sequent in the branch) such that* $\Delta_j \vdash y R z$ *or* $\Delta_j \vdash z R y$ *is provable in* $\mathrm{S}(\mathcal{T})$.

[8] Note that choosing between $x{:}\Box A$ and $x{:}\Box B$ constitutes a backtracking point in a backwards proof: if the chosen lwff does not allow us to close the proof, we must backtrack and try the other formula; cf. the derivations of the rule (K) and of its labelled equivalent in §9.2.

In other words, since it cannot be that y is z, y accesses z, or z accesses y, the independent worlds y and z diverge *from x as they generate two distinct, divergent, (sub-)chains of worlds that have x as their origin. By extension, we then say that* $\mathrm{S}(\mathcal{L})$ *is a* divergent *sequent system.* ∎

Observe that, given the absence of relational rules such as symmetry, euclideaness, or convergency, each system $\mathrm{S}(\mathcal{L})$ where $\mathcal{L} \in \{\mathrm{K}, \mathrm{T}, \mathrm{K4}, \mathrm{S4}\}$ is a divergent system. For example, in the following branch of a S(K4)-proof

$$\cfrac{\cfrac{\vdots \atop \Gamma, \Delta, xRy, xRz \vdash \Gamma', y{:}A_1, z{:}A_2}{\Gamma, \Delta, xRy \vdash \Gamma', y{:}A_1, x{:}\Box A_2}\ \Box\mathrm{R}}{\Gamma, \Delta \vdash \Gamma, x{:}\Box A_1, x{:}\Box A_2}\ \Box\mathrm{R} \atop {\vdots \atop S}$$

the atomic labels y and z are independent as y is different from z (by the condition on the application of $\Box$R) and neither of the two sequents

$$\Delta, xRy, xRz \vdash yRz \qquad \text{and} \qquad \Delta, xRy, xRz \vdash zRy$$

is provable in S(K4). Thus S(K4) is a divergent system.[9]

Proposition 8.2.9 ($\Box$-disjunction property) *Let* $\mathrm{S}(\mathcal{L}) = \mathrm{S}(\mathrm{K}) + \mathrm{S}(\mathcal{T})$ *be a divergent system, and consider a* $\mathrm{S}(\mathcal{L})$*-proof* Π *of the sequent*

$$\Gamma, \Delta \vdash \Gamma', x{:}\Box A_1, x{:}\Box A_2, \ldots, x{:}\Box A_n$$

where

(i) Γ *and* Γ' *are not x-branching with respect to* Δ *in* $\mathrm{S}(\mathcal{L})$.

Then there exists a $\mathrm{S}(\mathcal{L})$*-proof of* $\Gamma, \Delta \vdash \Gamma', x{:}\Box A_i$ *for some i with* $1 \leq i \leq n$.

Proof We prove the proposition for $n = 2$; the extension to the general case is straightforward. Let the sequent $\Gamma, \Delta \vdash \Gamma', x{:}\Box A_1, x{:}\Box A_2$ satisfy condition (i), and let Π be its $\mathrm{S}(\mathcal{L})$-proof.

Observe that by the permutability of the rules we can assume that Π does not contain any contraction of $x{:}\Box A_1$ or $x{:}\Box A_2$. If, for example, the last rule application in Π is a contraction of $x{:}\Box A_1$,

$$\cfrac{\Pi_1 \atop \Gamma, \Delta \vdash \Gamma', x{:}\Box A_1, x{:}\Box A_1, x{:}\Box A_2}{\Gamma, \Delta \vdash \Gamma', x{:}\Box A_1, x{:}\Box A_2}\ \mathrm{ClR} \quad ,$$

[9] Note that our proof-theoretical notion of divergent systems essentially corresponds to the semantic notion of *tree-frame* modal logics, i.e. modal logics whose Kripke frames are trees; see [140, §7] and [57, 204].

then we consider the proof Π_1 of the sequent $\Gamma, \Delta \vdash \Gamma'_1, x{:}\Box A_1, x{:}\Box A_2$ with $\Gamma'_1 = \Gamma' \cup \{x{:}\Box A_1\}$ and show that there is a proof of $\Gamma, \Delta \vdash \Gamma'_1, x{:}\Box A_i$ for $i = 1$ or 2. Similarly, if one of the two lwffs $x{:}\Box A_i$ is contracted somewhere in Π_1, then we simply permute this ClR over the rules below it.

We now proceed by induction on Π, and since the sequent cannot be an axiom we distinguish two cases, depending on the last rule in Π.

(*Case 1*) If the rule has principal formula other than $x{:}\Box A_i$, then, since Γ and Γ' are not x-branching with respect to Δ in $\mathrm{S}(\mathcal{L})$, we conclude by applying the induction hypothesis.

(*Case 2*) If the rule has principal formula $x{:}\Box A_i$, then it is either WlR or $\Box$R.

(*Case 2.1*) If the rule is an application of WlR with principal formula $x{:}\Box A_i$, then we conclude trivially; for example, if Π is

$$\dfrac{\begin{array}{c}\Pi_1\\ \Gamma, \Delta \vdash \Gamma', x{:}\Box A_1\end{array}}{\Gamma, \Delta \vdash \Gamma', x{:}\Box A_1, x{:}\Box A_2}\ \mathrm{WlR}$$

then Π_1 is the desired proof.

(*Case 2.2*) Suppose that the rule is an application of $\Box$R with principal formula $x{:}\Box A_1$. Then Π has the form

$$\dfrac{\begin{array}{c}\Pi_1\\ \Gamma, \Delta, xRy \vdash \Gamma', y{:}A_1, x{:}\Box A_2\end{array}}{\Gamma, \Delta \vdash \Gamma', x{:}\Box A_1, x{:}\Box A_2}\ \Box\mathrm{R} \tag{8.3}$$

and we distinguish two cases, depending on how $x{:}\Box A_2$ is introduced in Π_1.

(*Case 2.2.1*) If $x{:}\Box A_2$ is the principal formula of an application of WlR, then (8.3) has the form shown below on the left, and, since $x{:}\Box A_2$ is a parametric formula in Π_2, we can delete the application of WlR that introduces it to obtain the proof on the right:

$$\dfrac{\begin{array}{c}\dfrac{\begin{array}{c}\Pi_3\\ \Gamma_1, \Delta_1 \vdash \Gamma'_1\end{array}}{\Gamma_1, \Delta_1 \vdash \Gamma'_1, x{:}\Box A_2}\ \mathrm{WlR}\\ \Pi_2\\ \Gamma, \Delta, xRy \vdash \Gamma', y{:}A_1, x{:}\Box A_2\end{array}}{\Gamma, \Delta \vdash \Gamma', x{:}\Box A_1, x{:}\Box A_2}\ \Box\mathrm{R} \quad \rightsquigarrow \quad \dfrac{\begin{array}{c}\Pi_3\\ \Gamma_1, \Delta_1 \vdash \Gamma'_1\\ \Pi_2^\dagger\\ \Gamma, \Delta, xRy \vdash \Gamma', y{:}A_1\end{array}}{\Gamma, \Delta \vdash \Gamma', x{:}\Box A_1}\ \Box\mathrm{R}\ .$$

Note that for this transformation to be possible it must be the case that $x{:}\Box A_2$ is introduced by WlR in all branches of Π_1. If it is introduced by WlR in some branches and by $\Box$R in other branches, then we make the mode of introduction of $x{:}\Box A_2$ uniform by replacing all such introductions by weakening with introductions by $\Box$R. This is achieved by appropriately weakening active formulas so that we can introduce $x{:}\Box A_2$ by an application of $\Box$R. The rest of the proof is as before, modulo possible applications of weakening and AXr, and we then proceed like in the next case.

(*Case 2.2.2*) If $x{:}\Box A_2$ is the principal formula of an application of $\Box$R, then (8.3) has the form shown below on the left, and, since $x{:}\Box A_2$ is a parametric formula in Π_2, we can postpone its introduction, i.e., by Lemma 8.2.1, we permute the uppermost $\Box$R over the rules below it to obtain the proof on the right:

$$\dfrac{\dfrac{\begin{array}{c}\Pi_3\\ \Gamma_1, \Delta_1, xRz \vdash \Gamma_1', z{:}A_2\end{array}}{\Gamma_1, \Delta_1 \vdash \Gamma_1', x{:}\Box A_2}\ \Box R \atop {\Pi_2 \atop \Gamma, \Delta, xRy \vdash \Gamma', y{:}A_1, x{:}\Box A_2}}{\Gamma, \Delta \vdash \Gamma', x{:}\Box A_1, x{:}\Box A_2}\ \Box R \quad \rightsquigarrow \quad \dfrac{\dfrac{\begin{array}{c}\Pi_3\\ \Gamma_1, \Delta_1, xRz \vdash \Gamma_1', z{:}A_2\\ \Pi_2^\dagger\\ \Gamma, \Delta, xRy, xRz \vdash \Gamma', y{:}A_1, z{:}A_2\end{array}}{\Gamma, \Delta, xRy \vdash \Gamma', y{:}A_1, x{:}\Box A_2}\ \Box R}{\Gamma, \Delta \vdash \Gamma', x{:}\Box A_1, x{:}\Box A_2}\ \Box R \ .$$

(During the transformation it might be necessary to rename some labels to avoid possible variable clashes). Now we have a proof

$$\Pi_4 = \begin{cases} \quad \Pi_3 \\ \Gamma_1, \Delta_1, xRz \vdash \Gamma_1', z{:}A_2 \\ \quad \Pi_2^\dagger \end{cases} \qquad \text{of} \qquad \Gamma, \Delta, xRy, xRz \vdash \Gamma', y{:}A_1, z{:}A_2 \,,$$

where Γ and Γ' are not x-branching. We show that then there is a proof of either

$$\Gamma, \Delta, xRy \vdash \Gamma', y{:}A_1 \qquad \text{or} \qquad \Gamma, \Delta, xRz \vdash \Gamma', z{:}A_2 \,,$$

from which we conclude by an application of $\Box$R.

Since all our relational rules introduce rwffs only in the succedent of the conclusion (Fact 6.1.5), xRy and xRz must both be introduced by applications of WrL in Π_4. Moreover, and most importantly, since both y and z are arbitrary worlds accessible from x, and since $\mathrm{S}(\mathcal{L})$ is divergent, y and z are independent: it cannot be that y is equal to z, y accesses z, or z accesses y. This implies that $y{:}A_1$ and $z{:}A_2$ are independent as well, so that at least one of them is weak in Π_4. In other words, one of $y{:}A_1$ and $z{:}A_2$ is introduced by weakening in Π_4, or so are the formulas it is inferred from. By deleting the weakening(s) we obtain a proof Π_4' of either

$$\Gamma, \Delta, xRy, xRz \vdash \Gamma', y{:}A_1 \qquad \text{or} \qquad \Gamma, \Delta, xRy, xRz \vdash \Gamma', z{:}A_2 \,.$$

This is however not enough to conclude, since in both cases we must dispose of the additional rwff in the antecedent.

If Π_4' is a proof of the sequent $\Gamma, \Delta, xRy, xRz \vdash \Gamma', y{:}A_1$, then we dispose of xRz as follows (we proceed analogously when Π_4' is a proof of the other sequent and we need to dispose of xRy). If the application of WrL that introduces xRz is the last step in Π_4', then Π_4' has the form

$$\dfrac{\begin{array}{c}\Pi_5\\ \Gamma, \Delta, xRy \vdash \Gamma', y{:}A_1\end{array}}{\Gamma, \Delta, xRy, xRz \vdash \Gamma', y{:}A_1}\ \text{WrL}$$

and Π_5 is the desired proof. If the application of WrL that introduces xRz is not the last step in Π_4', then we can delete it provided that we modify the steps where xRz is active. That is, since xRz can only have been active in applications of $\Box$L with principal formula of the form $x{:}\Box C$ or in applications of relational rules to introduce $u{:}\Box C$ by $\Box$L for some u accessing x, we must replace these applications of $\Box$L by suitable steps, which depend on how the subformulas of $x{:}\Box$ or $u{:}\Box C$ are introduced. For example, suppose that Π'_4 has the form

$$\dfrac{\begin{array}{c}\Pi_6\\ \Delta_1, xR\,z \vdash xR\,z\end{array} \quad \dfrac{\begin{array}{c}\dfrac{\begin{array}{c}\Pi_8\\ \Gamma_2, \Delta_2 \vdash \Gamma'_2\end{array}}{\Gamma_2, \Delta_2, xR\,z \vdash \Gamma'_2}\,\text{WrL}\\ \Pi_7\\ \Gamma_1, \Delta_1, xR\,z \vdash \Gamma'_1\end{array}}{z{:}C, \Gamma_1, \Delta_1, xR\,z \vdash \Gamma'_1}\,\text{WlL}}{\begin{array}{c}x{:}\Box C, \Gamma_1, \Delta_1, xR\,z \vdash \Gamma'_1\\ \Pi_5\\ \Gamma, \Delta, xR\,y, xR\,z \vdash \Gamma', y{:}A_1\end{array}}\,\Box\text{L}$$

where $xR\,z$ is not active in Π_5 and Π_7. In this case, after having deleted the application of WrL that introduces $xR\,z$ in the right branch, we also delete Π_6 and the application of $\Box$L, and 'blow up' the weakening: we apply WlL to introduce $x{:}\Box C$ directly and transform the above to

$$\begin{array}{c}\Pi_8\\ \Gamma_2, \Delta_2 \vdash \Gamma'_2\\ \Pi_7^\dagger\\ \dfrac{\Gamma_1, \Delta_1 \vdash \Gamma'_1}{x{:}\Box C, \Gamma_1, \Delta_1 \vdash \Gamma'_1}\,\text{WlL}\\ \Pi_5^\dagger\\ \Gamma, \Delta, xR\,y \vdash \Gamma', y{:}A_1\end{array} \quad .$$

With such suitable changes we obtain the desired proof of $\Gamma, \Delta, xR\,y \vdash \Gamma', y{:}A_1$, from which we conclude by an application of $\Box$R. ■

Corollary 8.2.10 *Let* S($\mathcal{L}$) *be a divergent system, and let* Π *be an* S($\mathcal{L}$)*-proof of*

$$\Gamma, \Delta, xR\,y_1, \ldots, xR\,y_m \vdash \Gamma', y_1{:}B_1, \ldots, y_m{:}B_m, x{:}\Box A_1, \ldots, x{:}\Box A_n \tag{8.4}$$

where

(i) Γ *and* Γ' *are not* x*-branching with respect to* Δ *in* S($\mathcal{L}$)*, and*

(ii) $x, y_1, \ldots, y_m$ *are all distinct, and for each* j*, with* $1 \leq j \leq m$*,* y_j *does not occur in* Γ*,* Δ *or* Γ'*.*

Then there is a S($\mathcal{L}$)*-proof either of* $\Gamma, \Delta \vdash \Gamma', x{:}\Box A_i$ *for some* i *with* $1 \leq i \leq n$*, or of* $\Gamma, \Delta, xR\,y_j \vdash \Gamma', y_j{:}B_j$ *for some* j *with* $1 \leq j \leq m$*.* ■

This follows because, by (ii), each y_j is an arbitrary world accessible from x, and is independent from all other y_k's, so that the sequent (8.4) is 'equivalent' to (e.g. is obtained by m backwards applications of $\Box$R from) the sequent

$$\Gamma, \Delta \vdash \Gamma', x{:}\Box B_1, \ldots, x{:}\Box B_m, x{:}\Box A_1, \ldots, x{:}\Box A_n\,,$$

and we conclude analogously to Proposition 8.2.9.[10] Similarly we can show that:

Corollary 8.2.11 *Let* S($\mathcal{L}$) *be a divergent system, and let* Π *be a* S($\mathcal{L}$)*-proof of*

$$\Gamma, \Delta \vdash \Gamma', y_1{:}\Box B_1, \ldots, y_m{:}\Box B_m, x{:}\Box A_1, \ldots, x{:}\Box A_n$$

where

(i) Γ *and* Γ' *are not* x*-branching with respect to* Δ *in* S($\mathcal{L}$),

(ii) $x, y_1, \ldots, y_m$ *are all distinct, and*

(iii) $\Delta \vdash xRy_j$ *is provable in* S($\mathcal{L}$) *for each* j *where* $1 \leq j \leq m$.

Then there is a S($\mathcal{L}$)*-proof either of* $\Gamma, \Delta \vdash \Gamma', x{:}\Box A_i$ *for some* i *with* $1 \leq i \leq n$, *or of* $\Gamma, \Delta \vdash \Gamma', y_j{:}\Box B_j$ *for some* j *with* $1 \leq j \leq m$. ■

The intuition for Proposition 8.2.9 is that if in a backwards proof of $\Gamma, \Delta \vdash \Gamma', x{:}\Box A_1, x{:}\Box A_2, \ldots, x{:}\Box A_n$ there is no branching in x or in its predecessors, as required by the condition (i), then only one $x{:}\Box A_i$ (if any) will lead to axioms, i.e. all the other lwffs $x{:}\Box A_j$ are weak in the proof. The intuition for the corollaries is analogous; for example, since condition (ii) in Corollary 8.2.10 requires that each y_i is an arbitrary world accessible from x, at most one of $y_1{:}B_1, \ldots, y_m{:}B_m, x{:}\Box A_1, \ldots, x{:}\Box A_n$ leads to axioms and all the others are weak in the proof. In other words, the proposition and the corollaries state that if there is no branching in x or in its predecessors, then we only need to follow one chain of worlds that originates from x.

Note that the corollaries fail if we remove condition (ii), and that Corollary 8.2.11 does not generalize to hold for

$$\Gamma, \Delta \vdash \Gamma', y_1{:}B_1, \ldots, y_m{:}B_m, x{:}\Box A_1, \ldots, x{:}\Box A_n$$

even when $x, y_1, \ldots, y_m$ are all distinct. For example, to prove in S(K4)

$xRy_i, y_iRy_j \vdash y_i{:}\Box A \supset \bot, y_j{:}\Box A$ (which is equivalent to $\vdash x{:}\Box(\Box A \supset \Box\Box A)$)

we need both lwffs in the succedent.

The condition that Γ and Γ' are not x-branching with respect to Δ in S($\mathcal{L}$), condition (i) in the proposition and the corollaries, requires that in Π there is no application of $\supset$L with principal formula of the form $x{:}A \supset B$ or $u{:}A \supset B$ where $\Delta \vdash uRx$ is provable, and thus no implicit duplication (contraction) of the lwffs $x{:}\Box A_i$.[11] If this condition is not satisfied, then we can immediately find counter-examples to the proposition; for example, for p a propositional variable, although we have a proof of

$$x{:}(\Box A_1 \supset \bot) \supset (p \wedge \Box A_2) \vdash x{:}\Box A_1, x{:}\Box A_2\,, \tag{8.5}$$

[10] Note that our rules for rwffs are such that none of the y_j's can be a composite label built using Skolem function constants.

[11] Note that the implicit duplication resulting from an application of $\supset$L with principal formula $y{:}A \supset B$, where $\Delta \vdash xRy$ is provable, does not affect Proposition 8.2.9 and its corollaries, since in a backwards proof this $\supset$L can only occur after an application of $\Box$R with some $x{:}\Box A_i$ or some formula in Γ or Γ' as its principal formula.

neither of the two sequents

$$x{:}(\Box A_1 \supset \bot) \supset (p \wedge \Box A_2) \vdash x{:}\Box A_i \quad (1 \leq i \leq 2)$$

is provable. In this case, however, we can proceed as follows. By the permutability of the rules, we can find Π_1 and Π_2 such that we can transform each proof of (8.5) to:

$$\dfrac{\dfrac{\begin{array}{c}\Pi_1\\ x{:}\Box A_1 \vdash x{:}\Box A_1, x{:}\Box A_2, x{:}\bot\end{array}}{\vdash x{:}\Box A_1, x{:}\Box A_2, x{:}\Box A_1 \supset \bot}\supset\mathrm{R} \qquad \begin{array}{c}\Pi_2\\ x{:}p \wedge \Box A_2 \vdash x{:}\Box A_1, x{:}\Box A_2\end{array}}{x{:}(\Box A_1 \supset \bot) \supset (p \wedge \Box A_2) \vdash x{:}\Box A_1, x{:}\Box A_2}\supset\mathrm{L} \ . \quad (8.6)$$

Π_1 and Π_2 are proofs of $x{:}\Box A_1 \vdash x{:}\Box A_1, x{:}\Box A_2, x{:}\bot$ and $x{:}p \wedge \Box A_2 \vdash x{:}\Box A_1$, $x{:}\Box A_2$, and since both sequents satisfy condition (i) we can apply Proposition 8.2.9 to each of them separately. Indeed, there are proofs of $x{:}\Box A_1 \vdash x{:}\Box A_1, x{:}\bot$ and $x{:}p \wedge \Box A_2 \vdash x{:}\Box A_2$, and we can further transform (8.6) to a proof in which only one of $x{:}\Box A_1$ and $x{:}\Box A_2$ leads to axioms in each branch (and the other is weak):

$$\dfrac{\dfrac{\dfrac{\dfrac{\begin{array}{c}\text{axioms}\\ \vdots\\ x{:}\Box A_1 \vdash x{:}\Box A_1\end{array}}{x{:}\Box A_1 \vdash x{:}\Box A_1, x{:}\bot}\mathrm{WlR}}{x{:}\Box A_1 \vdash x{:}\Box A_1, x{:}\Box A_2, x{:}\bot}\mathrm{WlR}}{\vdash x{:}\Box A_1, x{:}\Box A_2, x{:}\Box A_1 \supset \bot}\supset\mathrm{R} \qquad \dfrac{\dfrac{\dfrac{\begin{array}{c}\text{axioms}\\ \vdots\\ x{:}\Box A_2 \vdash x{:}\Box A_2\end{array}}{x{:}\Box A_2 \vdash x{:}\Box A_1, x{:}\Box A_2}\mathrm{WlR}}{x{:}p, x{:}\Box A_2 \vdash x{:}\Box A_1, x{:}\Box A_2}\mathrm{WlL}}{x{:}p \wedge \Box A_2 \vdash x{:}\Box A_1, x{:}\Box A_2}\wedge\mathrm{L}}{x{:}(\Box A_1 \supset \bot) \supset (p \wedge \Box A_2) \vdash x{:}\Box A_1, x{:}\Box A_2}\supset\mathrm{L} \ .$$

In general, given a $\mathrm{S}(\mathcal{L})$-proof Π of

$$\Gamma, \Delta \vdash \Gamma', x{:}\Box A_1, \ldots, x{:}\Box A_n \, ,$$

where $\mathrm{S}(\mathcal{L})$ is a divergent system, we can permute the x-branching rules (i.e. the applications of $\supset$ L with principal formulas labelled with x or with predecessors of x in a chain) over the rules below them, and thereby transform Π so that it contains proofs of

$$\Gamma_1, \Delta \vdash \Gamma_1', x{:}\Box A_1, \ldots, x{:}\Box A_n \quad \text{and} \ \cdots \ \text{and} \quad \Gamma_m, \Delta \vdash \Gamma_m', x{:}\Box A_1, \ldots, x{:}\Box A_n$$

where Γ_j and Γ_j' are not x-branching with respect to Δ in $\mathrm{S}(\mathcal{L})$, for each j with $1 \leq j \leq m$. Then, for each

$$\Gamma_j, \Delta \vdash \Gamma_j', x{:}\Box A_1, \ldots, x{:}\Box A_n \quad (1 \leq j \leq m)$$

there exists an i, where $1 \leq i \leq n$, such that there is an $\mathrm{S}(\mathcal{L})$-proof of

$$\Gamma_j, \Delta \vdash \Gamma_j', x{:}\Box A_i \, .$$

Performing these transformations, it is straightforward to show that applications of ClR with principal formula of the form $x{:}\Box A$ are eliminable in proofs of theorems of $\mathrm{S}(\mathcal{L})$, so that we have:

Corollary 8.2.12 *Every sequent* $\vdash x_1{:}D$ *provable in a divergent system* S($\mathcal{L}$) *has a proof in which there are no applications of* ClR *with principal formula of the form* $x{:}\Box A$. ■

To illustrate this further, suppose that we have

$$\cfrac{\begin{array}{c}\Pi\\ \Gamma, \Delta \vdash \Gamma', x{:}\Box A, x{:}\Box A\end{array}}{\begin{array}{c}\Gamma, \Delta \vdash \Gamma', x{:}\Box A\\ \Pi_0\\ \vdash x_1{:}D\end{array}}\ \text{ClR} \quad .$$

The corollary follows because, by applying the above transformation to Π, we permute possible applications of x-branching rules over the rules below them. Thus, we transform Π so that it contains proofs $\Pi_1, \ldots, \Pi_m$ of

$$\Gamma_1, \Delta \vdash \Gamma'_1, x{:}\Box A, x{:}\Box A \quad \text{and} \quad \cdots \quad \text{and} \quad \Gamma_m, \Delta \vdash \Gamma'_m, x{:}\Box A, x{:}\Box A\,,$$

i.e. we transform Π to

$$\begin{array}{c}\begin{array}{ccc}\Pi_1 & & \Pi_m\\ \Gamma_1, \Delta \vdash \Gamma'_1, x{:}\Box A, x{:}\Box A & \cdots & \Gamma_m, \Delta \vdash \Gamma'_m, x{:}\Box A, x{:}\Box A\end{array}\\ \Pi_{m+1}\\ \cfrac{\Gamma, \Delta \vdash \Gamma', x{:}\Box A, x{:}\Box A}{\Gamma, \Delta \vdash \Gamma', x{:}\Box A}\ \text{ClR}\\ \Pi_0\\ \vdash x_1{:}D\end{array} \quad . \tag{8.7}$$

We can then permute the ClR upwards, i.e. we can further transform (8.7) to

$$\begin{array}{c}\begin{array}{ccc}\Pi_1 & & \Pi_m\\ \cfrac{\Gamma_1, \Delta \vdash \Gamma'_1, x{:}\Box A, x{:}\Box A}{\Gamma_1, \Delta \vdash \Gamma'_1, x{:}\Box A}\ \text{ClR} & \cdots & \cfrac{\Gamma_m, \Delta \vdash \Gamma'_m, x{:}\Box A, x{:}\Box A}{\Gamma_m, \Delta \vdash \Gamma'_m, x{:}\Box A}\ \text{ClR}\end{array}\\ \Pi^{\dagger}_{m+1}\\ \Gamma, \Delta \vdash \Gamma', x{:}\Box A\\ \Pi_0\\ \vdash x_1{:}D\end{array}$$

where Γ_j and Γ'_j are not x-branching with respect to Δ for each j such that $1 \leq j \leq m$. Then, for each such j we can find a proof of $\Gamma_j, \Delta \vdash \Gamma'_j, x{:}\Box A$, and thus eliminate the right contraction of $x{:}\Box A$ displayed in (8.7). By iterating this for all other right contractions of boxed formulas in Π_0, we obtain a proof of $\vdash x_1{:}D$ in which there are no applications of ClR with principal formula of the form $x{:}\Box A$. In doing so, we replace explicit right contractions of $x{:}\Box A$ with implicit ones, i.e. the contractions implicit in the applications of $\supset$L (with x-branching principal formulas).

We summarize the above results as follows. Given the absence of relational rules such as symmetry, euclideaness, or convergency, each system S($\mathcal{L}$) where $\mathcal{L} \in \{\text{K}, \text{T}, \text{K4}, \text{S4}\}$ is a divergent system. Hence, each such S($\mathcal{L}$) satisfies the $\Box$-disjunction property, and applications of ClR are eliminable in S($\mathcal{L}$)-proofs of $\vdash x_1{:}D$.

Table 8.1. Counter-examples to extensions of the $\Box$-disjunction property

System	Counter-example	
S(KB)	$\vdash x{:}\Box A, x{:}\Box \sim\Box\Box A$	(i.e. $\vdash x{:} \sim\Box A \supset \Box\Diamond \sim\Box A$)
S(K5)	$\vdash x{:}\Box \sim\Box A, x{:}\Box A$	(i.e. $\vdash \Diamond \sim A \supset \Box\Diamond \sim A$)
S(K2)	$\vdash x{:}\Box \sim\Box A, x{:}\Box \sim\Box \sim A$	(i.e. $\vdash x{:}\Diamond\Box A \supset \Box\Diamond A$)

Corollary 8.2.13 *The sequent systems* S(K), S(T), S(K4) *and* S(S4) *satisfy the* $\Box$-*disjunction property. This implies that every sequent* $\vdash x_1{:}D$ *provable in* S(K), S(T), S(K4) *or* S(S4) *has a proof in which there are no applications of* ClR *with principal formula of the form* $x{:}\Box A$. ■

The $\Box$-disjunction property for a divergent system $S(\mathcal{L})$ is related to the fact that the semantic condition

if x accesses y and x accesses z, then y is different from z
and it is not the case that y accesses z or that z accesses y (8.8)

is consistent with the other properties of the accessibility relation in the logic $\mathcal{L}$, but is 'purely negative' and thus not modally axiomatizable (see [168, Lemma 3.6.1] and [204, Lemma 9], and see also the discussion on tree-frame modal logics in [140, §7] as we remarked in Footnote 9 above). Thus, the addition of (8.8) to a first-order metalogic (in which modal formulas are translated) does not alter the set of provable theorems of $\mathcal{L}$.

Although the $\Box$-disjunction property holds also for modal sequent systems other than those we considered, e.g. for the divergent systems S(D) and S(KD4), it does not hold in general. For example, if we have that

if x accesses both y and z, then y accesses z or z accesses y, (8.9)

then the lwffs $x{:}\Box A_1$ and $x{:}\Box A_2$ in the succedent of a sequent are not necessarily independent, since their subformulas $y{:}A_1$ and $z{:}A_2$ may be not independent. That is, $x{:}\Box A_1$ and $x{:}\Box A_2$ are not weak in the proof. Thus, given (8.9), it is not surprising that attempts to extend Proposition 8.2.9 fail for non-divergent systems such as S(KB) = S(K) + {*symm*}, S(K5) = S(K) + {*eucl*} or S(K2) = S(K) + {*conv1*, *conv2*}, and that instances of the characteristic axiom schemas of the corresponding logics provide us with the counter-examples given in Table 8.1.

While the failure of the $\Box$-disjunction property for a non-divergent system $S(\mathcal{L})$ does not imply that we cannot eliminate applications of ClR with principal formula of the form $x{:}\Box A$ in $S(\mathcal{L})$ (in fact, the sequents in Table 8.1 are all provable without ClR), it is relatively easy, at least for some systems, to find theorems that do require ClR. So, for example, to prove $x{:}\Box\Box \sim B \supset (\Box B \supset \Box A)$ in S(K5) we need a right contraction of $x{:}\Box A$, as is suggested in the following proof

$$
\dfrac{\dfrac{\dfrac{\dfrac{\dfrac{\dfrac{\begin{array}{c}\Pi_1\\ \Delta \vdash xRy\end{array} \quad \dfrac{\begin{array}{c}\Pi_2\\ \Delta \vdash xRz\end{array} \quad \dfrac{\begin{array}{c}\Pi_3\\ \Delta \vdash yRz\end{array} \quad \dfrac{\begin{array}{c}\text{axiom(s)}\\ \vdots\\ z{:}B \vdash z{:}B\\ \vdots \; \mathrm{W}\\ z{:}B, \Delta \vdash y{:}A, z{:}A, z{:}B\end{array}}{z{:}\sim B, z{:}B, \Delta \vdash y{:}A, z{:}A}\,{\sim}\mathrm{L}}{y{:}\Box \sim B, z{:}B, \Delta \vdash y{:}A, z{:}A}\,\Box\mathrm{L}}{y{:}\Box \sim B, x{:}\Box B, \Delta \vdash y{:}A, z{:}A}\,\Box\mathrm{L}}{x{:}\Box\Box \sim B, x{:}\Box B, \Delta \vdash y{:}A, z{:}A}\,\Box\mathrm{L}}{x{:}\Box\Box \sim B, x{:}\Box B, xRy \vdash y{:}A, x{:}\Box A}\,\Box\mathrm{R}}{x{:}\Box\Box \sim B, x{:}\Box B \vdash x{:}\Box A, x{:}\Box A}\,\Box\mathrm{R}}{x{:}\Box\Box \sim B, x{:}\Box B \vdash x{:}\Box A}\,\mathrm{ClR}}{x{:}\Box\Box \sim B \vdash x{:}\Box B \supset \Box A}\,{\supset}\mathrm{R}}{\vdash x{:}\Box\Box \sim B \supset (\Box B \supset \Box A)}\,{\supset}\mathrm{R}
$$

where $\Delta = \{xRy, xRz\}$, so that Π_1 and Π_2 are trivial and Π_3 is

$$
\dfrac{\dfrac{\dfrac{}{xRy \vdash xRy}\,\mathrm{AXr}}{xRy, xRz \vdash xRy}\,\mathrm{WrL} \qquad \dfrac{\dfrac{}{xRz \vdash xRz}\,\mathrm{AXr}}{xRy, xRz \vdash xRz}\,\mathrm{WrL}}{xRy, xRz \vdash yRz}\,\textit{eucl} \quad .
$$

We conjecture that the failure of the eliminability of ClR can be related to, and perhaps proof-theoretically justify, the need [87, 120] for a superformula principle in non-analytic logics such as KB and K5. However, we will not discuss this further. Instead we will focus on the analysis of left contractions in S(K), S(T), S(K4) and S(S4), in order to establish bounds needed for our complexity analysis.

9 SUBSTRUCTURAL ANALYSIS OF S(K)

We begin our analysis of contractions in $\mathrm{S(K)}$ by introducing additional terminology. We call *contraction constituents* the active formulas of an application of a contraction rule, and define the *rank* of a contraction of $x{:}A$, in symbols $rank(x{:}A)$, to be the largest number of steps immediately preceding the conclusion of the contraction and containing at least one of the contraction constituents. Since we can always transform a backwards proof so that a contraction of $x{:}A$ immediately precedes the step introducing the second constituent, the rank measures how many steps stand between the introduction of the first and second constituent (so that the minimum possible rank of a contraction is 2).

9.1 ELIMINATING CONTRACTIONS IN $\mathrm{S(K)}$

Lemma 8.2.4 and Corollary 8.2.13 tell us that to prove a sequent $\vdash x_1{:}D$ in S(K) we do not need to apply the contraction rules CrL and ClR. We can eliminate the rule ClL as well.

Theorem 9.1.1 *Every sequent* $\vdash x_1{:}D$ *provable in* $\mathrm{S(K)}$ *has a proof in which there are no applications of the contraction rules.*

Proof We adapt and extend the proof for propositional classical logic given by Zeman in [238], and proceed by three nested inductions. The first induction is on the number of contractions in the S(K)-proof, the second on the grade of the contracted lwff, and the third on the rank of the contraction.

For the first induction, consider a proof Π of $\vdash x_1{:}D$ that contains $i+1$ contractions. Pick a 'highest' (e.g. uppermost in the leftmost branch) contraction in Π, i.e. consider a subproof of Π that ends with a contraction of $x{:}A$ and such that the proof above the contraction is contraction-free. By the permutability of the rules, we can assume, without loss of generality, that the contraction immediately precedes the rule that introduces the second instance of $x{:}A$. We show, by induction on the grade of $x{:}A$, how to eliminate this contraction to obtain a proof Π' of $\vdash x_1{:}D$ that contains i contractions.

($grade(x{:}A) = 0$) The base case, $grade(x{:}A) = 0$ is trivial: since neither of the contraction constituents $x{:}A$ can be introduced by a logical rule, and since only the first, at most, can be introduced by an axiom, the second $x{:}A$ must be introduced by a weakening, and we conclude by deleting this weakening and the contraction.

Now let $grade(x{:}A) = k+1$. We proceed by induction on the rank of the contraction, $rank(x{:}A)$, which in the base case is equal to 2.

($grade(x{:}A) = k+1$, $rank(x{:}A) = 2$) We consider the different rules introducing the second $x{:}A$, where, by Fact 8.2.2, we do not need to consider the case when one of the two constituents is introduced by an axiom, and where the case for $\Box$R follows trivially by Corollary 8.2.13.

($grade(x{:}A) = k+1$, $rank(x{:}A) = 2$, $A = B \supset C$, $\supset$L) Suppose that $A = B \supset C$ and that there is a contraction-free proof of $x{:}B \supset C, x{:}B \supset C, \Gamma, \Delta \vdash \Gamma'$. Since the contraction has rank 2, the first $x{:}B \supset C$ is introduced twice, once into the left and once into the right premise of the application of $\supset$L. We distinguish three cases.

($grade(x{:}A) = k+1$, $rank(x{:}A) = 2$, $A = B \supset C$, $\supset$L, *case 1*) If the second $x{:}B \supset C$ is introduced by WlL, or if both instances of the first $x{:}B \supset C$ are introduced by WlL, then we conclude by simply deleting the weakening(s) and the ClL. For example, we transform

$$\dfrac{\dfrac{\begin{matrix}\Pi_1\\ x{:}B \supset C, \Gamma, \Delta \vdash \Gamma'\end{matrix}}{x{:}B \supset C, x{:}B \supset C, \Gamma, \Delta \vdash \Gamma'}\ \text{WlL}}{\begin{matrix}x{:}B \supset C, \Gamma, \Delta \vdash \Gamma'\\ \Pi_0\\ \vdash x_1{:}D\end{matrix}}\ \text{ClL} \qquad \text{to} \qquad \begin{matrix}\Pi_1\\ x{:}B \supset C, \Gamma, \Delta \vdash \Gamma'\\ \Pi_0\\ \vdash x_1{:}D\end{matrix}\ .$$

($grade(x{:}A) = k+1$, $rank(x{:}A) = 2$, $A = B \supset C$, $\supset$L, *case 2*) If one of the instances of the first $x{:}B \supset C$ is introduced by WlL and the other by $\supset$L, then we have

$$\dfrac{\dfrac{\begin{matrix}\Pi_1\\ \Gamma, \Delta \vdash \Gamma', x{:}B\end{matrix}}{x{:}B \supset C, \Gamma, \Delta \vdash \Gamma', x{:}B}\ \text{WlL} \quad \dfrac{\begin{matrix}\Pi_2\\ x{:}C, \Gamma, \Delta \vdash \Gamma', x{:}B\end{matrix} \quad \begin{matrix}\Pi_3\\ x{:}C, x{:}C, \Gamma, \Delta \vdash \Gamma'\end{matrix}}{x{:}C, x{:}B \supset C, \Gamma, \Delta \vdash \Gamma'}\ \supset\text{L}}{\dfrac{x{:}B \supset C, x{:}B \supset C, \Gamma, \Delta \vdash \Gamma'}{\begin{matrix}x{:}B \supset C, \Gamma, \Delta \vdash \Gamma'\\ \Pi_0\\ \vdash x_1{:}D\end{matrix}}\ \text{ClL}}\ \supset\text{L}$$

which we can transform so that both instances of $x{:}B \supset C$ are introduced by $\supset$L, i.e. we replace the application of WlL with

$$\dfrac{\dfrac{\begin{array}{c}\Pi_1\\ \Gamma,\Delta \vdash \Gamma', x{:}B\end{array}}{\Gamma,\Delta \vdash \Gamma', x{:}B, x{:}B}\,\text{WlR} \quad \dfrac{\begin{array}{c}\Pi_1\\ \Gamma,\Delta \vdash \Gamma', x{:}B\end{array}}{x{:}C,\Gamma,\Delta \vdash \Gamma', x{:}B}\,\text{WlL}}{x{:}B \supset C,\Gamma,\Delta \vdash \Gamma', x{:}B}\,\supset\text{L} \quad .$$

Since $grade(x{:}B)$ and $grade(x{:}C)$ are both less than $grade(x{:}B \supset C)$, we can apply the induction hypothesis to

$$\Gamma,\Delta \vdash \Gamma', x{:}B, x{:}B \quad \text{and} \quad x{:}C, x{:}C, \Gamma, \Delta \vdash \Gamma'$$

to obtain contraction-free proofs Π_1' and Π_3' of

$$\Gamma,\Delta \vdash \Gamma', x{:}B \quad \text{and} \quad x{:}C,\Gamma,\Delta \vdash \Gamma',$$

and then conclude by an application of $\supset$L, i.e.

$$\begin{array}{c}\dfrac{\begin{array}{c}\Pi_1'\\ \Gamma,\Delta \vdash \Gamma', x{:}B\end{array} \quad \begin{array}{c}\Pi_3'\\ x{:}C,\Gamma,\Delta \vdash \Gamma'\end{array}}{x{:}B \supset C,\Gamma,\Delta \vdash \Gamma'}\,\supset\text{L}\\ \Pi_0\\ \vdash x_1{:}D\end{array} \quad .$$

($grade(x{:}A) = k+1$, $rank(x{:}A) = 2$, $A = B \supset C$, $\supset$L, *case 3*) If both instances of the first $x{:}B \supset C$ are introduced by $\supset$L, then the proof has the form

$$\begin{array}{c}\dfrac{\dfrac{\dfrac{\begin{array}{c}\Pi_1\\ \Gamma,\Delta \vdash \Gamma', x{:}B, x{:}B\end{array} \quad \begin{array}{c}\Pi_2\\ x{:}C,\Gamma,\Delta \vdash \Gamma', x{:}B\end{array}}{x{:}B \supset C,\Gamma,\Delta \vdash \Gamma', x{:}B}\,\supset\text{L} \quad \dfrac{\begin{array}{c}\Pi_3\\ x{:}C,\Gamma,\Delta \vdash \Gamma', x{:}B\end{array} \quad \begin{array}{c}\Pi_4\\ x{:}C, x{:}C,\Gamma,\Delta \vdash \Gamma'\end{array}}{x{:}C, x{:}B \supset C,\Gamma,\Delta \vdash \Gamma'}\,\supset\text{L}}{x{:}B \supset C, x{:}B \supset C,\Gamma,\Delta \vdash \Gamma'}\,\supset\text{L}}{x{:}B \supset C,\Gamma,\Delta \vdash \Gamma'}\,\text{ClL}\\ \Pi_0\\ \vdash x_1{:}D\end{array}$$

and we conclude by applying the induction hypothesis like in case 2.

($grade(x{:}A) = k+1$, $rank(x{:}A) = 2$, $A = B \supset C$, $\supset$R) Suppose that $A = B \supset C$ and that there is a contraction-free proof of $\Gamma,\Delta \vdash \Gamma', x{:}B \supset C, x{:}B \supset C$. If one of the two constituents is introduced by WlR, we conclude by deleting this WlR and the application of ClR. Therefore suppose now that both constituents are introduced by $\supset$R, i.e.

$$\begin{array}{c}\dfrac{\dfrac{\dfrac{\begin{array}{c}\Pi_1\\ \Gamma,\Delta, x{:}B, x{:}B \vdash \Gamma', x{:}C, x{:}C\end{array}}{\Gamma,\Delta, x{:}B \vdash \Gamma', x{:}C, x{:}B \supset C}\,\supset\text{R}}{\Gamma,\Delta \vdash \Gamma', x{:}B \supset C, x{:}B \supset C}\,\supset\text{R}}{\Gamma,\Delta \vdash \Gamma', x{:}B \supset C}\,\text{ClR}\\ \Pi_0\\ \vdash x_1{:}D\end{array} \quad .$$

We apply the induction hypothesis to $\Gamma, \Delta, x{:}B, x{:}B \vdash \Gamma', x{:}C, x{:}C$ to obtain a contraction-free proof Π_1' of $\Gamma, \Delta, x{:}B \vdash \Gamma', x{:}C$, and then conclude by $\supset$R, i.e.

$$\begin{array}{c} \Pi_1' \\ \dfrac{\Gamma, \Delta, x{:}B \vdash \Gamma', x{:}C}{\Gamma, \Delta \vdash \Gamma', x{:}B \supset C}\supset\mathrm{R} \\ \Pi_0 \\ \vdash x_1{:}D \end{array} .$$

($grade(x{:}A) = k+1$, $rank(x{:}A) = 2$, $A = \Box B$, $\Box$L) Suppose that $A = \Box B$ and that there is a contraction-free proof of $x{:}\Box B, x{:}\Box B, \Gamma, \Delta \vdash \Gamma'$. If one of the two constituents is introduced by WlL, then we conclude by deleting this WlL and the application of ClL. Therefore suppose now that both constituents are introduced by $\Box$L, i.e.

$$\begin{array}{c} \dfrac{\dfrac{\overline{xRy \vdash xRy}\,\mathrm{AXr}}{\Delta, xRy, xRz \vdash xRy}\,\vdots\,\mathrm{WrL} \qquad \dfrac{\dfrac{\overline{xRz \vdash xRz}\,\mathrm{AXr}}{\Delta, xRy, xRz \vdash xRz}\,\vdots\,\mathrm{WrL} \qquad \begin{array}{c}\Pi_1\\ y{:}B, z{:}B, \Gamma, \Delta, xRy, xRz \vdash \Gamma'\end{array}}{y{:}B, x{:}\Box B, \Gamma, \Delta, xRy, xRz \vdash \Gamma'}\,\Box\mathrm{L}}{\dfrac{x{:}\Box B, x{:}\Box B, \Gamma, \Delta, xRy, xRz \vdash \Gamma'}{x{:}\Box B, \Gamma, \Delta, xRy, xRz \vdash \Gamma'}\,\mathrm{ClL}}\,\Box\mathrm{L} \\ \Pi_0 \\ \vdash x_1{:}D \end{array} . \tag{9.1}$$

Since (9.1) is a proof of $\vdash x_1{:}D$ in S(K), and since $rank(x{:}\Box B) = 2$, the rwffs xRy and xRz must be present in the conclusion of the lowest $\Box$L (and thus in the conclusion of ClL as well); they are then active in two applications of $\Box$R in Π_0. Let $x{:}\Box A_1$ and $x{:}\Box A_2$ be the principal formulas of these applications of $\Box$R, and, without loss of generality, assume that $y{:}A_1, z{:}A_2 \in \Gamma'$.[1] Then $\Gamma' = \Gamma'' \cup \{y{:}A_1, z{:}A_2\}$, and (9.1) is

$$\begin{array}{c} \dfrac{\dfrac{\overline{xRy \vdash xRy}\,\mathrm{AXr}}{\Delta, xRy, xRz \vdash xRy}\,\vdots\,\mathrm{WrL} \qquad \dfrac{\dfrac{\overline{xRz \vdash xRz}\,\mathrm{AXr}}{\Delta, xRy, xRz \vdash xRz}\,\vdots\,\mathrm{WrL} \qquad \begin{array}{c}\Pi_1\\ y{:}B, z{:}B, \Gamma, \Delta, xRy, xRz \\ \vdash \Gamma'', y{:}A_1, z{:}A_2\end{array}}{y{:}B, x{:}\Box B, \Gamma, \Delta, xRy, xRz \vdash \Gamma'', y{:}A_1, z{:}A_2}\,\Box\mathrm{L}}{\dfrac{x{:}\Box B, x{:}\Box B, \Gamma, \Delta, xRy, xRz \vdash \Gamma'', y{:}A_1, z{:}A_2}{x{:}\Box B, \Gamma, \Delta, xRy, xRz \vdash \Gamma'', y{:}A_1, z{:}A_2}\,\mathrm{ClL}}\,\Box\mathrm{L} \\ \Pi_0 \\ \vdash x_1{:}D \end{array} .$$

This proof contains a contraction-free sub-proof Π_2 of

$$x{:}\Box B, x{:}\Box B, \Gamma, \Delta, xRy, xRz \vdash \Gamma'', y{:}A_1, z{:}A_2 ,$$

[1]Two remarks. First, if the two applications of $\Box$L have the same active rwff, say xRy, then we conclude analogously. Second, if, e.g., $y{:}A_1 = y{:}\Box A_3$ is the principal formula of a $\Box$R in Π_0, then $yRu \in \Delta$ and $u{:}A_3 \in \Gamma'$; if $z{:}A_2 = z{:} \sim A_4$ is the principal formula of a $\sim R$ in Π_0, then $z{:}A_4 \in \Gamma$. In both these cases we conclude analogously to the case we consider.

and we can transform Π_2 so that it contains contraction-free proofs of

$$x{:}\Box B, x{:}\Box B, \Gamma_1, \Delta, xRy, xRz \vdash \Gamma''_1, y{:}A_1, z{:}A_2$$
$$\text{and} \quad \cdots \quad \text{and}$$
$$x{:}\Box B, x{:}\Box B, \Gamma_m, \Delta, xRy, xRz \vdash \Gamma''_m, y{:}A_1, z{:}A_2$$

where for each j, Γ_j and Γ''_j are not x-branching with respect to Δ in S(K). Then, since y and z do not occur in Γ_j, Γ''_j or Δ, Corollary 8.2.10 tells us that for each

$$x{:}\Box B, x{:}\Box B, \Gamma_j, \Delta, xRy, xRz \vdash \Gamma''_j, y{:}A_1, z{:}A_2 \quad (1 \leq j \leq m)$$

there is a contraction-free S(K)-proof Π_3 of either

$$x{:}\Box B, x{:}\Box B, \Gamma_j, \Delta, xRy \vdash \Gamma''_j, y{:}A_1 \quad \text{or} \quad x{:}\Box B, x{:}\Box B, \Gamma_j, \Delta, xRz \vdash \Gamma''_j, z{:}A_2 .$$

Let, for example, Π_3 be the proof of the leftmost sequent. We have permuted possible applications in Π_1 of x-branching rules (i.e. applications of $\supset$L with principal formulas labelled with x or with predecessors of x) over the two consecutive applications of $\Box$L. Therefore, Π_3 contains these two applications of $\Box$L, i.e. Π_3 is

$$\dfrac{\dfrac{\overline{xRy \vdash xRy}\ \text{AXr}}{\Delta, xRy \vdash xRy}\ \vdots\,\text{WrL} \qquad \dfrac{\dfrac{\overline{xRy \vdash xRy}\ \text{AXr}}{\Delta, xRy \vdash xRy}\ \vdots\,\text{WrL} \qquad \begin{array}{c}\Pi_4 \\ y{:}B, y{:}B, \Gamma_j, \Delta, xRy \vdash \Gamma''_j, y{:}A_1\end{array}}{y{:}B, x{:}\Box B, \Gamma_j, \Delta, xRy \vdash \Gamma''_j, y{:}A_1}\ \Box\text{L}}{x{:}\Box B, x{:}\Box B, \Gamma_j, \Delta, xRy \vdash \Gamma''_j, y{:}A_1}\ \Box\text{L} .$$

Since $grade(y{:}B) < grade(x{:}\Box B)$, by the induction hypothesis, there is a contraction-free proof Π'_4 of

$$y{:}B, \Gamma_j, \Delta, xRy \vdash \Gamma''_j, y{:}A_1 ,$$

and thus we can replace Π_3 with

$$\dfrac{\dfrac{\overline{xRy \vdash xRy}\ \text{AXr}}{\Delta, xRy \vdash xRy}\ \vdots\,\text{WrL} \qquad \begin{array}{c}\Pi'_4 \\ y{:}B, \Gamma_j, \Delta, xRy \vdash \Gamma''_j, y{:}A_1\end{array}}{x{:}\Box B, \Gamma_j, \Delta, xRy \vdash \Gamma''_j, y{:}A_1}\ \Box\text{L} .$$

Proceeding analogously for each j such that $1 \leq j \leq m$, we obtain a contraction-free proof Π'_2 of either

$$x{:}\Box B, \Gamma, \Delta, xRy \vdash \Gamma'', y{:}A_1 \qquad \text{or} \qquad x{:}\Box B, \Gamma, \Delta, xRz \vdash \Gamma'', z{:}A_2 ,$$

from which we conclude by weakening. For example:

$$\begin{array}{c} \Pi'_2 \\ \dfrac{\dfrac{x{:}\Box B, \Gamma, \Delta, xRy \vdash \Gamma'', y{:}A_1}{x{:}\Box B, \Gamma, \Delta, xRy, xRz \vdash \Gamma'', y{:}A_1}\ \text{WrL}}{x{:}\Box B, \Gamma, \Delta, xRy, xRz \vdash \Gamma'', y{:}A_1, z{:}A_2}\ \text{WlR} \\ \Pi_0 \\ \vdash x_1{:}D \end{array} .$$

To summarize, analogously to Corollary 8.2.12, we have first copied $x{:}\Box B$ in each branch via the implicit contractions in applications of $\supset$ L (with principal formulas labelled with x or with predecessors of x), and then eliminated the explicit contraction.

This concludes the proof for the case $grade(x{:}A) = k+1$ and $rank(x{:}A) = 2$. Consider now the final case when $grade(x{:}A) = k+1$ and $rank(x{:}A) = j+1 > 2$. In this case, it is possible that the first contraction constituent was introduced by a weakening into one or more of the places which give the contraction a rank of $j+1$. We make the mode of introduction of that constituent uniform by replacing all such introductions by weakening with introductions by the proper logical rule; the rest of the proof is as before, modulo possible applications of weakening and AXr. For example, we transform the proof

$$
\begin{array}{c}
\dfrac{\dfrac{\begin{array}{c}\dfrac{}{xRy \vdash xRy}\,\text{AXr}\\ \vdots\ \text{WrL}\\ \Delta, xRy \vdash xRy\end{array} \qquad \begin{array}{c}\Pi_2\\ \dfrac{\Gamma_1,\Delta_1,xRy\vdash\Gamma_1'}{x{:}\Box B,\Gamma_1,\Delta_1,xRy\vdash\Gamma_1'}\,\text{WlL}\\ \Pi_1\\ y{:}B, x{:}\Box B,\Gamma,\Delta,xRy\vdash\Gamma'\end{array}}{x{:}\Box B, x{:}\Box B,\Gamma,\Delta,xRy\vdash\Gamma'}\,\Box\text{L}}{x{:}\Box B,\Gamma,\Delta,xRy\vdash\Gamma'}\,\text{ClL}\\
\Pi_0\\
\vdash x_1{:}D
\end{array}
$$

to

$$
\begin{array}{c}
\dfrac{\dfrac{\begin{array}{c}\dfrac{}{xRy \vdash xRy}\,\text{AXr}\\ \vdots\ \text{WrL}\\ \Delta, xRy \vdash xRy\end{array} \qquad \begin{array}{c}\dfrac{\begin{array}{c}\dfrac{}{xRy \vdash xRy}\,\text{AXr}\\ \vdots\ \text{WrL}\\ \Delta_1, xRy \vdash xRy\end{array} \qquad \begin{array}{c}\Pi_2\\ \dfrac{\Gamma_1,\Delta_1,xRy\vdash\Gamma_1'}{y{:}B,\Gamma_1,\Delta_1,xRy\vdash\Gamma_1'}\,\text{WlL}\end{array}}{x{:}\Box B,\Gamma_1,\Delta_1,xRy\vdash\Gamma_1'}\,\Box\text{L}\\ \Pi_1\\ y{:}B, x{:}\Box B,\Gamma,\Delta,xRy\vdash\Gamma'\end{array}}{x{:}\Box B, x{:}\Box B,\Gamma,\Delta,xRy\vdash\Gamma'}\,\Box\text{L}}{x{:}\Box B,\Gamma,\Delta,xRy\vdash\Gamma'}\,\text{ClL}\\
\Pi_0\\
\vdash x_1{:}D
\end{array}
.
$$

The rank of ClL is still $j+1$, but now we have a proof in which all highest introductions of the first contraction constituent are by the logical rule proper to that constituent. We consider the different cases for this proper rule.

($grade(x{:}A) = k+1$, $rank(x{:}A) = j+1$, $A = B \supset C$, $\supset$ L) We conclude straightforwardly by first permuting the uppermost $\supset$ L over the rule below it to obtain a proof with $rank(x{:}B \supset C) = j$, and then applying the induction hypothesis.

($grade(x{:}A) = k+1$, $rank(x{:}A) = j+1$, $A = B \supset C$, $\supset$ R) We conclude straightforwardly by first permuting the uppermost $\supset$ R over the rule below it to obtain a proof with $rank(x{:}B \supset C) = j$, and then applying the induction hypothesis.

($grade(x{:}A) = k+1$, $rank(x{:}A) = j+1$, $A = \Box B$, $\Box$L) We distinguish two cases, depending on whether or not the rwff xRz active in the uppermost $\Box$L appears in the premise of ClL; as before, the rwff xRy active in the lowest $\Box$L must appear in the premise of ClL since this is a proof in S(K).

($grade(x{:}A) = k+1$, $rank(x{:}A) = j+1$, $A = \Box B$, $\Box$L, *case 1*) In the first case, the rwffs active in the two applications of $\Box$L both appear in the premise of ClL, i.e.

$$
\dfrac{
\dfrac{
\begin{array}{c}\dfrac{}{xRy \vdash xRy}\,\text{AXr}\\ \vdots\ \text{WrL}\\ \Delta, xRy, xRz \vdash xRy\end{array}
\qquad
\begin{array}{c}
\dfrac{
\begin{array}{c}\dfrac{}{xRz \vdash xRz}\,\text{AXr}\\ \vdots\ \text{WrL}\\ \Delta_1, xRz \vdash xRz\end{array}
\qquad
\begin{array}{c}\Pi_2\\ z{:}B, \Gamma_1, \Delta_1, xRz \vdash \Gamma_1'\end{array}
}{x{:}\Box B, \Gamma_1, \Delta_1, xRz \vdash \Gamma_1'}\,\Box\text{L}\\
\Pi_1\\
y{:}B, x{:}\Box B, \Gamma, \Delta, xRy, xRz \vdash \Gamma'
\end{array}
}{x{:}\Box B, x{:}\Box B, \Gamma, \Delta, xRy, xRz \vdash \Gamma'}\,\Box\text{L}
}{
\begin{array}{c}x{:}\Box B, \Gamma, \Delta, xRy, xRz \vdash \Gamma'\\ \Pi_0\\ \vdash x_1{:}D\end{array}
}\,\text{ClL}
\quad .
$$

Since $x{:}\Box B$ and xRz are parametric in Π_2, the uppermost $\Box$L is permutable over the rule immediately below it (even if this rule were an application of $\Box$R). Performing this permutation, we obtain a proof in which $rank(x{:}\Box B) = j$, and we then conclude by applying the induction hypothesis.

($grade(x{:}A) = k+1$, $rank(x{:}A) = j+1$, $A = \Box B$, $\Box$L, *case 2*) In the second case, xRz does not appear in Δ, but is the active rwff of an application of $\Box$R occurring between the two applications of $\Box$L, i.e.

$$
\dfrac{
\dfrac{
\begin{array}{c}\dfrac{}{xRy \vdash xRy}\,\text{AXr}\\ \vdots\ \text{WrL}\\ \Delta, xRy \vdash xRy\end{array}
\qquad
\begin{array}{c}
\dfrac{
\begin{array}{c}\dfrac{}{xRz \vdash xRz}\,\text{AXr}\\ \vdots\ \text{WrL}\\ \Delta_1, xRz \vdash xRz\end{array}
\qquad
\begin{array}{c}\Pi_3\\ z{:}B, \Gamma_1, \Delta_1, xRz \vdash \Gamma_1'\end{array}
}{x{:}\Box B, \Gamma_1, \Delta_1, xRz \vdash \Gamma_1'}\,\Box\text{L}\\
\Pi_2\\
\dfrac{x{:}\Box B, \Gamma_2, \Delta_2, xRz \vdash z{:}C, \Gamma_2'}{x{:}\Box B, \Gamma_2, \Delta_2 \vdash x{:}\Box C, \Gamma_2'}\,\Box\text{R}\\
\Pi_1\\
y{:}B, x{:}\Box B, \Gamma, \Delta, xRy \vdash \Gamma'
\end{array}
}{x{:}\Box B, x{:}\Box B, \Gamma, \Delta, xRy \vdash \Gamma'}\,\Box\text{L}
}{
\begin{array}{c}x{:}\Box B, \Gamma, \Delta, xRy \vdash \Gamma'\\ \Pi_0\\ \vdash x_1{:}D\end{array}
}\,\text{ClL}
\quad . \qquad (9.2)
$$

As explained for Lemma 8.2.1, we cannot, in general, reduce the rank of ClL by permuting the uppermost $\Box$L over the rule below it, since, if Π_2 is empty, the uppermost $\Box$L is not permutable over the $\Box$R with active rwff xRz. We can however permute this $\Box$R over the lowest $\Box$L. To show this, we reason as follows. Since we are proving $\vdash x_1{:}D$, Π_0 must contain an application of $\Box$R with active rwff xRy, and, for this to be possible, y must be an arbitrary world accessible from x. In particular, $y \neq x$. Thus, since xRy and since there are no relational rules, $x{:}\Box C$ cannot follow from (be a subformula of) $y{:}B$. Then $x{:}\Box C$ must be a subformula of some formula in Γ or Γ', and Π_1 can be divided into two separate subproofs, Π_4 and Π_5. Π_5 introduces $y{:}B$ and xRy, and $x{:}\Box C$ is active in Π_4 or $x{:}\Box C \in \Gamma'$. (If $y{:}B$ and xRy are contained in

Γ_2 and Δ_2, then Π_5 is empty and $\Pi_4 = \Pi_1$.) Given this separation, we can transform (9.2) to a proof that has the form

$$
\cfrac{
\cfrac{
\begin{array}{c}
\cfrac{}{xRy \vdash xRy}\,\text{AXr} \\
\vdots\ \text{WrL} \\
\Delta_3, xRy, xRz \vdash xRy
\end{array}
\qquad
\begin{array}{c}
\cfrac{
\begin{array}{c}
\cfrac{}{xRz \vdash xRz}\,\text{AXr} \\
\vdots\ \text{WrL} \\
\Delta_1, xRz \vdash xRz
\end{array}
\qquad
\begin{array}{c}
\Pi_3 \\
z{:}B, \Gamma_1, \Delta_1, xRz \vdash \Gamma_1'
\end{array}
}{x{:}\Box B, \Gamma_1, \Delta_1, xRz \vdash \Gamma_1'}\,\Box\text{L} \\
\Pi_2 \\
x{:}\Box B, \Gamma_2, \Delta_2, xRz \vdash z{:}C, \Gamma_2' \\
\Pi_5 \\
y{:}B, x{:}\Box B, \Gamma_3, \Delta_3, xRy, xRz \vdash z{:}C, \Gamma_3'
\end{array}
}{x{:}\Box B, x{:}\Box B, \Gamma_3, \Delta_3, xRy, xRz \vdash z{:}C, \Gamma_3'}\,\Box\text{L}
}{
\begin{array}{c}
x{:}\Box B, x{:}\Box B, \Gamma_3, \Delta_3, xRy \vdash x{:}\Box C, \Gamma_3' \\
\Pi_4 \\
\cfrac{x{:}\Box B, x{:}\Box B, \Gamma, \Delta, xRy \vdash \Gamma'}{x{:}\Box B, \Gamma, \Delta, xRy \vdash \Gamma'}\,\text{ClL} \\
\Pi_0 \\
\vdash x_1{:}D
\end{array}
}\,\Box\text{R}
$$

and then to

$$
\cfrac{
\cfrac{
\cfrac{
\begin{array}{c}
\cfrac{}{xRy \vdash xRy}\,\text{AXr} \\
\vdots\ \text{WrL} \\
\Delta_3, xRy, xRz \vdash xRy
\end{array}
\qquad
\begin{array}{c}
\cfrac{
\begin{array}{c}
\cfrac{}{xRz \vdash xRz}\,\text{AXr} \\
\vdots\ \text{WrL} \\
\Delta_1, xRz \vdash xRz
\end{array}
\qquad
\begin{array}{c}
\Pi_3 \\
z{:}B, \Gamma_1, \Delta_1, xRz \vdash \Gamma_1'
\end{array}
}{x{:}\Box B, \Gamma_1, \Delta_1, xRz \vdash \Gamma_1'}\,\Box\text{L} \\
\Pi_2 \\
x{:}\Box B, \Gamma_2, \Delta_2, xRz \vdash z{:}C, \Gamma_2' \\
\Pi_5 \\
y{:}B, x{:}\Box B, \Gamma_3, \Delta_3, xRy, xRz \vdash z{:}C, \Gamma_3'
\end{array}
}{x{:}\Box B, x{:}\Box B, \Gamma_3, \Delta_3, xRy, xRz \vdash z{:}C, \Gamma_3'}\,\Box\text{L}
}{x{:}\Box B, \Gamma_3, \Delta_3, xRy, xRz \vdash z{:}C, \Gamma_3'}\,\text{ClL}
}{
\begin{array}{c}
x{:}\Box B, \Gamma_3, \Delta_3, xRy \vdash x{:}\Box C, \Gamma_3' \\
\Pi_4^\dagger \\
x{:}\Box B, \Gamma, \Delta, xRy \vdash \Gamma' \\
\Pi_0 \\
\vdash x_1{:}D
\end{array}
}\,\Box\text{R}
\quad .
$$

Now we have a proof in which $rank(x{:}\Box B) \leq j$, and we conclude by applying the induction hypothesis. ■

Theorem 9.1.1 provides the basis for showing that K is decidable (cf. §12). Furthermore, its proof is extensible: when we add relational rules to S(K), we only need to consider the new cases that are generated by these rules. In particular, we must just investigate the eliminability of ClL when it contracts lwffs of the form $x{:}\Box A$. By Corollary 8.2.13 and (the proof of) Theorem 9.1.1, it follows that in S(T), S(K4) and S(S4) we can eliminate ClR and all applications of ClL with principal formula other than $x{:}\Box A$.

Corollary 9.1.2 *Every sequent* $\vdash x_1{:}D$ *provable in* S(T), S(K4) *or* S(S4) *has a proof in which there are no contractions, except for applications of* ClL *with principal formula of the form* $x{:}\Box A$. ∎

Corollary 9.1.2 holds also for other systems extending S(K), e.g. S(D) and S(KD4). However, it turns out that, even for a simple system like S(T), applications of ClL with principal formula $x{:}\Box A$ are not always eliminable. For example, as we previously indicated in (6.1), the theorem $x{:} \sim\Box \sim(B \supset \Box B)$ cannot be proved in S(T) without one left contraction of $x{:}\Box \sim(B \supset \Box B)$. In the following chapters we show how to bound applications of ClL in S(T) and in the transitive systems S(K4) and S(S4).

9.2 S(K) AND SS(K)

Before moving on to the analysis of contractions in extension of S(K), we show that Theorem 9.1.1, together with Proposition 8.2.9, provides the basis for a proof-theoretical justification of the rules of the standard sequent system SS(K) given in Figure 6.3. We can derive a labelled equivalent of the standard rule

$$\frac{\Gamma \vdash A}{\Sigma, \Box\Gamma \vdash \Box A, \Sigma'}\ (\mathrm{K})$$

as follows:

$$\frac{y{:}\Gamma \vdash y{:}A}{x{:}\Sigma, x{:}\Box\Gamma \vdash x{:}\Box A, x{:}\Sigma'}\ \Box\mathrm{LR_K} \quad \rightsquigarrow \quad \begin{array}{c} \dfrac{y{:}\Gamma \vdash y{:}A}{y{:}\Gamma, xRy \vdash y{:}A}\ \mathrm{WrL} \\ \vdots\ \Box\mathrm{L}\ (\text{all with active rwff } xRy) \\ \dfrac{x{:}\Box\Gamma, xRy \vdash y{:}A}{x{:}\Box\Gamma \vdash x{:}\Box A}\ \Box\mathrm{R} \\ \vdots\ \mathrm{W} \\ x{:}\Sigma, x{:}\Box\Gamma \vdash x{:}\Box A, x{:}\Sigma' \end{array}$$

where y is different from x, the multisets of lwffs $x{:}\Sigma$ and $x{:}\Sigma'$ contain only formulas labelled with x, and if $y{:}\Gamma = \{y{:}B_1, \ldots, y{:}B_n\}$ then $x{:}\Box\Gamma = \{x{:}\Box B_1, \ldots, x{:}\Box B_n\}$.

Since all the formulas in $x{:}\Sigma$ and $x{:}\Sigma'$ are labelled with x, the rule $\Box\mathrm{LR_K}$, like (K), is a *transitional rule*: the conclusion represents a world x and the premise an arbitrary world y accessible from x, so that applying the rule means moving from one world to the other. Note also that if $x{:}\Sigma'$ contains some lwff $x{:}\Box B$, then an application of $\Box\mathrm{LR_K}$ explicitly performs the 'metatheoretical choice' in the $\Box$-disjunction property in Proposition 8.2.9 by actually selecting, via weakening, the lwff $x{:}\Box A$ as the principal formula of $\Box$R. Furthermore, like (K), the rule $\Box\mathrm{LR_K}$ constitutes a potential 'backtracking point' in a backwards proof since it does not permute over any other rule (it contains no parametric formulas). This is because the applications of $\Box$L in the derivation of $\Box\mathrm{LR_K}$ are not permutable over the application of $\Box$R.

We can go one step further, and transform S(K)-proofs so that we can replace particular sequences of applications of $\Box$L and $\Box$R with applications of $\Box\mathrm{LR_K}$. Formally, let $\widehat{\mathrm{S}}(\mathrm{K})$ be the sequent system obtained from S(K) by

(i) eliminating $\Box$L and $\Box$R,

(ii) eliminating AXr and WrL,

(iii) eliminating WlL and WlR, and

(iv) adding $\Box\mathrm{LR_K}$ and extending the axioms AXl and $\bot$L.

The intuition for (i) and (ii) is that $\Box$L, $\Box$R, AXr and WrL are embedded into $\Box\mathrm{LR_K}$; this amounts to eliminating relational reasoning, i.e. sequents in $\widehat{\mathrm{S}}(\mathrm{K})$ do not contain Δ. The intuition for (iii) and (iv) is that we embed WlL and WlR into AXl, $\bot$L and $\Box\mathrm{LR_K}$. Then we have:

Lemma 9.2.1 *$\vdash x_1{:}D$ is provable in $\widehat{\mathrm{S}}(\mathrm{K})$ iff $\vdash x_1{:}D$ is provable in $\mathrm{S}(\mathrm{K})$.* ■

The left-to-right direction follows by the derivability of $\Box\mathrm{LR_K}$ in $\mathrm{S}(\mathrm{K})$ (recall that the extension of the axioms is justified in §6.1). For the converse direction, we begin by showing that we can transform each $\mathrm{S}(\mathrm{K})$-proof Π of $\vdash x_1{:}D$ into a *block form* by

(i) eliminating 'detours', and

(ii) adjoining 'related rules',

where we say that an application of a rule (r_1) is *related* to an application of a rule (r_2) in a $\mathrm{S}(\mathrm{K})$ proof if the principal formulas of these applications of (r_1) and (r_2) have the same label.

We show how to adjoin (r_1) and (r_2) below. For (i), we define an application of a rule (r) to be a *detour* in Π when all of the active formulas of (r) are introduced in Π either by weakenings or by detours (i.e. none of them appears in the axioms of Π so that they are weak in Π). Note that since we are in $\mathrm{S}(\mathrm{K})$ if a $\Box$ rule constitutes a detour, then its active rwff must be introduced by WrL.[2]

We eliminate a detour (r) by 'blowing up' the weakenings of lwffs, i.e. by transforming them so that they introduce lwffs of the highest grade possible, and deleting the relational reasoning. For example, let (r) be $\supset$L and Π be

$$
\begin{array}{c}
\dfrac{
 \dfrac{\dfrac{\begin{array}{c}\dfrac{\begin{array}{c}\Pi_2\\ \Gamma_1,\Delta_1\vdash\Gamma_1'\end{array}}{\Gamma_1,\Delta_1,xRy\vdash\Gamma_1'}\,\mathrm{WrL}\\ \Pi_1\\ \Gamma,\Delta,xRy\vdash\Gamma'\end{array}}{\Gamma,\Delta,xRy\vdash\Gamma',y{:}A}\,\mathrm{WlR}}{\Gamma,\Delta\vdash\Gamma',x{:}\Box A}\,\Box\mathrm{R}
 \quad
 \begin{array}{c}\dfrac{\begin{array}{c}\Pi_4\\ \Gamma_2,\Delta_2\vdash\Gamma_2'\end{array}}{x{:}B,\Gamma_2,\Delta_2\vdash\Gamma_2'}\,\mathrm{WlL}\\ \Pi_3\\ x{:}B,\Gamma,\Delta\vdash\Gamma'\end{array}
}{x{:}\Box A\supset B,\Gamma,\Delta\vdash\Gamma'}\,{\supset}\mathrm{L}\\
\Pi_0\\
\vdash x_1{:}D
\end{array}
\quad . \tag{9.3}
$$

This application of $\supset$L constitutes a detour, since $x{:}B$ is introduced by weakening and the application of $\Box$R with principal formula $x{:}\Box A$ is itself a detour (its active formulas

[2] In the following chapters we extend the definitions of detour and related rules to consider extensions of $\mathrm{S}(\mathrm{K})$.

are introduced by weakening). Since $x{:}B$ is parametric in Π_3, we can eliminate this application of $\supset$L and transform (9.3) to

$$\dfrac{\begin{array}{c}\Pi_4\\ \Gamma_2,\Delta_2\vdash\Gamma_2'\\ \Pi_3^{\dagger}\\ \Gamma,\Delta\vdash\Gamma'\end{array}}{\begin{array}{c}x{:}\Box A\supset B,\Gamma,\Delta\vdash\Gamma'\\ \Pi_0\\ \vdash x_1{:}D\end{array}}\ \mathrm{WlL}\quad .$$

Alternatively, we can take the left branch as the proof of $\Gamma,\Delta\vdash\Gamma'$, provided that we similarly transform the rules in Π_1 where $xR\,y$ is active; e.g. if Π_1 contains

$$\dfrac{\dfrac{}{xRy\vdash xRy}\ \mathrm{AXr}\quad \begin{array}{c}\vdots\ \mathrm{WrL}\end{array}\quad \Delta_3,xRy\vdash xRy \qquad \dfrac{\begin{array}{c}\Pi_5\\ \Gamma_3,\Delta_3,xRy\vdash\Gamma_3'\end{array}}{y{:}C,\Gamma_3,\Delta_3,xRy\vdash\Gamma_3'}\ \mathrm{WlL}}{x{:}\Box C,\Gamma_3,\Delta_3,xRy\vdash\Gamma_3'}\ \Box\mathrm{L}$$

then we transform it to

$$\dfrac{\begin{array}{c}\Pi_5\\ \Gamma_3,\Delta_3,xRy\vdash\Gamma_3'\end{array}}{x{:}\Box C,\Gamma_3,\Delta_3,xRy\vdash\Gamma_3'}\ \mathrm{WlL}\quad .$$

Then we can eliminate both the application of $\Box$R and the one of $\supset$L. Note that we have performed a similar transformation in case 2.2.2 in the proof of Proposition 8.2.9. In fact, Proposition 8.2.9 tells us that if Π contains two applications of $\Box$R with principal formulas $x{:}\Box A$ and $x{:}\Box B$, then at least one of them is a detour (i.e. at least one of $x{:}\Box A$ and $x{:}\Box B$ is weak in Π), provided that we take care of x-branching rules.

By iterating these transformations, we obtain a S(K)-proof Π' of $\vdash x_1{:}D$ that is free from detours. As a consequence, each rwff $xR\,y$ active in an application of $\Box$R is active in (at least) one application of $\Box$L, and, vice versa, each $xR\,y$ active in an application of $\Box$L is active in one application of $\Box$R. Thus, in order to achieve (ii), we transform Π' by permuting rules so that, for each x, rules with principal formulas labelled with x, including weakenings, are applied as an uninterrupted sequence. Then, in particular, an application of $\Box$R with principal formula $x{:}\Box A$ and active formulas $y{:}A$ and $xR\,y$

- is immediately preceded by a (possibly empty, e.g. when $\Box$R introduces $x_1{:}D$) sequence of all rules in Π' with active formulas labelled with x, and

- is immediately followed by a (possibly empty) sequence of all applications of $\Box$L in Π' with principal formula labelled with x and with active rwff $xR\,y$.

For example, if Π' is

$$
\dfrac{
\begin{array}{c}
\dfrac{}{uRx \vdash uRx}\,\mathrm{AXr} \\
\vdots\ \mathrm{WrL} \\
\Delta, uRx, xRy \vdash uRx
\end{array}
\quad
\dfrac{
\begin{array}{c}
\dfrac{
\begin{array}{c}
\dfrac{}{xRy \vdash xRy}\,\mathrm{AXr} \\
\vdots\ \mathrm{WrL} \\
\Delta_1, xRy \vdash xRy
\end{array}
\quad
\begin{array}{c}
\Pi_2 \\
y{:}A, \Gamma_1, \Delta_1, xRy \vdash \Gamma_1'
\end{array}
}{x{:}\Box A, \Gamma_1, \Delta_1, xRy \vdash \Gamma_1'}\,\Box\mathrm{L} \\
\Pi_1 \\
x{:}\Box A, x{:}E, y{:}B, \Gamma, \Delta, uRx, xRy \vdash \Gamma', y{:}C
\end{array}
}{x{:}\Box A, x{:}E, \Gamma, \Delta, uRx, xRy \vdash \Gamma', y{:}B \supset C}\,\supset\mathrm{R}
}{
\begin{array}{c}
\dfrac{x{:}\Box A, u{:}\Box E, \Gamma, \Delta, uRx, xRy \vdash \Gamma', y{:}B \supset C}{x{:}\Box A, u{:}\Box E, \Gamma, \Delta, uRx \vdash \Gamma', x{:}\Box(B \supset C)}\,\Box\mathrm{R} \\
\Pi_0 \\
\vdash x_1{:}D
\end{array}
}\,\Box\mathrm{L}
$$

then we transform it into

$$
\dfrac{
\begin{array}{c}
\dfrac{}{uRx \vdash uRx}\,\mathrm{AXr} \\
\vdots\ \mathrm{WrL} \\
\Delta, uRx, xRy \vdash uRx
\end{array}
\quad
\dfrac{
\dfrac{
\begin{array}{c}
\dfrac{}{xRy \vdash xRy}\,\mathrm{AXr} \\
\vdots\ \mathrm{WrL} \\
\Delta, uRx, xRy \vdash xRy
\end{array}
\quad
\dfrac{
\begin{array}{c}
\Pi_2 \\
y{:}A, \Gamma_1, \Delta_1, xRy \vdash \Gamma_1' \\
\Pi_1^{\dagger} \\
y{:}A, x{:}E, y{:}B, \Gamma, \Delta, uRx, \\
xRy \vdash \Gamma', y{:}C
\end{array}
}{
\begin{array}{c}
y{:}A, x{:}E, \Gamma, \Delta, uRx, \\
xRy \vdash \Gamma', y{:}B \supset C
\end{array}
}\,\supset\mathrm{R}
}{x{:}\Box A, x{:}E, \Gamma, \Delta, uRx, xRy \vdash \Gamma', y{:}B \supset C}\,\Box\mathrm{L}
}{x{:}\Box A, x{:}E, \Gamma, \Delta, uRx \vdash \Gamma', x{:}\Box(B \supset C)}\,\Box\mathrm{R}
}{
\begin{array}{c}
x{:}\Box A, u{:}\Box E, \Gamma, \Delta, uRx \vdash \Gamma', x{:}\Box(B \supset C) \\
\Pi_0 \\
\vdash x_1{:}D
\end{array}
}\,\Box\mathrm{L} .
$$

Moreover, we can further permute rules so that the sequence of $\Box$ rules is immediately followed by a weakening of the rwff active in the sequence, and is immediately preceded by a (possibly empty) subsequence of weakenings.

By iterating these transformations, we obtain the desired proof in block form. Thus, we have:

Lemma 9.2.2 *Every proof of* $\vdash x_1{:}D$ *in* $\mathrm{S(K)}$ *can be transformed into block form.* ∎

The following is a more concrete example, using also the derived rules in Figure 6.2.

Example 9.2.3 Given the S(K)-proof

$$
\begin{prooftree}
\AxiomC{}
\RightLabel{AXr}
\UnaryInfC{$xRy \vdash xRy$}
\RightLabel{WrL}
\UnaryInfC{$xRy, xRz \vdash xRy$}
\AxiomC{}
\RightLabel{AXr}
\UnaryInfC{$xRy \vdash xRy$}
\AxiomC{}
\RightLabel{AXl}
\UnaryInfC{$y{:}A \vdash y{:}A$}
\RightLabel{WrL}
\UnaryInfC{$y{:}A, xRy \vdash y{:}A$}
\RightLabel{WlR}
\UnaryInfC{$y{:}A, xRy \vdash y{:}A, y{:}B$}
\RightLabel{$\Box$L}
\BinaryInfC{$x{:}\Box A, xRy \vdash y{:}A, y{:}B$}
\RightLabel{WlL}
\UnaryInfC{$x{:}\Box A, y{:}C, xRy \vdash y{:}A, y{:}B$}
\RightLabel{WrL}
\UnaryInfC{$x{:}\Box A, y{:}C, xRy, xRz \vdash y{:}A, y{:}B$}
\RightLabel{WlR}
\UnaryInfC{$x{:}\Box A, y{:}C, xRy, xRz \vdash z{:}B, y{:}A, y{:}B$}
\RightLabel{$\Box$L}
\BinaryInfC{$x{:}\Box A, x{:}\Box C, xRy, xRz \vdash z{:}B, y{:}A, y{:}B$}
\RightLabel{$\Box$R}
\UnaryInfC{$x{:}\Box A, x{:}\Box C, xRy \vdash x{:}\Box B, y{:}A, y{:}B$}
\RightLabel{$\vee$R}
\UnaryInfC{$x{:}\Box A, x{:}\Box C, xRy \vdash x{:}\Box B, y{:}A \vee B$}
\RightLabel{$\Box$R}
\UnaryInfC{$x{:}\Box A, x{:}\Box C \vdash x{:}\Box B, x{:}\Box(A \vee B)$}
\RightLabel{$\wedge$L}
\UnaryInfC{$x{:}\Box A \wedge \Box C \vdash x{:}\Box B, x{:}\Box(A \vee B)$}
\RightLabel{$\vee$R}
\UnaryInfC{$x{:}\Box A \wedge \Box C \vdash x{:}\Box B \vee \Box(A \vee B)$}
\RightLabel{$\supset$R}
\UnaryInfC{$\vdash x{:}(\Box A \wedge \Box C) \supset (\Box B \vee \Box(A \vee B))$}
\end{prooftree}
\tag{9.4}
$$

we transform it into the following S(K)-proof in block form

$$
\begin{prooftree}
\AxiomC{}
\RightLabel{AXr}
\UnaryInfC{$xRy \vdash xRy$}
\AxiomC{}
\RightLabel{AXl}
\UnaryInfC{$y{:}A \vdash y{:}A$}
\RightLabel{WlR}
\UnaryInfC{$y{:}A \vdash y{:}A, y{:}B$}
\RightLabel{$\vee$R}
\UnaryInfC{$y{:}A \vdash y{:}A \vee B$}
\RightLabel{WrL}
\UnaryInfC{$y{:}A, xRy \vdash y{:}A \vee B$}
\RightLabel{$\Box$L}
\BinaryInfC{$x{:}\Box A, xRy \vdash y{:}A \vee B$}
\RightLabel{$\Box$R}
\UnaryInfC{$x{:}\Box A \vdash x{:}\Box(A \vee B)$}
\RightLabel{WlL}
\UnaryInfC{$x{:}\Box A, x{:}\Box C \vdash x{:}\Box(A \vee B)$}
\RightLabel{WlR}
\UnaryInfC{$x{:}\Box A, x{:}\Box C \vdash x{:}\Box B, x{:}\Box(A \vee B)$}
\RightLabel{$\wedge$L}
\UnaryInfC{$x{:}\Box A \wedge \Box C \vdash x{:}\Box B, x{:}\Box(A \vee B)$}
\RightLabel{$\vee$R}
\UnaryInfC{$x{:}\Box A \wedge \Box C \vdash x{:}\Box B \vee \Box(A \vee B)$}
\RightLabel{$\supset$R}
\UnaryInfC{$\vdash x{:}(\Box A \wedge \Box C) \supset (\Box B \vee \Box(A \vee B))$}
\end{prooftree}
\quad . \tag{9.5}
$$

■

As displayed by Example 9.2.3, a S(K)-proof in block form consists of alternating sequences of propositional and modal ($\Box$) rules, where the principal formulas in each sequence all have the same label, and thus represent the same world. Moreover, in each sequent, Δ consists of a single rwff. Thus, given a proof in block form, we can replace each sequence of $\Box$ rules and the weakenings surrounding it with an application of $\Box LR_K$. These applications of $\Box LR_K$ 'absorb' all instances of AXr and WrL in the proof, so that the axiom and the rule can be eliminated. This allows us to eliminate all rwffs from sequents and proofs, i.e. to eliminate relational reasoning.

Then, to obtain the $\widehat{\text{S}}(\text{K})$-proof claimed by the right-to-left direction in Lemma 9.2.1, we just need to 'absorb' the uppermost weakenings of lwffs into the instances of the extended axioms AXl or $\bot$L.

Example 9.2.4 Consider again Example 9.2.3. Given (9.4), we first transform it to (9.5) in block form, and then to the following $\widehat{\text{S}}(\text{K})$-proof:

$$\cfrac{\cfrac{\cfrac{\cfrac{\cfrac{\cfrac{}{y{:}A \vdash y{:}A, y{:}B}\text{AX1}}{y{:}A \vdash y{:}A \vee B}\vee\text{R}}{x{:}\Box A, x{:}\Box C \vdash x{:}\Box B, x{:}\Box(A \vee B)}\Box\text{LR}_{\text{K}}}{x{:}\Box A \wedge \Box C \vdash x{:}\Box B, x{:}\Box(A \vee B)}\wedge\text{L}}{x{:}\Box A \wedge \Box C \vdash x{:}\Box B \vee \Box(A \vee B)}\vee\text{R}}{\vdash x{:}(\Box A \wedge \Box C) \supset (\Box B \vee \Box(A \vee B))}\supset\text{R} \quad . \tag{9.6}$$

■

It is a trivial matter to show the equivalence of $\widehat{\text{S}}(\text{K})$ and SS(K). Indeed, we just need to delete or add labels as required and appropriately rename the rules. For example, given (9.6), we immediately obtain a proof in SS(K) by deleting the labels and replacing $\Box\text{LR}_{\text{K}}$ with (K) and the propositional rules with their standard counterparts, i.e.

$$\cfrac{\cfrac{\cfrac{\cfrac{\cfrac{\cfrac{}{A \vdash A, B}(\text{AX})}{A \vdash A \vee B}(\vee\text{R})}{\Box A, \Box C \vdash \Box B, \Box(A \vee B)}(\text{K})}{\Box A \wedge \Box C \vdash \Box B, \Box(A \vee B)}(\wedge\text{L})}{\Box A \wedge \Box C \vdash \Box B \vee \Box(A \vee B)}(\vee\text{R})}{\vdash (\Box A \wedge \Box C) \supset (\Box B \vee \Box(A \vee B))}(\supset\text{R}) \quad .$$

Thus, we have:

Lemma 9.2.5 $\vdash x_1{:}D$ *is provable in* $\widehat{\text{S}}(\text{K})$ *iff* $\vdash D$ *is provable in* SS(K). *Furthermore, the proofs in the two systems differ only in the names of the rules and in the presence of labels, which can be eliminated or added as required.* ■

Lemma 9.2.1 and Lemma 9.2.5 establish the equivalence of S(K) and SS(K), in that we have:

Theorem 9.2.6 $\vdash x_1{:}D$ *is provable in* S(K) *iff* $\vdash D$ *is provable in* SS(K). ■

In particular, we can then view SS(K) as the result of our substructural analysis of rules of S(K) and proofs built using them. In the following chapters we show that analogous results hold for other sequent systems.

10 SUBSTRUCTURAL ANALYSIS OF S(T)

10.1 BOUNDING CONTRACTIONS IN S(T)

Lemma 8.2.4 and Corollary 9.1.2 tell us that the rules CrL and ClR and each application of ClL with principal formula other than $x{:}\Box A$ can be eliminated in S(T). It is however easy to show that we cannot eliminate left contractions of lwffs of the form $x{:}\Box A$ and retain completeness of S(T). As an example, consider the S(T)-theorem $x{:}\sim\Box\sim(B\supset\Box B)$ and the proof (6.1) we gave in §6.1 (and see also the discussion at the end of §6.3). If we eliminate ClL, then this theorem cannot be proved: after the initial $\sim$R, we can only apply $\Box$L with active rwff introduced by reflexivity, so that we obtain

$$\dfrac{\dfrac{\dfrac{}{\vdash xRx}\,\textit{refl} \qquad x{:}\sim(B\supset\Box B)\vdash}{x{:}\Box\sim(B\supset\Box B)\vdash}\,\Box\mathrm{L}}{\vdash x{:}\sim\Box\sim(B\supset\Box B)}\,\sim\mathrm{R} \quad .$$

We can then exploit the subformula property to show that $x{:}\sim(B\supset\Box B)\vdash$ is not provable (alternatively, we can argue semantically that $B\supset\Box B$ is not valid in T).

Table 10.1 contains additional S(T)-theorem schemas that require application of ClL; we also display there the overall number of contractions required in the proofs when C, D and E are propositional variables. For example, the following is a proof of the last theorem in the table,

$$x_1{:}\Box^3((C\supset\sim\Box\sim D)\wedge(D\supset\sim\Box\sim E)\wedge\sim E)\supset\Box\sim C\,, \tag{10.1}$$

where, for brevity, we use φ to denote the formula $(C \supset \sim\Box\sim D) \wedge (D \supset \sim\Box\sim E) \wedge \sim E$, and we use '$\ldots$' for the parts of sequents that are not relevant to the proof. Also, we do not explicitly display relational reasoning, and write Ψ_1 and Ψ_2 for the trivial proofs of the sequents $\ldots \vdash x_2{:}\sim C, x_2{:}C$ and $\ldots \vdash \ldots, x_3{:}\sim D, x_3{:}D$.

$$\frac{}{x_4{:}E \vdash x_4{:}E}\ \text{AXI}$$

$$\vdots\ \text{W}$$

$$x_4{:}\sim E, \ldots \vdash \ldots, x_4{:}\sim E$$

$$\vdots\ 2 \times \wedge\text{L}$$

$$x_4{:}\varphi, \ldots \vdash \ldots, x_4{:}\sim E$$

$$\vdots\ 1 \times \Box\text{L (with active rwff } x_3 R x_4)$$

$$\dfrac{\dfrac{\dfrac{x_3{:}\Box\varphi, \ldots, x_3 R x_4 \vdash \ldots, x_4{:}\sim E}{x_3{:}\Box\varphi, \ldots \vdash \ldots, x_3{:}\Box\sim E}\ \Box\text{R}}{\Psi_2 \qquad x_3{:}\sim\Box\sim E, x_3{:}\Box\varphi, \ldots \vdash \ldots}\ \sim\text{L}}{x_3{:}D \supset \sim\Box\sim E, x_3{:}\Box\varphi, \ldots \vdash \ldots, x_3{:}\sim D}\ \supset\text{L}$$

$$\vdots\ 2 \times \wedge\text{L}$$

$$x_3{:}\varphi, x_3{:}\Box\varphi, \ldots \vdash \ldots, x_3{:}\sim D$$

$$\vdots\ 2 \times \Box\text{L (with active rwff } x_2 R x_3)$$

$$x_2{:}\Box\varphi, x_2{:}\Box^2\varphi, \ldots \vdash \ldots, x_3{:}\sim D$$

$$\vdots\ 1 \times \Box\text{L (with active rwff } x_2 R x_2)$$

$$\dfrac{\dfrac{\dfrac{\dfrac{x_2{:}\Box^2\varphi, x_2{:}\Box^2\varphi, \ldots, x_2 R x_3 \vdash \ldots, x_3{:}\sim D}{x_2{:}\Box^2\varphi, \ldots, x_2 R x_3 \vdash \ldots, x_3{:}\sim D}\ \text{ClL}}{x_2{:}\Box^2\varphi, \ldots \vdash x_2{:}\sim C, x_2{:}\Box\sim D}\ \Box\text{R}}{\Psi_1 \qquad x_2{:}\sim\Box\sim D, x_2{:}\Box^2\varphi, \ldots \vdash x_2{:}\sim C}\ \sim\text{L}}{x_2{:}C \supset \sim\Box\sim D, x_2{:}\Box^2\varphi, \ldots \vdash x_2{:}\sim C}\ \supset\text{L}$$

$$\vdots\ 2 \times \wedge\text{L}$$

$$x_2{:}\varphi, x_2{:}\Box^2\varphi, x_1 R x_2 \vdash x_2{:}\sim C$$

$$\vdots\ 1 \times \Box\text{L (with active rwff } x_2 R x_2)$$

$$x_2{:}\Box\varphi, x_2{:}\Box^2\varphi, x_1 R x_2 \vdash x_2{:}\sim C$$

$$\vdots\ 2 \times \Box\text{L (with active rwff } x_1 R x_2)$$

$$x_1{:}\Box^2\varphi, x_1{:}\Box^3\varphi, x_1 R x_2 \vdash x_2{:}\sim C$$

$$\vdots\ 1 \times \Box\text{L (with active rwff } x_1 R x_1)$$

$$\dfrac{\dfrac{\dfrac{x_1{:}\Box^3\varphi, x_1{:}\Box^3\varphi, x_1 R x_2 \vdash x_2{:}\sim C}{x_1{:}\Box^3\varphi, x_1 R x_2 \vdash x_2{:}\sim C}\ \text{ClL}}{x_1{:}\Box^3\varphi \vdash x_1{:}\Box\sim C}\ \Box\text{R}}{\vdash x_1{:}\Box^3\varphi \supset \Box\sim C}\ \supset\text{R}$$

The theorem schemas of Table 10.1 can be instantiated to require more contractions. For example, a S(T)-proof of

$$x_1{:}\sim\Box\sim((\sim\Box\sim(C \supset \Box C) \supset D) \supset \Box D) \tag{10.2}$$

requires at least 3 contractions when we replace '$\Box\sim(C \supset \Box C)$' with '$\Box\sim((\sim\Box\sim(E \supset \Box E) \supset F) \supset \Box F)$', at least four contractions when we further replace

Table 10.1. Some S(T)-theorem schemas requiring application of ClL.

S(T)-theorem schema	#ClL's
$x_1{:} \sim\Box\sim(D \supset \Box D)$	1
$x_1{:} \sim\Box\Box\sim\Box(D \supset \Box\sim\Box\sim D)$	1
$x_1{:} \sim\Box\sim((\sim\Box\sim(C \supset \Box C) \supset D) \supset \Box D)$	2
$x_1{:} \sim\Box\sim((\sim\Box\sim(C \supset \Box C) \supset \Box\sim(E \supset \Box E)) \supset \Box D)$	2
$x_1{:} \sim\Box\sim(D \supset \Box\sim(D \supset \Box\sim(C \supset \Box C)))$	2
$x_1{:}\Box^3((C \supset\sim\Box\sim D) \wedge (D \supset\sim\Box\sim E) \wedge \sim E) \supset \Box\sim C$	2

The overall number #ClL's of left contractions in the proofs is for when C, D and E are propositional variables.

'$\Box \sim (E \supset \Box E)$' with '$\Box \sim ((\sim \Box \sim (G \supset \Box G) \supset H) \supset \Box H)$', and so on. Similarly, we can generalize (10.1) to

$$x_1{:}\Box^p((C \supset\sim\Box\sim D) \wedge (D \supset\sim\Box\sim E) \wedge \sim E) \supset \Box\sim C$$

where $p \geq 3$, and then modify it so that its proof needs more than 2 contractions, e.g. by replacing '$\wedge \sim E$' with '$\wedge (E \supset\sim\Box\sim F) \wedge \sim F$' and requiring that $p \geq 4$.

Note that (10.1) (and similarly for its generalization) is the only theorem of the ones in the table for which all contractions occur on the same branch of the proof, except of course for the theorems that require only one application of ClL. All the other theorems require a total of two or more contractions, but these occur in different branches; for example, the first of the two ClL's in a proof of (10.2) is 'shared' by two branches while the second occurs only in one branch.

Although applications of ClL with principal formula $x{:}\Box A$ are needed in S(T), if we can bound their use then we need not give up decidability. To this end, in Theorem 10.1.4 below we show that when proving a sequent $\vdash x_1{:}D$ in S(T) we need at most linearly (in the size of $x_1{:}D$) many applications of ClL in each branch of the proof. Leading up to this, we begin by showing that each $\vdash x_1{:}D$ provable in S(T) has a S(T)-proof in which we do not need to left-contract any lwff $x{:}\Box A$ more than once in each branch.

Lemma 10.1.1 *Every sequent $\vdash x_1{:}D$ provable in* S(T) *has a proof in which there are no contractions, except for applications of* ClL *with principal formula of the form $x{:}\Box A$. However,* ClL *need not be applied more than once with the same principal formula $x{:}\Box A$ in each branch.*

Proof (Sketch) We extend Theorem 9.1.1, where, by Corollary 9.1.2, we only need to consider the additional cases that arise in S(T) when ClL is applied with principal formula $x{:}\Box A$. In particular, we show that if ClL is applied with principal formula $x{:}\Box A$ more than once in a branch $\mathcal{B}$ of a proof of $\vdash x_1{:}D$, then we can transform $\mathcal{B}$ so that it contains at most one left contraction ot $x{:}\Box A$.

By the permutability of the rules, we can assume, without loss of generality, that if ClL is applied n times with principal formula $x{:}\Box A$ in $\mathcal{B}$, then these n contractions are performed consecutively immediately below the rule that introduces the n-th instance of $x{:}\Box A$. Thus, for example, we transform

$$\dfrac{\dfrac{\dfrac{\Pi_2}{x{:}\Box A, x{:}\Box A, \Gamma_2, \Delta_2 \vdash \Gamma_2'}}{x{:}\Box A, \Gamma_2, \Delta_2 \vdash \Gamma_2'}\,\mathrm{ClL}}{\begin{array}{c}\Pi_1\\ \dfrac{x{:}\Box A, x{:}\Box A, \Gamma_1, \Delta_1 \vdash \Gamma_1'}{x{:}\Box A, \Gamma_1, \Delta_1 \vdash \Gamma_1'}\,\mathrm{ClL}\\ \Pi_0\\ \vdash x_1{:}D\end{array}} \quad \text{to} \quad \begin{array}{c}\Pi_2\\ x{:}\Box A, x{:}\Box A, \Gamma_2, \Delta_2 \vdash \Gamma_2'\\ \Pi_1^{\dagger}\\ \dfrac{\dfrac{x{:}\Box A, x{:}\Box A, x{:}\Box A, \Gamma_1, \Delta_1 \vdash \Gamma_1'}{x{:}\Box A, x{:}\Box A, \Gamma_1, \Delta_1 \vdash \Gamma_1'}\,\mathrm{ClL}}{x{:}\Box A, \Gamma_1, \Delta_1 \vdash \Gamma_1'}\,\mathrm{ClL}\\ \Pi_0\\ \vdash x_1{:}D\end{array} .$$

For each branch $\mathcal{B}$ we proceed as follows. If lwffs of different grade are contracted in $\mathcal{B}$, then we pick a 'lowest' (e.g. lowest in the rightmost subbranch of $\mathcal{B}$) sequence of n applications of ClL with principal formula of greatest grade. If $n = 1$, we move on to the next sequence of contractions. If $n > 1$, we consider the first (highest) two applications in the sequence, and we either eliminate the uppermost one, or we transform it into an application of ClL with principal formula of smaller grade (this contraction is then eliminable when the contracted formula does not have the form $\Box C$). By iterating this transformation, we obtain the desired proof.

Therefore consider a branch

$$\begin{array}{c}\Pi_1\\ \dfrac{\dfrac{x{:}\Box A, x{:}\Box A, x{:}\Box A, \Gamma, \Delta \vdash \Gamma'}{x{:}\Box A, x{:}\Box A, \Gamma, \Delta \vdash \Gamma'}\,\mathrm{ClL}}{x{:}\Box A, \Gamma, \Delta \vdash \Gamma'}\,\mathrm{ClL}\\ \Pi_0\\ \vdash x_1{:}D\end{array} \tag{10.3}$$

where Π_1 does not contain applications of ClL with principal formula $x{:}\Box A$ or an lwff of grade greater than $grade(x{:}\Box A)$.[1] We eliminate the uppermost ClL in (10.3) by considering the two possible cases for the last rule in Π_1. This last rule has principal formula $x{:}\Box A$, and thus is either WlL (in which case we conclude trivially by deleting WlL and the uppermost ClL) or $\Box$L.

Suppose that the last rule in Π_1 is $\Box$L, i.e.

$$\dfrac{\begin{array}{c}\Pi_2\\ \Delta \vdash xRy\end{array} \quad \begin{array}{c}\Pi_3\\ y{:}A, x{:}\Box A, x{:}\Box A, \Gamma, \Delta \vdash \Gamma'\end{array}}{x{:}\Box A, x{:}\Box A, x{:}\Box A, \Gamma, \Delta \vdash \Gamma'}\,\Box\mathrm{L}$$

where either $xRy \in \Delta$ or $y = x$. If the second contraction constituent $x{:}\Box A$ is introduced in all subbranches of Π_3 by an application of WlL, we conclude trivially as above by deleting the weakenings and the contraction. If $x{:}\Box A$ is introduced by WlL in some subbranches and by $\Box$L in other subbranches, then we make the mode of

[1]Note that, by the permutability of the rules, if Γ contains m instances of $x{:}\Box A$, then the uppermost m rule applications in Π_0 might be applications of ClL with principal formula $x{:}\Box A$.

introduction of $x{:}\Box A$ uniform by replacing all such introductions by weakening with introductions by $\Box$L. This is achieved by appropriately weakening active formulas so that we can introduce $x{:}\Box A$ by an application of $\Box$L. The rest of the proof is as before, modulo possible applications of weakening and AXr.

Suppose therefore that the second constituent is introduced in Π_2 by an application of $\Box$L with active rwff $xR\,z$. There are four cases, depending on y and z.

(*Case 1*) In the first case, $y \neq x$ and $z \neq x$. Then we eliminate the uppermost ClL by induction on the length of Π_2, i.e. by induction on the rank of the uppermost ClL; we conclude analogously to the cases in Theorem 9.1.1 in which $grade(x{:}\Box A) = k+1$, and $rank(x{:}\Box A) = 2$ or $rank(x{:}\Box A) = j+1$.

(*Case 2*) In the second case, $y = x = z$, so that (10.3) has the form

$$
\dfrac{
\dfrac{
\begin{array}{c}
\dfrac{}{\vdash xRx}\,\textit{refl}\\
\vdots\ \text{WrL}\\
\Delta \vdash xRx
\end{array}
\qquad
\begin{array}{c}
\dfrac{
\begin{array}{c}
\dfrac{}{\vdash xRx}\,\textit{refl}\\
\vdots\ \text{WrL}\\
\Delta_1 \vdash xRx
\end{array}
\qquad
\begin{array}{c}
\Pi_2\\
x{:}A, x{:}\Box A, \Gamma_1, \Delta_1 \vdash \Gamma_1'
\end{array}
}{x{:}\Box A, x{:}\Box A, \Gamma_1, \Delta_1 \vdash \Gamma_1'}\,\Box\text{L}\\
\Pi_1\\
x{:}A, x{:}\Box A, x{:}\Box A, \Gamma, \Delta \vdash \Gamma'
\end{array}
}{x{:}\Box A, x{:}\Box A, x{:}\Box A, \Gamma, \Delta, \vdash \Gamma'}\,\Box\text{L}
}{
\begin{array}{c}
\dfrac{x{:}\Box A, x{:}\Box A, \Gamma, \Delta \vdash \Gamma'}{x{:}\Box A, \Gamma, \Delta \vdash \Gamma'}\,\text{ClL}\\
\Pi_0\\
\vdash x_1{:}D
\end{array}
}\,\text{ClL}
\quad .
$$

Since applications of $\Box$L with active rwff introduced by *refl* permute over every rule, we can transform this to

$$
\dfrac{
\dfrac{
\begin{array}{c}
\dfrac{}{\vdash xRx}\,\textit{refl}\\
\vdots\ \text{WrL}\\
\Delta \vdash xRx
\end{array}
\qquad
\dfrac{
\begin{array}{c}
\dfrac{}{\vdash xRx}\,\textit{refl}\\
\vdots\ \text{WrL}\\
\Delta \vdash xRx
\end{array}
\qquad
\begin{array}{c}
\Pi_2\\
x{:}A, x{:}\Box A, \Gamma_1, \Delta_1 \vdash \Gamma_1'\\
\Pi_1^{\dagger}\\
x{:}A, x{:}A, x{:}\Box A, \Gamma, \Delta \vdash \Gamma'
\end{array}
}{x{:}A, x{:}\Box A, x{:}\Box A, \Gamma, \Delta \vdash \Gamma'}\,\Box\text{L}
}{x{:}\Box A, x{:}\Box A, x{:}\Box A, \Gamma, \Delta, \vdash \Gamma'}\,\Box\text{L}
}{
\begin{array}{c}
\dfrac{x{:}\Box A, x{:}\Box A, \Gamma, \Delta \vdash \Gamma'}{x{:}\Box A, \Gamma, \Delta \vdash \Gamma'}\,\text{ClL}\\
\Pi_0\\
\vdash x_1{:}D
\end{array}
}\,\text{ClL}
\quad .
$$

Now we can conclude by replacing the uppermost application of ClL with principal formula $x{:}\Box A$ with an application of ClL with principal formula $x{:}A$, which has smaller grade, i.e.

$$\begin{array}{c}\dfrac{\dfrac{\begin{array}{c}\dfrac{}{\vdash xRx}\;\textit{refl}\\ \vdots\;\text{WrL}\\ \Delta \vdash xRx\end{array}\qquad \dfrac{\begin{array}{c}\Pi_2\\ x{:}A, x{:}\Box A, \Gamma_1, \Delta_1 \vdash \Gamma_1'\\ \Pi_1^{\dagger}\\ x{:}A, x{:}A, x{:}\Box A, \Gamma, \Delta \vdash \Gamma'\end{array}}{x{:}A, x{:}\Box A, \Gamma, \Delta \vdash \Gamma'}\;\text{ClL}}{x{:}\Box A, x{:}\Box A, \Gamma, \Delta \vdash \Gamma'}\;\Box\text{L}}{x{:}\Box A, \Gamma, \Delta \vdash \Gamma'}\;\text{ClL}\\ \Pi_0\\ \vdash x_1{:}D\end{array}\quad .$$

This new contraction is then eliminable when A does not have the form $\Box C$.

(*Case 3*) In the third case, $y = z$ and $z \neq x$, so that (10.3) has the form

$$\begin{array}{c}\dfrac{\dfrac{\dfrac{\begin{array}{c}\dfrac{}{\vdash xRx}\;\textit{refl}\\ \vdots\;\text{WrL}\\ \Delta \vdash xRx\end{array}\qquad \begin{array}{c}\dfrac{\begin{array}{c}\dfrac{}{xRz \vdash xRz}\;\text{AXr}\\ \vdots\;\text{WrL}\\ \Delta_1, xRz \vdash xRz\end{array}\qquad \begin{array}{c}\Pi_2\\ z{:}A, x{:}\Box A, \Gamma_1, \Delta_1, xRz \vdash \Gamma_1'\end{array}}{x{:}\Box A, x{:}\Box A, \Gamma_1, \Delta_1, xRz \vdash \Gamma_1'}\;\Box\text{L}\\ \Pi_1\\ x{:}A, x{:}\Box A, x{:}\Box A, \Gamma, \Delta \vdash \Gamma'\end{array}}{x{:}\Box A, x{:}\Box A, x{:}\Box A, \Gamma, \Delta \vdash \Gamma'}\;\Box\text{L}}{x{:}\Box A, x{:}\Box A, \Gamma, \Delta \vdash \Gamma'}\;\text{ClL}}{x{:}\Box A, \Gamma, \Delta \vdash \Gamma'}\;\text{ClL}\\ \Pi_0\\ \vdash x_1{:}D\end{array}\quad . \tag{10.4}$$

We cannot, in general, permute the uppermost $\Box$L over the rule below it, since, in Π_1, $xR\,z$ might be the active rwff of an application of $\Box$R that has a subformula of $x{:}A$ as its principal formula, e.g. when A is $\sim(B \supset \Box B)$ as is the case in the proof (6.1) of $\vdash x{:} \sim\Box \sim(B \supset \Box B)$. However, if there is a proof of

$$x{:}A, x{:}\Box A, x{:}\Box A, \Gamma, \Delta \vdash \Gamma'\,,$$

then there is also a proof of

$$x{:}A, x{:}\Box A, \Gamma, \Delta \vdash \Gamma'\,,$$

and we can delete the uppermost ClL in (10.4). The intuition behind this is that since we already have $x{:}A$, we only need one instance of $x{:}\Box A$ to infer that A holds at some world w that is a successor of x but is different from it (i.e., for some Δ_i in the branch, $xR\,w \in \Delta_i$ so that $\Delta_i \vdash xR\,w$ is provable); then, $w{:}A$ leads to axioms and the second $x{:}\Box A$ is weak. Formally, we show that given

$$\begin{array}{c}\dfrac{\begin{array}{c}\dfrac{}{xRz \vdash xRz}\;\text{AXr}\\ \vdots\;\text{WrL}\\ \Delta_1, xRz \vdash xRz\end{array}\qquad \begin{array}{c}\Pi_2\\ z{:}A, x{:}\Box A, \Gamma_1, \Delta_1, xRz \vdash \Gamma_1'\end{array}}{x{:}\Box A, x{:}\Box A, \Gamma_1, \Delta_1, xRz \vdash \Gamma_1'}\;\Box\text{L}\\ \Pi_1\\ x{:}A, x{:}\Box A, x{:}\Box A, \Gamma, \Delta \vdash \Gamma'\end{array}$$

there is a proof Π_1' such that either

$$\begin{array}{c} \Pi_2' \\ x{:}\Box A, \Gamma_1, \Delta_1, xRz \vdash \Gamma_1' \\ \Pi_1^\dagger \\ x{:}A, x{:}\Box A, \Gamma, \Delta \vdash \Gamma' \end{array} \quad \text{or} \quad \begin{array}{c} \Pi_2' \\ z{:}A, \Gamma_1, \Delta_1, xRz \vdash \Gamma_1' \\ \Pi_1^\dagger \\ x{:}A, x{:}\Box A, \Gamma, \Delta \vdash \Gamma' \end{array} \;.$$

We consider the possible forms of Π_2 and distinguish two subcases.

(*Case 3.1*) In the first subcase, $x{:}\Box A$ is introduced by an application of $\Box$L with active rwff xRz. Then, by the permutability of the rules, we can transform Π_2 so that this application of $\Box$L is its last rule, i.e.

$$\dfrac{\begin{array}{c} \dfrac{}{xRz \vdash xRz}\,\text{AXr} \\ \vdots\ \text{WrL} \\ \Delta_1, xRz \vdash xRz \end{array} \qquad \begin{array}{c} \Pi_3 \\ z{:}A, z{:}A, \Gamma_1, \Delta_1, xRz \vdash \Gamma_1' \end{array}}{z{:}A, x{:}\Box A, \Gamma_1, \Delta_1, xRz \vdash \Gamma_1'}\,\Box\text{L} \;.$$

Then we replace the uppermost application of ClL with principal formula $x{:}\Box A$ in (10.4) with a left contraction of $z{:}A$, i.e. we transform (10.4) to

$$\begin{array}{c} \dfrac{\dfrac{\begin{array}{c} \dfrac{}{\vdash xRx}\,\mathit{refl} \\ \vdots\ \text{WrL} \\ \Delta \vdash xRx \end{array} \qquad \begin{array}{c} \dfrac{\begin{array}{c} \dfrac{}{xRz \vdash xRz}\,\text{AXr} \\ \vdots\ \text{WrL} \\ \Delta_1, xRz \vdash xRz \end{array} \qquad \dfrac{\begin{array}{c} \Pi_3 \\ z{:}A, z{:}A, \Gamma_1, \Delta_1, xRz \vdash \Gamma_1' \end{array}}{z{:}A, \Gamma_1, \Delta_1, xRz \vdash \Gamma_1'}\,\text{ClL}}{x{:}\Box A, \Gamma_1, \Delta_1, xRz \vdash \Gamma_1'}\,\Box\text{L} \\ \Pi_1^\dagger \\ x{:}A, x{:}\Box A, \Gamma, \Delta \vdash \Gamma' \end{array}}{x{:}\Box A, x{:}\Box A, \Gamma, \Delta \vdash \Gamma'}\,\Box\text{L}}{x{:}\Box A, \Gamma, \Delta \vdash \Gamma'}\,\text{ClL} \\ \Pi_0 \\ \vdash x_1{:}D \end{array}$$

so that Π_2' is

$$\dfrac{\begin{array}{c} \Pi_3 \\ z{:}A, z{:}A, \Gamma_1, \Delta_1, xRz \vdash \Gamma_1' \end{array}}{z{:}A, \Gamma_1, \Delta_1, xRz \vdash \Gamma_1'}\,\text{ClL} \;.$$

(*Case 3.2*) In the second subcase, the active rwff in the application of $\Box$L that introduces $x{:}\Box A$ is not xRz but is xRw for some w different from z (and x, for otherwise we could permute rules and conclude like in case 2). Since (10.4) is a proof in S(T), the $\Box$-disjunction property (Proposition 8.2.9 and its corollaries) tells us that z and w diverge from x. Then $z{:}A$ and $x{:}\Box A$ are independent, in the sense that only one of them (if any) leads to axioms and the other is weak. In other words, at least one of $z{:}A$ and $x{:}\Box A$, or the formulas it is inferred from, must be introduced by weakening in Π_4. By deleting the weakening(s), we obtain the desired proof Π_2'.

(*Case 4*) In the fourth and last case, $y \neq x$ and $z = x$, so that (10.3) has the form

$$
\dfrac{\dfrac{\dfrac{\begin{array}{c}\dfrac{}{xRy \vdash xRy}\,\text{AXr}\\ \vdots\ \text{WrL}\\ \Delta, xRy \vdash xRy\end{array} \qquad \begin{array}{c}\dfrac{\begin{array}{c}\dfrac{}{\vdash xRx}\,\textit{refl}\\ \vdots\ \text{WrL}\\ \Delta_1 \vdash xRx\end{array} \qquad \begin{array}{c}\Pi_2\\ x{:}A, x{:}\Box A, \Gamma_1, \Delta_1 \vdash \Gamma_1'\end{array}}{x{:}\Box A, x{:}\Box A, \Gamma_1, \Delta_1 \vdash \Gamma_1'}\,\Box\text{L}\\ \Pi_1\\ y{:}A, x{:}\Box A, x{:}\Box A, \Gamma, \Delta, xRy \vdash \Gamma'\end{array}}{x{:}\Box A, x{:}\Box A, x{:}\Box A, \Gamma, \Delta, xRy \vdash \Gamma'}\,\Box\text{L}}{x{:}\Box A, x{:}\Box A, \Gamma, \Delta, xRy \vdash \Gamma'}\,\text{ClL}}{\begin{array}{c}x{:}\Box A, \Gamma, \Delta, xRy \vdash \Gamma'\\ \Pi_0\\ \vdash x_1{:}D\end{array}}\,\text{ClL}
\quad .
$$

Since applications of $\Box$L with active rwff introduced by *refl* permute over every rule, we can transform this to

$$
\dfrac{\dfrac{\dfrac{\begin{array}{c}\dfrac{}{\vdash xRx}\,\textit{refl}\\ \vdots\ \text{WrL}\\ \Delta \vdash xRx\end{array} \qquad \dfrac{\begin{array}{c}\dfrac{}{xRy \vdash xRy}\,\text{AXr}\\ \vdots\ \text{WrL}\\ \Delta, xRy \vdash xRy\end{array} \qquad \begin{array}{c}\Pi_2\\ x{:}A, x{:}\Box A, \Gamma_1, \Delta_1 \vdash \Gamma_1'\\ \Pi_1^{\dagger}\\ y{:}A, x{:}A, x{:}\Box A, \Gamma, \Delta, xRy \vdash \Gamma'\end{array}}{x{:}A, x{:}\Box A, x{:}\Box A, \Gamma, \Delta, xRy \vdash \Gamma'}\,\Box\text{L}}{x{:}\Box A, x{:}\Box A, x{:}\Box A, \Gamma, \Delta, xRy \vdash \Gamma'}\,\Box\text{L}}{x{:}\Box A, x{:}\Box A, \Gamma, \Delta, xRy \vdash \Gamma'}\,\text{ClL}
$$

$$
\dfrac{x{:}\Box A, x{:}\Box A, \Gamma, \Delta, xRy \vdash \Gamma'}{\begin{array}{c}x{:}\Box A, \Gamma, \Delta, xRy \vdash \Gamma'\\ \Pi_0\\ \vdash x_1{:}D\end{array}}\,\text{ClL}
$$

and then conclude like in case 3. ■

The intuition behind this result is that in each branch of a backwards proof of a S(T)-theorem we need at most two instances of each $x{:}\Box A$ in the antecedent of a sequent: one to infer that A holds at x itself, and the other to infer that A holds at a new world z that is a successor of x (where we can exploit the $\Box$-disjunction property to 'choose' the appropriate z).

We can refine this to characterize further the form of the contracted lwffs. Consider again (10.4), for which we argued that, in general, the uppermost application of $\Box$L is not permutable over the rule preceding it since xRz might be the active rwff of an application of $\Box$R introducing a subformula of $x{:}A$, e.g. when A is $\sim\Box B$. However, if xRz is 'independent' from $x{:}A$ (e.g. when $xRz \in \Delta$ or xRz is the active rwff of an application of $\Box$R with principal formula $x{:}\Box E \in \Gamma'$), then we can permute the uppermost application of $\Box$L over the rule preceding it, and eventually eliminate the application of ClL.

This suggests that in a S(T)-proof of a sequent $\vdash x_1{:}D$, we need a contraction of $x{:}\Box A$ when the lwff $x{:}A$ that is inferred from the first constituent yields the rwff xRz active in the application of $\Box$L that has the second constituent as its principal formula. For example, when given two instances of $x{:}\Box\sim\Box B$ we use the first one to infer (reasoning backwards via reflexivity) that $x{:}\ \sim\Box B$, and then use this $x{:}\ \sim\Box B$ to generate a new world z that is a successor of x such that using the rwff xRz we can infer $z{:}\ \sim\Box B$ and $z{:}\ \sim B$. The contraction then 'amounts to' the backwards introduction of the rwff xRz. But this is not the only possibility as we might use a contraction of $x{:}\Box A$ to generate additional worlds accessible from x. For example, the lowest contraction in the above proof of theorem (10.1) duplicates $x_1{:}\Box^3\varphi$ to generate world x_3. In any case, it follows that we can expand on the previous intuition that we need to contract $x{:}\Box A$ at most once in each branch by requiring that A contains a negative subformula of the form $\Box B$, e.g. when A is $\sim\Box B$, $\Box B \supset C$, $C \supset\sim\Box B$, or even $\Box^p((C \supset\sim\Box B) \wedge D)$.

Formally, we have the following refinement of Lemma 10.1.1, where, by Definition 8.2.5, we graphically denote the condition on A by requiring that there is some B such that $\Box B \Subset_- A$ so that each contracted lwff has the form $x{:}\Box A[\![\Box B]\!]_-$.

Lemma 10.1.2 *Every sequent $\vdash x_1{:}D$ provable in* S(T) *has a proof in which there are no contractions, except for applications of* ClL *with principal formula of the form $x{:}\Box A[\![\Box B]\!]_-$. However,* ClL *need not be applied more than once with the same principal formula $x{:}\Box A[\![\Box B]\!]_-$ in each branch.* ∎

To illustrate this result, observe that all theorems in Table 10.1 require contractions of lwffs of the form $x{:}\Box A[\![\Box B]\!]_-$. For example, to prove the theorem (10.2) where C and D are propositional variables, we need two applications of ClL: the first one with principal formula $x{:}\Box A[\![\Box B]\!]_- = x_1{:}\Box\sim((\sim\Box\sim(C \supset \Box C) \supset D) \supset \Box D)$ where $\Box B$ is $\Box D$, and the second one with principal formula $x{:}\Box A[\![\Box B]\!]_- = x_2{:}\Box\sim(C \supset \Box C)$ where $\Box B$ is $\Box C$.

This lemma allows us to restrict our contraction rule, and we do so in Theorem 10.1.4 after having introduced additional terminology.

Definition 10.1.3 *Given a sequent $S = \Gamma, \Delta \vdash \Gamma'$, we define pbs$(S)$ and nbs(S) to be the number of positive and negative boxed subformulas of S, respectively, i.e. pbs$(S) = |\{\Box A \mid \Box A \Subset_+ S\}|$ and nbs$(S) = |\{\Box A \mid \Box A \Subset_- S\}|$. In other words, pbs$(S)$ and nbs(S) are the sizes of the multisets of positive and negative boxed subformulas of Γ and Γ'.* ∎

For example, if A, B and C are propositional variables and

$$S = \vdash x_1{:}\Box\sim\Box A \supset (\Box\Box\Box B \supset \Box A),$$

then *pbs*$(S) = |\{\Box\sim\Box A, \Box\Box\Box B, \Box\Box B, \Box B\}| = 4$ and *nbs*$(S) = |\{\Box A, \Box A\}| = 2$.

Observe now that the examples in Table 10.1 tell us that there are sequents $S = \vdash x_1{:}D$ that require at least *pbs*(S) applications of ClL in S(T). For example, to prove $x_1{:}D$ = (10.2) we need at least *pbs*$(\vdash x_1{:}D) = 2$ contractions. Using Definition 10.1.3, we can sharpen our results to show that every $S = \vdash x_1{:}D$ provable in S(T) has a proof in which each branch contains at most *pbs*(S) applications of ClL.

We denote this bound by annotating each sequent with a *contraction index* s, which we set to *pbs*(S) at the start of a backwards proof of S, and which tells us how many contractions we are allowed to perform in each branch of the proof from this point (i.e. sequent) on. Then we can restrict the rule ClL to be

$$\frac{x{:}\Box A[\![\Box B]\!]_-, x{:}\Box A[\![\Box B]\!]_-, \Gamma, \Delta \vdash^{s-1} \Gamma'}{x{:}\Box A[\![\Box B]\!]_-, \Gamma, \Delta \vdash^{s} \Gamma'} \text{ ClL}s \;,$$

which explicitly requires that the contraction index s of the conclusion is greater than 0. The index is decremented at every contraction and is imported in the premises of branching rules, e.g.

$$\frac{\Gamma, \Delta \vdash^{s} \Gamma', x{:}A \quad x{:}B, \Gamma, \Delta \vdash^{s} \Gamma'}{x{:}A \supset B, \Gamma, \Delta \vdash^{s} \Gamma'} \supset\text{L} \quad \text{and} \quad \frac{\Delta \vdash xRy \quad y{:}A, \Gamma, \Delta \vdash^{s} \Gamma'}{x{:}\Box A, \Gamma, \Delta \vdash^{s} \Gamma'} \Box\text{L} \;.$$

Note that we do not import s into the left premise $\Delta \vdash xRy$ of $\Box$L since we can eliminate contractions of rwffs by Lemma 8.2.4.

Then we have:

Theorem 10.1.4 *Every sequent $S = \vdash x_1{:}D$ provable in* S(T) *has a proof in which there are no contractions, except for applications of* ClL *with principal formula of the form $x_i{:}\Box A[\![\Box B]\!]_-$. However,* ClL *need not be applied more than pbs(S) times in each branch. Hence, we can restrict* ClL *to be* ClLs *with s set to pbs(S) at the start of a backwards proof, i.e. $\vdash^{pbs(\vdash x_1:D)} x_1{:}D$.* ■

Before giving a formal proof, we provide some intuition for this result. Given a branch of a S(T)-proof of $S = \vdash x_1{:}D$, we can apply Lemma 10.1.2 to eliminate superfluous contractions, retaining at most one for each lwff of the form $x_i{:}\Box A[\![\Box B]\!]_-$ in the branch. It is nevertheless possible that some of these contractions are still superfluous.

Recall that we argued above that we need at most two instances of each $x_i{:}\Box A[\![\Box B]\!]_-$ in each branch: one to infer that $A[\![\Box B]\!]_-$ holds at x_i and thus possibly generate a new world x_j with $j \geq i+1$ where B holds, and the other to infer that $A[\![\Box B]\!]_-$ also holds at x_{i+1}. In Lemma 10.1.2, we have formalized this by allowing at most one contraction for each $x_i{:}\Box A[\![\Box B]\!]_-$ in each branch. However, the lemma allows us also to perform one contraction for each of the subformulas of $x_i{:}\Box A[\![\Box B]\!]_-$, provided that they have the required form $x_j{:}\Box C[\![\Box E]\!]_-$. This yields exponentially, instead of linearly, many contractions on a branch.

For example, let $x_i{:}\Box A[\![\Box B]\!]_-$ be $x_i{:}\Box\Box C[\![\Box B]\!]_-$, where B is a *box-free formula* as in Definition 2.1.1, i.e. where B is a formula that does not contain any $\Box$. Then Lemma 10.1.2 allows us to perform 3 contractions, one of $x_i{:}\Box\Box C[\![\Box B]\!]_-$, and two of the subformula $\Box C[\![\Box B]\!]_-$ but labelled differently at each ClL (once with label x_j and once with label x_k), i.e.

$$\dfrac{\dfrac{\begin{array}{c}\Pi_1\\ \Delta \vdash x_i R x_j\end{array} \quad \dfrac{\dfrac{\begin{array}{c}\Pi_3\\ \Delta \vdash x_i R x_k\end{array} \quad \dfrac{\begin{array}{c}\Pi_4\\ x_k{:}\Box C[\![\Box B]\!]_-, x_k{:}\Box C[\![\Box B]\!]_-, \Gamma_1, \Delta_1 \vdash \Gamma_1'\end{array}}{x_k{:}\Box C[\![\Box B]\!]_-, \Gamma_1, \Delta_1 \vdash \Gamma_1'}\,\mathrm{ClL}}{\begin{array}{c}x_i{:}\Box\Box C[\![\Box B]\!]_-, \Gamma_1, \Delta_1 \vdash \Gamma_1'\\ \Pi_2\\ x_j{:}\Box C[\![\Box B]\!]_-, x_j{:}\Box C[\![\Box B]\!]_-, x_i{:}\Box\Box C[\![\Box B]\!]_-, \Gamma, \Delta \vdash \Gamma'\end{array}}\,\Box\mathrm{L}}{x_j{:}\Box C[\![\Box B]\!]_-, x_i{:}\Box\Box C[\![\Box B]\!]_-, \Gamma, \Delta \vdash \Gamma'}\,\mathrm{ClL}}{x_i{:}\Box\Box C[\![\Box B]\!]_-, x_i{:}\Box\Box C[\![\Box B]\!]_-, \Gamma, \Delta \vdash \Gamma'}\,\Box\mathrm{L}}{\begin{array}{c}x_i{:}\Box\Box C[\![\Box B]\!]_-, \Gamma, \Delta \vdash \Gamma'\\ \Pi_0\\ \vdash x_1{:}D\end{array}}\,\mathrm{ClL}$$

where j and k are either i or $i+1$ as this is a proof in S(T).

We can improve the results of the above lemmas and eliminate the superfluous contractions to retain only linearly many in each branch, where, intuitively, the superfluous contractions are those that produce more than two instances of some of the boxed subformulas of the contracted lwff. To illustrate this, let us first extend Definition 8.2.5 and after consider an example.

Definition 10.1.5 *We inductively define that a subformula B of A* occurs locally positive [negative] in *A, in symbols $B \Subset^l_+ A$ [$B \Subset^l_- A$], as follows:*

- *if $B = A$, then $B \Subset^l_+ A$;*
- *if $B \supset C \Subset^l_- A$, then $B \Subset^l_+ A$ and $C \Subset^l_- A$;*
- *if $B \supset C \Subset^l_+ A$, then $B \Subset^l_- A$ and $C \Subset^l_+ A$.*

We will also write $A[\![B]\!]^l_+$ to specify that $B \Subset^l_+ A$, and $A[\![B]\!]^l_-$ to specify that $B \Subset^l_- A$, and say that $x{:}B$ occurs locally positive [negative] in $x{:}A$ iff B occurs locally positive [negative] in A. ■

That is, B occurs locally positive [negative] in A if B is a local positive [negative] subformula of A, where "local" means that B is not in the scope of a $\Box$ in A. Thus, for example, if A is $\Box C$ and C is $\sim \Box B$ (so that A is $\Box \sim \Box B$), then $\Box B \Subset^l_- C$ and $\Box B \Subset_- A$ but $\Box B \not\Subset^l_- A$. Intuitively, we then have that a principal formula $x_i{:}\Box A[\![B]\!]^l_+$ of an application of $\Box$L with active rwff $x_i R x_j$ 'means' that $x_j{:}A[\![B]\!]^l_+$ occurs on the left of $\vdash$, and thus that, by applying only propositional rules, we can obtain $x_j{:}B$ on the left of $\vdash$ as well. Graphically, this amounts to

$$\dfrac{\Delta \vdash x_i R x_j \quad \begin{array}{c} x_j{:}B, \Gamma_1, \Delta_1 \vdash \Gamma_1' \\ \Pi \\ x_j{:}A[\![B]\!]^l_+, \Gamma, \Delta \vdash \Gamma' \end{array}}{x_i{:}\Box A[\![B]\!]^l_+, \Gamma, \Delta \vdash \Gamma'}\,\Box\mathrm{L}$$

where Π is obtained by applying only propositional rules with principal and active formulas labelled with x_j, inferring $x_j{:}B$ from $x_j{:}A[\![B]\!]^l_+$; hence, $\Delta_1 = \Delta$.

As an example, consider now the case when B is a box-free formula, and the contracted lwff has the form $x_i{:}\Box\Box C\llbracket\Box B\rrbracket^l_-$, i.e. A is $\Box C$ and $\Box B$ occurs locally negative in C. Then we have

$$\dfrac{\dfrac{\begin{array}{c}\Pi_1\\ \Delta\vdash x_i R\, x_j\end{array}\quad \begin{array}{c}\Pi_2\\ x_j{:}\Box C\llbracket\Box B\rrbracket^l_-, x_i{:}\Box\Box C\llbracket\Box B\rrbracket^l_-, \Gamma, \Delta\vdash\Gamma'\end{array}}{x_i{:}\Box\Box C\llbracket\Box B\rrbracket^l_-, x_i{:}\Box\Box C\llbracket\Box B\rrbracket^l_-, \Gamma, \Delta\vdash\Gamma'}\;\Box\mathrm{L}}{\begin{array}{c}x_i{:}\Box\Box C\llbracket\Box B\rrbracket^l_-, \Gamma, \Delta\vdash\Gamma'\\ \Pi_0\\ \vdash x_1{:}D\end{array}}\;\mathrm{ClL}\quad . \tag{10.5}$$

Suppose now that x_j is x_i (for the other alternative, x_{i+1}, we proceed similarly) and that, following the results above, we need the contraction to generate a successor of x_i where both B and $\Box B$ hold. That is, suppose that we now contract $x_i{:}\Box C\llbracket\Box B\rrbracket^l_-$ to produce the second instance that we need, and after that use one instance to generate x_{i+1}, so that (10.5) has the form

$$\dfrac{\dfrac{\begin{array}{c}\overline{\vdash x_i R\, x_i}\;\mathit{refl}\\ \vdots\;\mathrm{WrL}\\ \Delta\vdash x_i R\, x_i\end{array}\quad \dfrac{\dfrac{\begin{array}{c}\overline{\vdash x_i R\, x_i}\;\mathit{refl}\\ \vdots\;\mathrm{WrL}\\ \Delta\vdash x_i R\, x_i\end{array}\quad \begin{array}{c}\dfrac{\begin{array}{c}\Pi_4\\ x_i{:}\Box C\llbracket\Box B\rrbracket^l_-, x_i{:}\Box\Box C\llbracket\Box B\rrbracket^l_-,\\ \Gamma_1, \Delta_1, x_i R\, x_{i+1}\vdash\Gamma_1', x_{i+1}{:}B\end{array}}{\begin{array}{c}x_i{:}\Box C\llbracket\Box B\rrbracket^l_-, x_i{:}\Box\Box C\llbracket\Box B\rrbracket^l_-,\\ \Gamma_1, \Delta_1\vdash\Gamma_1', x_i{:}\Box B\end{array}}\;\Box\mathrm{R}\\ \Pi_3\\ x_i{:}C\llbracket\Box B\rrbracket^l_-, x_i{:}\Box C\llbracket\Box B\rrbracket^l_-,\\ x_i{:}\Box\Box C\llbracket\Box B\rrbracket^l_-, \Gamma, \Delta\vdash\Gamma'\end{array}}{\begin{array}{c}x_i{:}\Box C\llbracket\Box B\rrbracket^l_-, x_i{:}\Box C\llbracket\Box B\rrbracket^l_-,\\ x_i{:}\Box\Box C\llbracket\Box B\rrbracket^l_-, \Gamma, \Delta\vdash\Gamma'\end{array}}\;\Box\mathrm{L}}{x_i{:}\Box C\llbracket\Box B\rrbracket^l_-, x_i{:}\Box\Box C\llbracket\Box B\rrbracket^l_-, \Gamma, \Delta\vdash\Gamma'}\;\mathrm{ClL}}{x_i{:}\Box\Box C\llbracket\Box B\rrbracket^l_-, x_i{:}\Box\Box C\llbracket\Box B\rrbracket^l_-, \Gamma, \Delta\vdash\Gamma'}\;\Box\mathrm{L}}{\begin{array}{c}x_i{:}\Box\Box C\llbracket\Box B\rrbracket^l_-, \Gamma, \Delta\vdash\Gamma'\\ \Pi_0\\ \vdash x_1{:}D\end{array}}\;\mathrm{ClL} \tag{10.6}$$

where Π_3 consists only of propositional rules, inferring $x_i{:}\Box B$ on the right of $\vdash$ from $x_i{:}C\llbracket\Box B\rrbracket^l_-$ on the left.

It is possible that one of the two contractions in (10.6) is superfluous, as we can see by considering the subproof Π_4. Since we have already eliminated the contractions of lwffs that are weakened later (i.e. above) in the branch, $x_i{:}\Box\Box C\llbracket\Box B\rrbracket^l_-$ must be introduced in Π_4 by an application of $\Box$L, and we distinguish two cases depending on the active rwff of this $\Box$L, namely $x_i R\, x_i$ or $x_i R\, x_{i+1}$. In both cases, the permutability of the rules allows us to assume that this application of $\Box$L is the last rule in Π_4.

In the first case, the application of $\Box$L has active rwff $x_i R\, x_i$ and active lwff $x_i{:}\Box C\llbracket\Box B\rrbracket^l_-$. But this means that we have produced a third instance of $x_i{:}\Box C\llbracket\Box B\rrbracket^l_-$, which we know to be redundant, so that we can eliminate one of the two contractions in (10.6). (In fact, we can permute downwards the $\Box$L introducing $x_i{:}\Box C\llbracket\Box B\rrbracket^l_-$, so that the contraction is eliminable by Lemma 10.1.2.)

In the second case, Π_4 has the form

$$\dfrac{\dfrac{\dfrac{}{x_i R\, x_{i+1} \vdash x_i R\, x_{i+1}}\,\mathrm{AXr}}{\Delta_1, x_i R\, x_{i+1} \vdash x_i R\, x_{i+1}}\,\vdots\,\mathrm{WrL} \qquad \begin{array}{c}\Pi_5 \\ x_i{:}\Box C[\![\Box B]\!]^l_-, x_{i+1}{:}\Box C[\![\Box B]\!]^l_-, \\ \Gamma_1, \Delta_1, x_i R\, x_{i+1} \vdash \Gamma_1', x_{i+1}{:}B\end{array}}{x_i{:}\Box C[\![\Box B]\!]^l_-, x_i{:}\Box\Box C[\![\Box B]\!]^l_-, \Gamma_1, \Delta_1, x_i R\, x_{i+1} \vdash \Gamma_1', x_{i+1}{:}B}\,\Box\mathrm{L}$$

In this case we cannot immediately eliminate one of the two applications of ClL in (10.6). Consider however the two lwffs $x_i{:}\Box C[\![\Box B]\!]^l_-$ and $x_{i+1}{:}\Box C[\![\Box B]\!]^l_-$. As we have already eliminated the contractions of lwffs that are weakened later in the proof (branch), both formulas must be introduced by $\Box$L in Π_5. Reason now on the active rwffs of these applications of $\Box$L. Looking at (10.6), we see that we have already inferred that $C[\![\Box B]\!]^l_-$ holds at x_i. Hence, if the active rwff of the $\Box$L introducing $x_i{:}\Box C[\![\Box B]\!]^l_-$ is $x_i R\, x_i$, then we produce a second, redundant, instance of $x_i{:}C[\![\Box B]\!]^l_-$, so that we can eliminate the uppermost contraction in (10.6). (As above, we can permute downwards the $\Box$L introducing $x_i{:}\Box C[\![\Box B]\!]^l_-$, so that this contraction is eliminable by Lemma 10.1.2.) This implies that $x_i{:}\Box C[\![\Box B]\!]^l_-$ must be introduced by an application of $\Box$L with active rwff $x_i R\, x_{i+1}$ and active lwff $x_{i+1}{:}C[\![\Box B]\!]^l_-$. So what about $x_{i+1}{:}\Box C[\![\Box B]\!]^l_-$? We have just argued that $C[\![\Box B]\!]^l_-$ holds at x_{i+1}. Thus we can argue similarly as above, possibly permuting rules in Π_5 appropriately, to conclude that the application of $\Box$L introducing $x_{i+1}{:}\Box C[\![\Box B]\!]^l_-$ must have active lwff $x_{i+2}{:}C[\![\Box B]\!]^l_-$. Note that we can then also replace the uppermost left contraction of $x_i{:}\Box C[\![\Box B]\!]^l_-$ with a left contraction of $x_{i+1}{:}\Box C[\![\Box B]\!]^l_-$. In any case, we see that given $x_i{:}\Box\Box C[\![\Box B]\!]^l_-$, we need at most two contractions, one for each of its two positive boxed subformulas $\Box\Box C[\![\Box B]\!]^l_-$ and $\Box C[\![\Box B]\!]^l_-$.

We are not yet done as it might be the case that B is not a box-free formula, but has subformulas that also require contractions. This is for example the case in the proof of

$$\vdash x_1{:} \sim\Box \sim(D \supset \Box \sim(D \supset \Box \sim(C \supset \Box C)))$$

where B has the form $\sim(D \supset \Box \sim(C \supset \Box C))$.

We can iterate the above argument to eliminate superfluous contractions of subformulas of B. Observe however that the above transformations yield two instances of B on the right of $\vdash$, one labelled with x_{i+1} and the other with x_{i+2}. These two instances of B, albeit labelled differently, might produce exponentially many contractions of subformulas of B. To avoid this, in Lemma 10.1.6 below, we show that we can transform the proof so that only one instance of B is needed in each branch.

A generalization of this informal argument proves Theorem 10.1.4.

Proof [of Theorem 10.1.4] (Sketch) Consider a branch $\mathcal{B}$ in a S(T)-proof of $S = \vdash x_1{:}D$, and assume that we have eliminated all contractions that are trivially superfluous because one of the two constituents is introduced by weakening in all subbranches of $\mathcal{B}$. We show that if $\mathcal{B}$ contains $pbs(S)+n$ applications of ClL with principal formula of the form $x_i{:}\Box A[\![\Box B]\!]_-$, where $n > 0$, then we can transform it to a branch $\mathcal{B}'$ containing at most $pbs(S)$ such contractions. To this end, we consider each of the $pbs(S) + n$ contractions in turn, starting from the lowest one in $\mathcal{B}$, and for each of them we proceed with a preliminary analysis in which we eliminate unnecessary contractions according to Lemma 10.1.2. In other words, we exploit the lemma to show that we need not contract each lwff $x_i{:}\Box A[\![\Box B]\!]_-$ more than once in each branch.

To eliminate the remaining superfluous contractions, if any, we proceed as follows. Let, for example, the branch have the form

$$
\dfrac{\begin{array}{c}\vdots\\ \Delta \vdash x_i R\, x_j\end{array} \qquad \begin{array}{c}\Pi_3\\ \dfrac{x_j{:}\Box C[\![\Box B]\!]_-, x_i{:}\Box A[\![\Box C[\![\Box B]\!]_-]\!]^l_+, \Gamma_2, \Delta_2, x_l R\, x_{l+1} \vdash \Gamma_2', x_{l+1}{:}B}{x_j{:}\Box C[\![\Box B]\!]_-, x_i{:}\Box A[\![\Box C[\![\Box B]\!]_-]\!]^l_+, \Gamma_2, \Delta_2 \vdash \Gamma_2', x_l{:}\Box B}\,\Box R\\ \Pi_2\\ \dfrac{x_j{:}\Box C[\![\Box B]\!]_-, x_j{:}\Box C[\![\Box B]\!]_-, x_i{:}\Box A[\![\Box C[\![\Box B]\!]_-]\!]^l_+, \Gamma_1, \Delta_1 \vdash \Gamma_1'}{x_j{:}\Box C[\![\Box B]\!]_-, x_i{:}\Box A[\![\Box C[\![\Box B]\!]_-]\!]^l_+, \Gamma_1, \Delta_1 \vdash \Gamma_1'}\,\mathrm{ClL}\\ \Pi_1\\ x_j{:}A[\![\Box C[\![\Box B]\!]_-]\!]^l_+, x_i{:}\Box A[\![\Box C[\![\Box B]\!]_-]\!]^l_+, \Gamma, \Delta \vdash \Gamma'\end{array}}{\dfrac{x_i{:}\Box A[\![\Box C[\![\Box B]\!]_-]\!]^l_+, x_i{:}\Box A[\![\Box C[\![\Box B]\!]_-]\!]^l_+, \Gamma, \Delta \vdash \Gamma'}{\begin{array}{c}x_i{:}\Box A[\![\Box C[\![\Box B]\!]_-]\!]^l_+, \Gamma, \Delta \vdash \Gamma'\\ \Pi_0\\ \vdash x_1{:}D\end{array}}\,\mathrm{ClL}}\,\Box L \tag{10.7}
$$

where $\Box C$ occurs locally positive in A; $\Box B$ is the outermost boxed formula that occurs negative in A and C; x_j is either x_i or x_{i+1}; Π_1 consists only of applications of propositional rules inferring $x_j{:}\Box C[\![\Box B]\!]_-$ from $x_j{:}A[\![\Box C[\![\Box B]\!]_-]\!]^l_+$, both on the left of $\vdash$; and in Π_2 we infer $x_l{:}\Box B$ on the right of $\vdash$ from $x_j{:}\Box C[\![\Box B]\!]_-$ on the left.

Let us now reason on the introduction of $x_i{:}\Box A[\![\Box C[\![\Box B]\!]_-]\!]^l_+$ in Π_3. Since we have already applied Lemma 10.1.2, this lwff cannot be introduced by an application of WlL or ClL. It must be introduced by an application of $\Box$L, which by permutability we can assume to be the last rule application in Π_3 , i.e.

$$
\dfrac{\Delta_2, x_l R\, x_{l+1} \vdash x_i R\, x_k \qquad \begin{array}{c}\Pi_4\\ x_j{:}\Box C[\![\Box B]\!]_-, x_k{:}A[\![\Box C[\![\Box B]\!]_-]\!]^l_+,\\ \Gamma_2, \Delta_2, x_l R\, x_{l+1} \vdash \Gamma_2', x_{l+1}{:}B\end{array}}{x_j{:}\Box C[\![\Box B]\!]_-, x_i{:}\Box A[\![\Box C[\![\Box B]\!]_-]\!]^l_+, \Gamma_2, \Delta_2, x_l R\, x_{l+1} \vdash \Gamma_2', x_{l+1}{:}B}\,\Box L
$$

where either x_k is one of x_i or x_{i+1}, or x_k and x_{i+1} diverge from x_i. We now distinguish five subcases, depending on x_j and x_k; the fifth subcase occurs when $x_j = x_{i+1}$ and x_k diverge from x_i.

(*Case 1*) In the first case, $x_j = x_i = x_k$, and we can eliminate one of the two contractions in (10.7), for either $x_i{:}A[\![\Box C[\![\Box B]\!]_-]\!]^l_+$ is weak in Π_4, or Π_4 contains rules inferring $x_i{:}\Box C[\![\Box B]\!]_-$ from $x_i{:}A[\![\Box C[\![\Box B]\!]_-]\!]^l_+$, both on the left of $\vdash$. Then we can permute these rules downwards and transform Π_3 to

$$\dfrac{\dfrac{}{\vdash x_i R\, x_i}\,refl \quad \begin{array}{c}\vdots\ \text{WrL}\\ \Delta_2, x_l R\, x_{l+1} \vdash x_i R\, x_i\end{array} \qquad \begin{array}{c}\Pi_6\\ x_i{:}\Box C[\![\Box B]\!]_-, x_i{:}\Box C[\![\Box B]\!]_-,\\ \Gamma_3, \Delta_2, x_l R\, x_{l+1} \vdash \Gamma_3', x_{l+1}{:}B\\ \Pi_5\\ x_i{:}\Box C[\![\Box B]\!]_-, x_i{:}A[\![\Box C[\![\Box B]\!]_-]\!]^l_+,\\ \Gamma_2, \Delta_2, x_l R\, x_{l+1} \vdash \Gamma_2', x_{l+1}{:}B\end{array}}{x_i{:}\Box C[\![\Box B]\!]_-, x_i{:}\Box A[\![\Box C[\![\Box B]\!]_-]\!]^l_+, \Gamma_2, \Delta_2, x_l R\, x_{l+1} \vdash \Gamma_2', x_{l+1}{:}B}\,\Box\text{L}$$

where Π_5 consists only of applications of propositional rules inferring $x_i{:}\Box C[\![\Box B]\!]_-$ from $x_i{:}A[\![\Box C[\![\Box B]\!]_-]\!]^l_+$, both on the left of $\vdash$. But this means that we have a third instance of $x_i{:}\Box C[\![\Box B]\!]_-$, which we know to be redundant. In fact, in this case we can further permute rules and transform the proof so that one of the two displayed contractions is eliminable by Lemma 10.1.2. Note also that, as in the example above, there are alternatives as to which contractions we eliminate.

(*Case 2*) In the second case, $x_j = x_i$ and $x_k = x_{i+1}$. If $x_{i+1}{:}A[\![\Box C[\![\Box B]\!]_-]\!]^l_+$ is weak in Π_4, then we can eliminate the contraction of $x_i{:}\Box A[\![\Box C[\![\Box B]\!]_-]\!]^l_+$ in (10.7). Let therefore Π_4 contain rules inferring $x_{i+1}{:}\Box C[\![\Box B]\!]_-$ from $x_{i+1}{:}A[\![\Box C[\![\Box B]\!]_-]\!]^l_+$, both on the left of $\vdash$. We can permute these rules downwards and transform Π_3 to

$$\dfrac{\dfrac{}{x_i R\, x_{i+1} \vdash x_i R\, x_{i+1}}\,\text{AXr} \quad \begin{array}{c}\vdots\ \text{WrL}\\ \Delta_2, x_l R\, x_{l+1} \vdash x_i R\, x_{i+1}\end{array} \qquad \begin{array}{c}\Pi_6\\ x_i{:}\Box C[\![\Box B]\!]_-, x_{i+1}{:}\Box C[\![\Box B]\!]_-,\\ \Gamma_3, \Delta_2, x_l R\, x_{l+1} \vdash \Gamma_3', x_{l+1}{:}B\\ \Pi_5\\ x_i{:}\Box C[\![\Box B]\!]_-, x_{i+1}{:}A[\![\Box C[\![\Box B]\!]_-]\!]^l_+,\\ \Gamma_2, \Delta_2, x_l R\, x_{l+1} \vdash \Gamma_2', x_{l+1}{:}B\end{array}}{x_i{:}\Box C[\![\Box B]\!]_-, x_i{:}\Box A[\![\Box C[\![\Box B]\!]_-]\!]^l_+, \Gamma_2, \Delta_2, x_l R\, x_{l+1} \vdash \Gamma_2', x_{l+1}{:}B}\,\Box\text{L}$$

where $x_i R\, x_{i+1}$ must occur in Δ_2 and Π_5 consists only of propositional rules. Consider now the two lwffs $x_i{:}\Box C[\![\Box B]\!]_-$ and $x_{i+1}{:}\Box C[\![\Box B]\!]_-$. As we have already eliminated the contractions of lwffs that are weakened later in the proof (branch), both formulas must be introduced by $\Box$L in Π_6. Reason now on the active rwffs of these applications of $\Box$L. Looking at (10.7), we see that we have already inferred that $C[\![\Box B]\!]_-$ holds at x_i. Hence, if the active rwff of the application of $\Box$L introducing $x_i{:}\Box C[\![\Box B]\!]_-$ is $x_i R\, x_i$, then we produce a second, possibly redundant, instance of $x_i{:}C[\![\Box B]\!]_-$, so that we can either eliminate the uppermost contraction in (10.7), or replace it with a contraction of (some subformula of) $x_i{:}C[\![\Box B]\!]_-$ (if, e.g., $\Box E[\![\Box B]\!]_-$ occurs locally positive in C).

This implies that $x_i{:}\Box C[\![\Box B]\!]_-$ must be introduced by an application of $\Box$L with active rwff $x_i R\, x_{i+1}$ and active lwff $x_{i+1}{:}C[\![\Box B]\!]_-$. So what about $x_{i+1}{:}\Box C[\![\Box B]\!]_-$? We have just argued that $C[\![\Box B]\!]_-$ holds at x_{i+1}. Thus we can argue similarly as above, possibly permuting rules in Π_6 appropriately, to conclude that the application of $\Box$L introducing $x_{i+1}{:}\Box C[\![\Box B]\!]_-$ must have active lwff $x_{i+2}{:}C[\![\Box B]\!]_-$.

(*Case 3*) In the third case, $x_j = x_{i+1}$ and $x_k = x_i$. Then we can permute rules so that we can conclude like in case 2.

(*Case 4*) In the fourth case, $x_j = x_{i+1} = x_k$. Then we introduce a third instance of $x_{i+1}{:}C[\![\Box B]\!]_-$, and we can conclude like in case 1.

(*Case 5*) In the fifth and last case, $x_j = x_{i+1}$ and x_k diverge from x_i. Then $x_j{:}\Box C[\![\Box B]\!]_-$ and $x_k{:}A[\![\Box C[\![\Box B]\!]_-]\!]^l_+$ are independent, and we can conclude easily.

This concludes the case analysis. We can now iterate these arguments for all other contractions of positive boxed formulas that occur in C (and thus in A) and that contain $\Box B$; for example for $\Box E$ where $x_i{:}\Box A[\![\Box C[\![\Box E[\![\Box B]\!]_-]\!]^l_+]\!]^l_+$. It then follows that we can transform (10.7) so that in each branch it contains at most one contraction for each of the positive boxed subformulas of $x_i{:}\Box A[\![\Box C[\![\Box B]\!]_-]\!]^l_+$.

Reason now on B, as the above transformations yield two instances of B on the right of $\vdash$, but labelled differently. For example, one instance of B is labelled with x_{i+1} and the other with x_{i+2}, and the proof has the form

$$\begin{array}{c} \Pi_2 \\ \dfrac{\Gamma_1, \Delta_1, x_{i+1} R\, x_{i+2} \vdash \Gamma_1', x_{i+1}{:}B, x_{i+2}{:}B}{\Gamma_1, \Delta_1 \vdash \Gamma_1', x_{i+1}{:}B, x_{i+1}{:}\Box B}\ \Box\mathrm{R} \\ \Pi_1 \\ \Gamma, \Delta \vdash \Gamma', x_{i+1}{:}B, x_{i+1}{:}\Box B \\ \vdots \\ \vdash x_1{:}D \end{array} \qquad (10.8)$$

The problem that we have to tackle is that these two instances of B, albeit labelled differently, might produce exponentially many contractions of subformulas of B.[2] We proceed as follows. We first permute rules and transform (10.8) to

$$\begin{array}{c} \Pi_3 \\ \dfrac{\Gamma_2, \Delta_2, x_{i+1} R\, x_{i+2} \vdash \Gamma_2', x_{i+1}{:}B, x_{i+2}{:}B}{\Gamma_2, \Delta_2 \vdash \Gamma_2', x_{i+1}{:}B, x_{i+1}{:}\Box B}\ \Box\mathrm{R} \\ \vdots \\ \vdash x_1{:}D \end{array}$$

where Γ_2 and Γ_2' are not x_{i+1}-branching with respect to Δ_2, and x_{i+2} does not occur in Γ_2, Δ_2 or Γ_2'. We can then apply Lemma 10.1.6 below, which tells us that given Π_3 there is a proof Π_3' of either $\Gamma_2, \Delta_2 \vdash \Gamma_2', x_{i+1}{:}B$ or $\Gamma_2, \Delta_2, x_{i+1} R\, x_{i+2} \vdash \Gamma_2', x_{i+2}{:}B$. That is, there is a proof of either $\Gamma_2, \Delta_2 \vdash \Gamma_2', x_{i+1}{:}B$ or $\Gamma_2, \Delta_2 \vdash \Gamma_2', x_{i+1}{:}\Box B$. We can thus iterate the above argument and eliminate superfluous contractions of subformulas of B.

It then follows that we can transform each S(T)-proof of $S = \vdash x_1{:}D$ so that in each branch we need at most one contraction for each boxed positive subformula of D, and thus at most $pbs(S)$ applications of ClL in each branch. ■

[2]This problem does not occur when $x_{i+1}{:}\Box B$ is not introduced by the application of $\Box$R displayed in (10.8), but by an application of WlR in all branches originating from $\Gamma, \Delta \vdash \Gamma', x_{i+1}{:}B, x_{i+1}{:}\Box B$. Then we can simply permute downwards this weakening so that we have a proof of $\Gamma, \Delta \vdash \Gamma', x_{i+1}{:}B$. We proceed analogously when $x_{i+1}{:}B$ is introduced by an application of WlR in all branches originating from $\Gamma, \Delta \vdash \Gamma', x_{i+1}{:}B, x_{i+1}{:}\Box B$.

Lemma 10.1.6 *If there is a* S(T)*-proof of* $\Gamma, \Delta \vdash \Gamma', x_i{:}A, x_i{:}\Box A$, *where* Γ *and* Γ' *are not* x_i*-branching with respect to* Δ, *then there is a* S(T)*-proof of either* $\Gamma, \Delta \vdash \Gamma', x_i{:}A$ *or* $\Gamma, \Delta \vdash \Gamma', x_i{:}\Box A$.

Proof Let Π be a S(T)-proof of $\Gamma, \Delta \vdash \Gamma', x_i{:}A, x_i{:}\Box A$, where Γ and Γ' are not x_i-branching with respect to Δ. We begin with two observations. If one of $x_i{:}A$ and $x_i{:}\Box A$ is introduced by weakening in all branches of Π, then we conclude trivially. If one of $x_i{:}A$ and $x_i{:}\Box A$ is introduced by WlL in some branches and by its proper logical rule in other branches, then we make the mode of introduction of the lwff uniform by replacing all its introductions by weakening with introductions by the proper logical rule. We proceed by induction on the structure of Π and distinguish three cases.

(*Case 1*) If the last rule application in Π has principal formula different from $x_i{:}A$ and $x_i{:}\Box A$, then we conclude by applying the induction hypothesis.

(*Case 2*) If the last rule application in Π has principal formula $x_i{:}\Box A$, then Π has the form

$$\dfrac{\begin{array}{c}\Pi_1\\ \Gamma, \Delta, x_i R\, x_{i+1} \vdash \Gamma', x_i{:}A, x_{i+1}{:}A\end{array}}{\Gamma, \Delta \vdash \Gamma', x_i{:}A, x_i{:}\Box A}\ \Box\text{R}$$

and we proceed by induction on the structure (grade) of A to show that there is a proof of either $\Gamma, \Delta \vdash \Gamma', x_i{:}A$ or $\Gamma, \Delta, x_i R\, x_{i+1} \vdash \Gamma', x_{i+1}{:}A$.

(*Case 2.1*) If A is a propositional variable, then, since Γ and Γ' are not x_i-branching with respect to Δ_1, at least one of $x_i{:}A$ and $x_{i+1}{:}A$ is weak in Π_1.

(*Case 2.2*) If A has the form $B \supset C$, then we can permute downwards first the rules introducing $x_i{:}A$ and $x_{i+1}{:}A$, and then the x_{i+1}-branching rules, so that Π_1 has the form

$$\dfrac{\dfrac{\begin{array}{c}\Pi_3\\ x_i{:}B, x_{i+1}{:}B, \Gamma_1, \Delta_1, x_i R\, x_{i+1} \vdash \Gamma_1', x_i{:}C, x_{i+1}{:}C\\ \Pi_2\\ x_i{:}B, x_{i+1}{:}B, \Gamma, \Delta, x_i R\, x_{i+1} \vdash \Gamma', x_i{:}C, x_{i+1}{:}C\end{array}}{x_i{:}B, \Gamma, \Delta, x_i R\, x_{i+1} \vdash \Gamma', x_i{:}C, x_{i+1}{:}B \supset C}\ \supset\text{R}}{\Gamma, \Delta, x_i R\, x_{i+1} \vdash \Gamma', x_i{:}B \supset C, x_{i+1}{:}B \supset C}\ \supset\text{R}$$

where Π_3 is such that Γ_1 and Γ_1' are not x_{i+1}-branching with respect to Δ_1. Nor are they x_i-branching (by the initial assumption). We then proceed by induction on the structure of B and apply the induction hypothesis on C to obtain a proof Π_3' of either $x_i{:}B, \Gamma_1, \Delta_1 \vdash \Gamma_1', x_i{:}C$ or $x_{i+1}{:}B, \Gamma_1, \Delta_1, x_i R\, x_{i+1} \vdash \Gamma_1', x_{i+1}{:}C$.

(*Case 2.2.1*) If B is a propositional variable, then at least one of $x_i{:}B$ and $x_{i+1}{:}B$ is weak in Π_3.

(*Case 2.2.2*) If B has the form $E \supset F$, then we conclude by applying the induction hypotheses.

(*Case 2.2.3*) If B has the form $\Box E$, then Π_3 has the form

$$\dfrac{\begin{array}{c}\Pi_4\\ \Delta_1, x_i R x_{i+1} \vdash x_i R x_j\end{array} \quad \begin{array}{c}\dfrac{\begin{array}{c}\Pi_6\\ \Delta_2 \vdash x_{i+1} R x_k\end{array} \quad \begin{array}{c}\Pi_7\\ x_k{:}E, \Gamma_2, \Delta_2 \vdash \Gamma_2'\end{array}}{x_{i+1}{:}\Box E, \Gamma_2, \Delta_2 \vdash \Gamma_2'}\,\Box\mathrm{L}\\ \Pi_5\\ x_j{:}E, x_{i+1}{:}\Box E, \Gamma_1, \Delta_1, x_i R x_{i+1} \vdash \Gamma_1', x_i{:}C, x_{i+1}{:}C\end{array}}{x_i{:}\Box E, x_{i+1}{:}\Box E, \Gamma_1, \Delta_1, x_i R x_{i+1} \vdash \Gamma_1', x_i{:}C, x_{i+1}{:}C}\,\Box\mathrm{L}\ .$$

Since x_j is a successor of x_i and x_k is a successor of x_{i+1}, we distinguish five subcases depending on x_j and x_k (the fifth one occurs when x_j and x_{i+1} diverge from x_i).

(*Case 2.2.3.1*) If $x_j = x_i$ and $x_k = x_{i+1}$, then we can permute rules so that Π_3 has the form

$$\dfrac{\begin{array}{c}\dfrac{}{\vdash x_i R x_i}\,\mathrm{AXr}\\ \vdots\ \mathrm{WrL}\\ \Delta_1, x_i R x_{i+1} \vdash x_i R x_i\end{array} \quad \dfrac{\begin{array}{c}\dfrac{}{\vdash x_{i+1} R x_{i+1}}\,\mathrm{AXr}\\ \vdots\ \mathrm{WrL}\\ \Delta_1 \vdash x_{i+1} R x_{i+1}\end{array} \quad \begin{array}{c}\Pi_7\\ x_{i+1}{:}E, \Gamma_2, \Delta_2 \vdash \Gamma_2'\\ \Pi_5^\dagger\\ x_i{:}E, x_{i+1}{:}E, \Gamma_1, \Delta_1, x_i R x_{i+1}\\ \vdash \Gamma_1', x_i{:}C, x_{i+1}{:}C\end{array}}{x_i{:}E, x_{i+1}{:}\Box E, \Gamma_1, \Delta_1, x_i R x_{i+1} \vdash \Gamma_1', x_i{:}C, x_{i+1}{:}C}\,\Box\mathrm{L}}{x_i{:}\Box E, x_{i+1}{:}\Box E, \Gamma_1, \Delta_1, x_i R x_{i+1} \vdash \Gamma_1', x_i{:}C, x_{i+1}{:}C}\,\Box\mathrm{L}$$

and we conclude by applying the induction hypotheses.

(*Case 2.2.3.2*) If $x_j = x_i$ and $x_k = x_{i+2}$, then we conclude easily since we can transform the proof (branch) so that at least one of $x_i{:}E$ and $x_{i+2}{:}E$ is weak.

(*Case 2.2.3.3*) If $x_j = x_{i+1} = x_k$, then we can permute rules and contract $x_{i+1}{:}\Box E$ (or $x_i{:}\Box E$; note that the contraction is then of course eliminable if no $\Box F$ occurs negative in E). That is, we can for example transform Π_3 to

$$\dfrac{\dfrac{\begin{array}{c}\dfrac{}{\vdash x_{i+1} R x_{i+1}}\,\mathit{refl}\\ \vdots\ \mathrm{WrL}\\ \Delta_1, x_i R x_{i+1} \vdash x_{i+1} R x_{i+1}\end{array} \quad \dfrac{\begin{array}{c}\dfrac{}{\vdash x_{i+1} R x_{i+1}}\,\mathit{refl}\\ \vdots\ \mathrm{WrL}\\ \Delta_1, x_i R x_{i+1} \vdash x_{i+1} R x_{i+1}\end{array} \quad \begin{array}{c}\Pi_7\\ x_{i+1}{:}E, \Gamma_2, \Delta_2 \vdash \Gamma_2'\\ \Pi_5^\dagger\\ x_{i+1}{:}E, x_{i+1}{:}E, \Gamma_1,\\ \Delta_1, x_i R x_{i+1}\\ \vdash \Gamma_1', x_i{:}C, x_{i+1}{:}C\end{array}}{\begin{array}{c}x_{i+1}{:}E, x_{i+1}{:}\Box E, \Gamma_1, \Delta_1,\\ x_i R x_{i+1} \vdash \Gamma_1', x_i{:}C, x_{i+1}{:}C\end{array}}\,\Box\mathrm{L}}{x_{i+1}{:}\Box E, x_{i+1}{:}\Box E, \Gamma_1, \Delta_1, x_i R x_{i+1} \vdash \Gamma_1', x_i{:}C, x_{i+1}{:}C}\,\Box\mathrm{L}}{x_{i+1}{:}\Box E, \Gamma_1, \Delta_1, x_i R x_{i+1} \vdash \Gamma_1', x_i{:}C, x_{i+1}{:}C}\,\mathrm{ClL}\ .$$

We then conclude by applying the induction hypotheses.

(*Case 2.2.3.4*) If $x_j = x_{i+1}$ and $x_k = x_{i+2}$, then we can similarly conclude by a contraction of $x_{i+1}{:}\Box E$.

(*Case 2.2.3.5*) If x_j and x_{i+1} diverge from x_i, then $x_j{:}E$ and $x_k{:}E$ are independent, and we can conclude easily.

(*Case 2.3*) If A has the form $\Box B$, then we conclude by the $\Box$-disjunction property (Proposition 8.2.9 and its corollaries).

(*Case 3*) If the last rule application in Π has principal formula $x_i{:}A$, then we proceed by induction on the structure (grade) of A, where A cannot be a propositional variable, as we have already considered the case where $x_i{:}A$ is introduced by weakening.

(*Case 3.1*) If A has the form $B \supset C$, then we conclude like in case 2.2.

(*Case 3.2*) If A has the form $\Box B$, then we again conclude by the $\Box$-disjunction property (Proposition 8.2.9 and its corollaries). ■

10.2 S(T) AND SS(T)

Like for S(K), our substructural analysis of S(T) provides a proof-theoretical justification of the rules of SS(T). Moreover, we can propagate our results to give a refined version of SS(T).

We follow the development for S(K) in §9.2, and, before defining an intermediate system $\widehat{\mathrm{S}}(\mathrm{T})$, which we relate to both S(T) and SS(T), we first derive a labelled equivalent $\Box\mathrm{L_T}$ of

$$\frac{A, \Box A, \Sigma \vdash \Sigma'}{\Box A, \Sigma \vdash \Sigma'}\ (\mathrm{T})$$

as follows:

$$\frac{x{:}A, x{:}\Box A, x{:}\Sigma \vdash x{:}\Sigma'}{x{:}\Box A, x{:}\Sigma \vdash x{:}\Sigma'}\ \Box\mathrm{L_T} \quad \rightsquigarrow \quad \dfrac{\dfrac{\overline{\vdash xRx}\ \mathit{refl} \qquad x{:}A, x{:}\Box A, x{:}\Sigma \vdash x{:}\Sigma'}{x{:}\Box A, x{:}\Box A, x{:}\Sigma \vdash x{:}\Sigma'}\ \Box\mathrm{L}}{x{:}\Box A, x{:}\Sigma \vdash x{:}\Sigma'}\ \mathrm{ClL}$$

where the multisets of lwffs $x{:}\Sigma$ and $x{:}\Sigma'$ contain only formulas labelled with x. Thus, $\Box\mathrm{L_T}$ is a *local* rule: the principal, active and parametric formulas all have the same label, so that, like in (T), the premise and the conclusion represent the same world.

Let $\widehat{\mathrm{S}}(\mathrm{T})$ be the system obtained by extending $\widehat{\mathrm{S}}(\mathrm{K})$ with $\Box\mathrm{L_T}$. Then, $\widehat{\mathrm{S}}(\mathrm{T})$ and SS(T), like $\widehat{\mathrm{S}}(\mathrm{K})$ and SS(K), do not contain structural rules, since the unavoidable left contractions of formulas of the form $\Box A$ are embedded in $\Box\mathrm{L_T}$ and (T). We show the equivalence of S(T), $\widehat{\mathrm{S}}(\mathrm{T})$ and SS(T) by transforming S(T)-proofs into a *block form*. As before, this is achieved by eliminating detours and adjoining related rules, and there are only a few minor changes with respect to our development for S(K).

As for S(K) and $\widehat{\mathrm{S}}(\mathrm{K})$, the right-to-left direction of the first such equivalence, i.e. if there is a $\widehat{\mathrm{S}}(\mathrm{T})$-proof of $\vdash x_1{:}D$ then there is a S(T)-proof, follows by transforming a proof in $\widehat{\mathrm{S}}(\mathrm{T})$ into one in S(T) by exploiting the derivability of $\Box\mathrm{LR_K}$ and $\Box\mathrm{L_T}$ in S(T). To show the other direction of this equivalence, we begin by extending the definition of detour as follows. An application of a rule (r) is a detour in a S(T)-proof Π of $\vdash x_1{:}D$ if

(i) all of the active formulas of (r) are introduced in Π either by weakenings or by detours (i.e. none of them appears in the axioms of Π so that they are weak in Π), or

(ii) (r) is an application of $\Box$L in which the active rwff is introduced by *refl* and the active lwff is introduced by weakening or by detours.

For example, the application of □L in the proof shown below on the left is a detour that we eliminate by ('blowing up' the application of WlL and) transforming the proof to the one on the right:

$$\dfrac{\dfrac{}{\vdash xRx}\,\mathit{refl} \qquad \dfrac{\begin{array}{c}\Pi_1\\ \Gamma,\Delta\vdash\Gamma'\end{array}}{x{:}A,\Gamma,\Delta\vdash\Gamma'}\,\text{WlL}}{\begin{array}{c}x{:}\Box A,\Gamma,\Delta\vdash\Gamma'\\ \Pi_0\\ \vdash x_1{:}D\end{array}}\,\Box\text{L} \quad \rightsquigarrow \quad \dfrac{\begin{array}{c}\Pi_1\\ \Gamma,\Delta\vdash\Gamma'\end{array}}{\begin{array}{c}x{:}\Box A,\Gamma,\Delta\vdash\Gamma'\\ \Pi_0\\ \vdash x_1{:}D\end{array}}\,\text{WlL}\ .$$

If the rule (r) in (i) is an application of ClL, then we simply delete it together with the corresponding weakenings; e.g. we transform

$$\dfrac{\dfrac{\begin{array}{c}\Pi_4\\ \Delta\vdash xRy\end{array} \qquad \dfrac{\dfrac{\begin{array}{c}\Pi_1\\ x{:}\Box(A\supset B),\Gamma,\Delta\vdash\Gamma'\end{array}}{x{:}\Box(A\supset B),\Gamma,\Delta\vdash\Gamma',y{:}A}\,\text{WlR} \qquad \begin{array}{c}\Pi_3\\ \dfrac{\Gamma_1,\Delta_1\vdash\Gamma_1'}{y{:}B,\Gamma_1,\Delta_1\vdash\Gamma_1'}\,\text{WlL}\\ \Pi_2\\ y{:}B,x{:}\Box(A\supset B),\Gamma,\Delta\vdash\Gamma'\end{array}}{y{:}A\supset B,x{:}\Box(A\supset B),\Gamma,\Delta\vdash\Gamma'}\,\supset\text{L}}{x{:}\Box(A\supset B),x{:}\Box(A\supset B),\Gamma,\Delta\vdash\Gamma'}\,\Box\text{L}}{\begin{array}{c}x{:}\Box(A\supset B),\Gamma,\Delta\vdash\Gamma'\\ \Pi_0\\ \vdash x_1{:}D\end{array}}\,\text{ClL}$$

$$\text{to} \quad \begin{array}{c}\Pi_1\\ x{:}\Box(A\supset B),\Gamma,\Delta\vdash\Gamma'\\ \Pi_0\\ \vdash x_1{:}D\end{array} \quad \text{or} \quad \begin{array}{c}\Pi_3\\ \Gamma_1,\Delta_1\vdash\Gamma_1'\\ \Pi_2^{\dagger}\\ x{:}\Box(A\supset B),\Gamma,\Delta\vdash\Gamma'\\ \Pi_0\\ \vdash x_1{:}D\end{array}$$

by considering the left or the right branch, respectively.

We adjoin related rules as for S(K), with the addition that we permute each left contraction of $x{:}\Box A$ so that it immediately precedes the application of □L introducing $x{:}\Box A$. By iterating these transformations, we obtain the desired S(T)-proof in block form, which consists of alternating sequences of local reasoning (ClL–□L pairs and propositional rules) and transitional reasoning (□R–□L sequences surrounded by weakenings). From this S(T)-proof in block form, we obtain a $\widehat{\text{S}}$(T)-proof by replacing the ClL–□L pairs with applications of $\Box\text{L}_\text{T}$, and replacing the sequences of □ rules and the weakenings surrounding them with applications of $\Box\text{LR}_\text{K}$. $\Box\text{L}_\text{T}$ thus 'absorbs' all applications of ClL and all of *refl*, so that we can eliminate all rwffs from sequents and proofs. From this $\widehat{\text{S}}$(T)-proof we obtain a proof in SS(T) by deleting the labels, replacing $\Box\text{LR}_\text{K}$ and $\Box\text{L}_\text{T}$ with (K) and (T), and renaming the propositional rules.

Example 10.2.1 Consider again the proof (6.1) of $\vdash x{:} \sim\Box\sim(B\supset\Box B)$ given in Example 6.1.6. We first transform it into block form simply by permuting two of the weakenings over the rules below them, i.e.

$$\dfrac{\dfrac{}{\vdash xRx}\,\mathit{refl} \qquad \dfrac{\dfrac{\dfrac{\dfrac{\dfrac{\dfrac{}{xRy \vdash xRy}\,\mathrm{AXr} \qquad \dfrac{\dfrac{\dfrac{\dfrac{\dfrac{}{y{:}B \vdash y{:}B}\,\mathrm{AXl}}{y{:}B \vdash y{:}B, y{:}\Box B}\,\mathrm{WlR}}{\vdash y{:}B, y{:}B \supset \Box B}\,{\supset}\mathrm{R}}{y{:} \sim(B \supset \Box B) \vdash y{:}B}\,{\sim}\mathrm{L}}{y{:} \sim(B \supset \Box B), xRy \vdash y{:}B}\,\mathrm{WrL}}{x{:}\Box \sim(B \supset \Box B), xRy \vdash y{:}B}\,\Box\mathrm{L}}{x{:}\Box \sim(B \supset \Box B) \vdash x{:}\Box B}\,\Box\mathrm{R}}{x{:}\Box \sim(B \supset \Box B), x{:}B \vdash x{:}\Box B}\,\mathrm{WlL}}{x{:}\Box \sim(B \supset \Box B) \vdash x{:}B \supset \Box B}\,{\supset}\mathrm{R}}{x{:} \sim(B \supset \Box B), x{:}\Box \sim(B \supset \Box B) \vdash}\,{\sim}\mathrm{L}}{\dfrac{\dfrac{x{:}\Box \sim(B \supset \Box B), x{:}\Box \sim(B \supset \Box B) \vdash}{x{:}\Box \sim(B \supset \Box B) \vdash}\,\mathrm{ClL}}{\vdash x{:} \sim\Box \sim(B \supset \Box B)}\,{\sim}\mathrm{R}}\,\Box\mathrm{L}$$

From this we obtain the $\widehat{\mathrm{S}}(\mathrm{T})$-proof shown below on the left, which then yields the SS(T)-proof on the right:

$$\dfrac{\dfrac{\dfrac{\dfrac{\dfrac{\dfrac{\dfrac{}{y{:}B \vdash y{:}B, y{:}\Box B}\,\mathrm{AXl}}{\vdash y{:}B, y{:}B \supset \Box B}\,{\supset}\mathrm{R}}{y{:} \sim(B \supset \Box B) \vdash y{:}B}\,{\sim}\mathrm{L}}{x{:}\Box \sim(B \supset \Box B), x{:}B \vdash x{:}\Box B}\,\Box\mathrm{LR_K}}{x{:}\Box \sim(B \supset \Box B) \vdash x{:}B \supset \Box B}\,{\supset}\mathrm{R}}{x{:} \sim(B \supset \Box B), x{:}\Box \sim(B \supset \Box B) \vdash}\,{\sim}\mathrm{L}}{\dfrac{x{:}\Box \sim(B \supset \Box B) \vdash}{\vdash x{:} \sim\Box \sim(B \supset \Box B)}\,{\sim}\mathrm{R}}\,\Box\mathrm{L_T} \qquad \dfrac{\dfrac{\dfrac{\dfrac{\dfrac{\dfrac{\dfrac{}{B \vdash B, \Box B}\,(\mathrm{AX})}{\vdash B, B \supset \Box B}\,({\supset}\mathrm{R})}{\sim(B \supset \Box B) \vdash B}\,({\sim}\mathrm{L})}{\Box \sim(B \supset \Box B), B \vdash \Box B}\,(\mathrm{K})}{\Box \sim(B \supset \Box B) \vdash B \supset \Box B}\,({\supset}\mathrm{R})}{\sim(B \supset \Box B), \Box \sim(B \supset \Box B) \vdash}\,({\sim}\mathrm{L})}{\dfrac{\Box \sim(B \supset \Box B) \vdash}{\vdash \sim\Box \sim(B \supset \Box B)}\,({\sim}\mathrm{R})}\,(\mathrm{T})$$

■

Formally, we have:

Lemma 10.2.2 *Lemma 9.2.1, Lemma 9.2.2, Lemma 9.2.5, and Theorem 9.2.6 extend as follows.*

1. *Every proof of* $\vdash x_1{:}D$ *in* S(T) *can be transformed into block form.*

2. *The following are equivalent:*

   (a) $\vdash x_1{:}D$ *is provable in* S(T).

   (b) $\vdash x_1{:}D$ *is provable in* $\widehat{\mathrm{S}}(\mathrm{T})$.

   (c) $\vdash D$ *is provable in* SS(T).

3. *The proofs in* $\widehat{\mathrm{S}}(\mathrm{T})$ *and* $\mathrm{SS}(\mathrm{T})$ *differ only in the names of the rules and in the presence of labels, which can be eliminated or added as required.* ■

We can thus view $\mathrm{SS}(\mathrm{T})$ as the result of our substructural analysis of the rules of $\mathrm{S}(\mathrm{T})$ and of the proofs built using them. However, up to now we have used only part of Theorem 10.1.4, which tells us also that for theoremhood we can restrict the left contraction rule in our labelled system $\mathrm{S}(\mathrm{T})$ to be

$$\frac{x{:}\Box A[\![\Box B]\!]_{-}, x{:}\Box A[\![\Box B]\!]_{-}, \Gamma, \Delta \vdash^{s-1} \Gamma'}{x{:}\Box A[\![\Box B]\!]_{-}, \Gamma, \Delta \vdash^{s} \Gamma'}\ \mathrm{ClL}s$$

with s set to $pbs(\vdash x_1{:}D)$ at the start of a backwards proof $\vdash x_1{:}D$.

We can propagate this refinement to the systems $\widehat{\mathrm{S}}(\mathrm{T})$ and $\mathrm{SS}(\mathrm{T})$. This is best achieved by undoing the embedding of contraction, so that by transforming into block form $\mathrm{S}(\mathrm{T})$-proofs in which ClL is restricted to be ClLs, and by introducing contraction indices in $\widehat{\mathrm{S}}(\mathrm{T})$ and $\mathrm{SS}(\mathrm{T})$ as well, we can replace $\Box\mathrm{L}_{\mathrm{T}}$ with the rules

$$\frac{x{:}\Box A[\![\Box B]\!]_{-}, x{:}\Box A[\![\Box B]\!]_{-}, x{:}\Sigma \vdash^{s-1} x{:}\Sigma'}{x{:}\Box A[\![\Box B]\!]_{-}, x{:}\Sigma \vdash^{s} x{:}\Sigma'}\ \mathrm{ClL}s2$$

and

$$\frac{x{:}A, x{:}\Sigma \vdash^{s} x{:}\Sigma'}{x{:}\Box A, x{:}\Sigma \vdash^{s} x{:}\Sigma'}\ \Box\mathrm{L}_{\mathrm{T}}2\ ,$$

and we can replace (T) with the rules

$$\frac{\Box A[\![\Box B]\!]_{-}, \Box A[\![\Box B]\!]_{-}, \Sigma \vdash^{s-1} \Sigma'}{\Box A[\![\Box B]\!]_{-}, \Sigma \vdash^{s} \Sigma'}\ (\mathrm{CLs}) \qquad \text{and} \qquad \frac{A, \Sigma \vdash^{s} \Sigma'}{\Box A, \Sigma \vdash^{s} \Sigma'}\ (\mathrm{T2})\ .$$

The rule (T2) is a non-contracting version of (T), while (CLs) is an explicit contraction rule that is applied only when the conclusion has contraction index greater than 0. Then, by Lemma 10.2.2, we have:

Theorem 10.2.3 *The following are equivalent:*

1. $\vdash x_1{:}D$ *is provable in* $\mathrm{S}(\mathrm{T})$.
2. $\vdash x_1{:}D$ *is provable in* $\mathrm{S}(\mathrm{T})$ *where* ClL *is restricted to be* ClLs *with* s *set to* $pbs(\vdash x_1{:}D)$ *at the start of a backwards proof of* $\vdash x_1{:}D$.
3. $\vdash x_1{:}D$ *is provable in* $\widehat{\mathrm{S}}(\mathrm{T})$.
4. $\vdash x_1{:}D$ *is provable in* $\widehat{\mathrm{S}}(\mathrm{T})$ *where* $\Box\mathrm{L}_{\mathrm{T}}$ *is replaced by* $\Box\mathrm{L}_{\mathrm{T}}2$ *and* ClLs2 *with* s *set to* $pbs(\vdash x_1{:}D)$ *at the start of a backwards proof of* $\vdash x_1{:}D$.
5. $\vdash D$ *is provable in* $\mathrm{SS}(\mathrm{T})$.
6. $\vdash D$ *is provable in* $\mathrm{SS}(\mathrm{T})$ *where* (T) *is replaced by* (T2) *and* (CLs) *with* s *set to* $pbs(\vdash x_1{:}D)$ *at the start of a backwards proof of* D. ■

11 SUBSTRUCTURAL ANALYSIS OF S(K4) AND S(S4)

Lemma 8.2.4 and Corollary 9.1.2 tell us that the rules CrL and ClR and each application of ClL with principal formula other than $x{:}\Box A$ can be eliminated in S(K4) and S(S4). It is however easy to show that we cannot eliminate left contractions of lwffs of the form $x{:}\Box A$ and retain completeness of S(K4) and S(S4). For instance, similar to the example for S(T) at the beginning of §10.1, we can exploit the subformula property to show that the left contraction of $x_1{:}\Box{\sim}\Box B$ in the proof (6.2) of the S(K4)-theorem

$$x_1{:}\Box{\sim}\Box B \supset \Box{\sim}\Box\Box B$$

is indispensable, in the sense that the theorem cannot be proved without it. Table 11.1 contains additional S(K4)-theorem schemas that require application of ClL; we also display there the condition under which each lwff is a theorem and the overall number of contractions required to prove it when C, D and E are propositional variables. (A negative number means, of course, that no contractions are required.) For S(S4), observe that when C, D and E are propositional variables the last theorem in Table 11.1 requires two contractions also in S(S4), and that, similarly, the formula

$$x_1{:}\Box(\Box(C \supset \Box(D \supset E)) \supset E) \supset \\ ((\Box(C \supset \Box(D \supset \Box(C \supset \Box(D \supset \Box(C \supset \Box(D \supset E))))))) \supset E) \quad (11.1)$$

(suggested by Nicola Olivetti in a private communication) requires 2 left contractions of $x_1{:}\Box(\Box(C \supset \Box(D \supset E)) \supset E)$.

Since contraction is not eliminable, S(K4) and S(S4) suffer from a drawback common also to other (unlabelled, labelled or prefixed) deduction systems for K4

Table 11.1. Some S(K4)-theorem schemas requiring application of ClL.

S(K4)-theorem schema	Condition	#ClL's
$x_1{:}\Box \sim\Box C \supset (\Box^p C \supset \Box D)$	$p \geq 1$	$p-2$
$x_1{:}\Box \sim\Box C \supset (\Box D \supset \Box \sim\Box^p(D \supset C))$	$p \geq 1$	$p-1$
$x_1{:}\Box \sim\Box C \supset \Box \sim\Box^p C$	$p \geq 1$	$p-1$
$x_1{:}\Box \sim\Box C \supset \Box(\sim\Box^p C \vee \Box D)$	$p \geq 1$	$p-2$
$x_1{:}\Box \sim\Box C \supset (\Box \sim\Box D \supset \Box \sim\Box^p C)$	$p \geq 1$	$p-2$
$x_1{:}\Box \sim\Box \sim\Box^p(C\wedge \sim C) \supset \Box D$	$p \geq 0$	p
$x_1{:}\Box((C \supset\sim\Box \sim D) \wedge (D \supset\sim\Box \sim E) \wedge \sim E) \supset \Box \sim C$		2

The displayed formulas are theorems of S(K4) only when the indicated condition on p is satisfied. The overall number #ClL's of left contractions in the proofs is for when C, D and E are propositional variables.

and S4, namely: backwards proof search in S(K4) or S(S4) may not terminate since *unbounded* applications of left contractions may give rise to *infinite* chains of worlds, and thus to infinite branches.[1] A typical example of this, which we discuss in more detail in Example 11.1.1 below, is the attempted proof in a system for K4 of the unprovable formula $\Box \sim\Box B \supset \Box \sim B$.

In order to guarantee termination of proof search, and thereby establish decidability of K4 and S4, we must therefore find a way of bounding applications of contraction and thereby stop the construction of infinite chains and branches. A common technique for doing so relies on the observation that each infinite chain in K4 and S4 (as well as in other decidable transitive modal logics, which, however, we do not discuss here) is also *periodic*: there exist worlds x_i and x_j in the chain such that x_j is accessible from x_i, and A holds at x_j iff A holds at x_i. That is, the chain is periodic after x_j, since then, for each A and each $l \geq 0$, the formula A holds at x_{j+l} iff it holds at x_{i+l}. The infinite branch thus contains two sequents that are 'essentially the same' (which Kleene, in the context of propositional intuitionistic logic [147, p. 480], calls *cognate sequents*). Therefore, infinite branches can be recognized and avoided by introducing *loop-checkers* (also called *repetition* or *redundancy checkers*), that dynamically test for periodicity by keeping a *history of the proof* during proof construction: if x_j is accessible from x_i and the same formulas hold at x_i and x_j, then, by connecting worlds appropriately, we can truncate the chain at x_j, and thus truncate the branch (and so eventually the proof) as well. As a result, we obtain a finite chain and thus a finite branch, where, moreover, no 'relevant information' is lost: if the original infinite branch allowed us to construct a counter-model for some end-sequent S, then so does the finite branch.

[1] See, e.g., [87, § 8] and [120, 238], and recall from Definition 8.2.8 that a chain is a sequence of worlds $x_1, x_2, x_3, \ldots$ where x_{i+1} is a successor of x_i. Note also that the same problems with unbounded contractions occur also in deduction systems for other transitive modal logics, as well as in systems for several other non-classical logics such as propositional intuitionistic logic [81, 136, 147].

Dynamic loop-checking may however be computationally expensive as we must carry along a history of the proof, and update and check it at every rule application.[2] In our systems we can replace dynamic loop-checking by a static counter-part: we give a-priori polynomial bounds on the number of applications of ClL in each branch, so that each branch is finite and proof search in S(K4) and S(S4) terminates. We establish these bounds by extending our results for S(K) and S(T), and by combining them with an adaptation of a result given by Ladner [153], who showed that there exists a polynomial bound on the length of branches in proofs built using a standard, unlabelled, tableaux system for S4.[3]

More specifically, exploiting the observations of §11.1 on infinite chains and periodicity, in the main section of this chapter, §11.2, we bound contractions in S(K4) and S(S4) by proceeding as follows. In Lemmas 11.2.1 and 11.2.2 we show that the number of applications of □R in each branch of a proof of a theorem is polynomially bounded in terms of the size (number of symbols) of the theorem itself. This provides us with polynomial bounds on the length of chains and branches, which we can exploit to give a polynomial (cubic) bound on the number of applications of ClL in each branch (Theorem 11.2.3), and which we can also combine with an extension of the results of the previous chapters to restrict the form of contracted lwffs (Theorem 11.2.5). We also discuss Conjecture 11.2.6, which surmises that we can improve our contraction bounds so that we can always find a proof of $S = \vdash x_1{:}D$ in S(K4) or S(S4) such that each branch of the proof contains at most quadratically many (in the size of S) applications of ClL.

In §11.3 and §11.4 we then show that our labelled sequent systems yield a proof-theoretical justification of the rules of the standard systems SS(K4) and SS(S4).

11.1 INFINITE CHAINS, INFINITE BRANCHES AND PERIODICITY

To provide intuition for our results, let us illustrate in more detail how unbounded applications of ClL in S(K4) and S(S4) may give rise to infinite chains of worlds, and thus to infinite branches, and how to show these infinite chains to be periodic.

Suppose that applications of ClL with principal formula of the form $x_i{:}\Box A$ are not bounded in S(K4) and S(S4); by Corollary 9.1.2, we know that we need not consider contractions of lwffs that do not have this form. We can further assume that, reasoning backwards, each application of such a ClL immediately precedes an application of □L with the same principal formula $x_i{:}\Box A$; otherwise, given a sequent $x_i{:}\Box A, \Gamma, \Delta \vdash \Gamma'$ we could construct an infinite branch simply by repeatedly applying ClL. An attempt

[2] Note however that efficient loop-checking procedures for sequent-based proof search in systems for S4 (and some other modal and non-classical logics) have been proposed, e.g. [127, 130], based on efficient representations of the histories of the proofs. Note also that, as shown in [127, 130], loop-checking is in fact required also for systems for T, where, however, loops occur locally inside a world when we repeatedly left-contract some boxed formula.

[3] Ladner does not explicitly give this bound, which is however implicit in his proof of the complexity of proof search in S4. Similar indirect applications of Ladner's results to various modal tableaux systems can be found in [55, 130] and [120, 160]; we discuss these in more detail in §13. Note also that we could have used similar results to bound applications of ClL in S(T) implicitly, as a consequence of bounds on the length of chains in S(T)-proofs, but preferred instead to give the constructive arguments of §10.

to prove such a sequent in S(K4) or S(S4) thus has the form

$$\dfrac{\dfrac{\Delta \vdash x_i R x_j \quad x_j{:}A, x_i{:}\Box A, \Gamma, \Delta \vdash \Gamma'}{x_i{:}\Box A, x_i{:}\Box A, \Gamma, \Delta \vdash \Gamma'}\,\Box\mathrm{L}}{x_i{:}\Box A, \Gamma, \Delta \vdash \Gamma'}\,\mathrm{ClL} \tag{11.2}$$

for some x_j such that we can prove $\Delta \vdash x_i R x_j$.

This contraction is, in general, not eliminable: if there is some B such that $\Box B \Subset_- A$, then $x_j{:}A[\![\Box B]\!]_-$ might be used to generate a new world x_{j+k} with $k \geq 1$ so that x_{j+k} is accessible from x_i by means of transitivity (i.e. there is some Δ_l in the branch such that we can prove $\Delta_l \vdash x_i R x_{j+k}$). If we delete the second instance of $x_i{:}\Box A[\![\Box B]\!]_-$ in (11.2), then we cannot infer that $A[\![\Box B]\!]_-$ holds at x_{j+k}, and, in general, we lose completeness.

For an example of such a situation, let $x_i{:}\Box A[\![\Box B]\!]_-$ be $x_1{:}\Box \sim \Box B$. Then from (11.2) we have

$$\dfrac{\dfrac{\Delta \vdash x_1 R x_2 \qquad \dfrac{\dfrac{x_1{:}\Box \sim\Box B, \Gamma, \Delta, x_2 R x_3 \vdash \Gamma', x_3{:}B}{x_1{:}\Box \sim\Box B, \Gamma, \Delta \vdash \Gamma', x_2{:}\Box B}\,\Box\mathrm{R}}{x_2{:} \sim\Box B, x_1{:}\Box \sim\Box B, \Gamma, \Delta \vdash \Gamma'}\,{\sim}\mathrm{L}}{x_1{:}\Box \sim\Box B, x_1{:}\Box \sim\Box B, \Gamma, \Delta \vdash \Gamma'}\,\Box\mathrm{L}}{x_1{:}\Box \sim\Box B, \Gamma, \Delta \vdash \Gamma'}\,\mathrm{ClL} \quad .$$

Now that we have a new world x_3 accessible from x_1, we can perform another 'ClL–□L sequence' with principal formula $x_1{:}\Box \sim\Box B$ to obtain

$$\dfrac{\dfrac{\Delta \vdash x_1 R x_2 \qquad \dfrac{\dfrac{\dfrac{\dfrac{\dfrac{\Delta, x_2 R x_3 \vdash x_1 R x_3 \qquad \dfrac{\dfrac{x_1{:}\Box \sim\Box B, \Gamma, \Delta, x_2 R x_3, x_3 R x_4 \vdash \Gamma', x_3{:}B, x_4{:}B}{x_1{:}\Box \sim\Box B, \Gamma, \Delta, x_2 R x_3 \vdash \Gamma', x_3{:}B, x_3{:}\Box B}\,\Box\mathrm{R}}{x_3{:} \sim\Box B, x_1{:}\Box \sim\Box B, \Gamma, \Delta, x_2 R x_3 \vdash \Gamma', x_3{:}B}\,{\sim}\mathrm{L}}{x_1{:}\Box \sim\Box B, x_1{:}\Box \sim\Box B, \Gamma, \Delta, x_2 R x_3 \vdash \Gamma', x_3{:}B}\,\Box\mathrm{L}}{x_1{:}\Box \sim\Box B, \Gamma, \Delta, x_2 R x_3 \vdash \Gamma', x_3{:}B}\,\mathrm{ClL}}{x_1{:}\Box \sim\Box B, \Gamma, \Delta \vdash \Gamma', x_2{:}\Box B}\,\Box\mathrm{R}}{x_2{:} \sim\Box B, x_1{:}\Box \sim\Box B, \Gamma, \Delta \vdash \Gamma'}\,{\sim}\mathrm{L}}{x_1{:}\Box \sim\Box B, x_1{:}\Box \sim\Box B, \Gamma, \Delta \vdash \Gamma'}\,\Box\mathrm{L}}{x_1{:}\Box \sim\Box B, \Gamma, \Delta \vdash \Gamma'}\,\mathrm{ClL}$$

where $\Delta, x_2 R x_3 \vdash x_1 R x_3$ follows by transitivity since $\Delta \vdash x_1 R x_2$ is provable.

It is easy to see that we can go on like this indefinitely, and by similar ClL–□L sequences construct an infinite chain of worlds

$$x_1 R x_2 R x_3 R x_4 R x_5 R \ldots$$

and thus an infinite branch. However, in constructing such a chain we would use only one formula, $x_1{:}\Box \sim \Box B$, completely neglecting the other lwffs in Γ and Γ'. But it is precisely the 'information' in these lwffs that allows us to eventually truncate the chain once we have generated some particular world.

The 'standard' intuition behind this is as follows. Suppose that we attempt to prove in S(K4) or S(S4) a non-provable sequent $S = \vdash x_1{:}D$. Then, one (or more) of the branches of this attempted proof might be infinite as we might be constructing in it an infinite chain of worlds. Since such a branch does not close, we need to find a way of stopping the construction of the chain. To this end, we show that the infinite chain must be periodic so that at some point formulas start to repeat themselves with larger labels, i.e. with labels with greater subscripts. Indeed, during the construction of this infinite chain/branch we will at some point obtain a sequent S' whose antecedent contains, either explicitly or implicitly as we discuss below, for some world x_n, all the positive boxed subformulas $x_n{:}\Box C_1, \ldots, x_n{:}\Box C_{pbs(S)}$ of the goal S. That is, $\Box C_l \Subset_+ S$ for each l where $1 \leq l \leq pbs(S)$.[4] For example, suppose that this S' has the form

$$x_n{:}\Box C_1, \ldots, x_n{:}\Box C_{pbs(S)}, \Gamma_n, \Delta_n \vdash \Gamma'_n, x_n{:}\Box B\,.$$

We can then continue the backwards proof by applying $\Box$R to generate world x_{n+1} to which we 'transfer' all C_l's by applying $\Box$L, i.e.

$$\dfrac{\begin{array}{c} x_{n+1}{:}C_1, \ldots, x_{n+1}{:}C_{pbs(S)}, \Gamma_{n+1}, \Delta_{n+1} \vdash \Gamma'_{n+1}, x_{n+1}{:}B \\ \vdots \\ x_n{:}\Box C_1, \ldots, x_n{:}\Box C_{pbs(S)}, \Gamma_n, \Delta_n, x_n R x_{n+1} \vdash \Gamma'_n, x_{n+1}{:}B \end{array}}{x_n{:}\Box C_1, \ldots, x_n{:}\Box C_{pbs(S)}, \Gamma_n, \Delta_n \vdash \Gamma'_n, x_n{:}\Box B}\ \Box\mathrm{R}$$

Suppose now that then, after a finite number of further steps, we obtain the sequent

$$S'' = x_{n+m}{:}\Box C_1, \ldots, x_{n+m}{:}\Box C_{pbs(S)}, \Gamma_{n+m}, \Delta_{n+m} \vdash \Gamma'_{n+m}, x_{n+m}{:}\Box B$$

where $m > 1$. We need not continue the attempted proof by creating a successor of x_{n+m} by applying $\Box$R with principal formula $x_{n+m}{:}\Box B$, since we already have

$$x_{n+1}{:}C_1, \ldots, x_{n+1}{:}C_{pbs(S)}, \ldots \vdash \ldots, x_{n+1}{:}B\,.$$

Hence we can simply 'short-circuit' the chain and 'connect' x_{n+m} with x_{n+1} in the diagram spawned by the failed branch. That is, we let x_{n+1} be the successor of x_{n+m}. Thus the chain contains $n + m$ worlds, and a loop. This provides us eventually also with a counter-example for the non-theorem $x_1{:}D$ we were trying to prove.

The following, more concrete, example shows that in our sequent systems we do not need all of $x_n{:}\Box C_1, \ldots, x_n{:}\Box C_{pbs(S)}$ to be explicitly contained in S'. Instead, it suffices that $S' = \Gamma_n, \Delta_n \vdash \Gamma'_n$ contains enough information to disclose the desired formulas. For example, for each $\Box C_l$ it suffices that there exists some x_k such that $\Delta_n \vdash x_k R x_n$ is provable and $x_k{:}\Box C_l \in \Gamma_n$. Then, if we generate x_{n+1}, we will be able to obtain $x_{n+1}{:}C_l$ by applying $\Box$L; the same holds for x_{n+m} in S''. (As we discuss below, a further advantage of this is that we may then be able to reduce the number of contractions in the branches.)

Example 11.1.1 We attempt to prove the non-provable sequent $\vdash x_1{:}\Box \sim \Box B \supset \Box \sim B$ in S(K4); see [87, p. 404] for a similar attempt to prove the contrapositive

[4] If $pbs(S) = 0$, then the situation is much simpler, as shown in Lemma 11.2.1 below.

$\Diamond A \supset \Diamond\Box A$ in Fitting's prefixed tableaux system, as well as a discussion of infinite chains in these systems. The first steps are mechanical, and we abuse notation by naming some of the sequents that appear in the proof in order to refer to them below:[5]

$$
\dfrac{
\dfrac{x_1 R x_2 \vdash x_1 R x_2}{}\,\text{AXr}
\qquad
\dfrac{
\dfrac{
\dfrac{
\dfrac{
\dfrac{\dfrac{\cdots \qquad \cdots}{x_1 R x_2, x_2 R x_3 \vdash x_1 R x_3}\,\mathit{trans}
\qquad
\dfrac{\begin{array}{l} S_2 = x_1{:}\Box \sim\Box B, x_1 R x_2, \\ x_2 R x_3 \vdash x_2{:} \sim B, x_3{:}B, x_3{:}\Box B \end{array}}{\begin{array}{l} x_3{:} \sim\Box B, x_1{:}\Box \sim\Box B, x_1 R x_2, \\ x_2 R x_3 \vdash x_2{:} \sim B, x_3{:}B \end{array}}\,{\sim}\text{L}}
{\begin{array}{l} x_1{:}\Box \sim\Box B, x_1{:}\Box \sim\Box B, x_1 R x_2, \\ x_2 R x_3 \vdash x_2{:} \sim B, x_3{:}B \end{array}}\,\Box\text{L}
}{x_1{:}\Box \sim\Box B, x_1 R x_2, x_2 R x_3 \vdash x_2{:} \sim B, x_3{:}B}\,\text{ClL}
}{S_1 = x_1{:}\Box \sim\Box B, x_1 R x_2 \vdash x_2{:} \sim B, x_2{:}\Box B}\,\Box\text{R}
}{x_2{:} \sim\Box B, x_1{:}\Box \sim\Box B, x_1 R x_2 \vdash x_2{:} \sim B}\,{\sim}\text{L}
}{
\dfrac{
\dfrac{
\dfrac{x_1{:}\Box \sim\Box B, x_1{:}\Box \sim\Box B, x_1 R x_2 \vdash x_2{:} \sim B}{x_1{:}\Box \sim\Box B, x_1 R x_2 \vdash x_2{:} \sim B}\,\text{ClL}
}{x_1{:}\Box \sim\Box B \vdash x_1{:}\Box \sim B}\,\Box\text{R}
}{\vdash x_1{:}\Box \sim\Box B \supset \Box \sim B}\,{\supset}\text{R}
}\,\Box\text{L}
\tag{11.3}
$$

Since $\Box \sim\Box B$ is the only positive boxed subformula of the goal S, i.e. $\mathit{pbs}(S) = 1$, we need not apply $\Box$R to S_2. Namely: although we cannot directly identify S_1 and S_2 with the desired sequents S' and S'', the missing lwffs $x_2{:}\Box \sim\Box B$ and $x_3{:}\Box \sim\Box B$ are implicit in S_1 and S_2, since both sequents contain in the antecedent the lwff $x_1{:}\Box \sim\Box B$, from which we can obtain $x_3{:} \sim\Box B$ and $x_4{:} \sim\Box B$.

We can, however, transform the proof to 'disclose' $x_2{:}\Box \sim\Box B$ and $x_3{:}\Box \sim\Box B$. We can do so by employing the admissible (by means of a *cut*) rule $\Box\text{L}_{\text{K4}}$,

$$
\dfrac{\Delta \vdash x_i R x_j \quad x_j{:}A, x_j{:}\Box A, \Gamma, \Delta \vdash \Gamma'}{x_i{:}\Box A, \Gamma, \Delta \vdash \Gamma'}\,\Box\text{L}_{\text{K4}} \quad \leadsto
$$

$$
\dfrac{
\dfrac{
\Delta \vdash x_i R x_j
\qquad
\dfrac{x_i{:}\Box A \vdash x_i{:}\Box\Box A \qquad \dfrac{\Delta \vdash x_i R x_j \quad x_j{:}A, x_j{:}\Box A, \Gamma, \Delta \vdash \Gamma'}{x_j{:}A, x_i{:}\Box\Box A, \Gamma, \Delta \vdash \Gamma'}\,\Box\text{L}}{x_j{:}A, x_i{:}\Box A, \Gamma, \Delta \vdash \Gamma'}\,\mathit{cut}
}{x_i{:}\Box A, x_i{:}\Box A, \Gamma, \Delta \vdash \Gamma'}\,\Box\text{L}
}{x_i{:}\Box A, \Gamma, \Delta \vdash \Gamma'}\,\text{ClL}
\; ,
\tag{11.4}
$$

where $x_i{:}\Box A \vdash x_i{:}\Box\Box A$ follows trivially by exploiting the transitivity of R.

[5]Note that if the uppermost $\Box$L had active rwff $x_1 R x_2$, then we would simply obtain a redundant sequent, $x_2{:} \sim\Box B, x_1{:}\Box \sim\Box B, x_1 R x_2, x_2 R x_3 \vdash x_2{:} \sim B, x_3{:}B$. Using the $\Box$-disjunction property and its corollaries we could then show that there is a proof of either $x_1{:}\Box \sim\Box B, x_1 R x_2, x_2 R x_3 \vdash x_2{:} \sim B, x_3{:}B$ or $x_2{:} \sim\Box B, x_1{:}\Box \sim\Box B, x_1 R x_2, \vdash x_2{:} \sim B$.

Then we can replace applications of □L and the contractions preceding them with applications of $\Box L_{K4}$, and transform the attempted proof (11.3) to

$$
\dfrac{
\dfrac{}{x_1 R x_2 \vdash x_1 R x_2}\;\text{AXr}
\qquad
\dfrac{
\dfrac{
\dfrac{
\dfrac{\dfrac{}{x_2 R x_3 \vdash x_2 R x_3}\;\text{AXr}}{x_1 R x_2, x_2 R x_3 \vdash x_2 R x_3}\;\text{WrL}
\qquad
\dfrac{\begin{array}{c} S'' = x_3{:}\Box \sim\Box B, x_1 R x_2, \\ x_2 R x_3 \vdash x_2{:} \sim B, x_3{:}B, x_3{:}\Box B \end{array}}{\begin{array}{c} x_3{:} \sim\Box B, x_3{:}\Box \sim\Box B, x_1 R x_2, \\ x_2 R x_3 \vdash x_2{:} \sim B, x_3{:}B \end{array}}\;{\sim}\text{L}
}{x_2{:}\Box \sim\Box B, x_1 R x_2, x_2 R x_3 \vdash x_2{:} \sim B, x_3{:}B}\;\Box\text{L}_{\text{K4}}
}{S' = x_2{:}\Box \sim\Box B, x_1 R x_2 \vdash x_2{:} \sim B, x_2{:}\Box B}\;\Box\text{R}
}{x_2{:} \sim\Box B, x_2{:}\Box \sim\Box B, x_1 R x_2 \vdash x_2{:} \sim B}\;{\sim}\text{L}
}{
\dfrac{
\dfrac{x_1{:}\Box \sim\Box B, x_1 R x_2 \vdash x_2{:} \sim B}{x_1{:}\Box \sim\Box B \vdash x_1{:}\Box \sim B}\;\Box\text{R}
}{\vdash x_1{:}\Box \sim\Box B \supset \Box \sim B}\;\supset\text{R}
}\;\Box\text{L}_{\text{K4}}
\,.
$$

We need not apply □R to S'' to generate a successor of the world $x_{n+m} = x_3$, since we already have a successor of the world $x_n = x_2$ where

$$x_n{:}\Box \sim\Box B, \ldots \vdash \ldots, x_n{:}B, x_n{:}\Box B\,.$$

It is x_3 itself. Thus, we can connect x_3 with itself and truncate the chain of worlds. This truncates and concludes the branch and the proof as well: $\vdash x_1{:}\Box \sim\Box B \supset \Box \sim B$ is not provable in S(K4), and we have the following counter-model

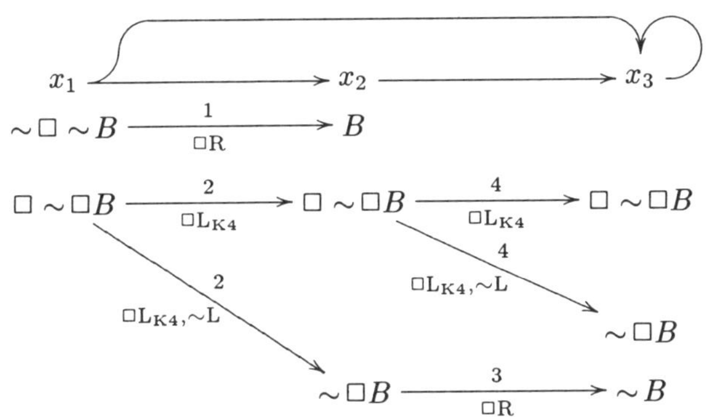

where the initial application of $\supset$ R yields $\sim\Box\sim B$ and $\Box\sim\Box B$ at x_1, and the numbered and indexed arrows represent applications of □R and $\Box L_{K4}$ together with the propositional reasoning following them; note that we represent the succedent and the antecedent of the premise of an application of $\Box L_{K4}$ by means of two arrows with the same number.

Although S_2 may contain different formulas and/or be a smaller sequent than S'' (it may contain fewer formulas when there is more than one positive boxed subformula), (11.3) contains enough information to build a similar counter-model to which the missing formulas can be easily added. In fact, we can associate (11.3) with the

diagram

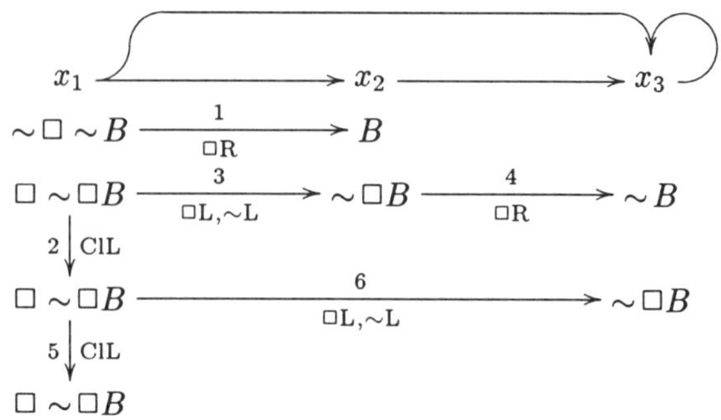

where the initial application of $\supset$ R yields $\sim\Box\sim B$ and $\Box\sim\Box B$ at x_1. (Looking at this diagram, we see that the third instance of $\Box\sim\Box B$ at x_1, and thus the second application of ClL, is redundant; we return to this below where we discuss superfluous contractions, e.g. in the examples and in §11.3.) ■

11.2 BOUNDING CONTRACTION IN S(K4) AND S(S4)

We now formalize the above intuitions and show that, when proving theorems in S(K4) or S(S4), in each branch we can bound applications of $\Box$R with principal formula $\Box B$ labelled with 'increasing' worlds in a chain.

Lemma 11.2.1 *Given a formula B and a sequent $S = \vdash x_1{:}D$ provable in* S(K4) *or* S(S4)*, there is a proof of S in the corresponding system such that in each branch $\Box$R is applied at most $pbs(S)+1$ times with principal formula $\Box B$ labelled with increasing worlds in a chain.*

Proof (Sketch) We begin by observing that Proposition 8.2.9 and its corollaries tell us that for fixed x_j we need not apply $\Box$R with principal formula $x_j{:}\Box B$ more than once (for otherwise we generate two divergent subchains of worlds that have x_j as their origin). Thus, we only need to consider the case where B is fixed and x_j varies over different, increasing, worlds in a chain. We distinguish two subcases.

If $pbs(S) = 0$, then, again by Proposition 8.2.9 and its corollaries, it immediately follows that we need at most one application of $\Box$R with principal formula $x_j{:}\Box B$.

If $pbs(S) > 0$, then consider the following branch

$$\begin{array}{c}
\Pi_2 \\
\dfrac{\Gamma_j, \Delta_j, x_j R\, x_{j+1} \vdash \Gamma'_j, x_{j+1}{:}B}{\Gamma_j, \Delta_j \vdash \Gamma'_j, x_j{:}\Box B}\ \Box\mathrm{R}\ (j) \\
\Pi_1 \\
\dfrac{\Gamma, \Delta, x_1 R\, x_2 \vdash \Gamma', x_2{:}B}{\Gamma, \Delta \vdash \Gamma', x_1{:}\Box B}\ \Box\mathrm{R}\ (1) \\
\Pi_0 \\
\vdash x_1{:}D
\end{array} \tag{11.5}$$

where $\Box$R (1) is the first application of $\Box$R with principal formula $x_j{:}\Box B$, which, for simplicity but without loss of generality, we have assumed to occur for $x_j = x_1$.

At each application of □R, we generate a new world x_{j+1} that is then available for applications of □L with principal formula $x_i{:}\Box C$, where $x_i{:}\Box C \Subset_+ S$ and $1 \leq i \leq j$ (or $1 \leq i \leq j+1$ in S(S4)). In particular, since we can prove $\Delta_j, x_j R\, x_{j+1} \vdash x_1 R\, x_{j+1}$, from $x_1{:}\Box C$ we can obtain $x_{j+1}{:}C$ by □L.

Now, for example, pick the highest l such that $0 < l \leq \mathit{pbs}(S)$ and $x_1{:}\Box^l C \in \Gamma$ (or $x_1{:}\Box^l C \Subset_+ S$), and suppose that we need C to close the branch at the leaves of (11.5). (In the general case, we consider instead $x_1{:}E \Subset_+ S$ with E containing l positive □'s and C being in the scope of the innermost such □.) Then we require a chain of $l+1$ worlds,

$$x_1 R\, x_2 R\, x_3 R \ldots R\, x_l R\, x_{l+1}\,,$$

generated by l applications of □R with, possibly, principal formulas $x_1{:}\Box B$, $x_2{:}\Box B$, ..., $x_l{:}\Box B$. Then, for example, (11.5) has the form

$$\dfrac{\dfrac{\dfrac{\dfrac{\Delta_l, x_l R\, x_{l+1} \vdash x_1 R\, x_{l+1} \qquad \begin{matrix}\Pi_2\\ x_{l+1}{:}C, x_l{:}\Box C, \Gamma_l, \Delta_l, x_l R\, x_{l+1} \vdash \Gamma_l', x_{l+1}{:}B\end{matrix}}{x_l{:}\Box C, x_l{:}\Box C, \Gamma_l, \Delta_l, x_l R\, x_{l+1} \vdash \Gamma_l', x_{l+1}{:}B}\,\Box L}{x_l{:}\Box C, \Gamma_l, \Delta_l, x_l R\, x_{l+1} \vdash \Gamma_l', x_{l+1}{:}B}\,\mathrm{ClL}}{x_l{:}\Box C, \Gamma_l, \Delta_l \vdash \Gamma_l', x_l{:}\Box B}\,\Box R\ (l)}{\begin{matrix}\Pi_1\\ \dfrac{x_1{:}\Box^l C, x_1{:}\Box \sim\Box B, \Gamma_1, \Delta_1, x_1 R\, x_2 \vdash \Gamma_1', x_2{:}B}{x_1{:}\Box^l C, x_1{:}\Box \sim\Box B, \Gamma_1, \Delta_1 \vdash \Gamma_1', x_1{:}\Box B}\,\Box R\ (1)\\ \Pi_0\\ \vdash x_1{:}D\end{matrix}}\ .$$

Once we have generated x_{l+1} we need not apply □R with principal formula $x_m{:}\Box B$ for $m \geq l+1$. This follows because, analogous with Example 11.1.1 and since we have picked the highest l, all the positive boxed subformulas of S hold at x_{l+1} (explicitly or implicitly, like in that example); e.g. Γ_l contains

$$x_{l+1}{:}\Box C, x_{l+1}{:}\Box\Box C, \ldots, x_{l+1}{:}\Box^l C, x_{l+1}{:}\Box \sim\Box B\,.$$

Therefore we can transform Π_2 so that it does not contain applications of □R with principal formula $x_m{:}\Box B$ for $m \geq l+1$. Note that we can perform another application of □R with principal formula $x_{l+1}{:}\Box B$; in doing so, we generate a world x_{l+2} where all the scopes of the positive boxed subformulas of S hold. We can thus combine the two subcases and conclude that, given B and a S(K4)-theorem or S(S4)-theorem $x_1{:}D$, we can find a proof of $S = \vdash x_1{:}D$ in the corresponding system where in each branch □R is applied at most $\mathit{pbs}(S) + 1$ times with principal formula $x_j{:}\Box B$. ■

Since each $S = \vdash x_1{:}D$ contains at most $\mathit{nbs}(S)$ negative boxed subformulas, and since ClR is eliminable by Corollary 9.1.2, in a proof of S in S(K4) or S(S4) there are at most $\mathit{nbs}(S)$ different candidates for the formula $x_j{:}\Box B$ in Lemma 11.2.1. It follows that if S is provable, then it has a proof where in each branch there are at most $\mathit{nbs}(S) \times (\mathit{pbs}(S) + 1)$ worlds accessible from x_1. In such a branch we need at most $(\mathit{nbs}(S) \times (\mathit{pbs}(S) + 1)) - 1$ applications of ClL with the same principal formula

$x_i{:}\Box A$, so that there is one instance of $x_i{:}\Box A$ for each world accessible from x_i. In other words, we have:

Lemma 11.2.2 *Given a proof of* $S = \vdash x_1{:}D$ *in* S(K4) *or* S(S4), *there is a proof of* S *in the corresponding system such that in each branch there are at most* $nbs(S) \times (pbs(S)+1)$ *applications of* $\Box$R, *so that chains contain at most* $1+nbs(S)\times(pbs(S)+1)$ *worlds. Hence, each branch is finite and contains at most* $(nbs(S)\times(pbs(S)+1))-1$ *applications of* ClL *with the same principal formula* $x_i{:}\Box A$. ∎

Since each $S = \vdash x_1{:}D$ contains at most $pbs(S)$ different formulas of the form $x_i{:}\Box A$ that can be contracted, it follows that in each branch there at most $((nbs(S) \times (pbs(S)+1)) - 1) \times pbs(S)$ left contractions. That is, from Lemma 11.2.2 we have:

Theorem 11.2.3 *Every sequent* $S = \vdash x_1{:}D$ *provable in* S(K4) *or* S(S4) *has a proof in the corresponding system in which there are no contractions, except for applications of* ClL *with principal formula of the form* $x_i{:}\Box A$. *However, in each branch of this proof* ClL *need not be applied with the same principal formula* $x_i{:}\Box A$ *more than* $(nbs(S) \times (pbs(S)+1)) - 1$ *times. Hence, in each branch there are at most* $((nbs(S) \times (pbs(S)+1)) - 1) \times pbs(S)$ *applications of* ClL. ∎

We can fairly straightforwardly extend our previous results to restrict the contraction rule: like for S(T), in S(K4) and S(S4) we only need to contract lwffs of the form $x_i{:}\Box A[\![\Box B]\!]_-$. (This is indeed the case for all theorem schemas in Table 11.1.)

Lemma 11.2.4 *Every sequent* $\vdash x_1{:}D$ *provable in* S(K4) *or* S(S4) *has a proof in the corresponding system in which there are no contractions, except for applications of* ClL *with principal formula of the form* $x_i{:}\Box A[\![\Box B]\!]_-$.

Proof (Sketch) We extend the proof of Theorem 9.1.1 for S(K) and the proof of Theorem 10.1.4 for S(T). By Corollary 9.1.2, given a proof of $\vdash x_1{:}D$, we consider the uppermost left contraction of an lwff of the form $x_i{:}\Box A$ where there is no B such that $\Box B \Subset_- A$, and show how to eliminate this contraction. The lemma follows by iterating the following argument.

We proceed by induction on the rank of the contraction, $rank(x_i{:}\Box A)$, where the permutability of the rules allows us to assume that the contraction immediately precedes the application of $\Box$L that introduces the second constituent.

$(rank(x_i{:}\Box A) = 2)$ If one of the two constituents is introduced by WlL, then we conclude by deleting this WlL and the application of ClL. Therefore suppose that both constituents are introduced by $\Box$L, i.e. consider

$$
\dfrac{\dfrac{\dfrac{\Pi_1}{\Delta \vdash x_i R x_j} \qquad \dfrac{\dfrac{\Pi_2}{\Delta \vdash x_i R x_k} \quad \dfrac{\Pi_3}{x_j{:}A, x_k{:}A, \Gamma, \Delta \vdash \Gamma'}}{x_j{:}A, x_i{:}\Box A, \Gamma, \Delta \vdash \Gamma'}\,\Box\mathrm{L}}{x_i{:}\Box A, x_i{:}\Box A, \Gamma, \Delta \vdash \Gamma'}\,\Box\mathrm{L}}{\dfrac{x_i{:}\Box A, \Gamma, \Delta \vdash \Gamma'}{\dfrac{\Pi_0}{\vdash x_1{:}D}}}\,\mathrm{ClL}
$$

.

Since Π_3 is contraction-free, we then proceed as follows, possibly applying Corollary 8.2.11 where we previously applied Corollary 8.2.10. If $x_j = x_i$ or $x_k = x_i$, then we conclude as in the proof of Theorem 10.1.4 for S(T). If $x_j \neq x_i$ and $x_k \neq x_i$, then we conclude as in the proof of Theorem 9.1.1 for S(K).

This concludes the proof for the case $rank(x_i{:}\Box A) = 2$. Consider now the case when $rank(x_i{:}\Box A) = j + 1 > 2$. As for S(K), in this case it is possible that the first $x_i{:}\Box A$ was introduced by WlL into one or more of the places which give the contraction a rank of $j + 1$. We make the mode of introduction of that $x_i{:}\Box A$ uniform by replacing all such introductions by WlL with introductions by □L; the rest of the proof is as before, modulo possible applications of weakening and AXr (cf. the example given in the proof of Theorem 9.1.1). Note that the rank is still $j + 1$. We distinguish two cases, depending on whether or not there is an application of □R between the two applications of □L.

($rank(x_i{:}\Box A) = j + 1$, *case 1*) If there is no application of □R between the two applications of □L, then we can permute rules to obtain a contraction of rank j, and conclude by applying the induction hypothesis.

($rank(x_i{:}\Box A) = j+1$, *case 2*) Suppose that there is an application of □R between the two applications of □L. That is, consider a branch of the form

$$\dfrac{\dfrac{\begin{matrix}\Pi_1\\ \Delta \vdash x_i R\, x_j\end{matrix} \qquad \begin{matrix}\dfrac{\dfrac{\begin{matrix}\Pi_4 & & \Pi_5\\ \Delta_2 \vdash x_i R\, x_k & & x_k{:}A, \Gamma_2, \Delta_2 \vdash \Gamma_2'\end{matrix}}{x_i{:}\Box A, \Gamma_2, \Delta_2 \vdash \Gamma_2'}\,\Box L}{\begin{matrix}\Pi_3\\ x_i{:}\Box A, \Gamma_1, \Delta_1, x_n R\, x_m \vdash x_m{:}C, \Gamma_1'\end{matrix}}\\ \overline{x_i{:}\Box A, \Gamma_1, \Delta_1 \vdash x_n{:}\Box C, \Gamma_1'}\;\Box R\\ \Pi_2\\ x_j{:}A, x_i{:}\Box A, \Gamma, \Delta \vdash \Gamma'\end{matrix}}{x_i{:}\Box A, x_i{:}\Box A, \Gamma, \Delta \vdash \Gamma'}\,\Box L}{\begin{matrix}x_i{:}\Box A, \Gamma, \Delta \vdash \Gamma'\\ \Pi_0\\ \vdash x_1{:}D\end{matrix}}\,\mathrm{ClL} \quad . \qquad (11.6)$$

Depending on x_n we distinguish three cases: x_n is different and not accessible from x_i (case 2.1), $x_n = x_i$ (case 2.2), x_n is accessible from x_i (case 2.3).

($rank(x_i{:}\Box A) = j + 1$, *case 2.1*) If x_n is not accessible from x_i and $x_n \neq x_i$, then $x_n R\, x_m \neq x_i R\, x_k$. It also follows that the provability of $\Delta_2 \vdash x_i R\, x_k$ implies that there exists a proof Π_6 of $\Delta \vdash x_i R\, x_k$, and we can permute rules to reduce the rank of the contraction, for example transforming (11.6) to

$$\begin{array}{c}\dfrac{\dfrac{\begin{array}{c}\Pi_1\\ \Delta \vdash x_i R x_j\end{array} \qquad \dfrac{\begin{array}{c}\Pi_6\\ \Delta\vdash x_iRx_k\end{array}\qquad \begin{array}{c} \Pi_5\\ x_k{:}A,\Gamma_2,\Delta_2\vdash\Gamma'_2\\ \Pi_3^\dagger \\ \dfrac{x_k{:}A,\Gamma_1,\Delta_1,x_nRx_m\vdash x_m{:}C,\Gamma'_1}{x_k{:}A,\Gamma_1,\Delta_1\vdash x_n{:}\Box C,\Gamma'_1}\,\Box R\\ \Pi_2^\dagger\\ x_j{:}A,x_k{:}A,\Gamma,\Delta\vdash\Gamma'\end{array}}{x_j{:}A,x_i{:}\Box A,\Gamma,\Delta\vdash\Gamma'}\,\Box L}{x_i{:}\Box A,x_i{:}\Box A,\Gamma,\Delta\vdash\Gamma'}\,\Box L}{x_i{:}\Box A,\Gamma,\Delta\vdash\Gamma'}\,\mathrm{ClL}\\ \Pi_0\\ \vdash x_1{:}D\end{array}$$

We conclude by applying the induction hypothesis.

($rank(x_i{:}\Box A) = j+1$, *case 2.2*) Suppose that $x_n = x_i$. If x_m and x_j (or x_m and x_k) diverge from x_i, then we can permute rules to reduce the rank of the contraction; else, we conclude similarly either to the proof of Theorem 10.1.4 for S(T), or to case ($grade(x{:}A) = k+1$, $rank(x{:}A) = j+1$, $A = \Box B$, $\Box$L, case 2) in the proof of Theorem 9.1.1 for S(K). Observe that it is possible that we need to apply Corollary 8.2.11 where we previously applied Corollary 8.2.10.

($rank(x_i{:}\Box A) = j+1$, *case 2.3*) Suppose that x_n is accessible but different from x_i (and x_m and x_j do not diverge from x_i, in which case we conclude like in case 2.2). Then x_n is either x_j (case 2.3.1), or a world different from x_j but still accessible from x_i (case 2.3.2).

($rank(x_i{:}\Box A) = j+1$, *case 2.3.1*) If $x_n = x_j$, then $x_j{:}\Box C$ must follow from some formula in Γ or Γ' ($x_j{:}\Box C$ cannot be a subformula of $x_j{:}A$ since we have assumed that there is no B such that $\Box B \in_{-} A$). We consider an example that points out the subtleties of the proof (the general case is dealt with similarly, by permutations of rules to reduce the rank of the contraction). The proof

$$\begin{array}{c}\dfrac{\dfrac{\begin{array}{c}\Pi_1\\ \Delta\vdash x_iRx_j\end{array}\qquad \dfrac{\dfrac{\begin{array}{c}\Pi_1\\ \Delta\vdash x_iRx_j\end{array}\qquad \dfrac{\dfrac{\dfrac{\begin{array}{c}\Pi_4\\ \Delta,x_jRx_k\vdash x_iRx_k\end{array}\qquad \begin{array}{c}\Pi_5\\ x_j{:}A,x_k{:}A,\Gamma_3,x_i{:}\Box\sim\Box C,\\ \Delta,x_jRx_k\vdash\Gamma',x_k{:}C\end{array}}{x_j{:}A,x_i{:}\Box A,\Gamma_3,x_i{:}\Box\sim\Box C,\Delta,x_jRx_k\vdash\Gamma',x_k{:}C}\,\Box L}{x_j{:}A,x_i{:}\Box A,\Gamma_3,x_i{:}\Box\sim\Box C,\Delta\vdash\Gamma',x_j{:}\Box C}\,\Box R}{x_j{:}A,x_i{:}\Box A,\Gamma_3,x_j{:}\sim\Box C,x_i{:}\Box\sim\Box C,\Delta\vdash\Gamma'}\,{\sim}L}{x_j{:}A,x_i{:}\Box A,\Gamma_3,x_i{:}\Box\sim\Box C,x_i{:}\Box\sim\Box C,\Delta\vdash\Gamma'}\,\Box L}{x_j{:}A,x_i{:}\Box A,\Gamma_3,x_i{:}\Box\sim\Box C,\Delta\vdash\Gamma'}\,\mathrm{ClL}}{x_i{:}\Box A,x_i{:}\Box A,\Gamma_3,x_i{:}\Box\sim\Box C,\Delta\vdash\Gamma'}\,\Box L}{x_i{:}\Box A,\Gamma_3,x_i{:}\Box\sim\Box C,\Delta\vdash\Gamma'}\,\mathrm{ClL}\\ \Pi_0\\ \vdash x_1{:}D\end{array}$$

is an instance of (11.6) where $x_nRx_m = x_jRx_k$, $\Gamma = \Gamma_3 \cup \{x_i{:}\Box\sim\Box C\}$, $\Delta_1 = \Delta \cup \{x_jRx_k\}$, $\Delta_2 = \Delta$, and $rank(x_i{:}\Box A) = 6$. By permuting rules, we can transform this proof to

$$
\begin{array}{c}
\dfrac{\begin{array}{c}\Pi_1\\ \Delta \vdash x_iRx_j\end{array} \quad \dfrac{\dfrac{\begin{array}{c}\Pi_1\\ \Delta \vdash x_iRx_j\end{array} \quad \dfrac{\dfrac{\dfrac{\begin{array}{c}\Pi_4\\ \Delta, x_jRx_k \vdash x_iRx_k\end{array} \quad \begin{array}{c}\Pi_5\\ x_j{:}A, x_k{:}A, \Gamma_3, x_i{:}\Box \sim\Box C,\\ \Delta, x_jRx_k \vdash \Gamma', x_k{:}C\end{array}}{x_j{:}A, x_i{:}\Box A, \Gamma_3, x_i{:}\Box \sim\Box C, \Delta, x_jRx_k \vdash \Gamma', x_k{:}C}\,\Box\mathrm{L}}{x_j{:}A, x_i{:}\Box A, \Gamma_3, x_i{:}\Box \sim\Box C, \Delta \vdash \Gamma', x_j{:}\Box C}\,\Box\mathrm{R}}{x_j{:}A, x_i{:}\Box A, \Gamma_3, x_j{:} \sim\Box C, x_i{:}\Box \sim\Box C, \Delta \vdash \Gamma'}\,\sim\mathrm{L}}{x_i{:}\Box A, x_i{:}\Box A, \Gamma_3, x_j{:} \sim\Box C, x_i{:}\Box \sim\Box C, \Delta \vdash \Gamma'}\,\Box\mathrm{L}}{x_i{:}\Box A, \Gamma_3, x_j{:} \sim\Box C, x_i{:}\Box \sim\Box C, \Delta \vdash \Gamma'}\,\mathrm{ClL}}{\dfrac{x_i{:}\Box A, \Gamma_3, x_i{:}\Box \sim\Box C, x_i{:}\Box \sim\Box C, \Delta \vdash \Gamma'}{x_i{:}\Box A, \Gamma_3, x_i{:}\Box \sim\Box C, \Delta \vdash \Gamma'}\,\mathrm{ClL}}\,\Box\mathrm{L} \\
\Pi_0 \\
\vdash x_1{:}D
\end{array}
\quad .
$$

Now $rank(x_i{:}\Box A) = 4$, and by the induction hypothesis we can eliminate the contraction (thus reducing also $rank(x_i{:}\Box \sim \Box C)$, which was temporarily increased). The intuition behind this is that since $\Box B \notin_- A$, only one of $x_j{:}A$ and $x_k{:}A$ contributes to the proof Π_1 when $x_k{:}C$ does (i.e. at most one of $x_j{:}A$ and $x_k{:}A$ leads to axioms and the other is weak).

($rank(x_i{:}\Box A) = j+1$, *case 2.3.2*) If x_n is a world different from x_j but accessible from x_i, then $x_n{:}\Box C$ must follow from (be a subformula of) some formula in Γ or Γ', and we conclude similarly to case 2.3.1. ∎

A straightforward combination of Theorem 11.2.3 and Lemma 11.2.4 yields:

Theorem 11.2.5 *Every sequent* $S = \vdash x_1{:}D$ *provable in* S(K4) *or* S(S4) *has a proof in the corresponding system in which there are no contractions, except for applications of* ClL *with principal formula of the form* $x_i{:}\Box A[\![\Box B]\!]_-$. *However, in each branch of this proof* ClL *need not be applied with the same principal formula* $x_i{:}\Box A[\![\Box B]\!]_-$ *more than* $(nbs(S) \times (pbs(S)+1)) - 1$ *times. Hence, in each branch there are at most* $((nbs(S) \times (pbs(S)+1)) - 1) \times pbs(S)$ *applications of* ClL. ∎

As for S(T), we can denote the bound on applications of ClL by annotating each sequent with a contraction index s, which we set to $((nbs(S) \times (pbs(S)+1)) - 1) \times pbs(S)$ at the start of a backwards proof, and restricting ClL in S(K4) and S(S4) to be

$$
\frac{x{:}\Box A[\![\Box B]\!]_-, x{:}\Box A[\![\Box B]\!]_-, \Gamma, \Delta \vdash^{s-1} \Gamma'}{x{:}\Box A[\![\Box B]\!]_-, \Gamma, \Delta \vdash^{s} \Gamma'}\ \mathrm{ClL}s\ ,
$$

which explicitly requires that $s > 0$. As for S(T), the index is decremented at every contraction, and is imported in the premises of branching rules.

Since in a proof of $S = \vdash x_1{:}D$ in S(K4) or S(S4) both $nbs(S)$ and $pbs(S)$ are bounded above by the *size* $n = |S|$ of S (i.e. the number of symbols in S's string representation), Lemma 11.2.2 and Theorem 11.2.5 tell us that each branch of the proof may contain chains that consist of $O(n^2)$ worlds, and thus it may contain $O(n^3)$ applications of ClL.

We show in §12 that this contraction upper-bound allows us to show that provability in the modal logics K4 and S4 is decidable in PSPACE. However, while it is true that in general we might need to contract $x_i{:}\Box A$ to infer $x_j{:}A$ from $x_i{:}\Box A$ by $\Box$L for each x_j accessible from x_i in a chain, it is often the case that many of these contractions are superfluous and we can dispose of them. It is thus reasonable to assume that this contraction bound is not 'optimal' and that there exists a better, smaller, bound; we conjecture that it is possible to establish a quadratic one, i.e. $O(|\vdash x_1{:}D|^2)$.

Conjecture 11.2.6 *Given a proof of $S = \vdash x_1{:}D$ in* S(K4) *or* S(S4), *there is a proof of S in the corresponding system such that in each branch there are at most $nbs(S) \times (pbs(S)+1)$ applications of* $\Box$R *and chains contain at most $1+nbs(S) \times (pbs(S)+1)$ worlds. Thus, in each branch there are at most $((nbs(S) \times (pbs(S) + 1)) - 1)$ applications of* ClL, *and we can restrict* ClL *to be* ClL*s with the contraction index s set to this number at the start of a backwards proof of $S = \vdash x_1{:}D$.* ■

The intuition behind this smaller bound is as follows. We need a contraction of $x_i{:}\Box A[\![\Box B]\!]_-$ only when it allows us to generate a new world in a chain by an application of $\Box$R with principal formula $x_j{:}\Box B$, where x_j is accessible from x_i. Hence, in the worst case we need such a contraction to generate each single world (except the first) in the longest chain that we ought to construct in the branches of a proof of $S = \vdash x_1{:}D$. Chains contain at most $1 + nbs(S) \times (pbs(S) + 1)$ worlds, and so we need at most $((nbs(S) \times (pbs(S) + 1)) - 1)$ applications of ClL. In other words, a contraction of $x_i{:}\Box A[\![\Box B]\!]_-$ 'corresponds' to the creation of a new world in a chain in some branch of a proof of $S = \vdash x_1{:}D$; since we need not build chains that are longer than $1 + nbs(S) \times (pbs(S) + 1)$ worlds, we do not need more than $((nbs(S) \times (pbs(S) + 1)) - 1)$ applications of ClL per branch. The following three observations expand on this.

First, the restriction to principal formulas of the form $x_i{:}\Box A[\![\Box B]\!]_-$ disposes of superfluous contractions of lwffs $x_i{:}\Box A$ where there is no B such that $\Box B \Subset_- A$.[6] However, in the multiplications in Theorems 11.2.3 and 11.2.5 we have made no distinction between contractions of $x_i{:}\Box A$ where there is some B such that $\Box B \Subset_- A$ and contractions of $x_i{:}\Box A$ where there is no such B.

Second, we have not been able to find theorems of S(K4) and/or S(S4) that require the $O(n^3)$ contractions of Theorem 11.2.5. On the other hand, we can fairly easily come up with theorems for which applications of ClL are bounded above in each branch by the number of Conjecture 11.2.6. For instance, the proof (6.2) of the S(K4)-theorem $x_1{:}\Box \sim \Box B \supset \Box \sim \Box\Box B$ that we have given in Example 6.1.6 contains only one contraction of $x_1{:}\Box \sim \Box B$, instead of the 21 allowed by Theorem 11.2.5, which also include left contractions of $x_2{:}\Box\Box B$ and $x_3{:}\Box B$ even when B is a propositional variable. Similar observations hold for the other S(K4)-theorems in Table 11.1 and for the S(S4)-theorem (11.1), as well as for their generalizations. For example, when

[6]Note that this is not the case in standard sequent systems for K4 and S4 as, e.g., the rule (K4) implicitly contracts boxed formulas on the left of $\vdash$ independent of their form. We return to this in §11.3 when we show the admissibility of (K4).

each C_i is a propositional variable, in both S(K4) and S(S4) a proof of

$$x_1{:}\Box(\bigwedge_{i=1}^{n}(C_i \supset \sim\Box\sim C_{i+1}) \wedge \sim C_n) \supset \Box\sim C_1\,, \tag{11.7}$$

which is a generalization of the last formula in Table 11.1 where $\bigwedge_{i=1}^{n}\varphi_i$ stands for the conjunction of the i formulas φ_i for $1 \le i \le n$, requires i contractions of

$$x_1{:}\Box(\bigwedge_{i=1}^{n}(C_i \supset \sim\Box\sim C_{i+1}) \wedge \sim C_n)\,,$$

namely, one contraction for each $\Box$ that occurs negative in it (i.e. one for each of its subformulas $\Box \sim C_{i+1}$).[7] Moreover, we can modify (11.7) to require additional contractions, and we can do so not only like we did for the corresponding example (10.1) in S(T), but also by replacing '$\wedge \sim C_n$' with

$$\text{`}\wedge \sim(C_n \wedge (\Box(\bigwedge_{j=1}^{m}(E_j \supset \sim\Box\sim E_{j+1}) \wedge \sim E_m) \supset \Box\sim E_1))\text{'}$$

so that we obtain a theorem of S(K4) and S(S4) such that for each subformula that has a positive $\Box$ as its main operator we need at most as many contractions as there are $\Box$'s that occur negative in its scope. That is, $O(|\vdash x_1{:}D|^2)$ left contractions.

The third and last observation supporting our conjecture is that we can transform backwards proofs in S(K4) and S(S4) so that each application of ClL that is not eliminable is followed by a particular derivation (sequence of rule applications) that has the effect of creating a new world in the chain of worlds that we are constructing in a branch.[8] All other contractions can be eliminated from the branch.

To illustrate this, recall that in each backwards proof we can always permute rules so that each left contraction of $x_i{:}\Box A[\![\Box B]\!]_-$ immediately precedes the application of $\Box$L introducing the second contraction constituent. Moreover, both constituents must be introduced by $\Box$L in at least one subbranch, for otherwise the application of ClL is trivially eliminable (since we can delete it together with the application of WlL introducing one of the two constituents). We can further transform a proof of $\vdash x_1{:}D$ so that before we apply a rule that has the second constituent as its principal formula, we 'decompose' the first one to generate a new world in the chain that we are building.

---

[7]The formula (11.7), suggested by Fabio Massacci in a private communication, is essentially a 'linearized' version of the S4-formula built by Halpern and Moses in [123] to show that there are formulas such that the models satisfying them contain exponentially many worlds (in the size of the formula). Hence, the models have exponential size, and the proofs built for such formulas in deduction systems for S4, including Halpern and Moses' tableaux system and our S(S4), require exponential size as well. Nonetheless, the decision problem for S4 is still in PSPACE as it is possible to build proofs in such a way that the entire model is not represented in memory and each branch in them has polynomial length in the size of the goal. We point to [82, 123, 153, 160] for additional details.

[8]At least one of the constituents must be introduced by such a derivation, but not necessarily both, as can be seen, e.g., when proving $\vdash x_1{:}\Box\sim(C \supset \Box C) \supset \Box D$ in S(K4).

That is, we can permute rules in the proof to obtain

$$\dfrac{\Delta_1 \vdash x_i R\, x_{i+j} \qquad \dfrac{\dfrac{x_i{:}\Box A[\![\Box B]\!]_-, \Gamma_2, \Delta_2, x_{i+j+k} R\, x_{i+j+k+1} \vdash \Gamma'_2, x_{i+j+k+1}{:}B}{x_i{:}\Box A[\![\Box B]\!]_-, \Gamma_2, \Delta_2 \vdash \Gamma'_2, x_{i+j+k}{:}\Box B}\,\Box\text{R}}{x_{i+j}{:}A[\![\Box B]\!]_-, x_i{:}\Box A[\![\Box B]\!]_-, \Gamma_1, \Delta_1 \vdash \Gamma'_1}\,\Pi_1}{\dfrac{x_i{:}\Box A[\![\Box B]\!]_-, x_i{:}\Box A[\![\Box B]\!]_-, \Gamma_1, \Delta_1 \vdash \Gamma'_1}{x_i{:}\Box A[\![\Box B]\!]_-, \Gamma_1, \Delta_1 \vdash \Gamma'_1}\,\text{ClL}}\,\Box\text{L}$$

$$\Pi_0 \qquad \vdash x_1{:}D$$

where $j > 0$ in S(K4) and $j \geq 0$ in S(S4), $k \geq 0$ in both S(K4) and S(S4), and $x_i{:}\Box A[\![\Box B]\!]_-$ is parametric in the subderivation Π_1 in which we infer $x_{i+j+k}{:}\Box B$ on the right of $\vdash$ from $x_{i+j}{:}A[\![\Box B]\!]_-$ on its left. In this transformed proof (branch), the application of $\Box$L that follows the contraction of $x_i{:}\Box A[\![\Box B]\!]_-$ is itself followed by subderivation ending with an application of $\Box$R with principal formula $x_{i+j+k}{:}\Box B$. Informally, if this is not the case, i.e. if there is no such application of $\Box$R, then it is as if there were no $\Box B$ such that $\Box B \Subset_- A$, and $x_{i+j+k}{:}\Box B$ is weak in the branch, so that we can eliminate the contraction like in Lemma 11.2.4. In other words, the sequence following a ClL generates at least one new world in a chain ('at least one' since each of the two constituents may be used to extend the chain). Hence, the worst case occurs when all the $1 + nbs(S) \times (pbs(S) + 1)$ worlds of a chain must be generated by contracting lwffs of the form $x_i{:}\Box A[\![\Box B]\!]_-$ for $((nbs(S) \times (pbs(S) + 1)) - 1)$ times.[9]

A formalization of these supporting observations would allow us to convert our conjecture into a fact, and thus reduce our cubic upper-bound on the number of applications of ClL into a quadratic one.

11.3 S(K4) AND SS(K4)

The standard sequent system SS(K4), see §6.1 and [87, 119, 120], is obtained from SS(K) by replacing (K) with the transitional rule

$$\dfrac{\Gamma, \Box\Gamma \vdash A}{\Sigma, \Box\Gamma \vdash \Box A, \Sigma'}\ (\text{K4}) \quad ,$$

which embeds multiple (implicit) contractions, namely one for each formula in Γ.

Recall that in §9.2 and §10.2 we have given a proof-theoretical justification of the rules of SS(K) and SS(T) by showing that their labelled equivalents can be derived

[9] A constructive proof of this would rely on the $\Box$-disjunction property and its corollaries to eliminate left contractions of $x_i{:}\Box A[\![\Box B]\!]_-$ that generate, by means of a sequence of rule applications such as the one described above, two worlds x_j and x_k that diverge from x_i. That is, left contractions that do not generate 'increasing' worlds in a chain are trivially eliminable. Mirroring the development for S(T), we could then also exploit Lemma 10.1.6 to dispose of redundant instances of the subformula B. To this end, note that the proof of Lemma 10.1.6 we have given in §10.1 does not depend on relational rules, except for case 2.2.3. Thus, the lemma holds also for modal systems other than S(T) provided that we check this case, which we can do fairly straightforwardly for S(K4) and S(S4) by exploiting the transitivity of R.

in our systems, and that there are intermediate systems equivalent to both ours and the standard ones. Further, based on our analysis of contractions in S(T), we have given a refined version of SS(T). For SS(K4) and SS(S4) things are, however, not so clear-cut: it is possible, at least to some extent, to use the above insights to similarly justify and refine SS(K4) and SS(S4) but it is more difficult to find the appropriate rules.

To illustrate this, observe that by a suitable number of applications of $\Box L_{K4}$ we can give a labelled equivalent $\Box LR_{K4}$ of the transitional rule (K4), i.e.

$$\frac{x_{i+1}{:}\Gamma, x_{i+1}{:}\Box\Gamma \vdash x_{i+1}{:}A}{x_i{:}\Sigma, x_i{:}\Box\Gamma \vdash x_i{:}\Box A, x_i{:}\Sigma'}\ \Box LR_{K4} \qquad \rightsquigarrow$$

$$\begin{array}{c} \dfrac{x_{i+1}{:}\Gamma, x_{i+1}{:}\Box\Gamma \vdash x_{i+1}{:}A}{x_{i+1}{:}\Gamma, x_{i+1}{:}\Box\Gamma, x_i R\, x_{i+1} \vdash x_{i+1}{:}A}\ \text{WrL} \\ \vdots\ \Box L_{K4}\ (\text{all with active rwff } x_i R\, x_{i+1}) \\ \dfrac{x_i{:}\Box\Gamma, x_i R\, x_{i+1} \vdash x_{i+1}{:}A}{x_i{:}\Box\Gamma \vdash x_i{:}\Box A}\ \Box R \\ \vdots\ \text{W} \\ x_i{:}\Sigma, x_i{:}\Box\Gamma \vdash x_i{:}\Box A, x_i{:}\Sigma' \end{array} \tag{11.8}$$

where the multisets of lwffs $x_i{:}\Sigma$ and $x_i{:}\Sigma'$ contain only formulas labelled with x_i, and if $x{:}\Gamma = \{x{:}B_1, \ldots, x{:}B_n\}$ then $x{:}\Box\Gamma = \{x{:}\Box B_1, \ldots, x{:}\Box B_n\}$.

The derivation (11.8) employs several applications of $\Box L_{K4}$ (one for each formula in $\Box\Gamma$), each of which is obtained by (a contraction and) a *cut* as in (11.4). Thus both $\Box L_{K4}$ and $\Box LR_{K4}$ are admissible, rather than derived, rules of S(K4).[10] As a consequence, although we can still define a system $\widehat{\mathrm{S}}$(K4) equivalent to SS(K4), showing the equivalence of S(K4) and $\widehat{\mathrm{S}}$(K4) requires more ingenuity.

Let $\widehat{\mathrm{S}}$(K4) be the system obtained from $\widehat{\mathrm{S}}$(K) by replacing $\Box LR_K$ with $\Box LR_{K4}$; since $\Box LR_{K4}$ embeds the unavoidable left contractions of boxed formulas, $\widehat{\mathrm{S}}$(K4), like SS(K4), does not contain structural rules. We have:

Lemma 11.3.1 *$\vdash x_1{:}D$ is provable in* S(K4) *iff $\vdash x_1{:}D$ is provable in* $\widehat{\mathrm{S}}$(K4). ■

The right-to-left direction follows by the admissibility of $\Box LR_{K4}$ in S(K4). For the left-to-right direction, we show that we can transform a S(K4)-proof Π of $\vdash x_1{:}D$ into a $\widehat{\mathrm{S}}$(K4)-proof of the same sequent.

For concreteness, we show the equivalence of S(K4), $\widehat{\mathrm{S}}$(K4) and SS(K4) by transforming S(K4)-proofs into a *block form*; as before, this is achieved by eliminating

[10]That $\Box L_{K4}$ and $\Box LR_{K4}$ are not derivable in S(K4) without *cut* follows from the subformula property. Note however that the applications of *cut* in (11.4), and thus in (11.8), are 'strictly controlled': the sequent $x_i{:}\Box\Gamma \vdash x_i{:}\Box A$ tells us precisely which formulas need to be cut in, namely the lwffs that we obtain by prefixing with another $\Box$ each lwff in $x_i{:}\Box\Gamma$. In order to derive $\Box LR_{K4}$ we thus need a form of 'superanalytic' *cut* in which only specific superformulas are allowed.

detours and adjoining related rules, and there are only a few changes with respect to our development for S(K).

We begin by eliminating detours in Π. We extend the definition for S(K) by saying that an application of a rule (r) is a detour in a S(K4)-proof Π of $\vdash x_1{:}D$ if

(i) all of the active formulas of (r) are introduced in Π either by weakenings or by detours (i.e. none of them appears in the axioms of Π so that they are weak in Π), or

(ii) (r) is an application of $\Box$L in which the active rwff is introduced by *trans* and the active lwff is introduced by weakening or by detours.

For example, the application of $\Box$L in

$$\dfrac{\begin{array}{c}\Pi_1\\ \Delta, x_i R x_j, x_j R x_k \vdash x_i R x_k\end{array} \quad \dfrac{\begin{array}{c}\Pi_2\\ \Gamma, \Delta, x_i R x_j, x_j R x_k \vdash \Gamma'\end{array}}{x_k{:}A, \Gamma, \Delta, x_i R x_j, x_j R x_k \vdash \Gamma'}\,\text{WlL}}{\begin{array}{c}x_i{:}\Box A, \Gamma, \Delta, x_i R x_j, x_j R x_k \vdash \Gamma'\\ \Pi_0\\ \vdash x_1{:}D\end{array}}\,\Box\text{L}$$

is a detour that we eliminate by ('blowing up' the application of WlL and) transforming the proof to

$$\dfrac{\begin{array}{c}\Pi_2\\ \Gamma, \Delta, x_i R x_j, x_j R x_k \vdash \Gamma'\end{array}}{\begin{array}{c}x_i{:}\Box A, \Gamma, \Delta, x_i R x_j, x_j R x_k \vdash \Gamma'\\ \Pi_0\\ \vdash x_1{:}D\end{array}}\,\text{WlL} \quad .$$

If the rule (r) in (i) is a contraction ClLs (or ClL), then we simply delete it together with the corresponding weakenings; e.g. we transform

$$\dfrac{\dfrac{\begin{array}{c}\Pi_1\\ \Delta \vdash x_i R x_j\end{array} \quad \dfrac{\begin{array}{c}\Pi_2\\ x_i{:}\Box \sim\Box B, \Gamma, \Delta \vdash^{s-1} \Gamma'\end{array}}{x_j{:} \sim\Box B, x_i{:}\Box \sim\Box B, \Gamma, \Delta \vdash^{s-1} \Gamma'}\,\text{WlL}}{x_i{:}\Box \sim\Box B, x_i{:}\Box \sim\Box B, \Gamma, \Delta \vdash^{s-1} \Gamma'}\,\Box\text{L}}{\begin{array}{c}x_i{:}\Box \sim\Box B, \Gamma, \Delta \vdash^{s} \Gamma'\\ \Pi_0\\ \vdash x_1{:}D\end{array}}\,\text{ClL}s$$

to

$$\begin{array}{c}\Pi_2'\\ x_i{:}\Box \sim\Box B, \Gamma, \Delta \vdash^{s} \Gamma'\\ \Pi_0\\ \vdash x_1{:}D\end{array}$$

where Π_2' is obtained from Π_2 simply by replacing the contraction index $s-1$ with s. By iterating these transformations, we obtain a proof that is free from detours.

Given a proof Π' free from detours, in §9.2 and §10.2 we showed that, for S(K) and S(T), a proof in block form is obtained from Π' by permuting rules so that, for each x_i, rules with principal formulas labelled with x_i are applied as an uninterrupted sequence. This does not straightforwardly generalize to S(K4). Consider, as a running example, the following simple proof where A is a propositional variable and $\Delta = \{x_1 R\, x_2, x_2 R\, x_3, x_3 R\, x_4\}$.

$$\dfrac{\dfrac{\dfrac{\dfrac{\dfrac{\dfrac{\dfrac{\begin{array}{c}\vdots \\ \Delta \vdash x_1 R\, x_3\end{array} \quad \begin{array}{c}\vdots \\ \Delta \vdash x_3 R\, x_4\end{array}}{\Delta \vdash x_1 R\, x_4}\,\textit{trans} \qquad \begin{array}{c}\dfrac{}{x_4{:}A \vdash x_4{:}A}\,\text{AX1} \\ \vdots\,\text{W} \\ x_4{:}A, \Delta \vdash x_4{:}A, x_4{:}B\end{array}}{x_1{:}\Box A, \Delta \vdash x_4{:}A, x_4{:}B}\,\Box\text{L}}{x_1{:}\Box A, \Delta \vdash x_4{:}A \vee B}\,\vee\text{R}}{x_1{:}\Box A, x_1 R\, x_2, x_2 R\, x_3 \vdash x_3{:}\Box(A \vee B)}\,\Box\text{R}}{x_1{:}\Box A, x_1 R\, x_2 \vdash x_2{:}\Box\Box(A \vee B)}\,\Box\text{R}}{x_1{:}\Box A \vdash x_1{:}\Box\Box\Box(A \vee B)}\,\Box\text{R}}{\vdash x_1{:}\Box A \supset \Box\Box\Box(A \vee B)}\,\supset\text{R} \quad . \qquad (11.9)$$

This proof is free from detours but we cannot transform it so that rules with principal formulas with the same label are applied as an uninterrupted sequence: we can permute the application of $\Box$L over that of $\vee$R but not over the applications of $\Box$R below it, since the rwffs active in the applications of $\Box$R are also (indirectly) active in $\Box$L by transitivity. This tells us that in S(K4) it is not enough to consider two rules to be related when their principal formulas have the same label; we must extend this by defining an application of $\Box$L and an application of $\Box$R to be related when there exists an rwff that is active in both rules via transitivity, as is the case in (11.9). Specifically, we define applications of $\Box$L and $\Box$R to be related when

- the active rwff $x_{i+j} R\, x_{i+j+1}$ of $\Box$R is also active in the applications of *trans* that introduce the active rwff $x_i R\, x_{i+j+1+k}$ of $\Box$L.

Note that when $j = k = 0$ this reduces to the case where the principal formulas of $\Box$L and $\Box$R have the same label. Thus, in (11.9), the application of $\Box$L is related to all three applications of $\Box$R, and we can transform (11.9) to

$$\dfrac{\dfrac{\dfrac{\dfrac{\dfrac{\dfrac{\begin{array}{c}\vdots \\ \Delta \vdash x_1 R\, x_3\end{array} \quad \begin{array}{c}\vdots \\ \Delta \vdash x_3 R\, x_4\end{array}}{\Delta \vdash x_1 R\, x_4}\,\textit{trans} \qquad \dfrac{\begin{array}{c}\dfrac{}{x_4{:}A \vdash x_4{:}A}\,\text{AX1} \\ \vdots\,\text{W} \\ x_4{:}A, \Delta \vdash x_4{:}A, x_4{:}B\end{array}}{x_4{:}A, \Delta \vdash x_4{:}A \vee B}\,\vee\text{R}}{x_1{:}\Box A, \Delta \vdash x_4{:}A \vee B}\,\Box\text{L}}{x_1{:}\Box A, x_1 R\, x_2, x_2 R\, x_3 \vdash x_3{:}\Box(A \vee B)}\,\Box\text{R}}{x_1{:}\Box A, x_1 R\, x_2 \vdash x_2{:}\Box\Box(A \vee B)}\,\Box\text{R}}{x_1{:}\Box A \vdash x_1{:}\Box\Box\Box(A \vee B)}\,\Box\text{R}}{\vdash x_1{:}\Box A \supset \Box\Box\Box(A \vee B)}\,\supset\text{R} \quad . \qquad (11.10)$$

The permutations that adjoin related rules affect also contractions: we obtain a proof in which related rules are applied as uninterrupted sequences, where, in particular, each left contraction of $x_i{:}\Box A$ immediately precedes the application of $\Box$L that introduces $x_i{:}\Box A$. However, this is still not enough: at applications of $\Box\mathrm{LR}_{\mathrm{K4}}$,

$$\frac{x_{i+1}{:}\Gamma, x_{i+1}{:}\Box\Gamma \vdash x_{i+1}{:}A}{x_i{:}\Sigma, x_i{:}\Box\Gamma \vdash x_i{:}\Box A, x_i{:}\Sigma'}\ \Box\mathrm{LR}_{\mathrm{K4}}\ ,$$

and, analogously, at applications of (K4), *each* formula in $x_i{:}\Box\Gamma$ is implicitly contracted. As a consequence, proofs in $\widehat{\mathrm{S}}$(K4) and SS(K4) might contain more contractions, albeit all of them implicit, than the corresponding proofs in S(K4). This is because sequents in $\widehat{\mathrm{S}}$(K4) and SS(K4) are committed to representing only one world; thus, when applying the transitional rule $\Box\mathrm{LR}_{\mathrm{K4}}$ or (K4), i.e. when moving from one world to another, we must bring along, by contraction, boxed formulas that might be inessential to the successful conclusion of the proof. On the other hand, the sequents that appear in S(K4)-proofs may contain formulas labelled differently, i.e. the sequents contain information about different worlds, and this allows us to reduce the number of required contractions since we can postpone applications of $\Box$L until they are really useful to the proof.

Thus, to obtain a $\widehat{\mathrm{S}}$(K4)-proof, we need to transform our S(K4)-proof further by adding the possibly missing contractions. Specifically, we transform our S(K4)-proof by

- eliminating applications of *trans*,

- replacing each application of $\Box$L, and the possible contraction below it, with a suitable number m of applications of $\Box\mathrm{L}_{\mathrm{K4}}$ (i.e. for each $\Box$L we implicitly add m contractions and m applications of *cut*), and

- possibly adding applications of WlL to introduce the additional lwffs required in the premises of the applications of $\Box\mathrm{L}_{\mathrm{K4}}$.

The number m depends on the rule introducing the active rwff of the particular application of $\Box$L that we are replacing: if the active rwff of $\Box$L follows just by applications of WrL and AXr, then we can replace $\Box$L with only one application of $\Box\mathrm{L}_{\mathrm{K4}}$; if the active rwff of $\Box$L is introduced by an application of *trans* with premise $\Delta \vdash x_i R x_j$, then we can replace $\Box$L with m applications of $\Box\mathrm{L}_{\mathrm{K4}}$, where m is the 'distance' $j - i$ between x_i and x_j, i.e. the number of applications of *trans* required to prove $\Delta \vdash x_i R x_j$. In both cases, $\Box\mathrm{L}_{\mathrm{K4}}$ absorbs the possible contraction preceding the application of $\Box$L. $\Box\mathrm{L}_{\mathrm{K4}}$ absorbs also the possible applications of WlL and WlR that surround $\Box$L, although this, as well as the absorption of the contractions, is not illustrated in our simple example, which illustrates only the second of the above cases: $m = 3$ and we transform (11.10) to

$$
\dfrac{\dfrac{\dfrac{\dfrac{\dfrac{\begin{array}{c}\vdots\\ \Delta \vdash x_1 R x_2\end{array} \quad \dfrac{\begin{array}{c}\vdots\\ \Delta \vdash x_2 R x_3\end{array} \quad \dfrac{\begin{array}{c}\vdots\\ \Delta \vdash x_3 R x_4\end{array} \quad \dfrac{\dfrac{\begin{array}{c}\dfrac{}{x_4{:}A \vdash x_4{:}A}\,\mathrm{AXl}\\ \vdots\ \mathrm{W}\\ x_4{:}A, \Delta \vdash x_4{:}A, x_4{:}B\end{array}}{x_4{:}A, x_4{:}\Box A, \Delta \vdash x_4{:}A, x_4{:}B}\,\mathrm{WlL}}{x_4{:}A, x_4{:}\Box A, \Delta \vdash x_4{:}A \lor B}\,{\lor}\mathrm{R}}{x_3{:}A, x_3{:}\Box A, \Delta \vdash x_4{:}A \lor B}\,\Box\mathrm{L_{K4}}}{x_2{:}A, x_2{:}\Box A, \Delta \vdash x_4{:}A \lor B}\,\Box\mathrm{L_{K4}}}{x_1{:}\Box A, \Delta \vdash x_4{:}A \lor B}\,\Box\mathrm{L_{K4}}}{x_1{:}\Box A, x_1 R x_2, x_2 R x_3 \vdash x_3{:}\Box(A \lor B)}\,\Box\mathrm{R}}{x_1{:}\Box A, x_1 R x_2 \vdash x_2{:}\Box\Box(A \lor B)}\,\Box\mathrm{R}}{x_1{:}\Box A \vdash x_1{:}\Box\Box\Box(A \lor B)}\,\Box\mathrm{R}}{\vdash x_1{:}\Box A \supset \Box\Box\Box(A \lor B)}\,{\supset}\mathrm{R}\ . \tag{11.11}
$$

Note that a S(K4)-proof so transformed may contain additional applications of WlL (in this case the weakening of $x_4{:}\Box A$), but it does not contain applications of *trans*: all the relational reasoning that is left consists of applications of WrL and AXr.

We can now further adjoin related rules, namely the applications of $\Box\mathrm{L_{K4}}$ and $\Box\mathrm{R}$ that have the same (implicit) active rwff and principal formula labelled with the same label. Specifically, we transform (11.11) to

$$
\dfrac{\dfrac{\dfrac{\dfrac{}{x_1 R x_2 \vdash x_1 R x_2}\,\mathrm{AXr} \quad \dfrac{\dfrac{\begin{array}{c}\vdots\\ x_1 R x_2, x_2 R x_3 \vdash x_2 R x_3\end{array} \quad \dfrac{\dfrac{\begin{array}{c}\vdots\\ \Delta \vdash x_3 R x_4\end{array} \quad \dfrac{\dfrac{\begin{array}{c}\dfrac{}{x_4{:}A \vdash x_4{:}A}\,\mathrm{AXl}\\ \vdots\ \mathrm{W}\\ x_4{:}A, \Delta \vdash x_4{:}A, x_4{:}B\end{array}}{x_4{:}A, x_4{:}\Box A, \Delta \vdash x_4{:}A, x_4{:}B}\,\mathrm{WlL}}{x_4{:}A, x_4{:}\Box A, \Delta \vdash x_4{:}A \lor B}\,{\lor}\mathrm{R}}{x_3{:}A, x_3{:}\Box A, \Delta \vdash x_4{:}A \lor B}\,\Box\mathrm{L_{K4}}}{\begin{array}{c}x_3{:}A, x_3{:}\Box A, x_1 R x_2,\\ x_2 R x_3 \vdash x_3{:}\Box(A \lor B)\end{array}}\,\Box\mathrm{R}}{\begin{array}{c}x_2{:}A, x_2{:}\Box A, x_1 R x_2,\\ x_2 R x_3 \vdash x_3{:}\Box(A \lor B)\end{array}}\,\Box\mathrm{L_{K4}}}{x_2{:}A, x_2{:}\Box A, x_1 R x_2 \vdash x_2{:}\Box\Box(A \lor B)}\,\Box\mathrm{R}}{x_1{:}\Box A, x_1 R x_2 \vdash x_2{:}\Box\Box(A \lor B)}\,\Box\mathrm{L_{K4}}}{x_1{:}\Box A \vdash x_1{:}\Box\Box\Box(A \lor B)}\,\Box\mathrm{R}}{\vdash x_1{:}\Box A \supset \Box\Box\Box(A \lor B)}\,{\supset}\mathrm{R}\ . \tag{11.12}
$$

We are almost done: we have finally obtained the desired S(K4)-proof in block form, which consists of alternating sequences of local reasoning (propositional rules) and transitional reasoning ('$\Box\mathrm{R}$–$\Box\mathrm{L_{K4}}$ pairs'). We can then transform this into a $\widehat{\mathrm{S}}$(K4)-proof by absorbing the uppermost weakenings of lwffs into instances of (extended) AXl, and replacing the transitional reasoning with applications of $\Box\mathrm{LR_{K4}}$; this latter step eliminates all rwffs and the remaining relational reasoning (applications of WrL and AXr) from sequents and proofs.

By similar proof transformations we can transform any S(K4)-proof of $\vdash x_1{:}D$ into a $\widehat{\mathrm{S}}$(K4)-proof of the same sequent, and thus conclude the (informal) justification of the left-to-right direction of Lemma 11.3.1. We can then transform this $\widehat{\mathrm{S}}$(K4)-proof into a SS(K4)-proof of $\vdash D$, where, as for $\widehat{\mathrm{S}}$(T) and SS(T), proofs in $\widehat{\mathrm{S}}$(K4) and SS(K4) differ only in the names of the rules, e.g. $\Box\mathrm{LR}_{\mathrm{K4}}$ and (K4), and in the presence of labels, which can be eliminated or added as required. For example, (11.12) yields the $\widehat{\mathrm{S}}$(K4)-proof shown below on the left, which then yields the SS(K4)-proof on the right:

$$
\dfrac{\dfrac{\dfrac{\dfrac{\dfrac{\dfrac{}{x_4{:}A, x_4{:}\Box A \vdash x_4{:}A, x_4{:}B}\,\mathrm{AXl}}{x_4{:}A, x_4{:}\Box A \vdash x_4{:}A \vee B}\,\vee\mathrm{R}}{x_3{:}A, x_3{:}\Box A \vdash x_3{:}\Box(A \vee B)}\,\Box\mathrm{LR}_{\mathrm{K4}}}{x_2{:}A, x_2{:}\Box A \vdash x_2{:}\Box\Box(A \vee B)}\,\Box\mathrm{LR}_{\mathrm{K4}}}{x_1{:}\Box A \vdash x_1{:}\Box\Box\Box(A \vee B)}\,\Box\mathrm{LR}_{\mathrm{K4}}}{\vdash x_1{:}\Box A \supset \Box\Box\Box(A \vee B)}\,\supset\mathrm{R}
\qquad
\dfrac{\dfrac{\dfrac{\dfrac{\dfrac{\dfrac{}{A, \Box A \vdash A, B}\,(\mathrm{AX})}{A, \Box A \vdash A \vee B}\,(\vee\mathrm{R})}{A, \Box A \vdash \Box(A \vee B)}\,(\mathrm{K4})}{A, \Box A \vdash \Box\Box(A \vee B)}\,(\mathrm{K4})}{\Box A \vdash \Box\Box\Box(A \vee B)}\,(\mathrm{K4})}{\vdash \Box A \supset \Box\Box\Box(A \vee B)}\,(\supset\mathrm{R})\ .
$$

Although the development is rather entangled, since it relies on applications of *cut* and thus on admissible but not derivable rules, admissibility suffices for producing the proof transformations needed to show the equivalence of S(K4), $\widehat{\mathrm{S}}$(K4) and SS(K4). However, admissibility does complicate the analyses of the resources (i.e. applications of contractions): Theorem 11.2.5 restricts contractions in the system S(K4), and it is thus reasonable to expect that the restriction propagates to $\widehat{\mathrm{S}}$(K4) and SS(K4), as was the case for S(T). Unfortunately, the above transformation of a S(K4)-proof Π into a $\widehat{\mathrm{S}}$(K4)-proof does not distinguish applications of ClL and ClLs in Π, and it is unclear how to modify this transformation so that it propagates the side conditions on the principal formulas of ClLs. Indeed, we have seen that in the transformations we may need to add, rather than remove, contractions. Thus, while we can show that $\widehat{\mathrm{S}}$(K4) is equivalent to SS(K4), there is no immediate refinement of the rules. For example, we might consider replacing (K4) with the rule

$$\dfrac{\Gamma_1, \Gamma_2, \Box\Gamma_2 \vdash A}{\Sigma, \Box\Gamma_1, \Box\Gamma_2 \vdash \Box A, \Sigma'}\ (\dagger)$$

where for every $\Box F \in \Box\Gamma_1$ there is no B such that $\Box B \Subset_- F$, while for every $\Box F \in \Box\Gamma_2$ there is some B such that $\Box B \Subset_- F$. However, this rule yields an incomplete system: we can easily show that the K4-theorem

$$\Box \sim\Box C \supset (\Box D \supset \Box \sim\Box\Box(D \supset C))\,,$$

which is an instance of the second theorem schema in Table 11.1 with $p = 2$, is not provable when (†) is the only modal rule.

We could add (†) together with the rule

$$\dfrac{\Box\Gamma_1, \Gamma_2, \Box\Gamma_2 \vdash A}{\Sigma, \Box\Gamma_1, \Box\Gamma_2 \vdash \Box A, \Sigma'}\ (\ddagger)$$

where Γ_1 and Γ_2 have the same side conditions as for (†), but the resulting system then contracts, implicitly, also the formulas $\Box F \in \Box\Gamma_1$ where there is no B such that $\Box B \Subset_- F$.

To conclude, we therefore content ourselves with the following theorem, which states that, by Lemma 11.3.1 and the results above, our labelled system S(K4) provides a proof-theoretical justification of the standard sequent system SS(K4).

Theorem 11.3.2 *The following are equivalent:*

1. $\vdash x_1{:}D$ *is provable in* S(K4).
2. $\vdash x_1{:}D$ *is provable in* S(K4) *where* ClL *is restricted to be* ClLs *with* S *set according to Theorem 11.2.5.*
3. $\vdash x_1{:}D$ *is provable in* $\widehat{\mathrm{S}}$(K4).
4. $\vdash D$ *is provable in* SS(K4). ■

11.4 S(S4) AND SS(S4)

The standard system SS(S4), see §6.1 and [87, 119, 120], is obtained by extending SS(K) with the rules

$$\frac{A, \Box A, \Sigma \vdash \Sigma'}{\Box A, \Sigma \vdash \Sigma'}\ (\mathrm{T}) \qquad \text{and} \qquad \frac{\Box\Gamma \vdash A}{\Sigma, \Box\Gamma \vdash \Box A, \Sigma'}\ (\mathrm{S4}) \ .$$

We have already shown in §10.2 how to justify (T) and the contraction embedded in it by deriving the rule $\Box\mathrm{L_T}$; the same derivation can be given in S(S4). In order to give a labelled equivalent $\Box\mathrm{LR_{S4}}$ of (S4), we first give an admissible rule $\Box\mathrm{L_{S4}}$,

$$\frac{\Delta \vdash x_i R x_j \quad x_j{:}\Box A, \Gamma, \Delta \vdash \Gamma'}{x_i{:}\Box A, \Gamma, \Delta \vdash \Gamma'}\ \Box\mathrm{L_{S4}} \quad \rightsquigarrow$$

$$\frac{x_i{:}\Box A \vdash x_i{:}\Box\Box A \qquad \dfrac{\Delta \vdash x_i R x_j \quad x_j{:}\Box A, \Gamma, \Delta \vdash \Gamma'}{x_i{:}\Box\Box A, \Gamma, \Delta \vdash \Gamma'}\ \Box\mathrm{L}}{x_i{:}\Box A, \Gamma, \Delta \vdash \Gamma'}\ cut \ ,$$

where $x_i{:}\Box A \vdash x_i{:}\Box\Box A$ follows trivially by the transitivity of R. The transitional rule

$$\frac{x_{i+1}{:}\Box\Gamma \vdash x_{i+1}{:}A}{x_i{:}\Sigma, x_i{:}\Box\Gamma \vdash x_i{:}\Box A, x_i{:}\Sigma'}\ \Box\mathrm{LR_{S4}}$$

is then obtained by a suitable number of weakenings and applications of $\Box\mathrm{L_{S4}}$ where, as before, the multisets of lwffs $x_i{:}\Sigma$ and $x_i{:}\Sigma'$ contain only formulas labelled with x_i, and if $x{:}\Gamma = \{x{:}B_1, \ldots, x{:}B_n\}$ then $x{:}\Box\Gamma = \{x{:}\Box B_1, \ldots, x{:}\Box B_n\}$.

If we now compare the derivation of $\Box\mathrm{L_{S4}}$ with the derivation (11.4) of $\Box\mathrm{L_{K4}}$, we see that for $\Box\mathrm{L_{S4}}$ it suffices to use *cut* and we do not need to apply ClL. The embedded application of *cut* implies however that $\Box\mathrm{L_{S4}}$ and $\Box\mathrm{LR_{S4}}$ are admissible, rather than derived, rules of S(S4). It follows that, like for K4, the refinements that result from our analysis cannot be directly propagated to SS(S4). To make a long story short, we can combine the definitions and results for systems for T and K4 in §10.2 and §11.3 to show, by proof transformations (and block forms), the following equivalence, where, however, we do not distinguish between applications of ClL and ClLs in S(S4).

Theorem 11.4.1 $\vdash x_1{:}D$ *is provable in* S(S4) *iff* $\vdash D$ *is provable in* SS(S4). ■

That is, our analysis of S(S4) provides a justification of the rules of the standard sequent system SS(S4). However, since the proof transformations are again based on admissible rules, there is no immediate way of exploiting our analysis to restrict the rules of SS(S4).

12 COMPLEXITY OF PROOF SEARCH IN K, T, K4 AND S4

In this chapter we show how bounds from substructural analysis can be combined with bounds for relational reasoning to provide decision procedures with space complexity upper-bounds. In particular, we show that contraction elimination for S(K) and bounded contraction for S(T), S(K4) and S(S4), combined with the soundness and completeness of our systems with respect to the corresponding Kripke semantics, tell us that the provability (validity) problems for the modal logics K, T, K4 and S4 are decidable in PSPACE.

For clarity, we first summarize the results of the previous chapters, then begin our complexity analysis with K, and after consider extensions to other modal logics.

12.1 SUMMARY OF OUR SUBSTRUCTURAL ANALYSIS

We can summarize the main results of the previous chapters as follows.

Theorem 12.1.1 *For each* $\mathcal{L} \in \{\mathrm{K}, \mathrm{T}, \mathrm{K4}, \mathrm{S4}\}$, *let the 'restricted' system* $\mathrm{Sr}(\mathcal{L})$ *be defined as follows.*

- $\mathrm{Sr}(\mathrm{K})$ *consists of the axioms* AXl, AXr *and* $\bot$L, *where* AXl *and* $\bot$L *are restricted to atomic lwffs, the logical rules* $\supset$L, $\supset$R, $\Box$L *and* $\Box$R, *and the weakening rules* WrL, WlL *and* WlR.

- $\mathrm{Sr}(\mathrm{T})$ *extends* $\mathrm{Sr}(\mathrm{K})$ *with the contraction rule*

$$\frac{x{:}\Box A[\![\Box B]\!]_{-}, x{:}\Box A[\![\Box B]\!]_{-}, \Gamma, \Delta \vdash^{s-1} \Gamma'}{x{:}\Box A[\![\Box B]\!]_{-}, \Gamma, \Delta \vdash^{s} \Gamma'} \; \mathrm{ClL}s \; (s > 0)$$

with s set to pbs(S) at the start of a backwards proof of $S = \vdash x_1{:}D$, and with the relational rule refl.

- $\mathrm{Sr(K4)}$ *extends* $\mathrm{Sr(K)}$ *with the contraction rule* ClL*s with s set appropriately set (according to Theorem 11.2.5 or Conjecture 11.2.6) at the start of a backwards proof of* $S = \vdash x_1{:}D$*, and with the relational rule trans.*
- $\mathrm{Sr(S4)}$ *extends* $\mathrm{Sr(K4)}$ *with the relational rule refl.*

The systems $\mathrm{S}(\mathcal{L})$ *and* $\mathrm{Sr}(\mathcal{L})$ *are equivalent for theoremhood, in the sense that an lwff* $x_1{:}D$ *is a theorem of* $\mathrm{S}(\mathcal{L})$ *iff it is a theorem of* $\mathrm{Sr}(\mathcal{L})$*. Hence,* $\mathrm{Sr}(\mathcal{L})$ *is a sound and complete system for theoremhood in each logic* $\mathcal{L} \in \{\mathrm{K}, \mathrm{T}, \mathrm{K4}, \mathrm{S4}\}$*.* ■

12.2 COMPLEXITY OF PROOF SEARCH IN K

We apply rules of $\mathrm{Sr(K)}$ backwards to build a proof top-down, starting with the end-sequent $S = \vdash x_1{:}D$ and working towards the axioms of the proof. We begin by showing that the search space for proofs is finite and after analyze space requirements.

Let the *size of a formula A* be the number of symbols in A's string representation, i.e. $|A|$, and let the *degree of an lwff x:A* be twice the size of A, i.e. $2 \times |A|$. We define the *degree of a sequent S* to be the sum of the degrees of the lwffs in S plus the number of rwffs in S.[1] The degree serves as a measure of the complexity of a sequent; by examining the rules of $\mathrm{Sr(K)}$ we can establish:

Fact 12.2.1 (Measure Decreasingness) *In each rule of* $\mathrm{Sr(K)}$*, each premise has degree smaller than that of the conclusion.* ■

It follows that the length of any branch in a proof of $S = \vdash x_1{:}D$ is $O(n)$, where $n = |S|$ is the *size of the sequent S* (defined, like for A, as the number of symbols in its string representation). We can also bound the distinct lwffs appearing in proofs. An examination of the rules of $\mathrm{Sr(K)}$ shows:

Fact 12.2.2 (Subformula Property) *Each premise of each rule of* $\mathrm{Sr(K)}$ *contains only labelled subformulas of the labelled formulas in the conclusion.* ■

Due to the rules $\Box$L and $\Box$R, subformulas may have different labels. However, new labels are only generated by applications of $\Box$R and an application of $\Box$L can only lead to axioms (the first premise of an application of $\Box$L is only provable) when the label y chosen has been generated by a previous application of $\Box$R. Since the number of possible applications of $\Box$R in a Sr(K)-proof of $S = \vdash x_1{:}D$ is bounded above by

[1] Note that this amounts to implicitly defining the degree of an rwff to be 1. Other definitions of sizes and degrees (of lwffs, rwffs and sequents) are possible.

$pbs(S) \leq |S|$, we can additionally bound the number of labels appearing in branches, and with this the possible lwffs and rwffs.

Fact 12.2.3 (Bounded Labels and Formulas) *The number of labels appearing on any branch of a* Sr(K)*-proof of* $S = \vdash x_1{:}D$ *is* $O(n)$*, where* $n = |S|$ *is the size of* S*. The number of different possible lwffs and rwffs is* $O(n^2)$ *and of these only* $O(n)$ *can occur in any sequent.* ■

It follows that provability is decidable. Since the length of each branch is bounded and so are the formulas appearing in each sequent, for any end-sequent we need only check a finite number of possible proofs.

To give a finer analysis of decidability, we distinguish between two different kinds of branching that arise in the search space for proofs.

- *Conjunctive branching*: applying rules with multiple premises builds a branching tree, where all branches must be proved.
- *Disjunctive branching*: more than one rule may be applicable and a given rule may be applicable in different ways.

Conjunctive branching, caused by rules like $\supset$L, leads to proofs that are exponential in size (although the length of the branches may be only polynomial). Disjunctive branching arises when more than one rule may be applied to a sequent or when a single rule can be applied to more than one formula in a sequent (e.g. weakening) or to a formula in more than one way (e.g. $\Box$L, where we can pick any suitable y that has been previously generated). Disjunctive branching is reflected by a branching point in the search space for proofs rather than in the proofs themselves.

To minimize space requirements in navigating the search space, we adapt a standard technique (see, e.g., [136, 138]). Rather than storing entire proofs, we store *a sequent and a stack*. The stack maintains information sufficient to reconstruct both kinds of branching points, and stack entry is a triple consisting of the name of the rule applied to a sequent, the principal formula of the rule, and an index. The stack allows us to reconstruct the sequent associated with a branching point by replaying the stack entries to that point. The index records sufficient information such that on return to the branching point we can generate the remaining branches. For example, for $\supset$L, the index is a bit indicating the first or second premise. For $\Box$L, the index also records the label y chosen.

A proof begins with an end-sequent $S = \vdash x_1{:}D$ and the empty stack. Each rule application generates a new sequent and appropriately extends the stack. (This extension represents a disjunctive branching point; we assume rules are ordered arbitrarily, e.g. alphabetically, and we apply them in order.) If the generated sequent is an axiom and the stack contains no conjunctive branching points that still need to be explored, then S is provable. Otherwise we pop entries off the stack until we find a conjunctive branching point that must be further explored and then generate the next branch (first incrementing the index on the stack to record this). Alternatively, if we arrive at a sequent that is not an axiom and no rule applies, then we pop stack entries and continue at the first available disjunctive branching point with the remaining choices. If no such branching point remains, then S is not provable.

This procedure terminates since, by Fact 12.2.1, the stack depth is $O(n)$ and branching is bounded (by n, the number of rules, and the maximum number of premises). Since a formula D of the modal logic K is provable iff the sequent $\vdash x_1{:}D$ is provable in $\mathrm{Sr}(\mathrm{K})$, and because the entire search space for proofs is navigated, if need be, on backtracking, this is a decision procedure for K.

The space required by this decision procedure is the sum of the space required to store any intermediate sequent arising in the proof and the space required for the stack (plus the space n required to store the end-sequent). Fact 12.2.1 tells us that every proof of a sequent $S = \vdash x_1{:}D$, of size n, is bounded in length by the degree of S, which is $O(n)$. We can represent any generated sequent in $O(n^2)$-space since by Fact 12.2.3 there are only $O(n)$ lwffs and rwffs, each lwff is a subformula of the end-sequent (Fact 12.2.2), and there are only n possible labels. We can reduce this to $O(n \log n)$-space by representing any subformula by an index into the end-sequent S, which requires only $log\, n$ bits, and using $log\, n$ bits to encode each label. The stack can also be stored in $O(n \log n)$-space since it contains $O(n)$ entries and each entry requires constant space for the rule name, $O(\log n)$-space for a pointer to the principle formula (and its associated label), and $O(\log n)$-space for the index. Putting the above together gives us:

Theorem 12.2.4 *Provability for* K *is decidable in* PSPACE, *namely in* $O(n \log n)$-*space.* ■

With minimal extensions, the same decision procedure and analysis applies to other modal logics. However, $\mathrm{Sr}(\mathrm{K})$ is a good starting point since it is comparatively simple in two respects. First, contraction is eliminated, not just bounded. Second, relational reasoning, which arises in proving the first premise of $\Box$L, is trivial: $\Delta \vdash xRy$ is provable precisely when $xRy \in \Delta$.

12.3 COMPLEXITY OF PROOF SEARCH IN T

$\mathrm{Sr}(\mathrm{T})$ extends $\mathrm{Sr}(\mathrm{K})$ with reflexivity, *refl*, and the bounded contraction rule $\mathrm{ClL}s$, where each sequent is annotated with a contraction index s that is initially set to be the number of positive boxes in the end-sequent $S = \vdash x_1{:}D$.[2]

The backwards application of contraction increases the measure of sequents, hence, as is, Fact 12.2.1 fails. We address this problem with two changes. First, we incorporate the contraction index in the measure: we define the degree of an lwff $x{:}A$ to be, as before, twice the number of symbols in the string representation of A, i.e. $2 \times |A|$, and we now define the degree of a sequent S with contraction index s lexicographically as the pair (s, Σ), where Σ is the sum of the degrees of the lwffs of S plus the number of

[2]Observe, however, that if we are only interested in bounding the space complexity of our decision procedure, as we are here, then it suffices to consider a left contraction rule

$$\frac{x{:}A, x{:}A, \Gamma, \Delta \vdash^{s-1} \Gamma'}{x{:}A, \Gamma, \Delta \vdash^{s} \Gamma'}$$

in which we do not impose any syntactic restriction on the contracted lwff. The absence of the restriction has no effect on overall space requirements, although more backtracking may be required.

rwffs in S. Second, taking advantage of the separation of derivations in our systems, we shall only consider whether *non-logic-specific* rules (i.e. logical and weakening rules) reduce the measure and separately analyze the space requirements for *logic-specific* (relational) reasoning. By inspecting the rules, the reader can check that ClLs and every non-logic-specific rule reduces this measure.

Relational reasoning in S(T) is also trivial. $\Delta \vdash xRy$ is provable iff $xRy \in \Delta$ or y is x. A proof of xRy is thus trivially constructible in space linear in the size of Δ with $O(|\Delta|)$ possible applications of WrL followed either by AXr or *refl*.

We analyze the overall space requirements as follows. Given an end-sequent $S = \vdash x_1{:}D$, the contraction index s is bounded by $n = |S|$. From our measure, we have that the length of any branch is $O(n \times s)$, i.e. $O(n^2)$, ignoring relational reasoning. It follows that there are $O(n^2)$ lwffs and rwffs in any sequent and $O(n^2)$ possible labels in any branch. Now, using the fact that we can represent any label in $O(log\, n)$-space and thus any lwff in $O(log\, n)$-space (as an index into the end-sequent paired with a label), we have that any generated sequent can be represented using $O(n^2 \log n)$-space. Moreover, the stack depth is $O(n^2)$ (the depth ignoring relational reasoning plus the space required for such reasoning) and each entry again requires $O(\log n)$-space. So the stack itself also requires $O(n^2 \log n)$-space. Thus we have:

Theorem 12.3.1 *Provability for* T *is decidable in* PSPACE, *namely in* $O(n^2 \log n)$-*space.* ∎

12.4 COMPLEXITY OF PROOF SEARCH IN K4 AND S4

Sr(K4) extends Sr(K) with *trans* and the bounded contraction rule ClLs, where, at the start of a proof of $S = \vdash x_1{:}D$, we set $s = ((nbs(S) \times (pbs(S) + 1)) - 1) \times pbs(S)$ according to Theorem 11.2.5 (or to $((nbs(S) \times (pbs(S) + 1)) - 1)$ according to Conjecture 11.2.6). Sr(S4) extends Sr(K4) with the relational rule *refl*.

Since the bounded contraction rule of Sr(K4) and Sr(S4) is the same as that of Sr(T), except for the initial value of the contraction index s, we can use the same lexicographic measure (s, Σ). The reader can again check that ClLs and every non-logic-specific rule of Sr(K4) and Sr(S4) reduces this measure.

Relational reasoning also has the same space requirement. A goal $\Delta \vdash xRy$ is provable in Sr(K4) if and only if xRy is in the transitive closure of Δ; this can be determined in time linear in Δ using depth-first search and the result can be translated to a proof of linear depth. A goal $\Delta \vdash xRy$ is provable in Sr(S4) if and only if xRy is in the reflexive-transitive closure of Δ, which is also provable in linear space.

Hence, the analysis of overall space requirements is similar to that for Sr(T), where now s is $O(n^3)$, so that the length of any branch ignoring relational reasoning, and thus the stack depth, is $O(n^4)$. We can thus conclude:

Theorem 12.4.1 *Provability for* K4 *and* S4 *is decidable in* PSPACE, *namely in* $O(n^4 \log n)$-*space.* ∎

Note that by Conjecture 11.2.6 we would have instead that s is $O(n^2)$ and the stack depth is $O(n^3)$, from which it follows that provability for K4 and S4 is decidable in $O(n^3 \log n)$-space. There are various other ways in which we could improve our bounds for S(K4) and S(S4), and also that for S(T). One of them is to find 'better' measures than our lexicographic ones. For example, by defining, based perhaps on an analysis extending ours for S(K4) and S(S4), a measure that is decreased by *sequences* of rule applications (instead of a measure like ours, which is decreased by each rule application in isolation). Another way, which we illustrate in more detail while discussing related work in the next chapter, is to translate formulas into some normal form and then consider only proofs of formulas pre-processed that way.

13 DISCUSSION

We have shown how our framework provides a basis for new proof-theoretical method for establishing decidability and bounding the complexity of some non-classical logics, and, as examples, we have given PSPACE decision procedures for the propositional modal logics K, T, K4 and S4. We establish these bounds by combining restrictions on the structural rules of our labelled sequent systems with an analysis of the accessibility relation of the corresponding Kripke frames. Furthermore, we have shown that as a by-product of our analysis we can obtain justifications (and in some cases refinements) of the rules of standard sequent systems.

We view these results as a first step towards the application of our method to the analysis of decidability and complexity of families of modal, relevance and other non-classical logics, and for implementing decision procedures for these logics. For example, our results for S(K) and S(K4) should extend fairly straightforwardly to the serial modal logics S(D) and S(KD4), respectively; moreover, based on the method presented here, [231] gives a $O(n \log n)$-space decision procedure for the basic positive relevance logic B^+. We return to the results of [231] in §13.1.4, after discussing related work.

13.1 RELATED WORK

Modal logics are typically shown to be decidable semantically, by showing that they possess the *finite model property* [58, 141, 217]. (The same technique can be applied also to other non-classical logics [175, 196].) However, for many modal logics, including for example K, T and S4, there are classes of satisfiable formulas such that every satisfying model contains exponentially many worlds, and thus has exponential size; see [123] and also our brief discussion in §11.2. Thus, it is necessary to analyze the complexity of particular decision procedures to get better results.

A number of other authors have shown that the logics we have considered are decidable, but often without analyzing the complexity of their decision procedures or giving specific upper-bounds. Most of these procedures are based on two ideas: (i) termination of the procedure is shown by dynamically checking for loops during proof search, e.g. [87, 120, 127, 130, 153], or (ii) a specialized deduction system is developed that restricts cut and contraction, and thus allows for the definition of a measure that is decreased at every rule application, e.g. [55, 120, 136, 137, 138]. We now make comparisons with related work based on these techniques. After, in §13.1.3, we consider work on empirical performance analysis of automated decision procedures.

13.1.1 Dynamic loop-checking

Ladner. In [153], Ladner shows that provability in K, T and S4 is PSPACE-complete and provides upper-bounds. Using semantic-based tableaux systems and building upon the decision procedures of Kripke [150], he proves that K is decidable in $O(n^2)$-space, T in $O(n^3)$-space, and S4 in $O(n^4)$-space, where n is the size of the goal.

There are a number of similarities between Ladner's work and ours, including the usual analogies between tableaux and sequent systems, e.g. he partitions the generated formulas into a positive and a negative set, in the same way that we partition them in the antecedent or succedent of a sequent. His rules for manipulating these sets have the subformula property and he analyzes space requirements in terms of the stack depth of his procedures and the information stored at each level of recursion. As remarked in §11, we have adapted some of his results in our analysis of S(K4) and S(S4).

There are also important differences, however. For example, since Ladner's formulas are unlabelled, he needs logic-specific rules for decomposing boxed formulas; e.g. he must distinguish not just positively and negatively occurring formulas, but also track and handle subformulas of boxed formulas specially, since these are used to 'restart' tableaux at appropriate points during proof construction.

Ladner's work is representative of approaches based on loop-checking, and his solution is to introduce a 'global stack' recording the history of restarts and then show that only finitely many different restarts are possible. This bounding of restarts plays a role analogous to our bounding of contractions, and it is possible to replace the global history with a 'restart index' analogous to our contraction indices. Note also that using the techniques of §12 for minimizing space requirements in navigating the search space, Ladner's upper-bounds for K, T and S4 can be reduced to $O(n \log n)$-space, $O(n^2 \log n)$-space and $O(n^3 \log n)$-space, respectively.

Halpern and Moses. The modal logics considered by Ladner can be used to model knowledge and belief. In [123] (but see also [82]), Halpern and Moses extend the results of Ladner to model not just what a single agent knows (or believes) but more generally what multiple agents know. Each agent's knowledge corresponds to an indexed modal operator, e.g. n different boxes are used to model the knowledge of n different agents. Halpern and Moses show that the complexity of the satisfiability problem is PSPACE-complete for the multi-modal (one or more agent) versions of the logics K, T and S4. The logics S5 and KD45 are also PSPACE-complete, but only when there are at least two agents, and are NP-complete in the single agent case. The authors also analyze modal operators that model common and distributed knowledge.

This work goes beyond ours with respect to both the kinds of modal operators considered and the study of modal logics based on the '5' axiom, $\Diamond A \supset \Box\Diamond A$, which says that the accessibility relation is euclidean (cf. *eucl* in Table 2.3). Thus, [123] raises the interesting question of whether our results could be extended to *multi-modal logics*, e.g. [51, 105, 121]. While we have not attempted this, we do not foresee any difficulties for the logics we have considered and conjecture that the complexity bounds are identical to the mono-modal case (at least when the n different modal operators do not 'interact'; interaction axioms in multi-modal logics are discussed in [14, 15, 16], together with some (un)decidability results.) Extending our results to modal logics whose accessibility relation is euclidean will probably be more problematic since the $\Box$-disjunction property in Proposition 8.2.9, which provided a basis for eliminating applications of ClR (and of ClL, too), holds only for divergent logics/systems (cf. Table 8.1). Hence, analyses of non-divergent logics must be based on results other than Proposition 8.2.9. The question of whether our techniques can be extended to treat common and distributed knowledge operators also remains to be explored.

Fitting. In [87], Fitting generalizes the procedures of Hughes and Cresswell [139, 141] to give systematic decision procedures for a wide variety of prefixed modal tableaux systems, including systems for the logics we consider here. Although our labelled sequent systems share characteristics with Fitting's systems (recall from §7 that the main representational difference is that in his systems the different properties of the accessibility relation are expressed procedurally as side conditions on the application of the same set of rules, while we use relational rules to extend a fixed base system), the analysis of decidability relies on different mechanisms. We show that contraction can be eliminated in S(K) and bounded in S(T), S(K4) and S(S4). Fitting avoids explicit repetitions (contractions) of formulas of the form $x{:}\Box A$ in the antecedent of a sequent (i.e., in his notation, formulas of the form $x\ \mathrm{T}\ \Box A$ in a tableau) by having a $\Box$L rule that does not delete $x{:}\Box A$ but prevents its further use until a new y accessible from x is introduced in the tableau; in other words, $x{:}\Box A$ is 'asleep' (i.e. is not reused) until a new y accessible from x is created, at which point $x{:}\Box A$ 'wakes up'. Termination of this 'asleep/wake-up' decision procedure, which is used also in other tableaux systems (see, e.g., [120]) is then argued by exploiting König's Lemma. Aside from this difference, this kind of procedure leads to implementations based on loop-checking, as opposed to measure-decreasing rule applications (as in our work and in the work based on specialized deduction systems we describe below).

13.1.2 Specialized deduction systems

To get more refined space upper-bounds, non-classical logics are sometimes recast as specialized deduction systems in which dynamic loop-checking is replaced by static (a priori) termination checks, or in which rules are measure decreasing so that loops cannot arise.

Cerrito and Cialdea Mayer. In [55], Cerrito and Cialdea Mayer show that loop-checking can be avoided by establishing polynomial bounds on the length of branches in unlabelled tableaux and sequent systems for K4 and S4, and then show, similar to what we did for our systems, that these bounds on the overall number of inferences indirectly bound applications of contraction. In particular, they then exploit Mints' translation of modal formulas into modal clauses [162] to give a contraction-free sequent system for S4 (radically different from the ones of Hudelmaier [137, 138], which we discuss below).

Hudelmaier. Mints' translation is used also in [138], where Hudelmaier gives the best currently known space upper-bounds for K, T and S4. These bounds are identical to ours for K, and better than ours for T and S4 (namely $O(n \log n)$-space and $O(n^2 \log n)$-space, respectively). While Hudelmaier also shows contraction elimination, his approach is otherwise radically different from ours and it is not clear how to extend it to other logics.

Mirroring his previous work [136] on propositional intuitionistic logic for which he also gave a $O(n \log n)$-space upper-bound, in [138] Hudelmaier considers first unlabelled modal sequent systems in which cut is admissible and contraction is built into the rules for the modal operator $\Box$.[1] Then, by introducing new modal operators and rules, whose semantics is however quite unclear, he produces equivalent systems in which all rules are measure decreasing. This, combined with the standard technique we have also employed to minimize space requirements in navigating the search space, allows him to establish his improved complexity bounds.

These improved bounds crucially depend on the assumption that formulas have been translated into Mints' clausal form [162]. We conjecture that if we employ the same translation then we can also provide the same space bounds for T and S4 (and probably also reduce the bound for K4). To illustrate this, let us first summarize some definitions and results of [162] and [137, 138].

Let a *modal literal* be a formula of the form p, $\sim p$, $\Box p$, or $\sim \Box \sim p$, where p is a propositional variable, and a *modal clause* be a disjunction of modal literals or an expression of the form $\Box C$ where C is a modal clause. It is then possible to prove that for any formula A there are a finite set of modal clauses Σ_A and a new propositional variable p such that A is S4-valid iff the formula $\bigwedge \Sigma_A \supset p$ is S4-valid, where $\bigwedge \Sigma_A$

[1]Contractions of formulas not of the form $\Box A$ can be easily shown to be eliminable in standard sequent systems, as done by Zeman in [238], whose proof we have adapted and extended for our systems. Thus, in standard systems, the contraction rule is 'fully' eliminated by building the necessary contractions of formulas of the form $\Box A$ into the modal rules, as we have shown in the previous chapters.

is the conjunction of the elements of Σ_A. Similar equivalence results hold for T and other modal logics.

What is particularly interesting for comparing Hudelmaier's method with ours is the translation itself, i.e. the way in which the set of modal clauses is obtained. This set is built by iterating two steps: (i) each subformula B of the original formula A that is not a modal literal is replaced with a new propositional variable p_B, and (ii) for each such B, clauses are asserted that establish the equivalence of p_B and B.

Consider, as an example, the formula

$$
{\sim}\Box{\sim}({\sim}B \vee \Box B) \tag{13.1}
$$

where B is a propositional variable. Proving this formula in S(T) or S(S4) requires one application of ClL, e.g.

$$
\dfrac{\dfrac{\dfrac{\dfrac{}{\vdash x_1Rx_1}\,\mathit{refl} \quad \dfrac{\dfrac{\dfrac{\dfrac{\dfrac{}{x_1Rx_2 \vdash x_1Rx_2}\,\mathrm{AXr} \quad \dfrac{\dfrac{\dfrac{\dfrac{\dfrac{}{x_2{:}B \vdash x_2{:}B}\,\mathrm{AXl}}{x_2{:}B, x_1Rx_2 \vdash x_1{:}{\sim}B, x_2{:}B, x_2{:}\Box B}\,\mathrm{W}}{x_1Rx_2 \vdash x_1{:}{\sim}B, x_2{:}B, x_2{:}{\sim}B, x_2{:}\Box B}\,{\sim}\mathrm{R}}{x_1Rx_2 \vdash x_1{:}{\sim}B, x_2{:}B, x_2{:}{\sim}B \vee \Box B}\,\vee\mathrm{R}}{x_2{:}{\sim}({\sim}B \vee \Box B), x_1Rx_2 \vdash x_1{:}{\sim}B, x_2{:}B}\,{\sim}\mathrm{L}}{x_1{:}\Box{\sim}({\sim}B \vee \Box B), x_1Rx_2 \vdash x_1{:}{\sim}B, x_2{:}B}\,\Box\mathrm{L}}{x_1{:}\Box{\sim}({\sim}B \vee \Box B) \vdash x_1{:}{\sim}B, x_1{:}\Box B}\,\Box\mathrm{R}}{x_1{:}\Box{\sim}({\sim}B \vee \Box B) \vdash x_1{:}{\sim}B \vee \Box B}\,\vee\mathrm{R}}{x_1{:}{\sim}({\sim}B \vee \Box B), x_1{:}\Box{\sim}({\sim}B \vee \Box B) \vdash}\,{\sim}\mathrm{L}}{x_1{:}\Box{\sim}({\sim}B \vee \Box B), x_1{:}\Box{\sim}({\sim}B \vee \Box B) \vdash}\,\Box\mathrm{L}}{x_1{:}\Box{\sim}({\sim}B \vee \Box B) \vdash}\,\mathrm{ClL}}{\vdash x_1{:}{\sim}\Box{\sim}({\sim}B \vee \Box B)}\,{\sim}\mathrm{R} \quad . \tag{13.2}
$$

This is not surprising, as (13.1) is equivalent to ${\sim}\Box{\sim}(B \supset \Box B)$, which we have already shown to require contraction in S(T), cf. the proof (6.1). It is also equivalent to ${\sim}\Box(B \wedge {\sim}\Box B)$, which similarly requires an application of ClL, namely an initial contraction of $\Box(B \wedge {\sim}\Box B)$ labelled with x_1.

Mints' translation on the other hand produces a set of modal clauses that is equivalent to (13.1) but does not require contraction. In Hudelmaier's notation, cf. [137, Lemma 3] or [138, Lemma 1], the translation transforms the sequent $\vdash {\sim}\Box{\sim}({\sim}B \vee \Box B)$ to the sequent $\Box{\sim}p, \Box(p \vee B), \Box(p \vee {\sim}\Box B) \vdash$, where p is a new propositional variable: it is the variable replacing the formula ${\sim}B \vee \Box B$.[2] In our notation, the translation transforms the sequent

$$
x_1{:}\Box{\sim}({\sim}B \vee \Box B) \vdash \tag{13.3}
$$

[2]More generally, the above transformation follows by the equivalence of the sequents $\Box{\sim}(C \vee D) \vdash$ and $\Box p, \Box({\sim}p \vee C), \Box(p \vee {\sim}D) \vdash$, where C and D are arbitrary modal formulas (in our example C is ${\sim}B$ and D is $\Box B$) and p is a new propositional variable replacing the formula ${\sim}(C \vee D)$.

to the sequent

$$x_1{:}\Box \sim p, x_1{:}\Box(p \vee B), x_1{:}\Box(p \vee \sim \Box B) \vdash , \tag{13.4}$$

and proving this new sequent in S(T) does not require contraction as is shown in the proof

$$\dfrac{\dfrac{}{\vdash x_1 R x_1}\,\mathit{refl} \qquad \dfrac{\overset{\Pi_1}{x_1{:}\Box \sim p, x_1{:}\Box(p \vee B), x_1{:}p \vdash} \qquad \overset{\Pi_2}{x_1{:}\Box \sim p, x_1{:}\Box(p \vee B), x_1{:} \sim \Box B \vdash}}{x_1{:}\Box \sim p, x_1{:}\Box(p \vee B), x_1{:}p \vee \sim \Box B \vdash}\,\vee\text{L}}{x_1{:}\Box \sim p, x_1{:}\Box(p \vee B), x_1{:}\Box(p \vee \sim \Box B) \vdash}\,\Box\text{L}$$

where Π_1 is

$$\dfrac{\dfrac{}{\vdash x_1 R x_1}\,\mathit{refl} \qquad \dfrac{\dfrac{\dfrac{}{x_1{:}p \vdash x_1{:}p}\,\text{AXl}}{x_1{:}\Box(p \vee B), x_1{:}p \vdash x_1{:}p}\,\text{WlL}}{x_1{:} \sim p, x_1{:}\Box(p \vee B), x_1{:}p \vdash}\,\sim\text{L}}{x_1{:}\Box \sim p, x_1{:}\Box(p \vee B), x_1{:}p \vdash}\,\Box\text{L}$$

and Π_2 is

$$\dfrac{\dfrac{\dfrac{\dfrac{}{x_1 R x_2 \vdash x_1 R x_2}\,\text{AXr} \qquad \dfrac{\dfrac{}{x_1 R x_2 \vdash x_1 R x_2}\,\text{AXr} \qquad \dfrac{\dfrac{\dfrac{\dfrac{\dfrac{}{x_2{:}p \vdash x_2{:}p}\,\text{AXl}}{x_2{:}p \vdash x_2{:}B, x_2{:}p}\,\text{WlR} \qquad \dfrac{\dfrac{}{x_2{:}B \vdash x_2{:}B}\,\text{AXl}}{x_2{:}B \vdash x_2{:}B, x_2{:}p}\,\text{WlR}}{x_2{:}p \vee B \vdash x_2{:}B, x_2{:}p}\,\vee\text{L}}{x_2{:}p \vee B, x_1 R x_2 \vdash x_2{:}B, x_2{:}p}\,\text{WrL}}{x_2{:} \sim p, x_2{:}p \vee B, x_1 R x_2 \vdash x_2{:}B}\,\sim\text{L}}{x_1{:}\Box \sim p, x_2{:}p \vee B, x_1 R x_2 \vdash x_2{:}B}\,\Box\text{L}}{x_1{:}\Box \sim p, x_1{:}\Box(p \vee B), x_1 R x_2 \vdash x_2{:}B}\,\Box\text{L}}{x_1{:}\Box \sim p, x_1{:}\Box(p \vee B) \vdash x_1{:}\Box B}\,\Box\text{R}}{x_1{:}\Box \sim p, x_1{:}\Box(p \vee B), x_1{:} \sim \Box B \vdash}\,\sim\text{L} \; .$$

The contraction that we required is thus 'factored out' by the translation that transforms (13.3) into the equivalent (13.4). The crucial step there is the introduction of the new propositional variable p, which allows us to 'separate' $\sim B$ from $\Box B$. That is, while in (13.3) the subformulas $\sim B$ and $\Box B$ both occur inside a positive boxed subformula that must be contracted, in (13.4) the propositional variable p allows us to split them into two different positive boxed subformulas that do not require contraction. More formally, we prove the equivalence of the sequents (13.3) and (13.4) as follows.

For the left-to-right direction of the equivalence, let us assume that we have a proof of (13.3). Then we obtain a proof the sequent (13.4) by an application of *cut* with the sequent

$$x_1{:}\Box \sim p, x_1{:}\Box(p \vee B), x_1{:}\Box(p \vee \sim \Box B) \vdash x_1{:}\Box \sim (\sim B \vee \Box B) ,$$

which we can easily prove without contractions.

The right-to-left direction is slightly trickier. Given a proof of (13.4), we obtain the desired proof of (13.3) by first substituting all occurrences of the variable p with the formula $\sim B \vee \Box B$ it originally replaced, and then cutting the two provable formulas $x_1{:}\Box(\sim B \vee \Box B \vee B)$ and $x_1{:}\Box(\sim B \vee \Box B \vee \sim\Box B)$ from the resulting sequent. That is:

$$\dfrac{\begin{array}{c}\Pi_1\\ \vdash x_1{:}\Box(\sim B \vee \Box B \vee \sim\Box B)\end{array} \qquad \dfrac{\begin{array}{c}\Pi_2\\ \vdash x_1{:}\Box(\sim B\vee\Box B\vee B)\end{array} \qquad \dfrac{x_1{:}\Box \sim p, x_1{:}\Box(p\vee B), x_1{:}\Box(p\vee\sim\Box B)\vdash}{\begin{array}{l} x_1{:}\Box\sim(\sim B\vee\Box B),\\ x_1{:}\Box(\sim B\vee\Box B\vee B),\\ x_1{:}\Box(\sim B\vee\Box B\vee\sim\Box B)\vdash\end{array}}\,\sigma}{\begin{array}{l}x_1{:}\Box\sim(\sim B\vee\Box B),\\ x_1{:}\Box(\sim B\vee\Box B\vee\sim\Box B)\vdash\end{array}}\,cut}{x_1{:}\Box\sim(\sim B\vee\Box B)\vdash}\,cut$$

where Π_1 and Π_2 follow easily (without contractions), and the double inference line labelled with σ stands for the substitution of $\sim B \vee \Box B$ for p.

We have thus made explicit the applications of cut that were implicit in the translation. But showing the admissibility of cut requires applications of contraction in labelled as well as in unlabelled modal sequent systems (see, e.g., [238], and, to some extent, also §6.3). Hence, this example shows how the translation into clausal form allows us to trade the explicit contraction in (13.2) for the contractions implicit in the applications of cut, which are in turn implicitly required in the translation itself.[3] While this is only an example, we believe that the intuition underlying it has a general character and could be turned into a more general, formal, argument. Pre-processing formulas of S(T) and S(S4) using Mints' translation would then allow us to lower our bounds for T and S4 to $O(n \log n)$-space and $O(n^3 \log n)$-space (or maybe even $O(n^2 \log n)$-space), respectively. Investigating this more formally, as well as the relationship between Hudelmaier's new modal operators and our contraction indices, remains as future work.

Massacci. As we observed in §7, an approach based on a specialized deduction system employing labels is that of Massacci [160, 120], who gives 'single-step' prefixed tableaux systems for several propositional modal logics, in which the single-step nature of the rules allows him to avoid an explicit characterization of the properties of the underlying accessibility relation.

Massacci gives decision procedures for his systems that replace loop-checking with a termination check based on bounding the length of prefixes and thus the length of branches. (Thus, his method is similar to our adaptation of Ladner's results to S(K4) and S(S4).) He then shows that provability in various modal logics, including the ones we consider, is in PSPACE. He also investigates the problem of deciding logical consequence, and shows that provability in K45, KD45 and S5 is co-NP-complete, while we ruled out analysis of these logics by assuming the $\Box$-disjunction property.

[3]That the need for contraction depends on the form of the formula to be proved can also be seen by transforming (13.3) to the sequent $\vdash x_1{:} \sim (\Box B \wedge \Box \sim \Box B)$, for which we can give a S(T)-proof that does not apply ClL. This transformation follows by the equivalence of $x_1{:}\Box(B \wedge \sim \Box B) \vdash$ and $x_1{:}\Box B \wedge \Box \sim \Box B \vdash$. Showing that we can 'push' a $\Box$ inside a conjunction requires cut like in the example above, and thus, implicitly, contraction.

Wallen. Other work on proof-theoretical decision procedures for modal and other non-classical logics (related to that of Fitting) is that of Wallen [232], who uses *matrix*-characterizations to investigate a range of logics, where duplication is achieved by increasing a *multiplicity* index associated with the formulas. While Wallen stresses the importance of 'computationally sensitive' characterizations of logics, he does not explicitly address the complexity bounds associated with the systems he investigates. Also, the central point of his work concerns identifying complementary terms that can be used as the basis of a search procedure, rather than analyzing the structural rules.

More specifically, in our systems, 'contraction' means reusing the same principal formula, while in the matrix-method the multiplicity of a formula φ denotes the total number of times φ is used, as principal or parametric formula, in a proof. In other words, the multiplicity of φ denotes not only explicit contractions of φ, but also what we call implicit contractions, i.e. the duplication of the parametric formulas in the premises of a multi-premise rule (e.g. $\supset$ L). It is then interesting to study how bounds on contractions relate to bounds on multiplicities. For example, in [54] Cerrito and Cialdea Mayer exploit their previous results on tableaux and sequent systems [55], which we discussed above, to perform precisely this kind of analysis for the propositional modal logics K, T, K4 and S4.

13.1.3 Empirical performance analysis

Much effort has been recently devoted to the empirical performance analysis of the several automated theorem provers that have been devised for (or specialized to) non-classical logics, and, in particular, multi-modal and description logics. Examples of such provers are documented (together with pointers to related publications and world-wide-web pages and interfaces) in the comparison sections in the latest proceedings of the conference on 'Automated Reasoning with Analytic Tableaux and Related Methods' ('TABLEAUX' [72, 165]). For instance, FaCT and DLP [134, 135] are two optimized provers for descriptions logics, KSAT and *SAT [110, 112, 113, 210] are provers for description, modal and temporal logics, based on extensions of the Davis-Putnam procedure, KRIS [11] is a tableaux-based prover for knowledge representation, MSPASS [101, 144, 143, 206, 207] is an enhancement of the first-order theorem prover SPASS [236] with a translator of formulas of modal or description logics into first-order logic with equality, and the Logics Workbench LWB [127, 128, 130] is a sequent-based prover for a number of propositional non-classical logics. As future work, we plan to increase the level of automation of our Isabelle implementations in order to evaluate, in the light of the above analyses, the practical efficiency of our decision procedures.[4]

[4] As we discussed in §7, in the approaches based on semantic embedding the emphasis is mostly on automated although not necessarily 'natural' theorem proving. Indeed, as shown in the publications referenced above, the optimized functional translation, combined with the first-order theorem prover SPASS, allows the use of resolution as an efficient decision procedure for several (multi-)modal logics. Investigating the relationships between our bounds on contractions and the bounds of the depth of terms in the optimized functional translation remains also as future work.

13.1.4 *Decidability and complexity of* B^+ *and other relevance logics*

In [231] we apply our proof-theoretical method to bound applications of the contraction and monotony rules of $\mathrm{S}(\mathrm{B}^+)$, see Figure 3.2 and §6.2, and thus give a $O(n \log n)$ upper-bound on the space complexity of the basic positive relevance logic B^+.

This bound is new: to our knowledge, no similar investigations have been carried out, and, in particular, we are not aware of any subrecursive bounds for the basic relevance logic B or any of its fragments. In fact, the analysis of decidability and complexity of relevance logics is a subtle issue (especially in comparison with other non-classical logics such as modal logics), and positive results have been presented alongside open problems and several negative results, including the undecidability of the principal relevance logics R and E; see [1, 2, 40, 175, 196, 198, 201, 224, 225, 226] and, in particular, [195], which also contains a detailed summary of work in this area.

Like for modal and other non-classical logics, the decidability of relevance logics is often established semantically by showing that they possess the finite model property. Although effective to establish decidability, this often yields poor complexity bounds. Moreover, the standard proofs of the finite model property fail for several relevance logics, although not for B and its fragments.

Proof-theoretical decision procedures have also been given for (deduction systems for) B and other relevance logics, and, although we are not aware of any analysis of the complexity of these procedures, some of this work is close to ours, at least in spirit. We now briefly discuss two of these approaches.

In [116], Gochet, Gribomont and Rossetto extend the matrix-method to B^+ and give a decision procedure that yields finite models for satisfiable formulas. However, as we already observed above, the matrix-method investigates the need for duplication, by means of explicit and implicit contractions, of indexed formulas, rather than explicitly bounding applications of the contraction rules as we do here.

In [195], Restall gives display calculi for a wide range of relevance and other substructural logics, and investigates their decidability by a proof-theoretical analysis. Similar to ours, this analysis focuses on controlling applications of the contraction rules. However, Restall must also control applications of other rules, which are specific to display calculi, such as the intensional display rules. (A similar analysis is required by the Gentzen-style systems of [1, 40, 108], which are also based on the introduction of new intensional operators.) While the display method establishes the decidability of several substructural logics, including the logic DW, which is a simple extension of B, no complexity results are given. As we remarked in §7, investigation of the connections and complementary aspects of the display method and ours remains as future work.

We plan to use our proof-theoretical method to investigate the complexity of other relevance logics, and thereby try to solve some of the decidability problems that are still open in this area. A first example includes extending our analysis to the basic relevance logic B, which we obtain by extending B^+ with negation. Since the addition of negation raises a number of interesting questions, we conclude this chapter by briefly remarking on this.

To begin with, since there are different ways of dealing with negation in relevance logics, e.g. [74, 78, 99, 196, 197], we must first choose the one that best suits our needs. For example, following §3 and §6.2, we could present B in terms of the cut-free labelled sequent system that we obtain by extending $S(B^+)$ with the logical rules for negation

$$\frac{\Gamma, \Delta \vdash \Gamma', a^*{:}A}{a{:}\neg A, \Gamma, \Delta \vdash \Gamma'} \neg L \quad \text{and} \quad \frac{a^*{:}A, \Gamma, \Delta \vdash \Gamma'}{\Gamma, \Delta \vdash \Gamma', a{:}\neg A} \neg R \ ,$$

and with the relational rules

$$\frac{\Delta \vdash R\,0\,a\,b}{\Delta \vdash R\,0\,b^*\,a^*}\ anti\ , \quad \frac{}{\vdash R\,0\,a\,a^{**}}\ {**}i \quad \text{and} \quad \frac{}{\vdash R\,0\,a^{**}\,a}\ {**}c \ .$$

Or, alternatively, we could follow Priest and Sylvan [188] and Restall [192], and develop a cut-free labelled sequent system for B in terms of the *reduced semantics* they propose. This would, for example, require us to distinguish two pairs of logical rules for relevant implication, depending on whether the lwff $a{:}A \to B$ they introduce has label $a \neq 0$ or $a = 0$. That is:

$$\frac{\Delta \vdash R\,a\,b\,c \quad \Gamma, \Delta \vdash \Gamma', b{:}A \quad c{:}B, \Gamma, \Delta \vdash \Gamma'}{a{:}A \to B, \Gamma, \Delta \vdash \Gamma'} \to L\ (a \neq 0)$$

$$\frac{b{:}A, \Gamma, \Delta, R\,a\,b\,c \vdash \Gamma', c{:}B}{\Gamma, \Delta \vdash \Gamma', a{:}A \to B} \to R\ (a \neq 0)$$

$$\frac{\Gamma, \Delta \vdash \Gamma', b{:}A \quad b{:}B, \Gamma, \Delta \vdash \Gamma'}{0{:}A \to B, \Gamma, \Delta \vdash \Gamma'} \to L_0 \qquad \frac{b{:}A, \Gamma, \Delta \vdash \Gamma', b{:}B}{\Gamma, \Delta \vdash \Gamma', 0{:}A \to B} \to R_0$$

where, in $\to$R, the labels b and c are distinct and do not occur in $\Gamma, \Delta \vdash \Gamma', a{:}A \to B$, while $\to R_0$ has the side condition that b does not occur in $\Gamma, \Delta \vdash \Gamma', 0{:}A \to B$.

This distinction between $a \neq 0$ and $a = 0$ in $a{:}A \to B$ has the effect of eliminating rwffs of the form $R\,0\,b\,c$, and thus also applications of *iden* (but not necessarily applications of other relational rules). This is an advantage, as it simplifies the solution of relational queries $\Delta \vdash R\,a\,b\,c$. It is however also a disadvantage, as it does not allow us anymore to define the partial order $\sqsubseteq$ in terms or R and the actual world 0, but requires us to introduce $\sqsubseteq$ explicitly, and reason explicitly about its properties. For example, *monl* becomes

$$\frac{\Delta \vdash a \sqsubseteq b \quad \Gamma, \Delta \vdash \Gamma', a{:}A}{\Gamma, \Delta \vdash \Gamma', b{:}A}\ monl$$

and $\sqsubseteq$ is formalized by the rules

$$\frac{}{\vdash a \sqsubseteq a} \quad \text{and} \quad \frac{\Delta \vdash a \sqsubseteq b \quad \Delta \vdash b \sqsubseteq c}{\Delta \vdash a \sqsubseteq c} \ .$$

When applying our method to extensions of B^+ we must thus examine these different possibilities in detail, as different choices may lead to different results.

14 CONCLUSIONS AND FURTHER RESEARCH

In this final chapter we summarize our main contributions and results, and discuss directions of future research.

Methodologically, the contribution of this book is the formalization of a framework for presenting families of non-classical logics in a uniform and modular way as labelled deduction systems. Moreover, we have shown that our systems lend themselves well to implementation in a Logical Framework such as Isabelle.

Technically, the contributions are as follows. We have given parameterized proofs of soundness and completeness of our labelled deduction systems with respect to the corresponding Kripke-style semantics, and of faithfulness and adequacy of their Isabelle encodings. Furthermore, we have analyzed structural and substructural properties of our systems, in particular normalization of derivations, the subformula property, and bounded contraction, and we have shown how we can exploit them to restrict proof search. As examples, we have given PSPACE decision procedures for the propositional modal logics K, T, K4 and S4, and we have discussed applications of our method to other non-classical logics such as the basic positive relevance logic B^+. Finally, we have exploited our substructural analysis to give justifications and partial refinements of the rules of standard sequent systems.

One of the main characteristics of our labelled presentations of non-classical logics is the identification of fixed bases from which we generate systems for families of logics by extension with separate theories formalizing the properties of the relations and/or of the domains of quantification in the corresponding semantics. This separation is of crucial importance. It allows us not only to prove metatheoretical results in a parameterized way, but also, and most importantly, to delineate the advantages and limitations of our systems. Specifically, we have shown that when the relational and domain theories are comprised of Horn rules, then our systems consist of the minimal deduction machinery necessary to present the corresponding logics, and allow separated derivations possessing a well-defined structure: derivations normalize, and normal derivations satisfy a subformula property. However, when we employ first-order theories to capture even larger families of logics, this separation is lost, and our approach does not seem to offer any advantages with respect to traditional semantic embeddings. Our investigation of tradeoffs in possible formalizations of deduction systems for non-classical logics can be developed further in several directions, and as a first step we plan to investigate in more detail the paraconsistent systems that we obtain when we restrict falsum to a local falsum.

Throughout the chapters we have pointed out a number of other directions for future research, and we now focus on four of the most interesting and promising ones.

First, we plan to extend our substructural analysis in order to investigate decidability and complexity of other non-classical logics; our work on B^+ is merely a first step in this direction. We also hope that, like for the example modal logics we have considered, this extended analysis will provide further insights in the understanding of other non-classical deduction systems that have been proposed in the literature.

Second, in order to present other logics than those considered here, we plan to extend our systems with other non-classical logical operators, such as operators for common and distributed knowledge and belief [82, 123] or 'graded modalities' [173]. This should be a straightforward task for several operators and logics. For example, as we observed in §13.1, common multi-modal logics are just special cases of the logics we presented in §3. But we also believe that there are logics, or combinations thereof, whose presentations in our framework will require complex theories (labelling algebras, in particular), thus endangering the naturalness and simplicity of our approach. This is in fact connected to the main 'problem' with our approach, and, similarly, with other approaches based on labelling, prefixing or embedding, namely the commitment to a Kripke-style (or some similar algebraic) semantics for the logics we want to present. While labels match our intuitions for a number of commonly considered non-classical logics, they do not do so for others. In the case of provability logic GL [36], for example, labels would closely match the semantics, which, however, does not fully accord with intuition; a Hilbert-style axiomatization, on the other hand, makes it clear that the Löb axiom is the central fact. In other words, there are logics for which a Hilbert-style presentation, although difficult to use, gives insights into intended applications that a labelled presentation does not. This establishes another, and, probably, the most important limitation of our approach: our labelled deduction framework is not a panacea for all problems that arise in connection with non-classical reasoning, since in some cases there are tradeoffs or even theoretical and practical limitations that must be taken into account.

Third, and closely related to multi-modal logics, we plan to use our labelled deduction systems to investigate combinations (and products) of logics. The recent research on *fibring*, a general methodology for combining logics proposed by Gabbay [91] and others [212, 237], and which can be based on labelled systems, e.g. [25, 64, 122], will provide a good starting point for our future work in this direction.

Finally, we plan to increase the level of automation of our Isabelle implementations in order to compare the practical efficiency of our decision procedures with existing theorem provers, and, most importantly, to consider the 'real' applications of non-classical logics mentioned in §1.

References

[1] A. R. Anderson and N. D. Belnap, Jr. *Entailment, The Logic of Relevance and Necessity*, volume 1. Princeton University Press, Princeton, New Jersey, 1975.

[2] A. R. Anderson, N. D. Belnap, Jr., and J. M. Dunn. *Entailment, The Logic of Relevance and Necessity*, volume 2. Princeton University Press, Princeton, New Jersey, 1992.

[3] C. Areces, P. Blackburn, and M. Marx. Hybrid logics. Characterization, interpolation and complexity. Technical Report CLAUS-Report Nr. 108, Computational Linguistics Department of the Universität des Saarlandes, Saarbrücken, Germany, 1999.

[4] A. Artosi, P. Benassi, G. Governatori, and A. Rotolo. Labelled proofs for quantified modal logic. In J. Alferes, L. Pereira, and E. Orlowska, editors, *Proceedings of Jelia'96*, LNCS 1126, pages 70–86. Springer, Berlin, 1996.

[5] Y. Auffray and P. Enjalbert. Modal theorem proving: An equational viewpoint. *Journal of Logic and Computation*, 2(3):247–297, 1992.

[6] A. Avron. Simple consequence relations. *Information and Computation*, 92:105–139, 1991.

[7] A. Avron. Using hypersequents in proof systems for non-classical logic. *Annals of Mathematics and Artificial Intelligence*, 4:225–248, 1991.

[8] A. Avron. The method of hypersequents in the proof theory of propositional non-classical logics. In W. Hodges, M. Hyland, C. Steinhorn, and J. Truss, editors, *Logic: From Foundations to Applications*, pages 1–32. Oxford Science Publications, Oxford, UK, 1996.

[9] A. Avron, F. Honsell, I. A. Mason, and R. Pollack. Using typed lambda calculus to implement formal systems on a machine. *Journal of Automated Reasoning*, 9:309–352, 1992.

[10] A. Avron, F. Honsell, M. Miculan, and C. Paravano. Encoding modal logics in logical frameworks. *Studia Logica*, 60(1):161–208, 1998.

[11] F. Baader and B. Hollunder. KRIS: Knowledge representation and inference system (system description). *ACM SIGART Bulletin*, 2:8–14, 1991.

[12] F. Baader and K. U. Schulz, editors. *Frontiers of Combining Systems*. Kluwer, Dordrecht, 1996.

[13] P. Balbiani and S. Demri. Prefixed tableaux systems for modal logics with enriched languages. In Pollack [184], pages 190–195.

[14] M. Baldoni. *Normal Multimodal Logics: Automatic Deduction and Logic Programming Extension*. PhD thesis, Università degli Studi di Torino, Italy, 1998.

[15] M. Baldoni. Normal multimodal logics with interaction axioms. In Basin et al. [17].

[16] M. Baldoni, L. Giordano, and A. Martelli. Tableau calculus for multimodal logics and some (un)decidability results. In de Swart [72], pages 44–59.

[17] D. Basin, M. D'Agostino, D. M. Gabbay, S. Matthews, and L. Viganò, editors. *Labelled Deduction*. Kluwer, Dordrecht, to appear.

[18] D. Basin and S. Matthews. Logical frameworks. Survey chapter to appear in the second edition of the *Handbook of Philosophical Logic*.

[19] D. Basin, S. Matthews, and L. Viganò. Labelled propositional modal logics: theory and practice. *Journal of Logic and Computation*, 7(6):685–717, 1997.

[20] D. Basin, S. Matthews, and L. Viganò. A new method for bounding the complexity of modal logics. In G. Gottlob, A. Leitsch, and D. Mundici, editors, *Proceedings of KGC'97*, LNCS 1289, pages 89–102. Springer, Berlin, 1997.

[21] D. Basin, S. Matthews, and L. Viganò. Labelled modal logics: quantifiers. *Journal of Logic, Language and Information*, 7(3):237–263, 1998.

[22] D. Basin, S. Matthews, and L. Viganò. Natural deduction for non-classical logics. *Studia Logica*, 60(1):119–160, 1998.

[23] D. Basin, S. Matthews, and L. Viganò. Modal logics K, T, K4, S4: labelled proof systems and new complexity results. *The Bulletin of Symbolic Logic*, 5(1):91–93, 1999.

[24] D. Basin and L. Viganò. A recipe for the complexity analysis of non-classical logics. In de Rijke and Gabbay [71].

[25] B. Beckert and D. M. Gabbay. Fibring semantic tableaux. In de Swart [72], pages 77–92.

[26] N. D. Belnap, Jr. Display logic. *Journal of Philosophical Logic*, 11:357–417, 1982.

[27] N. D. Belnap, Jr. Linear logic displayed. *Notre Dame Journal of Formal Logic*, 31(1):15–25, 1990.

[28] E. Bencivenga. Free logics. In Gabbay and Guenthner [94], pages 373–426.

[29] M. R. F. Benevides and T. S. E. Maibaum. A constructive presentation for the modal connective of necessity ($\Box$). *Journal of Logic and Computation*, 2(1):31–50, 1992.

[30] P. Blackburn. Internalizing labeled deduction. Technical Report CLAUS-Report Nr. 102, Computational Linguistics Department of the Universität des Saarlandes, Saarbrücken, Germany, 1998. To appear in *Journal of Logic and Computation*.

[31] P. Blackburn. Discipline as logic: treating labels as first class citizens. In Basin et al. [17].

[32] P. Blackburn, M. de Rijke, and Y. Venema. Modal logic. To appear.

[33] P. Blackburn and M. Tzakova. Hybrid completeness. *Logic Journal of the IGPL*, 6(4):625–650, 1998.

[34] P. Blackburn and M. Tzakova. Hybrid languages and temporal logic. *Logic Journal of the IGPL*, 7(1):27–54, 1999.

[35] G. S. Boolos. Don't eliminate cut. *Journal of Philosophical Logic*, 7:373–378, 1984.

[36] G. S. Boolos. *The logic of provability*. Cambridge University Press, Cambridge, UK, 1993.

[37] G. S. Boolos and R. C. Jeffrey. *Computability and Logic*. Cambridge University Press, Cambridge, UK, 1989.

[38] T. Borghuis. Interpreting modal natural deduction in type theory. In M. de Rijke, editor, *Diamonds and Defaults*, pages 67–102. Kluwer, Dordrecht, 1993.

[39] T. Borghuis. *Coming to Terms with Modal Logic*. PhD thesis, Department of Computer Science, University of Eindhoven, The Netherlands, 1994.

[40] R. T. Brady. Gentzenization and decidability of some contraction-less relevant logics. *Journal of Philosophical Logic*, 20:97–117, 1991.

[41] K. Broda, M. D'Agostino, and A. Russo. Transformation methods in LDS. In Ohlbach and Reyle [172], pages 335–376.

[42] K. Broda, M. Finger, and A. Russo. Labelled natural deduction for substructural logics. *Logic Journal of the IGPL*, 7(3):283–318, 199.

[43] K. Broda and D. M. Gabbay. CLDS for propositional intuitionistic logic. In Murray [165], pages 66–81.

[44] K. Broda and D. M. Gabbay. Labelled abduction (I). In Basin et al. [17].

[45] K. Broda and A. Russo. A unified compilation style labelled deductive system for modal and substructural logic using natural deduction. Technical report, Department of Computing, Imperial College, London, UK, 1997.

[46] R. Bull and K. Segerberg. Basic modal logic. In Gabbay and Guenthner [93], pages 1–88.

[47] R. Carnap. Modalities and quantification. *Journal of Symbolic Logic*, 11(2):33–64, 1946.

[48] R. Carnap. *Meaning and necessity*. University of Chicago Press, Chicago, 1947.

[49] M. Castilho, L. Fariñas del Cerro, O. Gasquet, and A. Herzig. Modal tableaux with propagation rules and structural rules. *Fundamenta Informaticae*, 32:281–297, 1997.

[50] M. Castilho, O. Gasquet, and A. Herzig. Modal tableaux for reasoning about actions and plans. In S. Steel and R. Alami, editors, *Proceedings of ECP'97*, LNAI 1348, pages 104–116. Springer, Berlin, 1997.

[51] L. Catach. Normal multimodal logic. In R. G. Smith and T. M. Mitchell, editors, *Proceedings of AAAI'88*, pages 491–495. Morgan Kaufmann, San Francisco, CA, 1988.

[52] L. Catach. TABLEAUX: A general theorem prover for modal logics. *Journal of Automated Reasoning*, 7(4):489–510, 1991.

[53] C. Cerrato. Modal sequents. In Wansing [234], pages 141–166.

[54] S. Cerrito and M. Cialdea Mayer. Hintikka multiplicities in matrix decision methods for some propositional modal logics. In Galmiche [100], pages 138–152.

[55] S. Cerrito and M. Cialdea Mayer. A polynomial translation of S4 into T and contraction-free tableaux for S4. *Logic Journal of the IGPL*, 5(2):287–300, 1997.

[56] S. Cerrito and M. Cialdea Mayer. Labelled tableaux for propositional linear time logic over finite frames. In Basin et al. [17].

[57] A. Chagrov and M. Zakharyaschev. *Modal Logic*. Clarendon Press, Oxford, UK, 1997.

[58] B. F. Chellas. *Modal Logic: an introduction*. Cambridge University Press, New York, NY, 1980.

[59] E. Clarke and D. Kozen, editors. *Logics of programs*. LNCS 164. Springer, Berlin, 1984.

[60] B. Copeland. On when a semantics is not a semantics : Some reasons for disliking the Routley-Meyer semantics for relevance logics. *Journal of Philosophical Logic*, 8:399–413, 1979.

[61] G. Corsi and S. Ghilardi. Semantical aspects of quantified modal logic. In C. Bicchieri and M. L. Dalla Chiara, editors, *Knowledge, Belief and Strategic Action*, pages 167–195. Cambridge University Press, Cambridge, UK, 1992.

[62] M. D'Agostino. Are tableaux an improvement on truth-tables? (Cut-free proofs and bivalence). *Journal of Logic, Language, and Information*, 1:235–252, 1992.

[63] M. D'Agostino and D. M. Gabbay. A generalization of analytic deduction via labelled deductive systems. Part I : basic substructural logics. *Journal of Automated Reasoning*, 13:243–281, 1994.

[64] M. D'Agostino and D. M. Gabbay. Fibred tableaux for multi-implication logics. In Miglioli et al. [161], pages 16–35.

[65] M. D'Agostino, D. M. Gabbay, and K. Broda. Tableau methods for substructural logics. In D'Agostino et al. [66], pages 397–467.

[66] M. D'Agostino, D. M. Gabbay, R. Hähnle, and J. Posegga, editors. *Handbook of Tableau Methods*. Kluwer, Dordrecht, 1999.

[67] M. D'Agostino, D. M. Gabbay, and A. Russo. Grafting modalities onto substructural implication systems. *Studia Logica*, 59:65–102, 1997.

[68] M. D'Agostino and M. Mondadori. The taming of the cut (classical refutations with analytic cut). *Journal of Logic and Computation*, 4(3):285–319, 1994.

[69] M. Dam. *Relevance logic and concurrent composition*. PhD thesis, Department of Computer Science, University of Edinburgh, UK, 1990.

[70] R. de Queiroz and D. M. Gabbay. Labelled natural deduction. In Ohlbach and Reyle [172], pages 173–250.

[71] M. de Rijke and D. M. Gabbay, editors. *Frontiers of Combining Systems 2*. Research Studies Press/Wiley, to appear.

[72] H. de Swart, editor. *Proceedings of TABLEAUX'98*. LNCS 1397. Springer, Berlin, 1998.

[73] K. Došen. Sequent-systems for modal logic. *Journal of Symbolic Logic*, 50(1):149–168, 1985.

[74] K. Došen. Negation as a modal operator. *Reports on Mathematical Logic*, 20:15–27, 1986.

[75] K. Došen. A historical introduction to substructural logics. In Došen and Schroeder-Heister [76], pages 1–30.

[76] K. Došen and P. Schroeder-Heister, editors. *Substructural Logics*. Oxford University Press, Oxford, UK, 1993.

[77] J. M. Dunn. Relevance logic and entailment. In Gabbay and Guenthner [94], pages 117–224.

[78] J. M. Dunn. Star and perp: Two treatments of negation. In J. Tomberlin, editor, *Philosophical Perspectives*, volume 7, pages 331–357. Ridgeview, Atascadero, CA, 1994.

[79] J. M. Dunn. Gaggle theory applied to modal, intuitionistic, and relevance logics. In I. Max and W. Stelzner, editors, *Logik und Mathematik (Frege Kolloquium '93)*, pages 335–368. de Gruyter, Berlin, 1995.

[80] J. M. Dunn. Positive modal logic. *Studia Logica*, 55:301–317, 1995.

[81] R. Dyckhoff. Contraction-free sequent calculi for intuitionistic logic. *Journal of Symbolic Logic*, 57(3):795–807, 1992.

[82] R. Fagin, J. Y. Halpern, Y. Moses, and M. Y. Vardi. *Reasoning about knowledge*. MIT Press, Cambridge, MA, 1995.

[83] L. Fariñas Del Cerro and A. Herzig. Modal deduction with applications in epistemic and temporal logics. In Gabbay et al. [96], pages 499–594.

[84] L. Fariñas Del Cerro and A. Herzig. Combining classical and intuitionistic logic. In Baader and Schulz [12], pages 93–102.

[85] A. Felty and D. Miller. Specifying theorem provers in a higher-order logic programming language. In E. Lusk and R. Overbeek, editors, *Proceedings of CADE-9*, LNCS 310, pages 61–80. Springer, Berlin, 1988.

[86] F. B. Fitch. Tree proofs in modal logic. *Journal of Symbolic Logic*, 31:152, 1966.

[87] M. Fitting. *Proof methods for modal and intuitionistic logics*. Kluwer, Dordrecht, 1983.

[88] M. Fitting. First-order modal tableaux. *Journal of Automated Reasoning*, 4(2):191–214, 1988.

[89] M. Fitting. Basic modal logic. In Gabbay et al. [95], pages 365–448.

[90] D. M. Gabbay. *Labelled Deductive Systems*, volume 1. Clarendon Press, Oxford, UK, 1996.

[91] D. M. Gabbay. *Fibring logics*. Clarendon Press, Oxford, UK, 1999.

[92] D. M. Gabbay and F. Guenthner, editors. *Handbook of Philosophical Logic*, volume I. Reidel, Dordrecht, 1983.

[93] D. M. Gabbay and F. Guenthner, editors. *Handbook of Philosophical Logic*, volume II. Reidel, Dordrecht, 1984.

[94] D. M. Gabbay and F. Guenthner, editors. *Handbook of Philosophical Logic*, volume III. Reidel, Dordrecht, 1986.

[95] D. M. Gabbay, C. J. Hogger, and J. A. Robinson, editors. *Handbook of Logic in Artificial Intelligence and Logic Programming*, volume I. Clarendon Press, Oxford, UK, 1993.

[96] D. M. Gabbay, C. J. Hogger, and J. A. Robinson, editors. *Handbook of Logic in Artificial Intelligence and Logic Programming*, volume IV. Clarendon Press, Oxford, UK, 1995.

[97] D. M. Gabbay and N. Olivetti. Algorithmic proof methods and cut elimination for implicational logics (Part I: Modal implication). *Studia Logica*, 61:237–280, 1998.

[98] D. M. Gabbay and N. Olivetti. *Goal-Directed Proof Theory*. Kluwer, Dordrecht, to appear.

[99] D. M. Gabbay and H. Wansing, editors. *What is negation?* Kluwer, Dordrecht, 1999.

[100] D. Galmiche, editor. *Proceedings of TABLEAUX'97*. LNCS 1227. Springer, Berlin, 1997.

[101] H. Ganzinger, U. Hustadt, C. Meyer, and R. A. Schmidt. A resolution-based decision procedure for extensions of K4. In K. Segerberg, M. Zakharyaschev, and M. de Rijke, editors, *Advances in Modal Logic*, volume 2. CSLI Publications, Stanford, CA, to appear.

[102] P. Gardner. A new type theory for representing logics. In A. Voronkov, editor, *Proceedings of LPAR'93*, LNAI 698, pages 146–157. Springer, Berlin, 1993.

[103] P. Gardner. Equivalences between logics and their representing type theories. *Mathematical Structures in Computer Science*, 5:323–349, 1995.

[104] J. Garson. Quantification in modal logic. In Gabbay and Guenthner [93], pages 249–307.

[105] O. Gasquet. Optimization of deduction for multi-modal logics. In L. Pólos and M. Masuch, editors, *Applied Logic: How, What and Why*, pages 61–77. Kluwer, Dordrecht, 1995.

[106] G. Gentzen. Untersuchungen über das logische Schließen I, II. *Mathematische Zeitschrift*, 39:176–210, 405–431, 1934–1935. English translation: Investigations into logical deduction, in M. Szabo, editor, *The collected papers of Gerhard Gentzen*, pages 68–131, North Holland, Amsterdam, 1969.

[107] C. Ghidini and L. Serafini. Distributed first order logics. In de Rijke and Gabbay [71].

[108] S. Giambrone. TW_+ and RW_+ are decidable. *Journal of Philosophical Logic*, 14:235–254, 1985.

[109] J.-Y. Girard. Linear logic. *Theoretical Computer Science*, 50:1–102, 1987.

[110] E. Giunchiglia, F. Giunchiglia, R. Sebastiani, and A. Tacchella. More evaluation of decision procedures for modal logics. In A. G. Cohn, L. Schubert, and S. Shapiro, editors, *Proceedings of KR'98*. Morgan Kauffman, San Francisco, CA, 1998.

[111] F. Giunchiglia. An epistemological science of common sense. *Artificial Intelligence*, 77(2):371–392, 1995.

[112] F. Giunchiglia and R. Sebastiani. A SAT-based decision procedure for ALC. In L. Carlucci Aiello, J. Doyle, and S. Shapiro, editors, *Proceedings of KR'96*, pages 304–314. Morgan Kaufmann, San Francisco, CA, 1996.

[113] F. Giunchiglia and R. Sebastiani. Building decision procedures for modal logics from propositional decision procedures – the case study of modal $K(m)$. *Information and Computation*, to appear.

[114] F. Giunchiglia and L. Serafini. Multilanguage hierarchical logics (or: how we can do without modal logics). *Artificial Intelligence*, 65:29–70, 1994.

[115] L. F. Goble. Gentzen systems for modal logic. *Notre Dame Journal of Formal Logic*, 15(3):455–461, 1974.

[116] P. Gochet, P. Gribomont, and D. Rossetto. Algorithms for relevant logic. *Logique et Analyse*, 150-151-152:329–346, 1995.

[117] R. Goldblatt. Semantical analysis of orthologic. *Journal of Philosophical Logic*, 3:19–35, 1974.

[118] R. Goldblatt. *Logics of time and computation*. CSLI Publications, Stanford, CA, 1992.

[119] R. P. Goré. *Cut–free sequent and tableau systems for propositional normal modal logics*. PhD thesis, University of Cambridge, UK, 1992.

[120] R. P. Goré. Tableau methods for modal and temporal logics. In D'Agostino et al. [66], pages 297–396.

[121] G. Governatori. Labelled tableaux for multi-modal logics. In P. Baumgartner, R. Hähnle, and J. Posegga, editors, *Proceedings of TABLEAUX'95*, LNAI 918, pages 79–94. Springer, Berlin, 1995.

[122] G. Governatori and D. M. Gabbay. Fibred modal tableaux. In Basin et al. [17].

[123] J. Y. Halpern and Y. O. Moses. A guide to completeness and complexity for modal logics of knowledge and belief. *Artificial Intelligence*, 54(3):319–379, 1992.

[124] D. Harel. Dynamic logic. In Gabbay and Guenthner [93], pages 497–604.

[125] R. Harper, F. Honsell, and G. Plotkin. A framework for defining logics. *Journal of the ACM*, 40(1):143–184, 1993.

[126] A. Herzig. *Raisonnement automatique en logique modale et algorithmes d'unification*. PhD thesis, Université Paul-Sabatier, Toulouse, France, 1989.

[127] A. Heuerding. *Sequent Calculi for Proof Search in Some Modal Logics*. PhD thesis, University of Bern, Switzerland, 1998.

[128] A. Heuerding, G. Jäger, S. Schwendimann, and M. Seyfried. The Logics Workbench LWB: A snapshot. *Euromath Bulletin*, 2(1):177–186, 1996.

[129] A. Heuerding and S. Schwendimann. On the modal logic K plus theories. In Kleine-Büning [148], pages 308–319.

[130] A. Heuerding, M. Seyfried, and H. Zimmermann. Efficient loop-check for backward proof search in some non-classical propositional logics. In Miglioli et al. [161], pages 210–225.

[131] J. R. Hindley and J. P. Seldin. *Introduction to combinators and lambda-calculus*. Cambridge University Press, New York, NY, 1993.

[132] J. Hintikka. *Knowledge and belief*. Cornell University Press, Ithaca, NY, 1962.

[133] F. Honsell and M. Miculan. A natural deduction approach to dynamic logics. In S. Berardi and M. Coppo, editors, *Proceedings of TYPES'95*, LNCS 1158, pages 165–182. Springer, Berlin, 1996.

[134] I. Horrocks. *Optimizing Tableaux Decision Procedures for Description Logics*. PhD thesis, University of Manchester, UK, 1997.

[135] I. Horrocks and P. F. Patel-Schneider. FaCT and DLP. In de Swart [72], pages 27–30.

[136] J. Hudelmaier. A $O(n \ log \ n)$-space decision procedure for intuitionistic propositional logic. *Journal of Logic and Computation*, 3(1):63–75, 1993.

[137] J. Hudelmaier. A contraction-free sequent calculus for S4. In Wansing [234], pages 3–15.

[138] J. Hudelmaier. Improved decision procedures for the modal logics K, T and S4. In Kleine-Büning [148], pages 320–334.

[139] G. E. Hughes and M. J. Cresswell. *An introduction to modal logic*. Routledge, London, UK, 1972.

[140] G. E. Hughes and M. J. Cresswell. *A companion to modal logic*. Methuen, London, UK, 1984.

[141] G. E. Hughes and M. J. Cresswell. *A new introduction to modal logic*. Routledge, London, UK, 1996.

[142] I. Humberstone. Interval semantics for tense logic: Some remarks. *Journal of Philosophical Logic*, 8:171–196, 1979.

[143] U. Hustadt and R. A. Schmidt. An empirical analysis of modal theorem provers. To appear in *Journal of Applied Non-Classical Logics*.

[144] U. Hustadt and R. A. Schmidt. On evaluating decision procedures for modal logics. In Pollack [184], pages 202–207.

[145] B. Jónsson and A. Tarski. Boolean algebras with operators. *American Journal of Mathematics*, 73–4:891–939, 127–162, 1951-52.

[146] S. Kanger. *Provability in logic*. Almqvist and Wiksell, Stockholm, 1957.

[147] S. C. Kleene. *Introduction to metamathematics*. North Holland, Amsterdam, 1952.

[148] H. Kleine-Büning, editor. *Proceedings of CSL'95*. LNCS 1092. Springer, Berlin, 1996.

[149] M. Kracht. Power and weakness of the modal display calculus. In Wansing [234], pages 93–121.

[150] S. Kripke. Semantical analysis of modal logic I – normal modal propositional calculi. *Zeitschrift für mathematische Logik und Grundlagen der Mathematik*, 9:67–96, 1963.

[151] S. Kripke. Semantical considerations on modal logics. *Acta Philosophica Fennica*, 16:83–94, 1963.

[152] S. Kripke. *Naming and necessity*. Harvard University Press, Cambridge, MA, 1980.

[153] R. E. Ladner. The computational complexity of provability in systems of modal propositional logics. *SIAM Journal of Computing*, 6(3):46–65, 1977.

[154] E. J. Lemmon. Algebraic semantics for modal logics I, II. *Journal of Symbolic Logic*, 31(1), 1966.

[155] V. Lifschitz, editor. *Formalizing common sense: Papers by John McCarthy*. Lawrence Erlbaum Associates, Pittsburgh, Pennsylvania, 1990.

[156] Z. Manna and A. Pnueli. The modal logic of programs. In H. A. Maurer, editor, *Proceedings of the 6th Colloquium on Automata, Languages and Programming*, LNCS 71, pages 385–410. Springer, Berlin, 1979.

[157] S. Martini and A. Masini. A computational interpretation of modal proofs. In Wansing [234], pages 213–241.

[158] A. Masini. 2-sequent calculus: a proof theory of modalities. *Annals of Pure and Applied Logic*, 58:229–246, 1992.

[159] F. Massacci. Strongly analytic tableaux for normal modal logics. In A. Bundy, editor, *Proceedings of CADE-12*, LNAI 814, pages 723–737. Springer, Berlin, 1994.

[160] F. Massacci. Single step tableaux for modal logics: Methodology, computations, algorithms. *Journal of Automated Reasoning*, to appear.

[161] P. Miglioli, U. Moscato, D. Mundici, and M. Ornaghi, editors. *Proceedings of TABLEAUX'96*. LNAI 1071. Springer, 1996.

[162] G. Mints. Gentzen-type systems and resolution rules. Part I: propositional logic. In P. Martin-Löf and G. Mints, editors, *Proceedings of COLOG'88*, LNCS 417, pages 198–231. Springer, 1990.

[163] G. Mints. *A short introduction to modal logic*. CSLI Publications, Stanford, CA, 1992.

[164] C. G. Morgan. Methods for automated theorem proving in nonclassical logics. *IEEE Transactions on Computers*, 25(8):852–862, 1976.

[165] N. V. Murray, editor. *Proceedings of TABLEAUX'99*. LNCS 1617. Springer, Berlin, 1999.

[166] R. P. Nederpelt, J. H. Geuvers, and R. C. de Vrijer, editors. *Selected papers on Automath*. Elsevier, Amsterdam, 1994.

[167] A. Nerode and R. Shore. *Logic for applications*. Springer, Berlin, 1993.

[168] A. Nonnengart. *A resolution-based calculus for temporal logics*. PhD thesis, Universität des Saarlandes, Saarbrücken, Germany, 1995.

[169] H. J. Ohlbach. *A resolution calculus for modal logics*. PhD thesis, Universität Kaiserslautern, Kaiserslautern, Germany, 1988.

[170] H. J. Ohlbach. Semantics based translation methods for modal logics. *Journal of Logic and Computation*, 1(5):691–746, 1991.

[171] H. J. Ohlbach. Translation methods for non-classical logics: an overview. *Bulletin of the IGPL*, 1(1):69–90, 1993.

[172] H. J. Ohlbach and U. Reyle, editors. *Logic, Language and Reasoning (Essays in Honour of Dov Gabbay)*. Kluwer, Dordrecht, 1999.

[173] H. J. Ohlbach, R. A. Schmidt, and U. Hustadt. Translating graded modalities into predicate logic. In Wansing [234], pages 253–291.

[174] M. Ohnishi and K. Matsumoto. Gentzen method in modal calculi. *Osaka Mathematical Journal*, 9:113–130, 1957.

[175] H. Ono. Decidability and finite model property of substructural logics. In J. Ginzburg, Z. Khasidashvili, C. Vogel, J.-J. Lévy, and E. Vallduví, editors, *The Tbilisi Symposium on Language, Logic and Computation*, pages 263–274. CSLI Publications, Stanford, CA, 1998.

[176] E. Orlowska. Relational interpretation of modal logics. In H. Andréka, J. D. Monk, and I. Németi, editors, *Algebraic logic*. North Holland, Amsterdam, 1991.

[177] E. Orlowska. Relational proof system for relevant logics. *Journal of Symbolic Logic*, 57(4):1425–1440, 1992.

[178] E. Orlowska. Relational proof systems for modal logics. In Wansing [234], pages 55–78.

[179] L. C. Paulson. The foundation of a generic theorem prover. *Journal of Automated Reasoning*, 5:363–397, 1989.

[180] L. C. Paulson. *ML for the Working Programmer*. Cambridge University Press, Cambridge, UK, 1991.

[181] L. C. Paulson. *Isabelle: a generic theorem prover*. LNCS 828. Springer, Berlin, 1994. URL of Isabelle: http://www.cl.cam.ac.uk/Research/HVG/Isabelle/.

[182] J. Perzanowski. The deduction theorem for the modal propositional calculi formalized after the manner of Lemmon, Part I. *Reports on Mathematical Logic*, 1:1–12, 1973.

[183] F. Pfenning. Structural cut elimination. In D. Kozen, editor, *Proceedings of LICS'95*, pages 156–166. IEEE Computer Society Press, 1995. Extended version to appear in *Information and Computation*.

[184] M. E. Pollack, editor. *Proceedings of IJCAI'97*. Morgan Kaufmann, San Francisco, CA, 1997.

[185] G. Pottinger. Normalization as a homomorphic image of cut-elimination. *Annals of Mathematical Logic*, 12:323–357, 1977.

[186] D. Prawitz. *Natural deduction, a proof-theoretical study*. Almqvist and Wiksell, Stockholm, 1965.

[187] D. Prawitz. Ideas and results in proof theory. In J. E. Fensted, editor, *Proceedings of the Second Scandinavian Logic Symposium*, pages 235–307. North Holland, Amsterdam, 1971.

[188] G. Priest and R. Sylvan. Simplified semantics for basic relevant logics. *Journal of Philosophical Logic*, 21:217–232, 1992.

[189] J.-P. Queille and J. Sifakis. Specification and verification of concurrent systems in CESAR. In *Proceedings of the Fifth Internationl Symposium in Programming*, LNCS 137, pages 337–351. Springer, Berlin, 1982.

[190] W. Rautenberg. Modal tableau calculi and interpolation. *Journal of Philosophical Logic*, 12:403–423, 1983.

[191] J. Renz and B. Nebel. On the complexity of qualitative spatial reasoning: A maximal tractable fragment of the region connection calculus. *Artificial Intelligence*, 108(1/2):69–123, 1999.

[192] G. Restall. Simplified semantics for relevant logics (and some of their rivals). *Journal of Philosophical Logic*, 22:481–511, 1993.

[193] G. Restall. Display logic and gaggle theory. *Reports on mathematical logic*, 29:133–146, 1995.

[194] G. Restall. Four-valued semantics for relevant logics (and some of their rivals). *Journal of Philosophical Logic*, 24:139–160, 1995.

[195] G. Restall. Displaying and deciding substructural logics 1: Logics with contraposition. *Journal of Philosophical Logic*, 27(2):179–216, 1998.

[196] G. Restall. *An Introduction to Substructural Logics*. Routledge, London, UK, 1999.

[197] G. Restall. Negation in relevant logics (how I stopped worrying and learned to love the Routley star). In Gabbay and Wansing [99], pages 53–76.

[198] J. Riche and R. K. Meyer. Kripke, Belnap, Urquhart and relevant decidability & complexity. In G. Gottlob, E. Grandjean, and K. Seyr, editors, *Proceedings of CSL'98*, LNCS 1584, pages 224–240. Springer, Berlin, 1999.

[199] R. Routley and R. K. Meyer. The semantics of entailment – III. *Journal of Philosophical Logic*, 1:192–208, 1972.

[200] R. Routley and R. K. Meyer. The semantics of entailment – I. In Leblanc, editor, *Truth, Syntax and Modality*, pages 199–243. North Holland, Amsterdam, 1973.

[201] R. Routley, V. Plumwood, R. K. Meyer, and R. T. Brady. *Relevant logics and their rivals*. Ridgeview, Atascadero, CA, 1982.

[202] A. Russo. Generalizing propositional modal logic using Labelled Deductive Systems. In Baader and Schulz [12], pages 57–73.

[203] A. Russo. *Modal logics as Labelled Deductive Systems*. PhD thesis, Department of Computing, Imperial College, London, UK, 1996.

[204] H. Sahlqvist. Completeness and correspondence in first and second order semantics for modal logic. In S. Kanger, editor, *Proceedings of the Third*

Scandinavian Logic Symposium, pages 110–143. North Holland, Amsterdam, 1975.

[205] R. A. Schmidt. *Optimised modal translation and resolution*. PhD thesis, Universität des Saarlandes, Saarbrücken, Germany, 1997.

[206] R. A. Schmidt. Resolution is a decision procedure for many propositional modal logics. In M. Kracht, M. de Rijke, H. Wansing, and M. Zakharyaschev, editors, *Advances in Modal Logic*, volume 1, pages 189–208. CSLI Publications, Stanford, CA, 1998.

[207] R. A. Schmidt. Decidability by resolution for propositional modal logics. *Journal of Automated Reasoning*, 22(4):379–396, 1999.

[208] M. Schmidt-Schauß and G. Smolka. Attributive concept descriptions with complements. *Artificial Intelligence*, 48:1–26, 1991.

[209] H. Schnelle and N. O. Bernsen. *Logic and linguistics*. Lawrence Erlbaum Associates, Pittsburgh, Pennsylvania, 1989.

[210] R. Sebastiani and A. Villafiorita. SAT-based decision procedures for normal modal logics: a theoretical framework. In F. Giunchiglia, editor, *Proceedings of AIMSA'98*, LNAI 1480, pages 583–597. Springer, Berlin, 1998.

[211] J. Seligman. The logic of correct description. In M. de Rijke, editor, *Advances in Intensional Logic*, pages 107–136. Kluwer, Dordrecht, 1997.

[212] A. Sernadas, C. Sernadas, and C. Caleiro. Fibring of logics as a categorial construction. *Journal of Logic and Computation*, 9(2):149–179, 1999.

[213] J. R. Shoenfield. *Mathematical logic*. Addison Wesley, Reading, MA, 1967.

[214] G. F. Shvarts. Gentzen style systems for $K45$ and $K45D$. In A. Meyer and M. Taitslin, editors, *Proceedings of the Symposium on Logical Foundations of Computer Science*, LNCS 363, pages 245–256. Springer, Berlin, 1989.

[215] D. F. Siemens Jr. Fitch–style rules for many modal logics. *Notre Dame Journal of Formal Logic*, 18(4):631–636, 1977.

[216] A. Simpson. *The proof theory and semantics of intuitionistic modal logic*. PhD thesis, University of Edinburgh, UK, 1993.

[217] E. Spaan. *Complexity of Modal Logics*. PhD thesis, University of Amsterdam, The Netherlands, 1993.

[218] C. Stirling. Modal logics for communicating systems. *Theoretical Computer Science*, 49(1):311–347, 1987.

[219] P. F. Syverson. Knowledge, belief and semantics in the analysis of cryptographic protocols. *Journal of Computer Security*, 1(3):317–334, 1992.

[220] A. S. Troelstra. *Metamathematical investigation of intuitionistic arithmetic and analysis*. LNM 344. Springer, Berlin, 1982.

[221] A. S. Troelstra and H. Schwichtenberg. *Basic proof theory*. Cambridge University Press, Cambridge, UK, 1996.

[222] M. Tzakova. Tableau calculi for hybrid logics. In Murray [165], pages 66–81.

[223] A. Urquhart. Semantics for relevant logics. *Journal of Symbolic Logic*, 37(1):159–169, 1972.

[224] A. Urquhart. The undecidability of entailment and relevant implication. *Journal of Symbolic Logic*, 49(4):1059–1073, 1984.

[225] A. Urquhart. The complexity of decision procedures in relevance logic. In J. M. Dunn and A. Gupta, editors, *Truth or Consequences*, pages 61–76. Kluwer, Dordrecht, 1990.

[226] A. Urquhart. The complexity of decision procedures in relevance logic II. Manuscript, 1997.

[227] J. van Benthem. Correspondence theory. In Gabbay and Guenthner [93], pages 167–248.

[228] J. van Benthem. *Modal logic and classical logic*. Bibliopolis, Napoli, 1985.

[229] D. van Dalen. Intuitionistic logic. In Gabbay and Guenthner [94], pages 225–339.

[230] D. van Dalen. *Logic and structure*. Springer, Berlin, 1994.

[231] L. Viganò. An $O(n \log n)$-space decision procedure for the relevance logic B^+ (extended version). To appear as a technical report of the Institut für Informatik of the Albert-Ludwigs-Universität Freiburg. Available at http://www.informatik.uni-freiburg.de/~luca.

[232] L. A. Wallen. *Automated deduction in non-classical logics*. MIT Press, Cambridge, MA, 1990.

[233] H. Wansing. Sequent calculi for normal modal propositional logics. *Journal of Logic and Computation*, 4(2):125–142, 1994.

[234] H. Wansing, editor. *Proof theory of modal logic*. Kluwer, Dordrecht, 1996.

[235] H. Wansing. *Displaying Modal Logic*. Kluwer, Dordrecht, 1998.

[236] C. Weidenbach, B. Gaede, and G. Rock. SPASS & FLOTTER, version 0.42. In M. McRobbie and J. Slaney, editors, *Proceedings of CADE-13*, LNAI 1104, pages 141–145. Springer, Berlin, 1996.

[237] A. Zanardo, A. Sernadas, and C. Sernadas. Fibring: Completeness preservation. Technical report, Section of Computer Science, Department of Mathematics, Instituto Superior Técnico, Lisboa, Portugal, 1999.

[238] J. J. Zeman. *Modal logic: the Lewis-modal systems*. Oxford University Press, Oxford, UK, 1973.

[239] J. Zucker. The correspondence between cut-elimination and normalization I, II. *Annals of Mathematical Logic*, 7:1–112, 113–155, 1974.

Symbols

A

B

C

D

E

F

M

N

T

U

W